Lecture Notes in Mechanical Engineering

Lecture Notes in Mechanical Engineering (LNME) publishes the latest developments in Mechanical Engineering—quickly, informally and with high quality. Original research or contributions reported in proceedings and post-proceedings represents the core of LNME. Volumes published in LNME embrace all aspects, subfields and new challenges of mechanical engineering.

To submit a proposal or request further information, please contact the Springer Editor of your location:

Europe, USA, Africa: Leontina Di Cecco at Leontina.dicecco@springer.com
China: Ella Zhang at ella.zhang@cn.springernature.com
India: Priya Vyas at priya.vyas@springer.com
Rest of Asia, Australia, New Zealand: Swati Meherishi at swati.meherishi@springer.com

Topics in the series include:

- Engineering Design
- Machinery and Machine Elements
- Mechanical Structures and Stress Analysis
- Automotive Engineering
- Engine Technology
- Aerospace Technology and Astronautics
- Nanotechnology and Microengineering
- Control, Robotics, Mechatronics
- MEMS
- Theoretical and Applied Mechanics
- Dynamical Systems, Control
- Fluid Mechanics
- Engineering Thermodynamics, Heat and Mass Transfer
- Manufacturing Engineering and Smart Manufacturing
- Precision Engineering, Instrumentation, Measurement
- Materials Engineering
- Tribology and Surface Technology

Indexed by SCOPUS, EI Compendex, and INSPEC.

All books published in the series are evaluated by Web of Science for the Conference Proceedings Citation Index (CPCI).

To submit a proposal for a monograph, please check our Springer Tracts in Mechanical Engineering at https://link.springer.com/bookseries/11693.

Mnaouar Chouchane · Moez Abdennadher ·
Nizar Aifaoui · Fakher Chaari · Slim Bouaziz ·
Zouhaier Affi · Mohamed Haddar ·
Lotfi Romdhane · Abdelmajid Benamara
Editors

Design and Modeling of Mechanical Systems - VI

Proceedings of the 10th Conference on Design and Modeling of Mechanical Systems, CMSM'2023, December 18–20, 2023, Hammamet, Tunisia - Volume 2: Materials Engineering and Manufacturing

Springer

Editors
Mnaouar Chouchane
National Engineering School of Monastir (ENIM)
Monastir, Tunisia

Nizar Aifaoui
Preparatory Institute for Engineering Studies of Monastir (IPEIM)
Monastir, Tunisia

Slim Bouaziz
National Engineering School of Sfax (ENIS)
Sfax, Tunisia

Mohamed Haddar
National Engineering School of Sfax (ENIS)
Sfax, Tunisia

Abdelmajid Benamara
National Engineering School of Monastir (ENIM)
Monastir, Tunisia

Moez Abdennadher
Preparatory Institute for Engineering Studies of Sfax (IPEIS)
Sfax, Tunisia

Fakher Chaari
National Engineering School of Sfax (ENIS)
Sfax, Tunisia

Zouhaier Affi
National Engineering School of Monastir (ENIM)
Monastir, Tunisia

Lotfi Romdhane
College of Engineering
American University of Sharjah (AUS)
Sharjah, United Arab Emirates

ISSN 2195-4356 ISSN 2195-4364 (electronic)
Lecture Notes in Mechanical Engineering
ISBN 978-3-031-65006-2 ISBN 978-3-031-65007-9 (eBook)
https://doi.org/10.1007/978-3-031-65007-9

This Springer imprint is published by the registered company Springer Nature Switzerland AG
The registered company address is: Gewerbestrasse 11, 6330 Cham, Switzerland

Preface

The tenth edition of the "International Congress Design and Modeling of Mechanical Systems" CMSM'2023 was held in Hammamet, Tunisia, from 18th to 20th of December 2023. The CMSM congress has been held in a Tunisian city every two years since its first edition in 2005. The CMSM congress brings together specialists from universities and industrial companies to present their recent research findings and to discuss and exchange research experiences with other attendees of the CMSM Congress. The CMSM Congress is jointly organized by two Tunisian research laboratories: The Mechanical Engineering Laboratory (LGM) of the National Engineering School of Monastir and the Mechanical, Modeling and Manufacturing Laboratory (LA2MP) of the National Engineering School of Sfax. The tenth edition of the CMSM Congress has been attended by about 250 participants who participated actively in the plenary sessions and in the sessions devoted to specialized topics. Five plenary conferences and about 156 papers and 17 posters have been presented during the three days of the congress.

This book is the sixth volume of the "Lecture Notes in Mechanical Engineering" series "Design and Modeling of Mechanical Systems". From the 173 papers and posters presented in the CMSM'2023, 91 papers are selected to be included in the two volumes of this book. The papers are classified into the following 8 topics:

1. Design and Analysis of Mechanical Systems
2. Numerical Modeling and Analysis of Structures and Systems
3. Mechanical Vibration Analysis and Applications
4. Industrial Engineering
5. Materials Science and Engineering
6. Composite and Bio-Materials
7. Surface Finishing and Coating
8. Manufacturing Engineering and Additive Manufacturing.

This first volume contains topics one to four. The second volume contains topics five to eight.

All the papers included in this volume have undergone rigorous reviewing by two or three reviewers. Authors have been provided by the comments of the reviewers and requested to submit revised papers. The review process contributed significantly to the improvement of the quality of the papers included in this volume. The editors would like to thank all the authors for submitting their recent research work to the CMSM'2023 congress and for considering the comments of the reviewers to revise their papers. The hard work of the reviewers is also highly appreciated. The editors are also grateful to the organizing committee of the CMSM'2023 and to the editing team of Springer publication for providing support for the publication of this proceedings.

December 2023

Mnaouar Chouchane
Moez Abdennadher
Nizar Aifaoui
Fakher Chaari
Slim Bouaziz
Zouhaier Affi
Mohamed Haddar
Lotfi Romdhane
Abdelmajid Benamara

Conference Plan

- Please indicate a conference webpage address and/or a Call for papers.

https://cmsm.tn/CMSM2023/

- Please indicate in which city and on which day/s the conference is taking place

Mehari Hotel, Yasmine Hammamet, Hammamet, Tunisia

- Please list Conference Organizers and Boards

Organizing Committee

KHEMILI Imed	LMS, ISSATS, TN
MZALI Slah	LGM, ENIM, TN
FTOUTOU Ezzeddine	LGM-ENIM, TN
BEN AHMED Mondher	ISET Mahdia, TN
BOUKTHIR Hafedh	ISET Mahdia, TN
GHORBEL Ahmed	ENIS, LA2MP, TN
GUIZANI Amir	ENIS, LA2MP, TN

Scientific Committee

Nizar Aifaoui	LGM, IPEIM, Tunisia
Dominique Deneux	LAMIH, UPHF, France
Hakim Naceur	INSA Hauts de France, France
Thierno Diallo	ISAE-Supméca, France
ALI KHALFALLAH	ISSAT Sousse, Tunisia
Habib SAHLAOUI	LMMP-ENSIT, Tunisia
Sami Bennour	LMS-ENISo, Tunisia
Rabi Ben SGHAIER	ISSAT Sousse, Tunisia
Noureddine BOUHADDI	FEMTO-ST Institute, UMR CNRS 6174, Université de Franche-Comté, Besançon, France
Naoufel Ben Moussa	LMMP-IPEIB, Tunisia
MONDHER BEN AHMED	ISET, MAHDIA, Tunisia
Imed KHEMILI	LMS, ISSAT Sousse, Tunisia
Belgacem MAMEN	Abbès laghrour university of khenchela, Algeria
Mohamed El Badaoui	Université Jean-Monnet, France

Mohamed Guedri	National Higher Engineering School of Tunis, University of Tunis, Tunisia
Jerome Antoni	LVA - INSA Lyon, France
Zied Driss	National School of Engineers of Sfax, Tunisia
Faida Mhenni	ISAE-Supmeca, France
AMINE AMMAR	ENSAM, France
Mohamed Amen Gahbiche	LGM, Tunisia
Mnaouar Chouchane	LGM, ENIM, University of Monastir, Tunisia
Khaled ELLEUCH	ENIS, Tunisia
Tarek BENAMEUR	LGM, ENIM, Tunisia
Abdelfattah MLIKA	ENISo-University of Sousse, Tunisia
Wacef Salem	LGM - ENIM, Tunisia
Haykel Marouani	LGM, ENIM, Tunisia
Farhat ZEMZEMI	ENISo, Sousse, Tunisia
Mohamed KHARRAT	LASEM - ENIS, Tunisia
Stephan Heyns	University of Pretoria, South Africa
Khalil hajlaoui	University KSA, Saudi Arabia
Sami Chatti	ISSAT Sousse, Tunisia
Marc Bonnet	ENSTA, UMA, France
Riccardo Rubini	University of Modena and Reggio Emilia, Italy
Abdallah Bouabidi	M2EM-ENIG/ISSIG, Tunisia
Fafa Ben Hatira	INSAT, Tunisia
MAHER BAILI	ENIT, France
Tarek Hassine	LGM, ENIM, Tunisia
Myriam LAROUSSI	LGM, ENIM, Tunisia
Jamel Mars	ISSIG, Gabès, Tunisia
KAMEL MEHDI	IPEIEM, UTM, Tunisia
Maroua Ghali	LGM, ENIM, Tunisia
Ezzeddine Ftoutou	LGM, ENIM, Tunisia
Foued MZALI	University of Monastir, Tunisia
Mohamed Amine Ben Abdallah	Sheffield Hallam University, UK
Mohamed GUERICH	ESILV, France
Anouar NASR	LGM, ENIM, Tunisia
Amira SALEM	LGM, ENIM, Tunisia
Fakher Chaari	ENIS, Tunisia
Zouhaier AFFI	LGM, ENIM, Tunisia
Ridha Ennetta	University of Gabès, Tunisia
SLIM BOUAZIZ	LA2MP, ENIS, Tunisia
Ated BEN KHALIFA	LGM, ENIM, Tunisia
Mourad Turki	ENIM, Tunisia
Moez Trigui	LGM, ENIM, Tunisia
Amira Amamou	LGM, ENIM, Tunisia

Slah Mzali	LGM, ENIM, Tunisia
Walid Larbi	CNAM, Paris, France
Sami Chatti	LGM, ENIM, Tunisia
Mehdi Tlija	King Saud University, Saudi Arabia
Sameh Meherzi	ENIM, Tunisia
Anis Korbi	ISSAT Kairouan, LGM, Tunisia
Salah Mezlini	LGM, ENIM, Tunisia
Zied Ktari	LGM, ENIM, Tunisia
Saoussen Belghith	LGM, ENIM, Tunisia
Fatma Elwasli	LGM, ENIM, Tunisia
Maryam Brahem	LGM, ENIM, Tunisia
Hadhami Ben Slama	LGM, ENIM, Tunisia
RAMZI BEN HMIDA	LGM, ENIM, Tunisia
Tarak BOURAOUI	ENIM/LGM/ATM - Université de Monastir (TUNISIE), Tunisia
Riadh Bahloul	LGM, ENIM, Tunisia
Mortadha GRAA	LGM, ENIM, Tunisia
Samir LAHOUAR	LGM, ENIM, Tunisia
Amani Amamou	LGM, ENIM, Tunisia
Safa BOUDHAOUIA	LGM, ENIM, Tunisia
Foued Mzali	ENIM - University of Monastir, Tunisia
Moez Abdennadher	LA2MP, ENIS, Tunisia

Contents

Materials Science and Engineering

Materials Science and Engineering

Evaluation of Parameters for Magnesium Fabrication by Powder Metallurgy Route

Mohammed Ibrahim Fathi[1,2](✉), Mohamed Ali Terres[3], and Jawdat Ali Yagoob[4]

[1] National School of Engineer Tunis, University of Tunis El Manar, Tunis, Tunisia
mohammedibrahim@uokirkuk.edu.iq

[2] Department of Medical Basic Science, College of Dentistry, University of Kirkuk, Kirkuk, Iraq

[3] Mechanical, Material and Process Laboratory (LR99ES05) ENSIT - University of Tunis, 5 AV Taha Hussein Montfleury, Tunis, Tunisia

[4] Engineering Technical Collage Kirkuk- Northern Technical University, Kirkuk, Iraq

Abstract. The key benefits of magnesium and its alloys are their high biocompatibility and specific strength. Powder metallurgy is a contemporary method for the preparation of materials based on magnesium. The micro-hardness of pure magnesium (Mg) samples is evaluated in this study after they were produced through the powder metallurgy process. Magnesium powder's purity and microstructure were examined using electron microscopy and x-ray diffractometry. The cold compaction of powdered magnesium was examined. At laboratory temperature, magnesium powder with a mean particle size of 30 μm was compressed using pressures of 400, 450, 500, 550, 600, and 650 MPa. The impact of compacting pressure was examined using %porosity and green density. It was found that a pressure of 650 MPa produced the best density of 1.715 gm/cm^3 and %porosity of 1.323. The sintering process was conducted at three different temperatures (455, 525, and 585 °C) and three different sintering time (30, 45, 60) min. It was found that 525 °C for 30 min produced the best results, with a density of 1.725 g/cm3, porosity of 0.75%, and micro-hardness of 97.77%.

Keywords: Evaluation · Magnesium · Powder metallurgy · Sintering process · Microhardness

1 Introduction

With a density of 1.74 g/cm^3, magnesium (Mg) is one of the light metals and is found in large quantities in both the sea and the crust of the earth. Mg has high strength-to-weight ratios due to their low density. Because of this wonderful quality, it is appealing for lightweight [1–3]. For instance, Mg and its alloys are good choice for reducing weight for carrying convenience in customer electronics and tools for power; in the automotive industry, weight reduction through the use of Mg components can significantly reduce CO_2 (Carbon dioxide gas) emissions and fuel consumption [1]. Since 1999, the global production of Mg and its alloys has been rising steadily. Cars built in Germany have already made use of some Mg components [1]. Due to their appropriate mechanical

M. Chouchane et al. (Eds.): CMSM 2023, LNME, pp. 3–9, 2024.
https://doi.org/10.1007/978-3-031-65007-9_1

properties and good biocompatibility, Mg and its some alloys are also utilized by means of biodegradable materials [4–10]. Mg and its alloys have attracted a lot of attention by means of potential replacements for conventional orthopedic implantation constituents in recent years [11–13]. What's more, these substances are entirely absorbed by the body once bone tissue regenerates since they are biodegradable [14]. Additionally, it was observed that Magnesium has effection on lipid profile [15]. The synthesis and characterization of pure Mg produced using the cold powder metallurgy method is the main goal of this work. The micro hardness of sintered Mg specimens, their density and porosity have been studied.

2 Material and Experimental Methods

The mean size of the Mg powder particle used in this study was approximately 30 μm, and it has an irregular blocky shape as illustrated in (Fig. 1). The purity of base Mg powder was 99.8%. It was provided by Jingan Chemicals & Alloy Company - China.

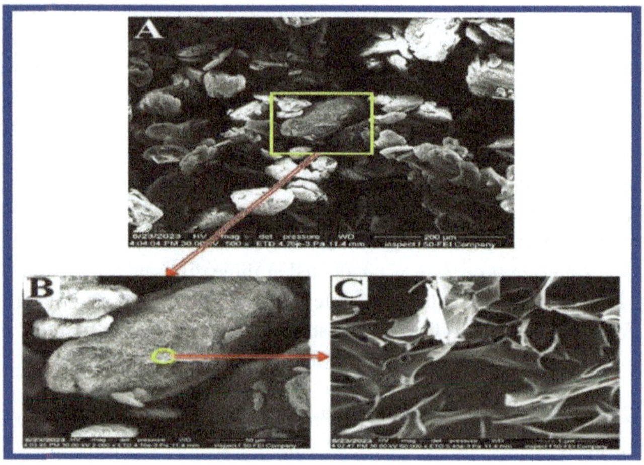

Fig. 1. FESEM Image for pure magnesium powder at different magnification powers

Lab XRD-6000 Shimadzu- Japan XRD machine was used for phase identification analysis for Mg powder. XRD has been used to ensure that the supplied powder is Mg powder and does not contain any other impurities as clearly displayed in (Fig. 2).

A 304 stainless steel vessel was charged with Mg powder and 11 mm diameter balls made of chromium iron with a ratio of weight 1:10, respectively. The powder was milled for three hours at 121 rpm with aid of the ball milling machine that was designed and manufactured for this purpose. The balls Vickers micro-hardness (HV) was (855) evaluated by Metkon micro-hardness device with 500 gm applied force. Alloy steel die and punch and other components shown in (Fig. 3) were used to create cylindrical samples with 15 mm diameter and with a height suitable for each test.

At room temperature, compaction of the Mg powder into compacts was achieved by applying varying uniaxial pressures of (400, 450, 500, 550, 600, and 650) MPa.

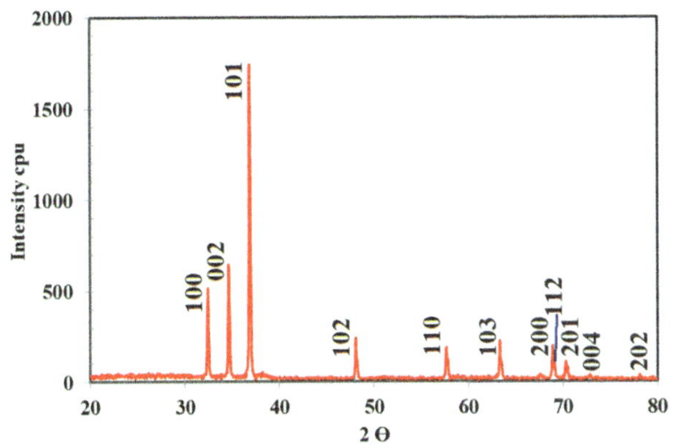

Fig. 2. XRD Analysis diagram for pure magnesium powder

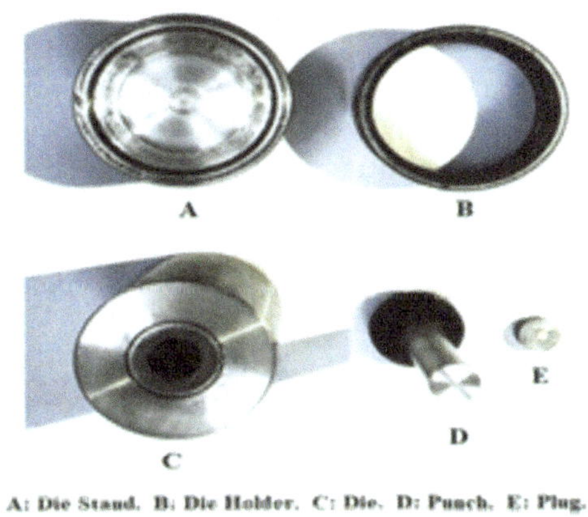

Fig. 3. Main parts pressing tool.

Green samples density (ρg) considered by as stated Eq. (1) [14] the mass of samples was weighted by electronic type balance with 0.1mg accurateness. The volume of samples was measured with the aid of determining the major dimensions.

$$Pg = Mc/Vg \tag{1}$$

Percent porosities (%P) are measured according to the Eq. (2) [14].

$$\%\text{P} = (1 - \rho g/\rho \text{Th}) \times 100 \tag{2}$$

Next, from the start of the sintering to room temperature, the samples are existed in an electric furnace of CARBOLITE-UK type while an argon gas stream is continuously applied. For thirty minutes, samples was soaked at the appropriate temperatures (455, 525, and 585 °C), several samples at each temperature. The rate of heating is set at 10 °C/min. Then samples were allowed to cool inside the furnace gradually to room temperature. The samples' parallel faces was wet ground with the utility of SiC emery papers with grits of 240, 600, 1000, 2000, and 3000, polished with a 5 μm suspension of alumina solution, cleaned with distilled water, and dried in a oven at 120 °C for 20 min. Vickers micro-hardness for sintered compacts (SCs) was determined using the same technique described above. For microstructure analyses, optical microscopy Italy made invert microscope - was employed.

3 Results and Discussion

Determining the best pressing pressure for Mg samples preparation is crucial. Pressing pressure and green density were found to be directly correlated (Table 1). Greatest (ρg) result was under the highest applied pressure (650 MPa) with the maximum 1.715 g/Cm^3 was achieved. Figure 4 illustrates how density and compression pressure are proportionally related. Hence the decision was that 650MPa is the ideal applying pressure. As a result, it was used later in the preparation of all samples.

Table 1. Density and Porosity of pure magnesium against applied pressure

Sample No	Pressing Pressure MPa	Green Density g/cm^3	Porosity %
1	400	1.58	9.09
2	450	1.615	7.077
3	500	1.622	6.674
4	550	1.681	3.28
5	600	1.699	2.244
6	650	1.715	1.323

On other hand the porosity decreased by increasing of the compacting pressure and the powder particle's deformation increased with increasing compaction pressure. The best porosity (1.323%) obtained at 650 MPa. Figure 5 illustrates how % porosity and compression pressure are related.

The best density value of 1.725 g/cm^3 (that is nearest to the theoretical density of Mg, % porosity of 0.75% and microhardness of 97.77 were obtained at a sintering temperature of 525 °C and 30 min of sintering time. (Table 2) displays the density, % porosity and Microhardness calculated at each sintering temperature and time.

Optical microscope photo in Fig. 6 A and B for non-etched microstructure is discovered the how the porosity was restricted at Mg particles boundaries. At some positions the powder particles boundaries are clearly shown while at other positions the boundaries between the nearby particles are partially disappeared due to the diffusion process that occurred at sintering.

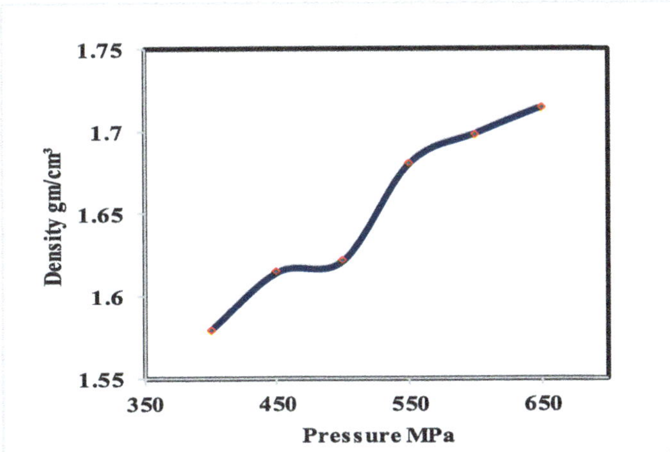

Fig. 4. Relationship between green density and applied compaction pressure

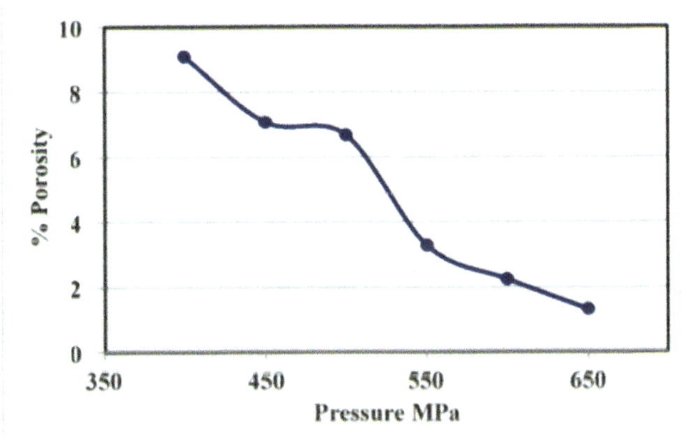

Fig. 5. Relation between % porosity and applied pressing pressure

Table 2. Density, %Porosity and Microhardness of pure Mg at each sintering temperature and time.

Pressure MPa	Temperature °C	Times min	Density g/cm3	Porosity %	HV
650	455	60	1.715	1.32	64.47
650	525	30	1.725	0. 75	97.77
650	585	45	1.695	2.5	80.56
650	455	30	1.722	0.92	75.46
650	525	45	1.7247	0.765	90.52
650	585	60	1.684	3.1	79.68
650	455	45	1.721	0.98	73.27
650	525	60	1.7246	0.77	87.14
650	585	30	1.708	1.37	86.9

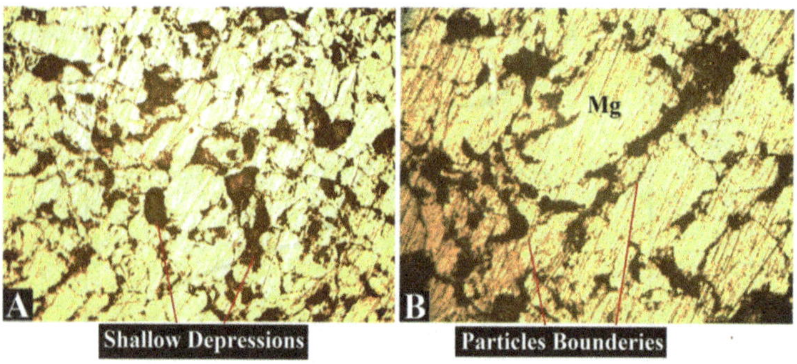

Fig. 6. **A and B** Microstructure of the Mg sample sintered at 525 °C 500X

4 Conclusion

1. This investigation is accomplished efficaciously from the view point of fabrication of Mg compacts initially for the reason of the capability to protecting the compacted samples from impurity that may be imported from powder metallurgy processing beginning from milling to sintering process.
2. The presence of the porosity inside the sintered samples was with very low value that is strong evidence to the success of pressing and sintering processes.
3. All three factors have a distinct impact on the overall quality of the result.
4. Maximum micro-hardness which extended to (~ 98 HV) is comparable to its value for cast Mg products.

References

1. Abbott, T.: Magnesium: industrial and research developments over the last 15 years, Corrosion (2014).https://doi.org/10.5006/1474
2. Cao, F., Shi, Z., Hofstetter, J., Uggowitzer, P.J., Song, G., Liu, M., Atrens, A.: Corrosion of ultra-highpurity Mg in 3.5% NaCl solution saturated with Mg(OH)2, Corros. Sci. **75**, 78–99 (2013)
3. Cao, F., Shi, Z., Song, G.-L., Liu, M., Atrens, A.: Corrosion behaviour in salt spray and in 3.5% NaCl solution saturated with Mg(OH)2 of as-cast and solution heat-treated binary Mg–X alloys: X = Mn, Sn, Ca, Zn, Al, Zr, Si, Sr, Corros. Sci. **76**, 60–97 (2013)
4. Cao, F., Shi, Z., Song, G.-L., Liu, M., Dargusch, M.S., Atrens, A.: Influence of hot rolling on the corrosion behavior of several Mg–X alloys. Corros. Sci. **90**, 176–191 (2015)
5. Dewidar, M.M., Yoon, H.C., Lim, J.K.: Mechanical properties of metals for biomedical applications using powder metallurgy process. A Rev. Metals Mater. Int. 12(3), 193–206
6. Shi, Z., Cao, F., Song, G.-L., Liu, M., Atrens, A.: Corrosion behaviour in salt spray and in 3.5% NaCl solution saturated with Mg(OH)2 of as-cast and solution heat-treated binary Mg–RE alloys: RE = Ce, La, Nd, Y, Gd, Corros. Sci. **76**, 98–118 (2013)
7. Song, G., Atrens, A.: Understanding magnesium corrosion—a framework for improved alloy performance. Adv. Eng. Mater. **5**, 837–858 (2003). Electrochemical Properties Powder Technol. **206**, 233–238 (2011)
8. Song, G., Atrens, A.: Recent insights into the mechanism of magnesium corrosion and research suggestions. Adv. Eng. Mater. **9**, 177–183 (2007)
9. Song, G.L., Atrens, A.: Corrosion mechanisms of magnesium alloys. Adv. Eng. Mater. **1**, 11–33 (1999)
10. Song, G., Hapugoda, S., St John, D.: Degradation of the surface appearance of magnesium and its alloys in simulated atmospheric environments. Corros. Sci. **49**, 1245–1265 (2007)
11. Song, G., StJohn, D.: Corrosion of magnesium alloys in commercial engine coolants. Mater. Corros. **56**, 15–23 (2005)
12. Xu, L., Pan, F., Yu, G., Yang, L., Zhang, E., Yang, K.: In vitro and in vivo evaluation of the surface bioactivity of a calcium phosphate coated magnesium alloy. Biomaterials **30**, 1512–1523 (2009)
13. Xin, Y., Liu, C., Huo, K., Tang, G., Tian, X.: Corrosion behavior of ZrN/Zr coated biomedical AZ91 magnesium alloy. Surf. Coat. Technol. **203**, 2554–2557 (2009)
14. Xin, Y., Liu, C., Huo, K., Tang, G., Tian, X., Chu, P.K.: Influence of heat treatment on degradation behavior of bio-degradable die-cast AZ63 magnesium alloy in simulated body fluid. Mater. Sci. Eng. A **456**, 350–357 (2007)
15. Yagoob, J.A., Abbass, M.K.: Characterization of cobalt based CoCrMo alloy fabricated by powder metallurgy route. In: 2nd International Conference for Engineering, Technology and Science of Al-Kitab University 4–5 December-2018 (2018)
16. Abdulla, K.N., Ahmed, M.S.: The effect of oral magnesium supplements on lipid profile and oxidative stress in adult albino female rats. Medico-Legal Update **20**, 596–599 (2020)

Study of the Effects of Gamma Irradiation on the Mechanical Properties of NBR Material

Slah Chayoukhi[1](✉), Mohamed Abid[2], and Arbi Mejri[3]

[1] Laboratoire Mecanique, Productique et Energetique (LMPE), Ecole Nationale Supérieure d'ingénieurs de Tunis. Université de Tunis, B.P. 56, Bab Menara, 1008 Tunis, Tunisia
slah.chayoukhi@fst.utm.tn

[2] Laboratory of Electromechanical Systems, National Engineering School of Sfax, Sfax, Tunisia

[3] Laboratoire de Dosimétrie, Centre National des Sciences et Technologies Nucléaires (CNSTN), Pôle TechnologiqueBP 72, 2020 Sidi Thabet, Tunisia

Abstract. This article presents an experimental study of the mechanical properties of nitrile-butadiene rubber (NBR) when irradiated with gamma rays. NBR was irradiated with gamma rays at different absorbed radiation energy doses (0, 30, 45, 60, 75 and 90 kGy). Tensile experiments and hardness test analyses show an improvement in the elastic behavior of NBRs and a decrease in their plastic behavior with increasing absorbed radiation energy doses. Scanning electron microscopy (SEM) shows the NBR surface layer to be porous and cracked as a result of gamma irradiation. The experimental study was completed by a numerical simulation with Abaqus software using hyperelastic models (Mooney Rivlin model, Ogden model, Yeoh model, Arruda-Boyce model and Neo-Hookean model). The simulation uses input data obtained from traction results. Curves of stress-strain for pristine and irradiated samples, obtained from numerical and experimental studies, were compared. Parameters of the hyperelastic deformation energy function are computed for each of the five models. So, the Ogden model was found to produce the most accurate curve, with a mean square error less than 0.5%. The Ogden model is therefore the most appropriate for simulating the mechanical behavior of irradiated nitrile-butadiene rubber. The results obtained provide a good basis for describing the performance of this material under gamma rays.

Keywords: Nitrile butadiene rubber · gamma irradiation · Abaqus · hyperelastic models · tensile test

1 Introduction

Nitrile-butadiene rubbers (NBR) are used over a wide temperature range and are known for their cost-effectiveness, high oil and fuel compatibility, good elasticity and excellent adherence [1–4].

The mechanical properties of NBR rubber can be affected by gamma irradiation. Indeed, tensile strength, modulus of elasticity, elongation at failure and hardness are all found to modify as irradiation dose γ increases [5–7]. Numerous researchers [8, 9] have studied the effect of gamma irradiation on elastic modulus, tensile strength and hardness.

M. Chouchane et al. (Eds.): CMSM 2023, LNME, pp. 10–18, 2024.
https://doi.org/10.1007/978-3-031-65007-9_2

They reported that the behavior of rubber when irradiated is complex. Therefore, it is necessary to use simulations to partially minimize or avoid experimental testing.

This article examines the mechanical behavior of NBR exposed to different doses of gamma radiation. Uniaxial tensile and hardness measurements were first carried out on pristine and irradiated NBR samples. The results are shown and commented on. Next, morphological properties were explored by scanning electron microscopy (SEM). Finally, a simulation was carried out using Abaqus.

2 Materials and Methods

In the present research, we used NBR (nitrile-butadiene rubber) marketed under the name of the product Europrene® N 3345. Several sheets of the material were irradiated with gamma rays at different doses (0 to 90 kGy). The effect of absorbed radiation energy doses on the mechanical properties of NBR was studied using tensile and hardness measurements. Tensile tests were carried out at room temperature on samples manufactured in accordance with ASTM D412 type C, using a universal machine (LLOYD LR5K). Hardness measurements were performed according to ASTM D 2240, using a Shore A hardness tester. Scanning electron microscopy (SEM) was used to visualize the effects of gamma irradiation on the morphology of the NBR material.

3 Results and Discussion

Results of tensile tests performed on samples of virgin and gamma-irradiated NBR, at absorbed radiation energy doses ranging from 00 to 90 kGy, are presented in Table 1. Figure 1 shows the corresponding stress-strain curves.

Table 1. Tensile properties of irradiated NBR samples

Gamma dose (kGy)	00	30	45	60	75	90
Elastic modulus (MPa)	13.15	13.26	14.67	17.50	18.06	18.01
Tensile strength (MPa)	4.93	4.72	4.51	4.63	4.73	5.02
Strain at break (%)	243	233	219	218	217	224
Strain at ultimate strength (%)	232.46	224.78	211.45	201.47	198.17	215.96
Stress at 100% strain (relative values %)	101	97	96	101	103	107
Stress at 200% strain (relative values %)	101	98	97	102	103	107
Toughness (106 J/m3)	8.05	7.23	6.64	6.54	6.6	6.8
Crosslink density (10-4mol/cm3)	2.73	2.77	2.65	3.05	3.17	3.38
Strain at break (%)	243	233	219	218	217	224

Figure 2 shows the influence of irradiation on the modulus of elasticity of NBR material. It is clear that the stiffness of NBR increases with increasing irradiation dose.

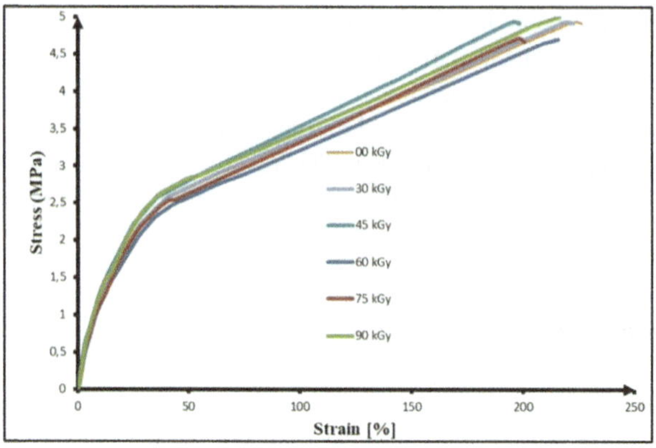

Fig. 1. NBR stress-strain curves for different doses

In fact, an increase in Young's modulus of up to 37 is due to new cross-linking reactions and a reduction in distance of atomic separation in irradiated specimens [1, 8–10]. In addition, irradiation significantly affected the tensile resistance of the material due to crosslinking/degradation [11]. The variation in tensile resistance with gamma irradiation dose is shown in Fig. 3. It can be seen that tensile strength varies slightly with increasing gamma dose. It gradually decreases to a minimum value at an absorbed radiation energy dose of 45 kGy, then slowly increases again. This is due to modifications in polymer chain scission and increased cross-linking density resulting from high absorbed radiation energy doses.

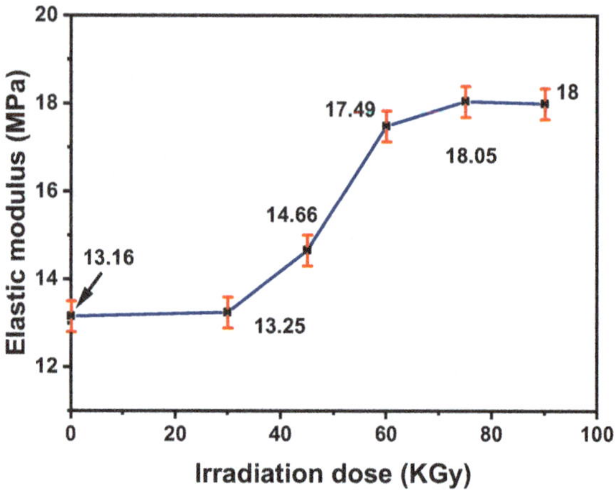

Fig. 2. NBR elastic modulus at different γ-doses.

The effect of radiation dose on NBR failure was investigated by examining the evolution of strain at failure versus absorbed radiation energy dose. The results show that strain at failure decreases with increasing absorbed dose. Indeed, a rapid decrease

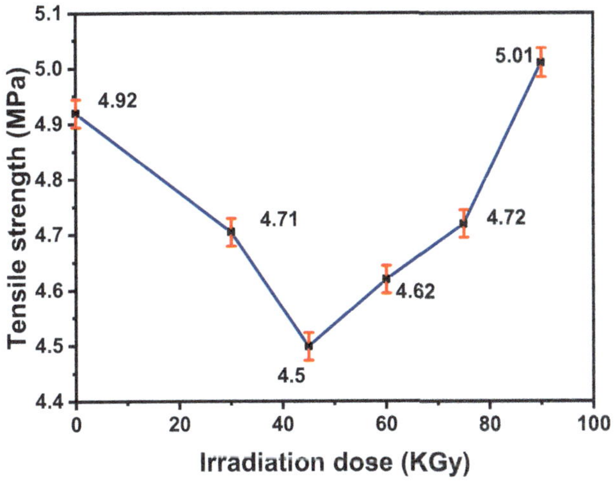

Fig. 3. Tensile strength as a function of irradiation dose

is observed at doses below 45 kGy, with a gradual decrease at higher doses (Fig. 4). It reaches 16.9% at 90 kGy. This can be explained by the supplementary bonding of filler and polymer due to the elevated free-radical formation [12]. It is also due to additional cross-linking reactions during irradiation [13].

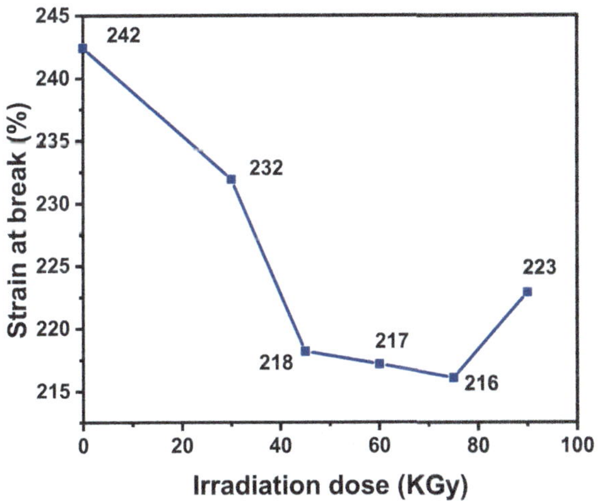

Fig. 4. Strain at break versus absorbed dose

A material's toughness is defined by the energy of deformation at failure per volume (J/m^3). It corresponds to the material's capacity to deform plastically without breaking. The toughness of virgin and gamma-irradiated NBR was calculated and plotted versus gamma dose (Fig. 5). Toughness decreases progressively with increasing gamma dose.

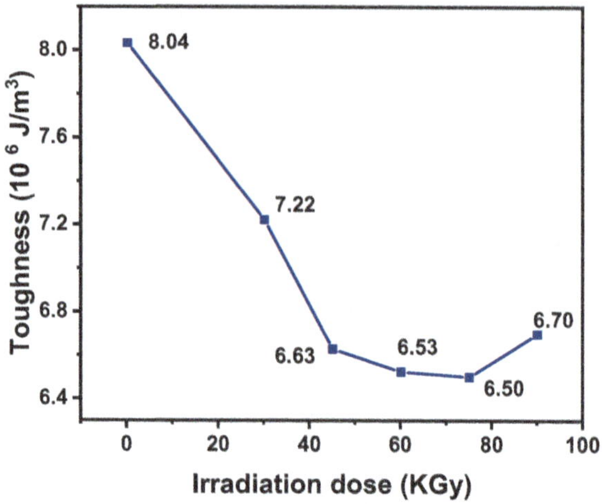

Fig. 5. Toughness versus absorbed dose

The effect of absorbed radiation energy dose on material hardness was examined by means of Shore A hardness measurements. The results presented in Table 2 and Fig. 6 indicate an increase in NBR hardness with increasing gamma dose. This is coherent with the observation made above for the elastic modulus parameter, and shows that NBR material is becoming stiffer with rising irradiation dose.

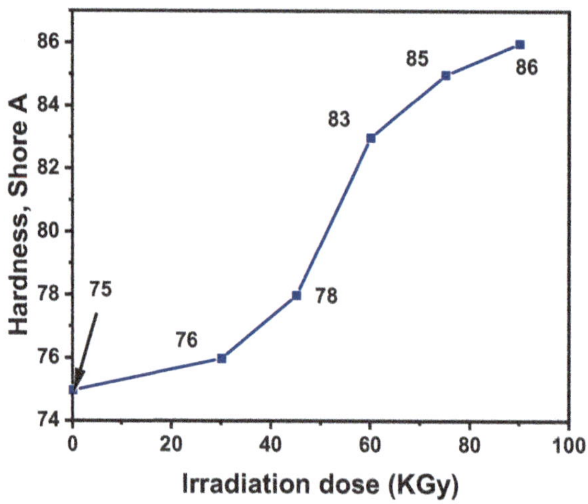

Fig. 6. NBR hardness as a function of γ-irradiation dose

The influence of gamma rays on the morphology of the NBR was examined by scanning electron microscopy (SEM). Figure 7 shows the surface morphology of virgin and irradiated specimens. Visibly porous surfaces and cracks in the NBR surface layer are the result of gamma irradiation. Polymer fibers randomly arranged in the pristine

Table 2. Hardness shore A of irradiated NBR samples.

Absorbed radiation energy dose (kGy)	00	30	45	60	75	90
Hardness (Sh-A)	75	76	78	83	85	86

specimen. In the irradiated samples, however, the fibers disappeared as pores formed on the surface. The result could be that the polymer fibers agglomerate to form semi-compact zones and pores.

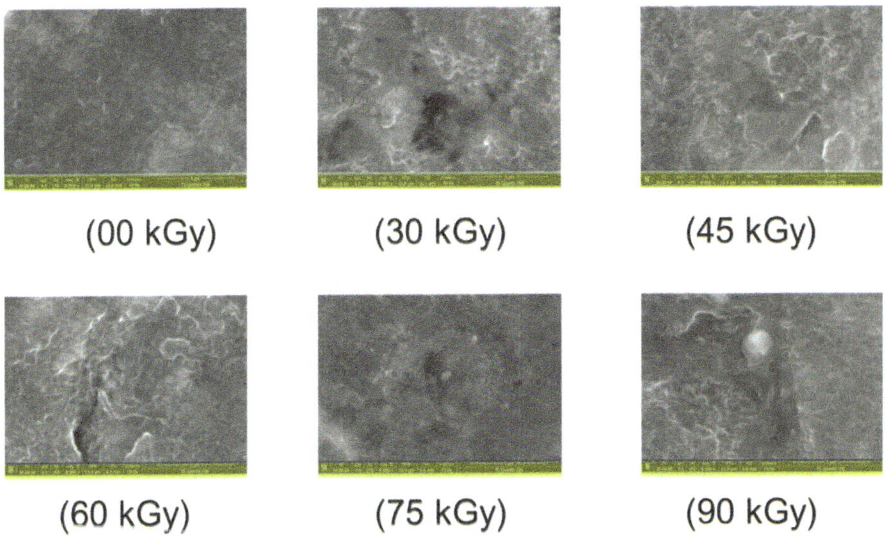

Fig. 7. SEM images.

4 Numerical Study

The principal purpose of the numerical analysis was to identify a suitable model to describe the hyperelastic properties of NBR rubber under gamma irradiation.

The tensile behavior of the material, before and after irradiation, was simulated based on the assumption of a hyperelastic material property model, using as input data those found by uniaxial tensile tests. The models employed (Mooney Rivlin, Ogden, Yeoh, Arruda-Boyce and Neo-Hookean models) express rubber stress and deformation as functions of strain energy.

The numerical study uses ABAQUS software to calculate the non-linear deformations of NBR material before and after irradiation. After selecting the hyperelastic models mentioned above, the results were assessed by reanalyzing the experimental data, verifying model stability, establishing stress-strain curves and identifying material-specific constants and best-fit.

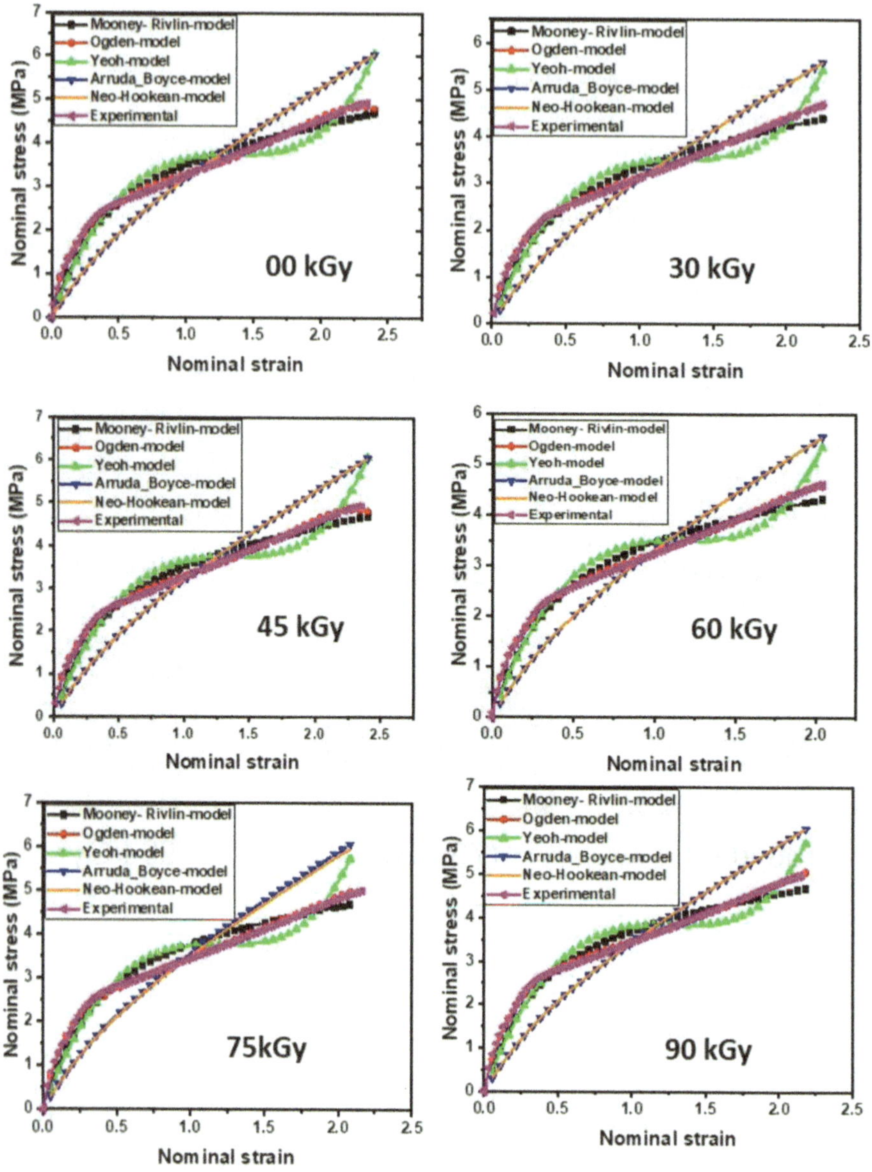

Fig. 8. Numerical and experimental Stress/strain curves.

The numerical results presented in Fig. 8 have made it possible to: i) confront experimental results with numerical plots for each model; ii) choose the model best corresponding to the properties of the NBR material; iii) determine the coefficients of NBR material irradiated with different doses of absorbed radiation energy.

The fitting quality of the hyperelastic models was evaluated by the RMS error, whose calculated values are shown in Table 3. The Ogden model shows a good fit, with the lowest root-mean-square error of less than 0.50%. Also, the adequacy of this model increases with increasing radiation dose. The RMS error decreases from 0.50% for pure NBR to 0.09% for NBR irradiated at 90 kGy. This result highlights the outstanding simulation of material properties.

Table 3. Mean square error corresponding to five hyperelastic models

	Mooney-Rivlin	Ogden	Yeoh	Arruda-Boyce	Neo-Hookean
00 kGy	2.53	0.50	5.27	31.8	31.8
30 kGy	3.06	0. 24	4.65	28.36	28.36
45 kGy	3.69	0. 23	4.86	33.69	33.69
60 kGy	3.53	0. 14	5.98	35.73	35.73
75 kGy	4.94	0.16	6.38	44.48	38.48
90 kGy	4.18	0.09	5.12	40.91	40.91

5 Conclusion

In this study, the mechanical characteristics of nitrile-butadiene rubber (NBR) irradiated with gamma rays were investigated experimentally and numerically. NBR samples were irradiated with different doses of absorbed radiation energy (0 to 90 kGy) at 10 meV. The results obtained lead to the following main conclusions:

- The elastic properties of NBR material improve with increasing absorbed dose;
- The plastic properties have changed, resulting in decreasing strain-at-failure and fracture toughness. Nonetheless, tensile strength was affected to a small extent by irradiation;
- Hardness improves as gamma radiation dose increases;
- Simulation showed that the Ogden model has root mean square errors, compared with experimental results, of less than 0.51% for all samples.

References

1. Jebur, Q.H., Jweeg, M.J., Al-Waily, M., Ahmad, H.Y., Resan, K.K.: Hyperelastic models for the description and simulation of rubber subjected to large tensile loading. Arch. Mater. Sci. Eng. **108**, 75–85 (2021). https://doi.org/10.5604/01.3001.0015.0256
2. Gajewskiam, M., Szczerbab, R., Jemioło, S.: Modelling of elastomeric bearings with application of yeoh hyperelastic material model. Procedia Eng. **111**, 220–227 (2015). https://doi.org/10.1016/j.proeng.2015.07.080

3. Jafari, F., Kashani, M.R., Hosseini, S.M., Pourhossaini, M.R.: Effects of modified poly (tetrafluoroethylene) on the physico-mechanical and tribological properties of carbonblack filled nitrile-butadiene rubber. Appl. Polym. Sci. **138**, 50061 (2021). https://doi.org/10.1002/app.50061
4. Eyssa, H.M., Abulyazied, D.E., Abdulrahman, M., Youssef, H.A.: Mechanical and physical properties of nanosilica/nitrile butadiene rubber composites cured by gamma irradiation. Egyptian J. Pet. **27**, 383–392 (2018). https://doi.org/10.1016/j.ejpe.2017.06.004
5. Markovic, G., Cincovic, M.S.M., Jovanovic, V., Jovanovic, S.S., Simendic, J.B.: Gamma irradiation aging of NBR/CSM rubber nanocomposites. Compos. Part B: Eng. **43**, 609–615 (2012). https://doi.org/10.1016/j.compositesb.2011.11.056
6. Jia, D., et al.: Constitutive modeling of γ-irradiated silicone rubber foams under compression and shear loading. Polym. Degrad. Stab. **183**, 109410 (2021). https://doi.org/10.1016/j.polymdegradstab.2020.109410
7. Hegazi, E.M., Eyssa, H.M., Abd-El-Megeed, A.A.: Effect of nanofiller on the ageing of rubber seal materials under gamma irradiation. J. Compos. Mater. **53**, 2065–2076 (2019). https://doi.org/10.1177/0021998318819178
8. Raslan, H.A., Abu-Ganima, H.G., Hassan, M.M., Abdel-Aala, S.E., Mahmoud, W.H.: Effect of gamma irradiation on the properties of short glass fibre-reinforced elastomers based on nitrile butadiene rubber and devulcanised waste rubber. Int. J. Environ. Analy. Chem. **100**, 1763–1776 (2019). https://doi.org/10.1080/03067319.2019.1657853
9. Nakhaei, M.R., Naderi, G., Ghoreishy, M.H.R.: Fracture mechanisms and failure analysis of PA6/NBR/graphene nanocomposites by essential work of fracture. Iranian Polym. J. **30**, 975–987 (2021). https://doi.org/10.1007/s13726-021-00950-9
10. Kohol, M.M., Asgari, M., Shakur, H.R.: Effect of gamma irradiation on the structural, mechanical and optical properties of polytetrafluoroethylene sheet. Radiat. Phys. Chem. **145**, 11–18 (2018). https://doi.org/10.1016/j.radphyschem.2017.12.007
11. Cheremisinoff, N.P.: Product design and testing of polymeric materials. J. Polym. Sci. **29**, 1365–1366 (2020). https://doi.org/10.1002/pola.1991.080290918
12. Atef, S., El-Nashar, D.E., Ashour, A.H., El-Fiki, S., El-Kameesy, S.U., Medhat, M.: Effect of gamma irradiation and lead content on the physical and shielding properties of PVC/NBR polymer blends. Polym. Bull. **77**, 5423–5438 (2019). https://doi.org/10.1007/s00289-019-03022-4
13. Murat, Ş, Davut, A., Bağdagül, K.: The effect of ionizing radiation on the mechanical properties of nbr elastomers reinforced by lignin. Radiat. Phys. Chem. **168**, 108626 (2020). https://doi.org/10.1016/j.radphyschem.2019.108626

Evaluation of Passivation and Corrosion Behavior of AISI 316L SS with Low Ferrite Content Using EIS Test in Artificial Seawater Solution

Atef Chokri(✉), Amir Ben Rhouma, and Habib Sahlaoui

Laboratory of Mechanics, Materials and Processes (LMMP) LR 99ES05, University of Tunis (UT), National High School of Engineering of Tunis (ENSIT), 5 Avenue Taha Hussien B.P 56, 1008 Bab Menara, Tunis, Tunisia

chokriatef@gmail.com, {amir.benrhouma, habib.sahlaoui}@ensit.rnu.tn

Abstract. In this study, the electrochemical behavior of AISI 316L Austenitic Stainless Steel with low ferrite content, aged at temperatures ranging from 550 to 700 °C for ranging times of 50 to 80.000 h, was followed by Electrochemical Impedance Spectroscopy (EIS). This technique is becoming increasingly tool used to characterize and determine the properties of the passive film formed on the surface of stainless steels. The obtained results were comprehensively discussed and then used to develop a specific models based on Equivalent Electrical Circuits (EEC). These models were exploited to predict the formation of the passivity film during the aging of AISI 316L austenitic steels. A good agreement between the experimental results and those obtained by the models was observed.

Keywords: AISI 316L · Aging · Corrosion · Passive film · Electrochemical Impedance Spectroscopy · Equivalent Electrical Circuits

1 Introduction

Austenitic stainless steels (ASS) are widely known for their many industrial applications. Among other things, they have good corrosion resistance due to their relatively high chromium content (≥ to 13% by weight) allowing the creation of a Cr_2O_3 protective film on the surface [1]. Furthermore, these materials are characterized by a good toughness in a wide range of temperatures [2] and a high resistance to pitting and stress corrosion [3]. Therefore, these materials are suggested for high temperature application and more severe environments such as offshore platforms, nuclear power plants [4], chemical industries and medical devices [5]. The localized corrosion resistance of a stainless steel is closely related to the structure of the passivity film. Its growth and stability depend on the potential difference between the metal and the electrolyte that will generate an electric field on the surface.

M. Chouchane et al. (Eds.): CMSM 2023, LNME, pp. 19–28, 2024.
https://doi.org/10.1007/978-3-031-65007-9_3

The phenomenon of passivity is characterized by the presence of a very thin film (thickness of the order of 10 to 20 nm), called passive film or passive layer, on the surface of the metal alloy, which considerably reduces the kinetics of corrosion susceptibility. The passive film, strongly attached to the metal substrate, prevents, or renders insignificant the contact between the alloy and the aggressive agents of the surrounding environment. For the phenomenon of passivity to be established, and maintained, it is necessary that the iron-chromium alloy has a minimum chromium content of the order of 13%. Indeed, there is a threshold effect beyond which the phenomenon of passivity can be established and it cannot exist below this threshold. In addition, if the passive layer were to be destroyed locally because of accidental damage (scratches for example), it has the property of being able to be reconstituted in many environments, this reflects the self-passivation property of the metal alloy in this environment [6]. Electrochemical impedance spectroscopy (EIS) represents the most widely used technique in the characterization of the kinetic parameters and electrochemical properties of the passive film [7]. Some study were interested to the diffusion–migration phenomenon and their analytical modeling. The model most used to illustrate the formation of passivity film is that based on the Electric Equivalent Circuit (EEC) proposed by Okamoto and Shibata. [8].

The passive film of stainless steels is of the "bilayer" type, characterized by the existence of an electric field [9]. Layers are differentiated by their hydration rate. Thus, the inner layer, in contact with the alloy, is composed of oxides. The outer layer, in contact with the electrolyte, is essentially composed of hydroxides and is the most hydrated by the presence of water molecules bound to the surface.

This contribution aims to evaluate the effectiveness of models based on Electric Equivalent Circuit (EEC) to describe the behavior of the passive film with diffusion-migration impedance consideration.

2 Experimental Details and Theoretical Aspects

2.1 Material and Aging Conditions

The nominal composition of the AISI 316L SS used in this study is given in Table 1. The samples were cut from a 30 mm thick hot-rolled plate using an electrical discharge machining (EDM) cutting. The samples were then subjected to annealing heat treatment at 1070 °C for 1 h followed by water quenching.

The annealed samples were subjected to aging heat treatments in a temperature range between 550 to 700 °C for short and long durations ranging from 50 h to 80,000 h followed by air-cooling.

Table 1. Nominal composition (wt. %) of AISI 316L ASS used in this study, Fe balance.

C	S	P	Si	Mn	Ni	Cr	Mo	Cu	N	B(ppm)
0.023	0.008	0.029	0.4	1.7	12.1	17.4	2.44	0.17	0.078	38

2.2 Test Method

The impedance measurements were carried out with a VoltaLab 40 PGZ301 potentiostat controlled by Voltamaster4 software for impedance measurements, and a three-electrode cell. This cell includes the sample as the test electrode, a saturated calomel electrode (SCE) as the reference electrode and a platinum electrode as the auxiliary electrode. The synthetic seawater used in all experiments was prepared by dissolving ten inorganic salts into deionized water according to ASTM141–86 the compositions are shown in Table 2. This electrolyte is usually used in the conduct of electrochemical tests for the determination of resistance characteristics to localized corrosion. The tests were carried out with scanning frequencies ranging from 100 kHz to 100 MHz and with a disturbing signal of amplitude 10 mV. The simulation of the EEC and the optimization of its parameters were carried out on the ZsimpWin software.

Table 2. Chemical compositions of artificial seawater (g.L^{-1}).

NaCl	MgC_{12}	Na_2SO_4	CaC_{12}	KCl	$NaHCO_3$	KBr	$H3BO_3$	SrC_{12}	NaF
24.534	10.676	1.094	1.160	0.695	0.201	0.101	0.027	0.025	0.003

For the interpretation of the electrochemical behavior clearly, the EIS plots were fitted according to the results given by the equivalent circuit shown in Fig. 1. The dissolution of atoms at the alloy/electrolyte interface generates the formation of a flow of metal ions (M +). This phenomenon is modeled in the Electric Equivalent Circuit (ECC) by a capacitance $CPE_{dl}(C_{dl}, n_{dl})$ in parallel with a charge transfer resistance R_t. The film thus formed covers the surface of the alloy to prevent the diffusion of ions to the interface and therefore increase corrosion resistance. This is modeled in the ECC by a capacitance $CPE_{ox}(C_{ox}, n_{ox})$ in parallel with a resistance R_{ox}. .

The effective capacitance is calculated using Eqs. (1) and (2) [10].

$$C_{ox} = Q_{ox}^{1/n_{ox}} \left(\frac{R_s \cdot R_{ox}}{R_s + R_{ox}} \right)^{(1-n_{ox})/n_{ox}} \tag{1}$$

$$C_{dl} = Q_{dl}^{1/n_{dl}} \left(\frac{R_{ox} \cdot R_t}{R_{ox} + R_t} \right)^{(1-n_{dl})/n_{dl}} \tag{2}$$

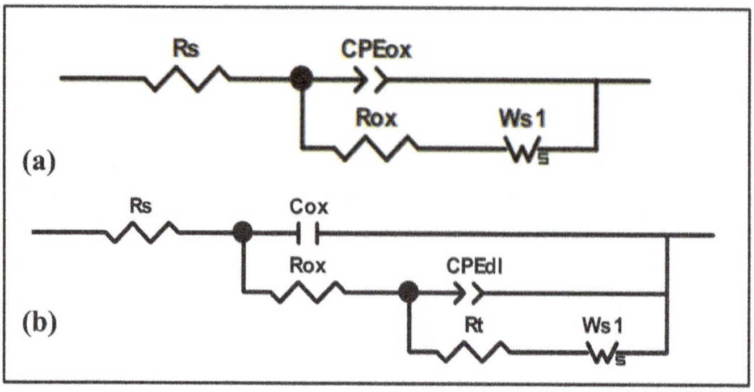

Fig. 1. Equivalent Electrical Circuits (EEC) used to adjust the experimental data obtained: (a) no sensitized state and (b) sensitized state

3 Results and Discussion

Electrochemical impedance spectroscopy is used to evaluate corrosion resistance and analyze the corrosion mechanism. Figure 2 schematizes the transfer phenomenon at the interface passive metal/solution and the passive film formation. The formation of the passive film is carried out in three steps: (1) metal atoms diffuse, with uncontrolled kinetics, from the metal lattice to the passive film. (2) Local transfer of the ions. (3) Charge transfer at the interface. The total impedance (Z_T) is given by the following equality (3) [11–13]:

$$Z_T = Z_{Metal/Film} + Z_{Film} + Z_{Film/Solution} \quad (3)$$

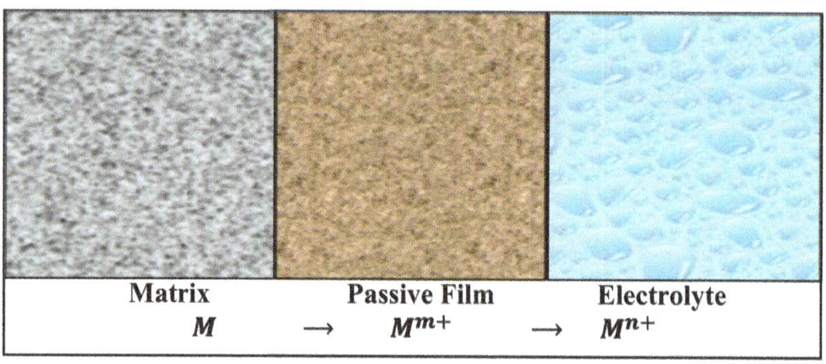

Fig. 2. Schematic description of the passive film formation phenomenon.

During the last decade, numerous analytical and empirical models have been proposed based on the Electrochemical Impedance Spectroscopy analysis of different systems. In all EEC models, two-time constants (Q, α) are used to describe the capacitive behaviors of the outer and inner layer [14, 15].

As the impedance data in the case of a solid electrolyte interface generally shows frequency dispersion attributed to "capacitive dispersion", the capacitance considered in the EEC is expressed in terms of constant phase element (CPE) [16]. The capacitance was replaced by a constant-phase element (CPE) to describe a surface with both capacitive and resistive properties. The impedance spectra fitting is given by Eq. (4):

$$Z_{CPE} = \frac{1}{Y_0(j \cdot \omega)^n} \tag{4}$$

Were, j is the imaginary number ($j = \sqrt{-1}$), ω is the angular frequency ($\omega = 2\pi f$), n represent the CPE power adjustable between 0 and 1as given by Yuan et al. [17].

Generally, CPE is related to the electrode surface heterogeneity such as roughness and/or porosity at the surface [18].

Figure 3 shows the Nyquist plots of AISI 316L SS at considered aging conditions. It can be observed that all the curves represent capacitive arcs indicating the formation of the passive films on the surface. Moreover, consistent with literature results, large arc radii mean high resistance to electrochemical dissolution and significant stability of the passive film [19–22].

Electrochemical impedance spectroscopy (EIS) makes it possible to evaluate the passivation behavior and corrosion resistance of the AISI 316L SS. To analyze the change in the electrochemical properties of the oxide layer for a no sensitized state, the equivalent circuit shown in Fig. 1.a was used. Nyquist diagrams, illustrated by Fig. 3, give the impedance spectra of the oxide layers formed on the surface of the AISI 316L ASS samples aged between 550 and 700 ∘C. They exhibited a large capacitive loop, corresponding to the R_{ox} coupled with CPE_{ox}, and a low frequency tail, attributed to the Warburg impedance W_s (Fig. 3 a). The same EEC was used to adjust the experimental results obtained for the repassived state (Fig. 3 b). The fitting (extracted parameters) shown in Table 3 and Table 4.

This CEE shows a suitable fit for a non-sensitized state. At high frequencies, the behavior is related to the charge transfer resistance, then at low frequencies, the phenomena related to the diffusion processes dominate and we observe the Warburg impedance. An element $R_t//CPE_{dl}(C_{dl}, n_{dl})$ additional was introduced in parallel (Fig. 3.c). The two parts of the equivalent electrical circuit were nested to consider both the presence of oxygen and the gradual modification of the passive film. The calculated values of the parameters are listed in Table 5. For the different experimental conditions, the values of n_{ox} characterizing capacitive behavior are relatively similar and not depend on aging time, ($n_{ox} \approx 0{,}9$). The system shows that the behavior is attributed to surface heterogeneity [23] or to distributed time constant for charge transfer reactions [24]. The total impedance Bode diagram (|Z| vs frequency) and the Bode phase angle diagram (θ vs Frequency) are shown in Fig. 4. The total impedance diagrams show that the impedance gradually increases over time and goes from 3.8 to 4.5 (logarithmic value) for the sensitized states (Fig. 4b). These results show the evolution and growth of the protective film.

In Bode phase diagrams, phase angles were always less than 90°. The different frequencies showed, in all states, a similar contribution to the corrosion impedance. Only differences in phase diagram were observed for repassived states at high temperatures and long holding time. The maximum peak angles occurring in the high frequency zone can be attributed to the formation of a protective corrosion product film [25–28].

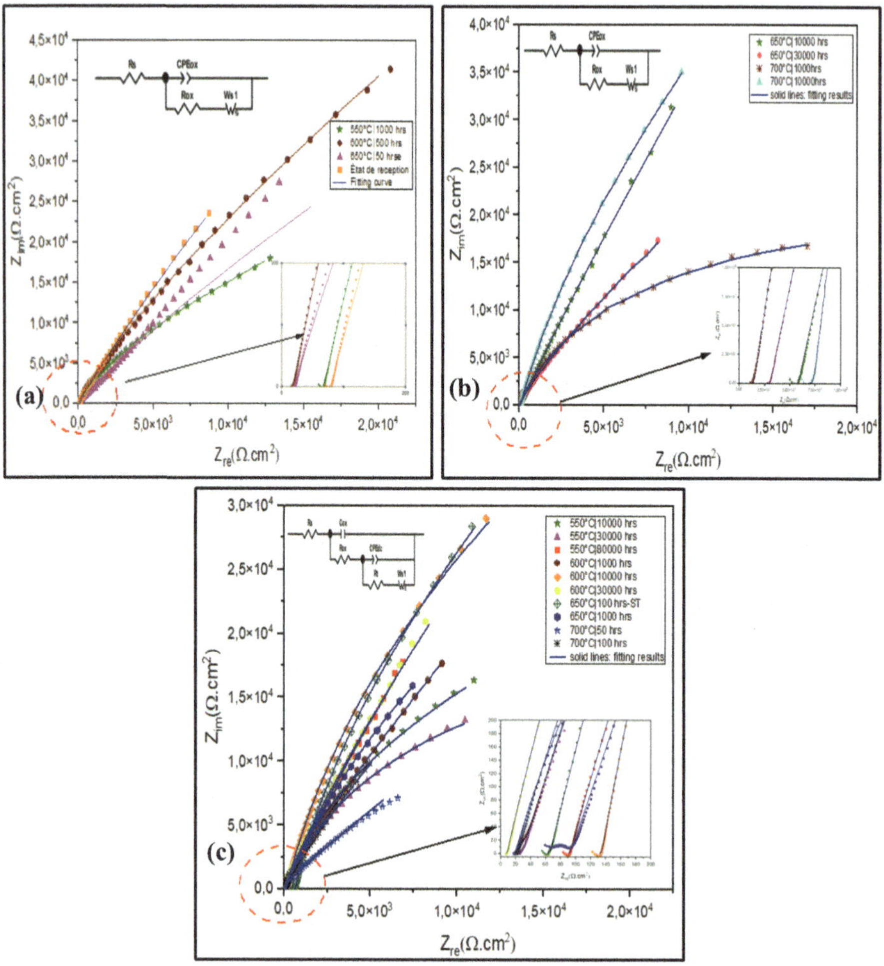

Fig. 3. Impedance spectra of oxide layers formed on the surface of AISI 316L SS aged between 550 and 700 ∘C. The symbols are the measured data, and the lines are the results of the adjustment: (**a**) As receiving and non-sensitized states, (**b**) repassived state and (**c**) sensitized state.

Table 3. Parameters fitted for the equivalent circuits corresponding to the EIS of the unsensitized state (Fig. 1a).

$T symbol (symbol °C)$	t(h)	R_s $(\Omega \cdot cm^2)$	$R_{ox} \cdot 10^4$ $(\Omega \cdot cm^2)$	$Q_{ox} \cdot 10^{-5}$ $(\Omega^{-1} \cdot s^{n_{ox}})$	n_{ox}	$W_{S1} \cdot 10^{-5}$ $(\Omega^{-1} \cdot s^{0,50})$	$\sum \chi^2 \cdot 10^{-4}$
As received state		78.51	9.593	4.52	0.85	1.56	6.96
550	1000	67.05	2.129	3.94	0.86	4.45	4.48
600	500	17.59	1.294	1.92	0.88	1.22	11.4
650	50	20.49	2.879	2.24	0.83	3.40	27.4

Table 4. Parameters fitted for equivalent circuits corresponding to the EIS of the repassived state (Fig. 1b).

$T symbol (symbol °C)$	$t(h)$	R_s $(symbol \Omega \cdot cm^2)$	R_{ox} 10^4 $(symbol \Omega \cdot cm^2)$	Q_{ox} 10^{-5} $(symbol \Omega^{-1} \cdot s^{n_{ox}})$	n_{ox}	W_{S1} 10^{-6} $(symbol \Omega^{-1} \cdot s^{0.5})$	$\sum \chi^2 10^{-4}$
650	10000	58.23	4.893	2.88	0.85	2.65	6.47
650	30000	30.02	1.856	3.98	0.85	21.96	3.40
700	1000	13.18	3.648	3.77	0.88	197.2	36.5
700	10000	71.48	2.429	3.65	0.91	31.75	6.11

Table 5. Parameters fitted for the equivalent circuits corresponding to the EIS of the sensitized state (Fig. 1c).

T (°C)	t (h)	R_s $(\Omega.cm^2)$	R_{OX} $(\Omega.cm^2)$	C_{OX} $(\mu F \cdot cm^{-2})$	$Q_{dl} 10^{-5}$ $(\Omega^{-1} s^{n_{dc}})$	**n$_{dl}$**	$R_t \cdot 10^3$ **(Ω.cm2)**	$W_{S1} . 10^{-5}$ $(\Omega^{-1} s^{0,5})$	$\sum \chi^2 \cdot 10^{-4}$
550	10000	62.48	5.234	3.53	5.12	0.85	36.04	7.60	5.87
550	30000	24.91	9.336	4.83	6.08	0.80	42.26	4.89	3.72
550	80000	90.7	3.701	2.77	4.00	0.83	5.337	1.23	3.62
600	1000	19.78	20.39	2.15	2.78	0.79	19.79	1.51	6.07
600	10000	132.1	4.767	4.88	3.59	0.87	131.9	6.40	1.63
600	30000	8.276	1.817	7.78	4.06	0.82	10.85	1.39	1.12
650	100	578.2	44.36	0.04	4.05	0.90	91.62	2.03	1.08
650	1000	64.27	29.36	0.18	7.13	0.82	114.2	1.45	7.01
700	50	18.88	4.665	5.89	4.69	0.82	2.661	10.5	4.39
700	100	21.83	5.220	5.35	4.39	0.82	3.267	3.30	3.33

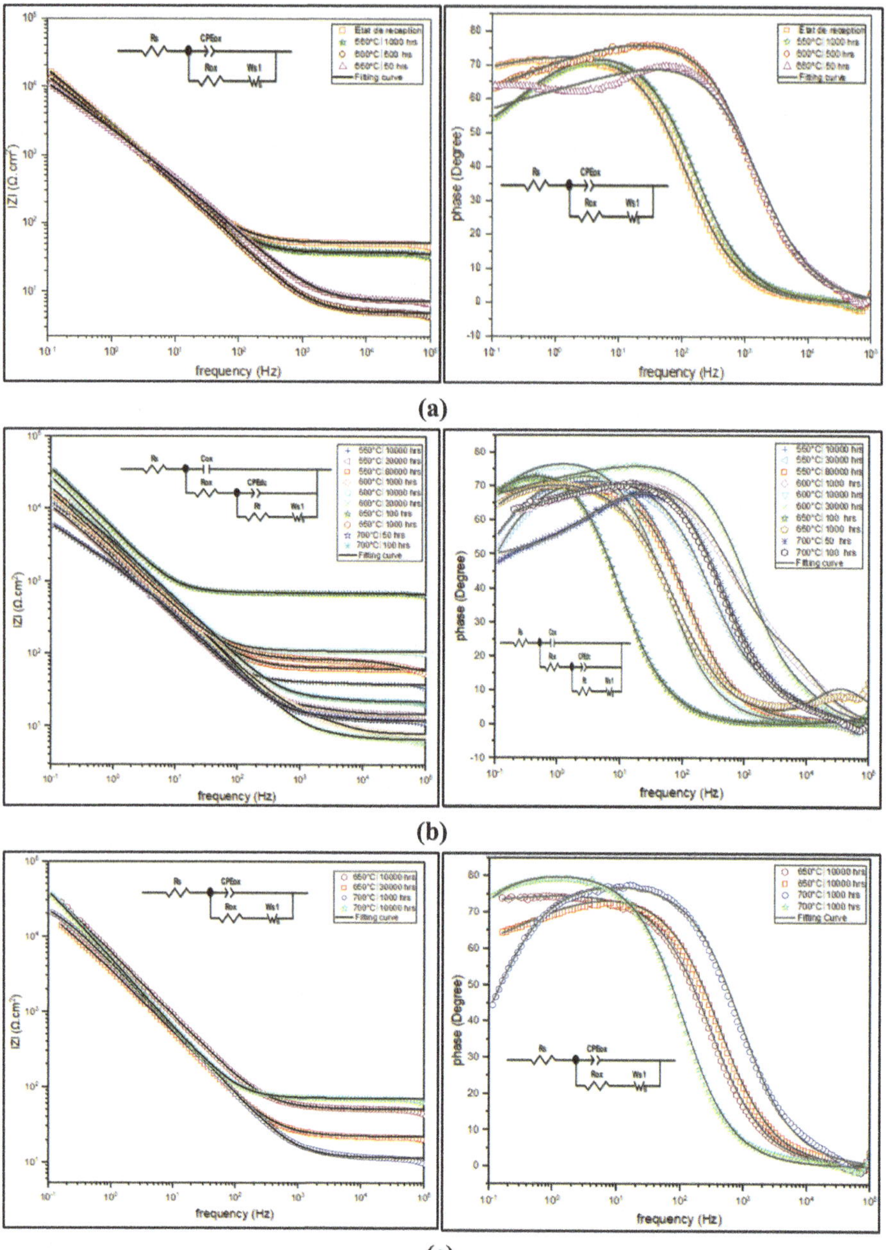

Fig. 4. Bode plots of EIS spectra for corrosion of oxide layers formed on the surface of AISI 316L SS aged between 550 and 700 ∘C **(a)** As received and non-sensitized state, **(b)** sensitized state and **(c)** repassived state.

4 Conclusion

The passivation film formed during aging of AISI 316L austenitic stainless steels was monitored and characterized by Electrochemical Impedance Spectroscopy (EIS). The results obtained were then exploited to establish empirical models based on Electrical Circuits Equivalents (CEE) to illustrate the formation of the passivity film. The circuit parameters were then determined for the as received, sensitized and unsensitized states with a good accuracy.

References

1. Ke, Y., Yibin, R.: Nickel-free austenitic stainless steels for medical applications. Sci. Technol. Adv. Mater. **11**, 014105 (2010). https://doi.org/10.1088/1468-6996/11/1/014105
2. Talha, M., Behera, C.K., Sinha, O.P.: A review on nickel-free nitrogen containing austenitic stainless steels for biomedical applications. Mater. Sci. Eng. C **33**, 3563–4357 (2013). https://doi.org/10.1016/j.msec.2013.06.002
3. Mohammadi Zahrani, E., Saatchi, A., Alfantazi, A.: Pitting of 316L stainless steel in flare piping of a petrochemical plant. Eng. Fail. Anal. **17**, 810–817 (2010). https://doi.org/10.1016/j.engfailanal.2009.10.015
4. Wang, Z., Di-Franco, F., Seyeux, A.: Passivation-induced physicochemical alterations of the native surface oxide film on 316L austenitic stainless steel. J. Electrochem. Soc. **166**, C3376–C3388 (2019). https://doi.org/10.1149/2.0321911jes
5. Cindra, M., Fonseca, I.N., Bastos, E.: Characterization of oxides of stainless steel UNS S30400 formed in offshore environment. Corros. Sci. **55**, 3439 (2012). https://doi.org/10.1016/j.corsci.2011.09.029
6. Sidhom, H., Ben Rhouma, A., Braham, C., Ledion, J.: Préparation des surfaces et tenue à la corrosion localisé des aciers inoxydables austénitiques, Mater.Tech. **86**, 31–37 (1998). https://doi.org/10.1016/j.jallcom.2017.02.273
7. André, J., Antoni, L., Petit, J.P.: Electrical contact resistance between stainless steel bipolar plate and carbon felt in PEFC: a comprehensive study. Int. J. Hydrogen Energy **34**, 3125–3133 (2009). https://doi.org/10.1016/j.ijhydene.2009.01.089
8. Wang, C., Yu, J.: Comparison of the corrosion and passivity behavior between CrMnFeCoNi and CrFeCoNi coatings prepared by argon arc cladding. J Mater. Res. Technol. **9**(4), 8482–8496 (2020). https://doi.org/10.1016/j.jmrt.2020.05.093
9. Kim, J.J., Young, Y.M.: Study on the passive film of type 316 stainless steel. Int. J. Electrochem. Sci. **8**, 11847–11859 (2013). https://doi.org/10.1016/S14523981(23)13227-5
10. Hirschorna, B., Orazema, M.E., Tribollet, B., Vivierb, V.: Determination of effective capacitance and film thickness from constant-phase-element parameters. Electrochim. Acta **55**, 6218–6227 (2010). https://doi.org/10.1016/j.electacta.2009.10.065
11. Horvat-Radošević, V., Kvastek, K., Hodko, D., Pravdić, V.: Impedance of anodically passivated Fe80B20 over potentials from passive state to oxygen evolution. Electrochim. Acta **39** (1), 119 (1994). https://doi.org/10.1016/0013-4686(94)85018-6
12. Armstrong, R.D., Edmondson, K.: The impedance of metals in the passive and transpassive regions Electrochim. Acta **18**, 937 (1973). https://doi.org/10.1016/0013-4686(73)87025-2
13. Yin, Q., Kelsall, G.H.: Mathematical models for time-dependent impedance of passive. Electrodesal J. Electrochem. Soc. **148**(3), A200 (2001). https://doi.org/10.1149/1.1345956
14. Pan, J., Thierry, D., Leygraf, C.: Electrochemical impedance spectroscopy study of the passive oxide film on titanium for implant application. Electrochim. Acta **41**, 11431153 (1996). https://doi.org/10.1016/00134686(95)00465-3

15. Wang, Z.B., Hu, H.X., Zheng, Y.G., Ke, W., Qiao, Y.X.: Comparison of the corrosion behavior of pure titanium and its alloys in fluoride-containing sulfuric acid. Corros. Sci. **103**, 50–65 (2016). https://doi.org/10.1016/j.corsci.2015.11.003
16. Jorcin, J.B., Orazem, M.E., Pebere, N.: CPE analysis by local electrochemical impedance spectroscopy. Electrochim. Acta **51**, 1473–1479 (2006)
17. Yuan, S.J., Pehkonen, S.O.: Surface characterization and corrosion behavior of 70/30 Cu–Ni alloy in pristine and sulfide-containing simulated seawater. Corros. Sci. **49**(3), 1276–1304 (2007). https://doi.org/10.1016/j.corsci.2006.07.003
18. Ariyosh, K., Siroma, Z.: Electrochemical impedance spectroscopy part 1: fundamentals J-stage: October 31 (2022). https://doi.org/10.5796/electrochemistry.22-66071
19. Wang, L., Dou, Y., Han, S., Wu, J., Cui, Z.: Influence of sulfide on the passivation behavior and surface chemistry of 2507 super duplex stainless steel in acidified artificial seawater. Appl. Surf. Sci. **504**, 144340 (2020). https://doi.org/10.1016/j.apsusc.2019.144340
20. Yang, Y.Y., et al.: Enhancements of passive film and pitting resistance in chloride solution for 316LX austenitic stainless steel after SN alloying. Acta Metall. Sin. (Engl. Lett.) **32**, 98–106 (2019). https://doi.org/10.1007/s40195-018-0855-9
21. Shin, H., Mansfeld, F.: Concerning the use of the Kramers-Kronig transforms for the validation of impedance data. Corros. Sci. **28**(9), 933–938 (1988)
22. Wang, L., Liang, J., Li, H., Cheng, L., Cui, Z.: Quantitative study of the corrosion evolution and stress corrosion cracking of high strength aluminum alloys in solution and thin electrolyte layer containing Cl. Corros. Sci. **178**, 109076 (2021)
23. Al-Ali, A., Elwakil, A., Maundy, B., Westwick, D.: A generic impedance modeling technique. Int. J. Electr. Commun. (2020)
24. Macdonald, J.R.: Power-law exponents and hidden bulk relation in the impedance spectroscopy of solids. J. Electroanal. Chem. **378**, 17–29 (1994). https://doi.org/10.1016/0022-0728(94)87053-5
25. Wang, Y.Z., Beccaria, A.M., Poggi, G.: The effect of temperature on the corrosion behaviour of a 70/30 Cu-Ni commercial alloy in seawater. Corros. Sci. **36**, 1277–1288 (1994). https://doi.org/10.1016/0010-938X(94)90181-3
26. Ai, H., Shen, M., Sun, H.: Effects of O2- additive on corrosion behavior of Fe–Cr–Ni alloy in molten fluoride salts. Corros. Sci. **150**, 175–182 (2019) https://doi.org/10.1016/j.corsci.2019.01.040
27. Ben Rhouma, A., Amadou, T., Sidhom, H., Braham, C.: Correlation between microstructure-and intergranular corrosion behavior of low delta-ferrite content AISI 316L aged in the range 550–700°C. J. Alloys Comp. **708**, 871–886 (2017). https://doi.org/10.1016/j.jallcom.2017.02.273
28. Huajie, T., Jianlin, S., Jinshan, P.: Electrochemical and localized corrosion behavior of 316L SS/copper/316LSS sandwich composite in chloride-containing environment. Corros Sci. **209**, 110719 (2022). https://doi.org/10.1016/j.corsci.2022.110719

Experimental and Analytical Study on the Thermomechanical Behavior of PC/ABS Blend

Fatma Hentati[1(✉)], Ridha Mnif[2], Naila Hfaiedh[3], Johan Petit[4], and Lorenzo Faelli[3]

[1] Laboratory of Applied Mechanics and Engineering (LR-MAI), National Engineering School of Tunis ENIT, 1002 Tunis, Tunisia
fatma.hentati1@gmail.com

[2] Laboratory of Electromechanical Systems (LASEM), National Engineering School of Sfax, University of Sfax, BP 1173, 3038 Sfax, Tunisia

[3] Léonard de Vinci Pôle Universitaire, Research Center, 92916 Paris, Courbevoie, France
naila.hfaiedh@devinci.fr

[4] Laboratory of Energetic Mechanic and Electromagnetism, Paris Nanterre University, 50, Rue de Sèvres, 92410 Ville d'Avray, France
johannpetit@parisnanterre.fr

Abstract. In this study, we examine the small and large strain tensile properties of the PC/ABS blend through experimental analysis and the application of predictive analytical models. Our investigation includes an investigation of the effects of temperature and strain rate on the tensile response, with strain rates ranging from 1.25 10^{-4} s^{-1} to 1.25 10^{-1} s^{-1} and temperatures ranging from 20 °C to 120 °C. The experimental findings prove that the tensile behavior of the material blend is sensitive to both strain rate and temperature. As temperature increases, the yield strength and strain of the material blend decrease significantly, while the modulus experiences a slight decrease. Following an analysis of material parameters using various constitutive models such as G'Sell and Jonas, and DSGZ, a phenomenological model named the DSGZ modified model by "Zhu et al." was selected to characterize the tensile behavior of PC/ABS. In comparison with other models, the "Zhu et al." model demonstrates a higher level of accuracy. The proposed model exhibits proficient capability in simulating and predicting tensile behavior across various strain rates. However, regarding the influence of temperature, the predicted responses notably deviate from the experimental data.

Keywords: PC/ABS blend · Tensile behavior · Strain rates and temperature dependence · predictive analytics models

1 Introduction

The PC/ABS blend offers exceptional properties derived from both components, combining the high impact strength and technical performance of PC with the superior toughness of ABS, all at a more cost-effective price point [1–3]. Thanks to their mechanical and thermal properties and processability, they have attracted considerable attention in the

M. Chouchane et al. (Eds.): CMSM 2023, LNME, pp. 29–42, 2024.
https://doi.org/10.1007/978-3-031-65007-9_4

automotive industry [4, 5]. An abundance of research within the field of engineering has been dedicated to understanding the deformation characteristics of thermoplastic polymers. For example, Yin et al. [6] explored the tensile deformation behaviors of PC, ABS, and PC/ABS blends by subjecting them to progressively increasing strain rates at room temperature. Among the various constitutive models employed for this purpose, G'Sell and Jonas [7] designed a concise phenomenological model specifically crafted to describe semi-crystalline polymers. Duan et al. [8] introduced a constitutive model that combines four models: the Johnson-Cook model, the G'Sell Jonas model, the Matsuoka model, and the Brooks model. This model is applicable to both glassy and semi-crystalline thermoplastics, explicitly demonstrating the dependency of true stress on true strain, true strain rate, and absolute temperature. Recent investigations by researchers such as Cao et al. [9] and Yu et al. [10] have explored into the tensile and compressive behavior of Polycarbonate (PC), exploring variations in strain rates and temperatures. Additionally, both studies introduced nonlinear aspects of PC's mechanical behavior, including apparent yielding and softening under load, while also examining the effects of strain rate and temperature. Simultaneously, Wang and co-workers [11] employed the DSGZ model to explore the compression behavior of PC/ABS across a wide range of strain rates and temperatures. They investigated both monotonic and cyclic loading at room temperature. Results indicated that the model struggled to accurately replicate polymer deformation at high strain rates when using identical material coefficients for both low and high strain rate deformations. In parallel, Zhu and co-workers [12] proposed a new constitutive model derived from the DSG model to forecast the tensile and compressive behavior of two thermoplastics: PEEK and PC. Their model considers the effect of strain, strain rate, and temperature.

This study aims to analyze the influence of temperature and strain rate on the tensile behavior of PC/ABS blends, while also evaluating the effectiveness of various models in predicting their tensile behavior. A comparative investigation is carried out among several commonly employed constitutive models, including G'Sell, DSGZ, and the model presented by "Zhu et al.". Experimental results were employed to quantify the parameters of these constitutive models.

2 Experimental Set up

2.1 Material Preparation and Samples

The commercial PC/ABS blend is known as T45PG and produced by BAYER, has a density of 1.1 g/cm3. Prior to processing, the pellets were oven dried to remove moisture. Injection molding was performed using an FCSHN 125SV injection molding machine with a temperature range of 235–260 °C and a screw diameter of 34 mm. Specimen dimensions and geometry were in accordance with ISO 572 and ISO 825 standards [13]. The analysis temperatures included room temperature, 50 °C, 80 °C, and 120 °C.

2.2 Tensile Tests

Tensile tests were conducted following standard procedures using an Instron 3R universal tensile machine equipped with a load cell, enabling precise force measurements in the range of 2.5 kN. To minimize testing errors, three specimens were tested for each strain rate for each condition. The tensile test was carried out several times for each of three specimens at the fixed position of extensometer arms (20 mm gauge length). The testing setup also included a self-assembly environmental chamber. Tensile tests on the selected PC/ABS samples for this study were performed at engineering strain rates of 1.25 10^{-4}, 1.25 10^{-3}, 1,25 10^{-2}, and 1.25 10^{-1} s^{-1}.

3 Typical Monotonic Tensile Analysis

Figure 1 depicts the typical deformation curve at a strain rate of 1.25 10^{-2} s^{-1} and a room temperature of 20 °C. The experimental curve illustrates the variation of true stress as a function of true strain. Notably, the deformation processes of PC/ABS can be segmented into four distinct phases: elastic viscoelasticity, yielding, strain softening, and strain hardening [11, 14].

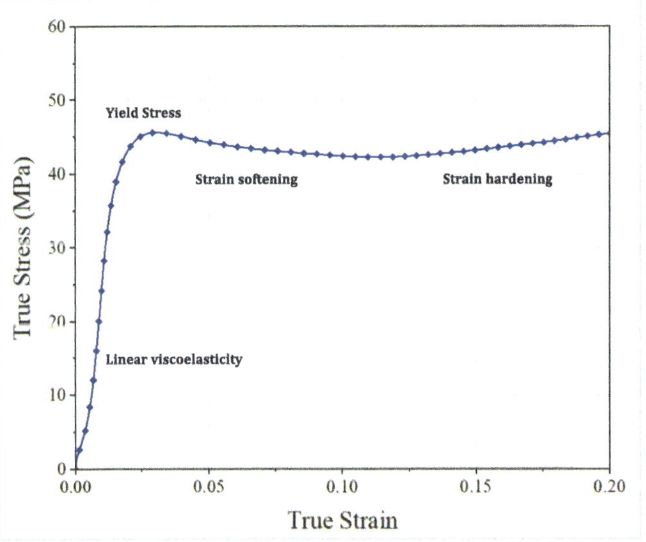

Fig. 1. Experimental true stress-strain curve of PC/ABS blend at room temperature and strain rate of 1.25 × 10^{-2} s^{-1}.

Polymer materials are subjected to diverse mechanical loading conditions, with temperature and strain rate playing significant roles. This experimental section explores the impact of these parameters on the tensile behavior. As a precautionary measure regarding the extensometer, deformation was limited to 20%.

As depicted in Fig. 1, the mechanical properties of PC/ABS blend, including Young's modulus E and Yield stress σ_y, are 1490 MPa and 45 MPa, respectively. Both Yield stress and Young's modulus exhibit sensitivity to variations in strain rates and temperature. The thermomechanical behaviors of the PC/ABS will be discussed in the following section, emphasizing its response to these parameters' changes.

3.1 Strain Rate Effect on the Tensile Behavior

Figure 2a represents the variation of the true stress as a function of true strain by ranging the strain rates from 1.25 10^{-4}, 1.25 10^{-3}, 1,25 10^{-2}, and 1.25 10^{-1} s^{-1} at room temperature of 20 °C. The gradual increase in strain rate is accompanied by a clear rise in true stress. The optimal true stress yield value is achieved at 52 MPa for a strain rate of 1.25 10^{-1} s^{-1}. Notably, when the strain rate abruptly increases from 1.25 10^{-4} to 1.25 10^{-1} s^{-1} during loading, there is an immediate and significant increase in true stress. These results corroborate the findings of Yin, Z. N. et al. [6], who examined the tensile deformation behavior of PC, ABS, and PC/ABS blends across various strain rates. The alignment of these results may be explained by the observed phenomenon: as the strain rate increases, the molecular mobility of polymer chains decreases, leading to increased stiffness. Concurrently, Wang, H. et al. [11] conducted a study examining the mechanical behavior of PC/ABS blends under both monotonic and cyclic loading, while accounting for variations in strain rates and temperatures.

Analysis of the curves depicted in Fig. 2b reveals that the strain rate significantly influences both the elastic modulus and the yield stress. The yield stress increases notably and more rapidly at high strain rates, ranging from 1.25 10^{-2} to 1.25 10^{-1} s^{-1}, compared to low strain rates, particularly evident at 1.25 10^{-2} s^{-1}. The increase in yield stress and elastic modulus values at high strain rates at room temperature may be attributed to the significant phenomenon of strain hardening and increased toughness [18, 19].

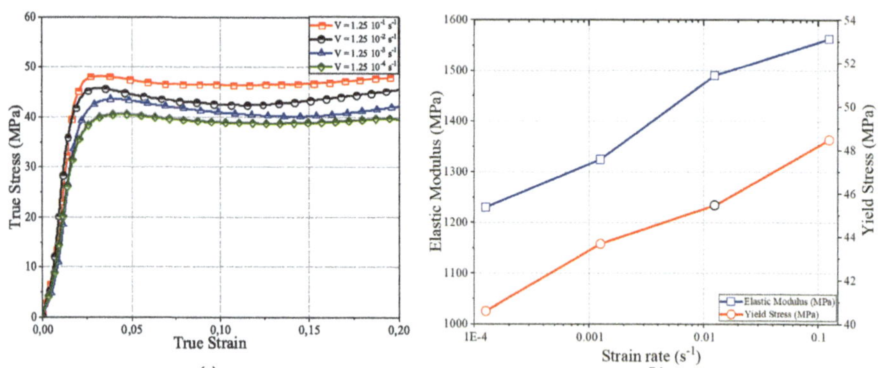

(a) (b)

Fig. 2. (**a**) Experimental true stress-strain curves of PC/ABS blend over a wide range of strain rates, (b) Elastic Modulus and Yield Stress as a function of the strain rate at room temperature.

3.2 Temperature Effect on the Tensile Behavior

Figure 3a displays the tensile true stress–true strain curves at a constant strain rate of 1.25 10^{-2} s^{-1} for different temperatures. It is observed that the strain hardening rate decreases as the temperature approaches the glass transition temperature [21]. In Fig. 3b, it is evident that the yield stress and elastic modulus decrease with increasing temperature. Both parameters show a clear dependency on temperature and strain rate. As a result, an increase in strain rate or a decrease in temperature leads to higher yield stress [22]. However, with increasing temperature, true strain notably decreases [21]. This phenomenon is explained to the reduced molecular mobility of polymer chains, making them stiffer [6, 23].

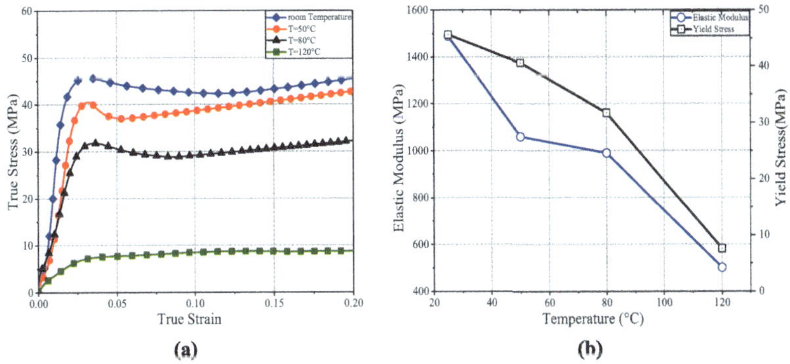

Fig. 3. (a) Experimental true stress-strain curves of PC/ABS blend over a wide range of temperatures, (b) Elastic Modulus and Yield Stress as a function of the temperatures at strain rate 1.25 10^{-2} s^{-1}.

The material blend's yield stress and elastic modulus decrease with increasing temperature, as illustrated in Fig. 3.b. Both parameters exhibit a dependence on temperature and strain rate. Therefore, an increase in strain rate or a decrease in temperature leads to higher yield stress [24]. Notably, as temperature increases, true strain noticeably decreases [23, 25].

4 Analytical Results

4.1 Description of the Constitutive Models' and Identification of Their Parameters

- **G'Sell Model**

  G'Sell and Jonas have suggested a phenomenological model for the stress-strain behavior of semi-crystalline polymers [7, 24]. Their model is presented as follow:

$$\sigma = K\,(1 - e^{\frac{\varepsilon}{\varepsilon_0}})\, e^{\frac{\gamma}{2}\varepsilon^2}\,(\dot{\varepsilon})^m \tag{1}$$

where σ is equivalent stress, ε is equivalent strain, $\dot{\varepsilon}$ is equivalent strain rate, K, ε_0, γ and m are material parameters. Two concepts are combined into one equation in the model: viscoelasticity and viscoplasticity. Plastic strain-hardening, the strain rate-dependence of the viscous behavior, and the initial non-linear elastic behavior in semicrystalline polymers have all been modeled by three functions.

The function $(1 - e^{\frac{\varepsilon}{\varepsilon_0}})$ refers to initial nonlinear elastic behavior, the function $e^{\frac{\gamma}{2}\varepsilon^2}$ describes plastic strain hardening, and the function $(\dot{\varepsilon})^m$ defines viscous behavior. However, it is noteworthy that this model currently does not capture the intrinsic strain-softening behavior exhibited by glassy polymers, as long as the model is not temperature dependent.

In this section, the model of G'Sell and Jonas was used to validate its applicability by testing the tensile properties of glassy PC/ABS polymers. The Table 1 depicts the model parameters. For the identified parameters, the reference conditions are: room temperature of 20 °C and the reference strain rate of 1.25 10^{-2} s^{-1}.

Origin software was used to identify all the parameters of the G'Sell Model and to generate the predictive curves.

Table 1. G'Sell and Jonas model's parameters identification.

K	**ε0**	γ	**m**
73.07	0.0135	3.17	0.087

- **DSGZ model**

  A comprehensive constitutive model called the DSGZ was introduced by Duan et al. [10] and it was derived from four different constitutive models. This approach extends its applicability to semi-crystalline thermoplastics, not just glassy polymers. The formulation of the DSGZ model is as follows:

$$\sigma(\varepsilon, \dot{\varepsilon}, T) = K\left\{f(\varepsilon) + \left[\frac{\varepsilon\, e^{\left(1-\frac{\varepsilon}{C_3 . h(\dot{\varepsilon}, T)}\right)}}{C_3\, h(\dot{\varepsilon}, T)} - f(\varepsilon)\right] e^{[\ln(g(\dot{\varepsilon}, T)) - C_4]\,\varepsilon}\right\} h(\dot{\varepsilon}, T) \quad (2)$$

$$f(\varepsilon) = \left(e^{-C_1\varepsilon} + \varepsilon^{C_2}\right)\left(1 - e^{-\alpha\varepsilon}\right) \quad (3)$$

$$h(\dot{\varepsilon}, T) = (\dot{\varepsilon})^m\, e^{\frac{a}{T}} \quad (4)$$

where σ, ε, $\dot{\varepsilon}$ and T are the stress, strain, strain rate and temperature, respectively.

$g(\dot{\varepsilon}, T)$ is the dimensionless form of $h(\dot{\varepsilon}, T)$.

K, C_1, C_2, C_3, C_4, a, m and α represent the eight material parameters in DSGZ model.

$\frac{\varepsilon\, e^{\left(1-\frac{\varepsilon}{C_3\, h(\dot{\varepsilon}, T)}\right)}}{C_3\, h(\dot{\varepsilon}, T)}$ describes shift and strain softening behaviors at small strains a function of strain rate and temperature.

$f(\varepsilon)$ represents the strain hardening behaviors at large strains.

$e^{[\ln(g(\dot{\varepsilon},T))-C_4]\,\varepsilon}$ is used to correlate deformation behaviors at small and large strains. $h(\dot{\varepsilon}, T)$ represents the value of stress sensitivity of strain rate and temperature.

The DSGZ model has demonstrated its capability to describe the mechanical behavior of both PC and ABS [8, 26], which are components of PC/ABS. Hence, the DSGZ model is supposed to be suitable for predicting the mechanical behavior of PC/ABS. For the identified parameters showed in Table 2, the reference conditions are: room temperature of 20 °C and the reference strain rate of 1.25 10^{-2} s^{-1}. Origin software was used to identify all the parameters of the G'Sell Model and to generate the predictive curves.

Table 2. DSGZ model's parameters identification.

C_1	C_2	C_3	C_4	K	α	m	a
−1.84	−0.15	−8,51 10^{24}	85.64	23.56	150.62	0.35	634

- **"Zhu et al." model**

The partial derivative of f (ε, ˙ε, T) with respect to ε can be obtained in Eq. 5:

$$f(\varepsilon, \dot{\varepsilon}, T) = K_1\, \varepsilon^n\, e^{-\frac{\varepsilon}{\mu\, h(\dot{\varepsilon},T)}} \tag{5}$$

where: ε,$\dot{\varepsilon}$ and T are the strain, strain rate and temperature, respectively.

K_1, n and μ are material parameters.

Another term g (ε) refers to the corresponding term in DSGZ model to predict the strain hardening behavior at large strains (larger than $\varepsilon_{critical}$) after the strain softening behavior at small strains.

$\varepsilon_{critical}$ defines the critical strain between the strain softening and the strain hardening behaviors. As the strain approaches a value close to $\varepsilon_{critical}$, the transition in flow stress from small-strain to large-strain cases occurs smoothly.

$$g\ (\varepsilon, \dot{\varepsilon}, T) - K_2\left(e^{-C_1\,\varepsilon} + \varepsilon^{C_2}\right)\left(1\ -\ e^{-\alpha\,\varepsilon}\right)h(\dot{\varepsilon}, T) \tag{6}$$

where, ε is the true strain, ˙ε is the strain rate, T is the absolute temperature.

K_2, C_1, C_2 and α are material parameters.

The value of g $(\varepsilon, \dot{\varepsilon}, T)$ increases with the rise in strain, aligning with the strain-hardening behavior observed at large strains.

The constitutive model is then expressed as follow:

$$\sigma\ (\varepsilon, \dot{\varepsilon}, T)\ =\ f(\varepsilon, \dot{\varepsilon}, T)\quad if\ \ 0 \le \varepsilon < \varepsilon_{critical}$$

$$\sigma\ (\varepsilon, \dot{\varepsilon}, T)\ =\ g(\varepsilon, \dot{\varepsilon}, T)\quad if\ \ \varepsilon > \varepsilon_{critical}$$

In the "Zhu et al." modified DSGZ model, two terms, u $(\varepsilon, \dot{\varepsilon}, T)$ and v $(\varepsilon, \dot{\varepsilon}, T)$, are suggested to allow a smooth transition between strain hardening and softening behaviors. The two transition functions in this model are:

$$u\ (\varepsilon, \dot{\varepsilon}, T) = \frac{1}{1 + k\ e^{w\,\varepsilon - \lambda\, h(\dot{\varepsilon},T)}} \tag{7}$$

$$v\ (\varepsilon, \dot{\varepsilon}, T) = \frac{1}{1 + k\ e^{\lambda\ h(\dot{\varepsilon}, T) - w\ \varepsilon}} \tag{8}$$

It is observed that the parameter k is exclusively associated with the $u\ (\varepsilon, \dot{\varepsilon}, T)$ function [14].

ε is the true strain, $\dot{\varepsilon}$ is the strain rate and T is the absolute temperature.

k, w and λ are material parameters.

Hence, according to the methodology described above, the proposed model called "Zhu et al." is described in Eq. 9 as follow:

$$\sigma\ (\varepsilon, \dot{\varepsilon}, T) = f(\varepsilon, \dot{\varepsilon}, T)u(\varepsilon, \dot{\varepsilon}, T) + g(\varepsilon, \dot{\varepsilon}, T)v(\varepsilon, \dot{\varepsilon}, T) \tag{9}$$

k, w, λ, n, μ, C_1, C_2, α, K_1, K_2, m and a represent the twelve material parameters in this new modified model. These material parameters are presented in Table 3 in order to validate the new modified constitutive model. The reference conditions are: room temperature of 20 °C and the reference strain rate of 1.25 10^{-2} s^{-1}. Origin software was used to identify all the parameters of the G'Sell Model and to generate the predictive curves.

Table 3. Zhu model's parameters identification.

k	K_1	K_2	ω	λ	μ	n	C_1	C_2	a	m	α
0.85	50379	3.02	380	0.35	0.0038	2.87	−2.08	−1.07	750	-0.54	0.8

4.2 Comparison Between the Experimental and the Predictive Models Results

In the current study, the root mean squared error ($RMSE$) and coefficient of determination R^2 are employed to assess each model's prediction accuracy. A high R^2 value and a low $RMSE$ value indicate good precision in predictions. Their expressions are as follows:

$$R^2 = \frac{\sum_{i=1}^{N} \left(E_i - \overline{E}\right)^2 - \sum_{i=1}^{N} (E_i - P_i)^2}{\sum_{i=1}^{N} \left(E_i - \overline{E}\right)^2} \tag{10}$$

$$RMSE = \sqrt{\frac{1}{N} \sum\nolimits_{i=1}^{N} (E_i - P_i)^2} \tag{11}$$

The values E_i and P_i refers to the experimental and predicted values, respectively. $\overline{E}$ and $\overline{P}$ denote the mean values of E_i and P_i, respectively, and N is the number of data used in the evaluation process.

The Fig. 4(a) depicts the experimental data and the prediction curves for the G'Sell and Jonas, DSGZ and the "Zhu et al." models.

The model predicted by G'Sell and Jonas shows disagreement with the experimental curve across all regions illustrated in the experimental curve: linear viscoelasticity, yielding, strain softening, and strain hardening. Notably, the strain-softening behavior is not detected. While the curves exhibit proximity mainly in the strain-hardening region (plastic deformation), it's essential to mention that this region is not a focal point in our study. This clear disagreement between predicted and experimental results aligns with warnings from scholars [9, 10], who alerted that the G'Sell and Jonas model is limited to predicting semi-crystalline polymers.

A number of features of the curve, including the presence of the maximum stress seen in the early phases of deformation, the following strain softening, and the hardening phenomena seen at large strains, are demonstrated to be precisely reproduced by the DSGZ model. Therefore, the shift and strain-softening behaviors at small strains as well as the strain-hardening behaviors at large strains may be effectively described by the DSGZ model, which can be used to consistently depict the flow stress behavior across a variety of thermoplastics.

Comparing the prediction curves of the two constitutive models, G'Sell and DSGZ, with the experimental findings, the "Zhu et al." model proves a much better agreement. The strain softening, strain hardening, yield behavior, linear and non-linear elastic, and yield behavior are all well-represented by the "Zhu et al." model. This detailed phase-by-phase identification is essential to improve material constitutive models' prediction precision. When it comes to predicting the tensile behavior of PC/ABS material, the "Zhu et al." model exceeds the G'Sell and Jonas and DSGZ constitutive models in terms of precision and efficiency.

A significant agreement is seen in Fig. 4(a) between the experimental data under constant temperature and strain rate conditions and the predicted results derived from the "Zhu et al." model. There is a little difference between the shapes of the predicted and experimental data, which show the same trend. This confirms that the new model accurately illustrates the impact of the strain on the flow stress.

A quantitative assessment of the predictive accuracy for G'Sell and Jonas, DSGZ, and the "Zhu et al." models was conducted, and the results are depicted in Fig. 4(b) by using the R^2 and *RMSE* values. The prediction precision of the G'Sell and Jonas model maintains a high level with an RMSE of 2.6 but reveals a relatively low level with an R^2 of 0.93639. These results indicate that the model is ineffective and unsuitable for predicting the behavior of PC/ABS blend, thereby contributing significantly to its poor precision in the current case. The DSG predictive model performs at an average level with an RMSE of 1.75 and a relatively high R^2 of 0.97113. These results reveal that the model may be beneficial for predicting the PC/ABS blend's tensile behavior. However, even if the DSGZ model is adequate for thermoplastics, it could be further improved.

Comparing the newly developed constitutive model to the models of G'Sell and Jonas, and DSGZ, the values of R^2 and *RMSE* displayed better results. According to Fig. 4(b), "Zhu et al." provides greater prediction accuracy, as proved by the better value of R^2 and the smaller value of *RMSE* of 0.9939 and 0.806, respectively. This demonstrates that this model, out of the three constitutive models, is the most accurate and can accurately predict the tensile behavior of PC/ABS material. The findings corroborated with the researches presented in Ref. [10, 11, 14], which suggested that the adjusted constitutive model may precisely predict the mechanical properties of the materials under investigation at different temperatures and strain rates.

To conclude, the developed "Zhu et al." model can describe the trend of yield stress by varying the strain rates and temperatures. However, this model still has some limitations when it is used to predict the responses of the material under various temperatures.

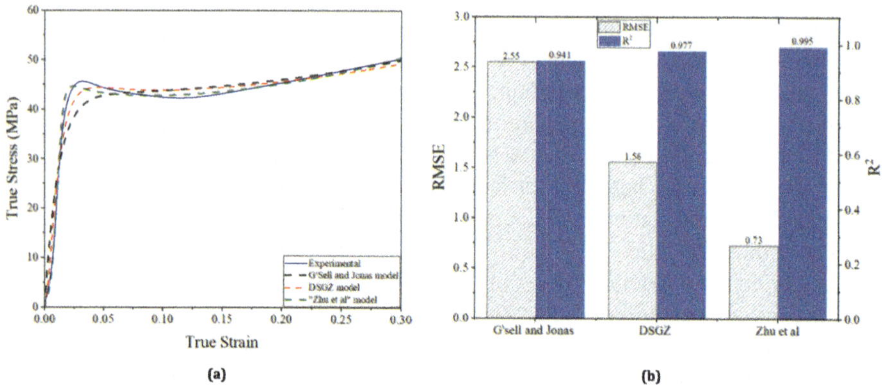

Fig. 4. Comparison of the experimental and the predicted models results provided by G'Sell and Jonas, DSGZ and the Zhu model at room temperature and stress train of 1.25 10^{-2} s^{-1} via: (a) True stress-strain curves, (b) RMSE and Rsquare values.

4.3 Prediction of the Strain Rate Influence on the Tensile Mechanical Behavior of PC/ABS Blend

Tensile tests on PC/ABS samples were conducted at varying strain rates 1.25 10^{-4}, 1.25 10^{-3}, 1.25 10^{-2}, and 1.25 10^{-1} s^{-1}.

Figure 5 indicates that the most accurate predictions occur at 1.25 10^{-2} s^{-1} and room temperature, aligning with the reference conditions. The maximum tensile true stress is achieved at a strain rate of 1.25 10^{-1} s^{-1}, revealing distinct shapes in the curves of the experimental and predicted data.

Figure 5 displays the predicted and experimental results, demonstrating acceptable agreement at various strain rates and a temperature of 20 °C. The flow stress, including the peak stress, increases with the rise in strain rate. Notably, the peak point shifts to the right with higher strain rates, indicating a delayed onset of yield behavior. These findings align with the work of Wang H [11], Zhu H et al. [12], Cao K et al. [26], Yang M et al. [27], and Dreistadt C et al. [28]. They have developed a constructive model effectively predicting the mechanical behavior of glassy and semi-crystalline materials under various values of strain rates and temperatures.

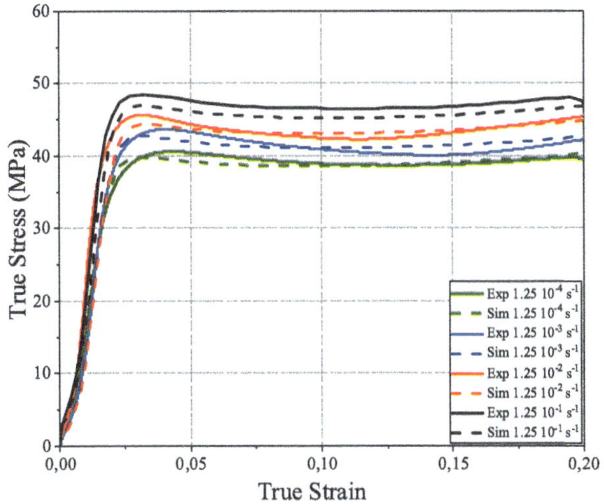

Fig. 5. True stress – strain curves and predicted results for tensile behavior of PC/ABS blend under different strain rates at room temperature.

4.4 Prediction of the Temperature Effect on the Tensile Mechanical Behavior

Tensile tests on PC/ABS samples were conducted with temperature ranging from room temperature (~20 °C) to 50 °C, 80 °C, and 120 °C at a strain rate of 1.25 10^{-2} s^{-1}. The results suggest that the current model demonstrates strong consistency with experimental data over a wide temperature range. As illustrated in Fig. 6 at a strain rate of 1.25 10^{-2} s^{-1}, both the yield stress and peak stress exhibit a decrease with increasing temperature. The peak point shifts to the left at higher temperatures, indicating an earlier onset of yield behavior. The deviation in PC/ABS behavior is observed at lower temperatures, specifically at 20 °C and 50 °C, confirming the strain-rate effect on yield properties.

These findings highlight the efficience of the Zhu model to predict the impact of strain rate on tensile true stress. Additionally, they are fully consistent with the work of Duan Y et al. [8], Wang H et al. [11], Zhu H et al. [12], Richeton J et al. [22], and Yang M et al. [28]. The predicted results at a constant strain rate underscore the influence of temperature on flow stress. Both experimental true stress-strain curves and predictions demonstrate that the new Zhu et al. model effectively describes the effect of temperature on flow stress with high precision.

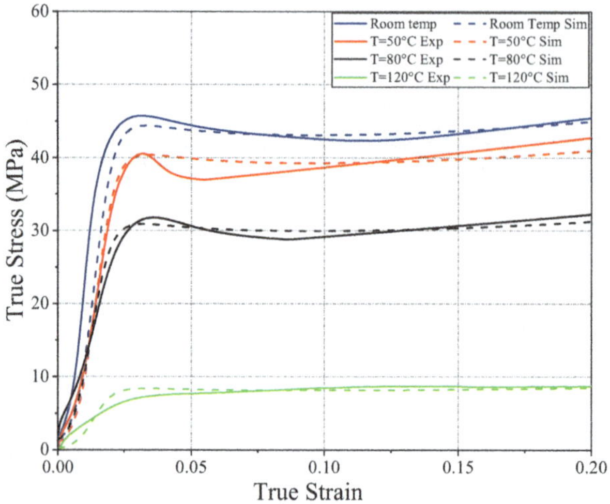

Fig. 6. True stress-strain curves and predicted results for tensile behavior of PC/ABS blend under different temperatures at strain rate of 1.25 10^{-2} s^{-1}.

5 Conclusion

This paper aims to evaluate the impact of temperature and strain rate on the tensile behavior of the PC/ABS blend. Experimental findings indicate that the tensile behavior of the material is sensitive to both strain rate and temperature. With increasing temperature, the yield strength and strain of the material mixture decrease significantly, while the Young's modulus exhibits a slight decline.

The DSGZ model and a new phenomenological model proposed by "Zhu et al." were selected to describe the tensile behavior of PC/ABS, considering various constitutive models. Comparative analysis between G'Sell, DSGZ, and Zhu et al. models indicates that the predictive accuracy of the "Zhu et al." model surpasses that of G'Sell and DSGZ. The Zhu et al. model proves its ability to predict tensile behavior at different strain rates. However, for the effect of temperature, the predicted responses deviate significantly from the experimental data. Nevertheless, it's important to acknowledge that the "Zhu et al." constitutive model has its limitations. Potential improvement could involve incorporating a mathematical term or introducing a transition function to enhance its predictive.

References

1. Belarbi, A., Dawood, M., Acun, B.: Sustainability of fiber-reinforced polymers (FRPs) as a construction material. In: Sustainability of Construction Materials, pp. 521–538 (2016)
2. Praveen, K.M., Pious, C.V., Thomas, S., Grohens, Y.: Non-thermal plasma technology for polymeric materials. Appl. Comp. Nanostr. Mater. Biomed. Fields 1–21 (2019)
3. Suarez, H., Barlow, J.W., Paul, D.R.: Mechanical properties of ABS/polycarbonate blends. J. Appl. Polym. Sci. **29**(11), 3253–3259 (1984)
4. Amoako, K., Gbyli, R.: Improving the hemocompatibility of biomedical polymers. In: Hemocompatibility of Biomaterials for Clinical Applications, pp. 223–252 (2018)

5. Greco, R., Astarita, M.F., Dong, L., Sorrentino, A.: Polycarbonate/ABS blends: processability, thermal properties, and mechanical and impact behavior. Advances in Polymer Technology: Journal of the Polymer Processing Institute **13**(4), 259–274 (1994)
6. Yin, Z.N., Wang, T.J.: Investigation of tensile deformation behavior of PC, ABS, and PC/ABS blends from low to high strain rates. Appl. Math. Mech. **33**(4), 455–464 (2012)
7. G'sell, C., Jonas, J.J.: Determination of the plastic behaviour of solid polymers at constant true strain rate. J. Mater. Sci. **14**(3), 583–591 (1979)
8. Duan, Y., Saigal, A., Greif, R., Zimmerman, M.A.: A uniform phenomenological constitutive model for glassy and semicrystalline polymers. Polym. Eng. Sci. **41**(8), 1322–1328 (2001)
9. Mulliken, A.D., Boyce, M.C.: Mechanics of the rate-dependent elastic–plastic deformation of glassy polymers from low to high strain rates. Int. J. Solids Struct. **43**(5), 1331–1356 (2006)
10. Yu, P., Yao, X., Han, Q., Zang, S., Gu, Y.: A visco-elastoplastic constitutive model for large deformation response of polycarbonate over a wide range of strain rates and temperatures. Polymer **55**(25), 6577–6593 (2014)
11. Wang, H., Zhou, H., Huang, Z., Zhang, Y., Qiao, H., Yu, Z.: Experimental investigation and modeling of the mechanical behavior of PC/ABS during monotonic and cyclic loading. Polym. Test. **50**, 216–223 (2016)
12. Mathiesen, D., Vogtmann, D., Dupaix, R.B.: Characterization and constitutive modeling of stress-relaxation behavior of Poly (methyl methacrylate) (PMMA) across the glass transition temperature. Mech. Mater. **71**, 74–84 (2014)
13. Hentati, F., Barhoumi, N., Khlifi, K.: Characterization of the surface properties and adhesion behavior of electroplated PC/ABS using nano-indentation and scratch tests. J. Adhes. Sci. Technol. **37**(10), 1724–1735 (2023)
14. Zhu, H., Ou, H., Popov, A.: A new phenomenological constitutive model for thermoplastics. Mech. Mater. **157**, 103817 (2021)
15. Vignjevic, R., Djordjevic, N., De Vuyst, T., Gemkow, S.: Modelling of strain softening materials based on equivalent damage force. Comput. Methods Appl. Mech. Eng. **335**, 52–68 (2018)
16. Hentati, F., Hadriche, I., Masmoudi, N., Bradai, C.: Experimental design to enhance the surface appearance, the internal structure, and the shear stresses of injected and metallized polycarbonate/acrylonitrile–butadiene-styrene parts. J. Appl. Polym. Sci. **137**(7), 48384 (2020)
17. Richeton, J., Ahzi, S., Daridon, L., Rémond, Y.: Modeling of strain rates and temperature effects on the yield behavior of amorphous polymers. J. de Physique IV (Proc.) **110**, 39–44 (2003)
18. Fang, Q.Z., Wang, T.J., Beom, H.G., Zhao, H.P.: Rate-dependent large deformation behavior of PC/ABS. Polymer **50**(1), 296–304 (2009)
19. G'Sell, C., Hiver, J.M., Dahoun, A., Souahi, A.: Video-controlled tensile testing of polymers and metals beyond the necking point. J. Mater. Sci. **27**, 5031–5039 (1992)
20. Boyer, R.F.: Dependence of mechanical properties on molecular motion in polymers. Polym. Eng. Sci. **8**(3), 161–185 (1968)
21. Kastelic, J.R., Baer, E.: Crazing, yielding, and fracture in polycarbonate and polyethylene terephthalate at low temperatures. J. Macromol. Sci. Part B Phys. **7**(4), 679–703 (1973)
22. Haward, R.N., Thackray, G.L.: The use of a mathematical model to describe isothermal stress-strain curves in glassy thermoplastics. Proc. Royal Soc. London. Ser. A. Math. Phys. Sci. **302**(1471), 453–472 (1968)
23. Duan, Y., Saigal, A., Greif, R., Zimmerman, M.A.: Analysis of multiaxial impact behavior of polymers. Polym. Eng. Sci. **42**(2), 395–402 (2002)
24. Cao, K., Wang, Y., Wang, Y.: Effects of strain rate and temperature on the tension behavior of polycarbonate. Mater. Des. **38**, 53–58 (2012)

25. Yang, M., et al.: Temperature and strain rate sensitivity of yield strength of amorphous polymers: Characterization and modeling. Polymer **251**, 124936 (2022)
26. Dreistadt, C., Bonnet, A.S., Chevrier, P., Lipinski, P.: Experimental study of the polycarbonate behaviour during complex loadings and comparison with the Boyce, Parks and Argon model predictions. Mater. Des. **30**(8), 3126–3140 (2009)

A Modified Frölich Magnetic Hysteresis Model for a ε-Fe2O3 Iron Ferrite

Khaoula Hergli[1,2](✉) and Haykel Marouani[3]

[1] Gafsa Preparatory Institute for Engineering Studies, University of Gafsa, Gafsa, Tunisia
kaoula.hergli@gmail.com
[2] LGM, University of Monastir, Monastir, Tunisia
[3] National Engineering School of Monastir, LGM, University of Monastir, Monastir, Tunisia
haykel.marouani@enim.rnu.tn

Abstract. As bulk magnetic behaviors are different from those of nanomaterials, the well-known classical models are not appropriate to simulate the magnetization-induction curves. In this work, a novel formulation for modeling the hysteresis loop of an iron ferrite nanomaterial is presented. Three adapted models are tested, based on the Rayleigh, the Potter and the Frölich expressions. Results show a non-satisfying fitting curve. An improved model, based on the Frölich one, is then expressed. The added model coefficients facilitate an accurate non-linearity representation of nanomaterials magnetic response, thus resulting in a good agreement between simulated and measured hysteresis loop. To accomplish this goal, the appropriate values of these parameters have to be identified. Then, a constrained optimization approach is used. The generalized reduced gradient is adopted for the optimization process. The new model is verified. Thus, its numerical hysteresis loop shows a better fitting compared to the experimental curve. The root mean square error is reduced from 2.519 to 2.022. Furthermore, it demonstrates better magnetic response description of the ε-Fe2O3 nano-iron-ferrite material than the other models.

Keywords: Magnetism · hysteresis · nanomaterial · modeling · Frölich

1 Introduction

The term 'magnetic materials' refers to a broad category of materials with a wide range of engineering and medical applications with electromechanical systems. The adaptability of magnetic materials in both engineering and medical domains highlights their importance in modern technology and their potential to revolutionize numerous fields. Magnetic materials are used in the generation and distribution of electricity and, in most cases, in devices that use this electricity. Magnetic hysteresis loop measurements are of eminent interest to assess to magnetic properties such as magnetic induction B (T), magnetic field H (A.m-1), magnetic moment μ (J.T-1), magnetization M (A.m-1 or J.T-1m-3), magnetic susceptibility χ (-) and magnetic flux φ (T.m^2 or Wb). The mathematical modeling of hysteresis loops is also fundamental to design and evaluate the performance of complex magnetic systems subject to an altering magnetic field, thermal

M. Chouchane et al. (Eds.): CMSM 2023, LNME, pp. 43–51, 2024.
https://doi.org/10.1007/978-3-031-65007-9_5

and mechanical stresses [1, 2]. Various models have been proposed to represent hysteresis behavior accurately such as Rayleigh, Frölich, Potter, Preisach and Jiles-Atherton models [3]. Some of them are extended to describe the magnetomechanical behavior [4–7]. The identification of the model's parameters requires, in many cases, the development of an efficient and a sophisticated identification technique based on genetic algorithm, particle swarm optimization, [8].

Nowadays, the miniaturized actuators are based on isotropic magnetic nanoparticles and submicron anisotropic magnetic nanomaterials, which opens up new possibilities for improving functionality, efficiency and precision in a wide range of applications, revolutionizing a variety of industries. Due to their numerous potential applications in biomedicine and technology, magnetic nanoparticles have been used in various medical applications, such as cell isolation, immunoassay, diagnostic testing and drug delivery [9–11]. However, a magnetic nanomaterial becomes improved with a variety of new properties, different to those observed for its bulk counterpart, which requires the development of novel experimental and theoretical techniques that assist on the design and optimization of nanomaterial for specific applications.

Indeed, epsilon-phase iron oxide ε-Fe2O3 has been a material of distinctive interest and is getting attracting attention due to its specific magnetic properties, which are highly dependent on the structure, size and shape of nanoparticles, temperature, magnetic anisotropy… This maghemite constitutes a remarkable phase, presenting an important coercive field of up to 20 kOe at room temperature, significant ferromagnetic resonance, magnetic transition at low temperature [12, 13] and coupled magnetoelectric structures that are not observed in any other simple metal oxide phase. This particularity provides it great potential applicability in many areas, with high-density magnetic recording media, information storage and magnetic resonance imaging. Chirita and Jiri et al. [14, 15] had presented a brief overview of the structural, magnetic properties of the ε-Fe2O3 and its various applications. Therefore, studynig its magnetic properties is crucial for modeling its magnetic behavior accurately.

In the present work, the hysteresis loop of a nano-iron-ferrite material is presented (named ε-Fe2O3). The purpose is to model its magnetic behavior using the well-known previously cited models and described for massive materials. None of them gives a satisfying result which push us to develop a new mathematical model based on the Frölich formulation. The manuscript is organized as follows: Sect. 2 reviews the experimental hysteresis loop of the ε-Fe2O3. Section 3 describes our attempts to model it by using the existing models. Section 4 presents the new modified Frölich model. Finally, Sect. 5 is the conclusion that summarize the contribution and highlights the benefits of the proposed approach.

2 Experimental Data

The development of iron ferrite with specific magnetic properties is a scientific challenge. One of the actual key-study areas, for instance, is the production of iron ferrites with a significant coercive field. Unfortunately, the low magneto-crystalline anisotropy of iron ferrites makes it difficult to increase the coercive field value. This has not prevented some researchers [16] from creating an iron oxide nanocrystal in a silica matrix that displays a massive coercive field value of 2T (1.6 × 106 A/m) at ambient temperature. This amount is the highest value for a magnetic material based on metal oxide that has been reported [16]. This nanocrystal is made of the iron oxide phase ε-Fe2O3. This accomplishment was possible because of the significant magneto-crystalline anisotropy of ε-Fe2O3 combined with the creation of a unique magnetic domain by appropriately sized nanocrystals. Figure 1 shows its magnetization M as a function of the magnetic induction B at room temperature [17].

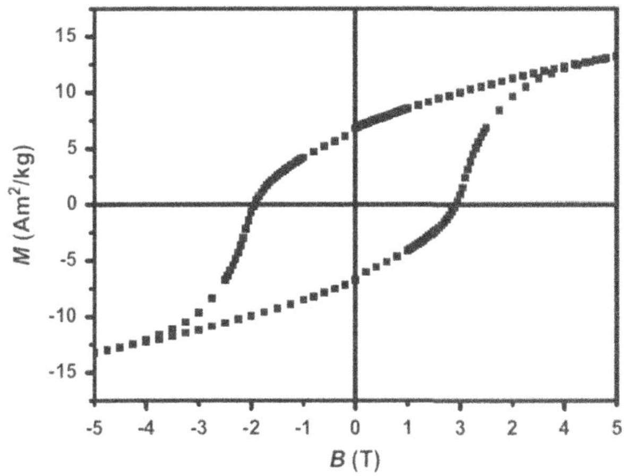

Fig. 1. Hysteresis loop of ε-Fe2O3 at room temperature (Tučková et al. 2014)

The main magnetic properties are resumed in Table 1. Magnetization at saturation is the state reached when an increase in induction cannot further increase the magnetization of the material. Coercive magnetization (or coercivity) is the value of magnetization that brings to zero the magnetic induction. Saturation induction is the point where no more magnetizations will give rise to any additional magnetic induction. Remanence is the capacity of a material to preserve magnetization, equal to the magnetic flux density of the material after the removal of the magnetizing field.

Table 1. Magnetic properties.

Quantity	Symbol	Unit	Value
Magnetization at saturation	Ms	A.m^2.Kg-1	13.086
Coercive magnetization	M0	A.m^2.Kg-1	6.804
Saturation induction	Bs	T	5
Remanence	Br	T	1.903

3 Hysteresis Models

Hysteresis modeling is a multidisciplinary field that involves physics, mathematics, materials science, and computational methods. The choice of model depends on the specific material, the desired level of accuracy, and the application. Often, a combination of theoretical and computational approaches is used to develop comprehensive hysteresis models that permit an effective study of the nanomaterial magnetic behavior.

In this section, three mathematical models are used to simulate the experimental hysteresis loop: Rayleigh, Potter and Frölich. The root mean square error RMSE (Eq. 1) is used to evaluate the fitting performance of each model.

$$RMSE = \sqrt{\frac{\sum_{i=1}^{n} (N_i - E_i)^2}{n}} \tag{1}$$

where N_i, E_i and n are the numerically estimated values, the experimentally measured values and the sampling points number, respectively.

3.1 Rayleigh Model

The classic Rayleigh model describes the H-B curve [18]. According to the literature the model is adapted for modeling high-coercivity materials We adapt the mathematical model to describe the B-M curve for the ε-Fe2O3 iron ferrite nanomaterial by rearranging the x and y-axis accordingly. The model is fully determined by 3 experimental values (B_S, M_s and M_0) as expressed by the following expressions:

$$M(B) = (\alpha_R + \beta_R B_s)B + sign(\gamma)\frac{1}{2}\beta_R\left(B^2 - B_s^2\right) \tag{2}$$

$$\alpha_R = \frac{M_s}{B_s} - \beta_R B_s \tag{3}$$

$$\beta_R = \frac{2M_0}{B_s^2} \tag{4}$$

where α_R and β_R are model constants. Sign(γ) takes the value + 1 and -1 for the ascending and descending branch of the cycle, respectively.

Figure 2 presents a comparison of the resulting curve and the experimental one. The Rayleigh model shape proves that it is not totally adequate to the ε-Fe_2O_3 iron ferrite nanomaterial behavior. The RMSE is equal to 2.188.

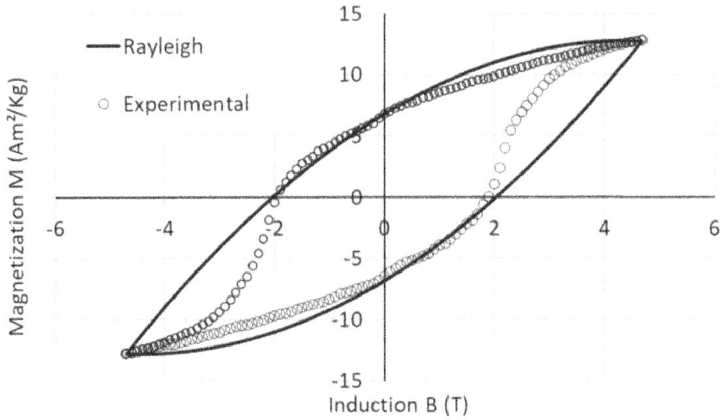

Fig. 2. Comparison of experimental hysteresis loop and Rayleigh model

3.2 Frölich Model

The Frölich model H-B is more adapted for the low hysteresis loss materials [19]. It is determined by 4 experimental values (B_r, B_s, M_s and M_0). The adapted expression B-M is expressed as follows:

$$M(B) = \frac{B - sign(\gamma)B_r}{\alpha_F + \beta_F|B - sign(\gamma)B_r|} \tag{5}$$

$$\alpha_F = \frac{B_R}{\frac{1}{M_0} - \frac{1}{M_S}} \tag{6}$$

$$\beta_F = \frac{1}{M_S} \tag{7}$$

where α_F and β_F are model constants.

Figure 3 presents a comparison of the resulting curve and the experimental one. The Frölich model shape proves that it is more adequate to the ε-Fe_2O_3 iron ferrite nanomaterial behavior, even with a RMSE greater than the Rayleigh's RMSE and equal to 2.519.

3.3 Potter Model

The H-B Potter model is determined by only 3 experimental values (B_r, M_s and M_0). The adapted expression B-M is expressed as follows:

$$M(B) = M_S\left(sign(\gamma) - sign(\gamma)\left(1 + tanh\left(\frac{B_r - sign(\gamma)B}{B_r}tanh^{-1}\left(\frac{M_S}{M_0}\right)\right)\right)\right) \tag{8}$$

Figure 4 presents a comparison of the resulting curve and the experimental one. The Potter model shape proves that it is not totally adequate to the ε-Fe_2O_3 iron ferrite nanomaterial behavior. The RMSE is equal to 2.495.

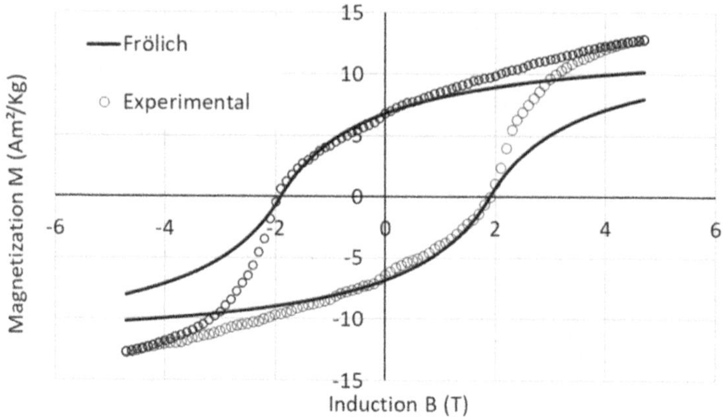

Fig. 3. Comparison of experimental hysteresis loop and Frölich model

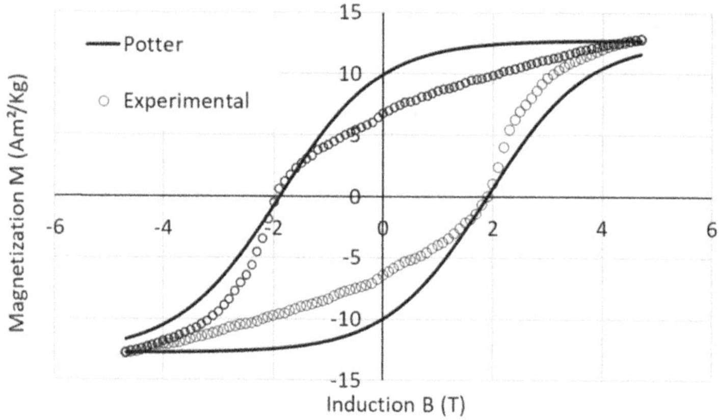

Fig. 4. Comparison of experimental hysteresis loop and Potter model

4 Modified Frölich Modeling

The Frölich model looks to be more adequate to fit the experimental hysteresis loop of the nanomaterial. A modification is adopted to improve the numerical approximation by adding two 'corrective' coefficients C_1 and C_2 as expressed in the following expression:

$$M(B) = \frac{B - sign(\gamma)B_r}{C_1\alpha_F + C_2\beta_F|B - sign(\gamma)B_r|} \tag{9}$$

In order to identify C_1 and C_2 values, a constrained optimization approach is used. The objective function is the RMSE value and the variables to be constrained are C_1 and C_2. The optimization problem is summarized as follows:

$$\begin{cases} \min RMSE \\ subject\ to \begin{cases} 0 \leq C_1 \leq 1 \\ 0 \leq C_2 \leq 1 \end{cases} \end{cases} \tag{10}$$

To solve the optimization problem, the generalized reduced gradient is used. It represents an iterative algorithm aims to find the optimal solution for a given objective function while keeping the constraints satisfied. This method is an extension of the reduced gradient method [20, 21]. Its aptitude to navigate through nonlinearities and convey an optimized solution makes it a valued tool in various domains. The identified constants are resumed in Table 2.

Table 2. Constant values of the modified Frölich model.

Constant	Value
C1	1.00
C2	0.81

Figure 4 presents a comparison of the resulting curve and the experimental one. The modified-Frölich model shape proves a better fitting than the classical formulation. The RMSE is improved by 20% and is equal to 2.022 (Fig. 5).

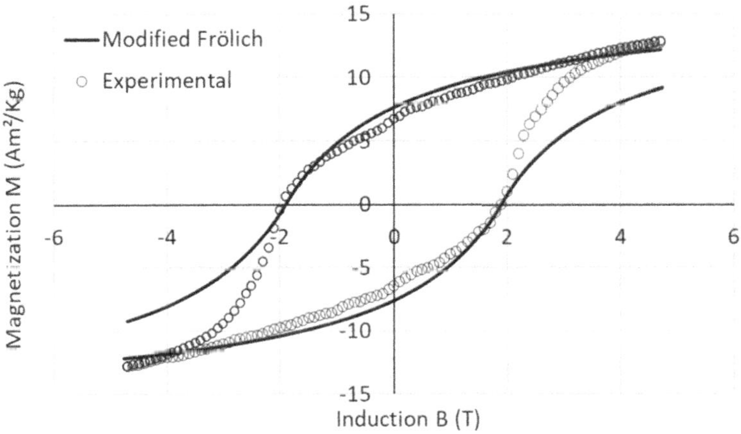

Fig. 5. Comparison of experimental hysteresis loop and modified Frölich model

5 Conclusion

A new nanomaterial, the ε-Fe_2O_3, is added to the family of iron ferrite nanomaterials that consist of maghemite (γ-Fe_2O_3), magnetite (Fe_3O_4) … It exhibits the highest coercive value that has been reported (i.e., 2T at ambient temperature). In this work, we focus on the modeling of the experimental magnetic hysteresis loop (Magnetisation – Induction). Since the nanomaterial behaviour differs from the bulk behaviour, the classical models are not adequate. Thus, accurate hysteresis modeling is essential for understanding and predicting the behavior of various materials and devices. In this paper, we firstly reformulate the Rayleigh, the Potter and the Frölich models. Results show that they are not perfectly fitting the experimental loop and the RMSE ranges from 2.519 to 2.188. Secondly, a new formulation, based on the Frölich model, is presented by adding two new constants. The generalized reduced gradient is then used to identify these constants. Final results show an improvement of RMSE by 20% compared to the classical Frölich formulation with a better fitting. Future improvement, by using more sophisticated hysteresis models (such as Preisach and Jiles-Atherton) and more performant identification techniques (such as genetic algorithms and particle swarm optimization) are planned.

References

1. You-Shu, Z., Chien-hong, L.: A constitutive model of coupled magneto-thermo-mechanical hysteresis behavior for giant magnetostrictive materials. Mech. Mater. **148** (2020)
2. Daniel L., Domenjoud M.: Effects of plastic strain and reloading stress on the magneto-mechanical behavior of electrical steels: experiments and modeling. Mech. Mater. **176** (2023)
3. Hergli, K., Marouani, H., Zidi, M.: Hysteresis identification models: a review. In: Lecture Notes in Mechanical Engineering, pp. 189–198, Springer (2018)
4. Hergli, K., Marouani, H., Zidi, M.: Numerical determination of Jiles-Atherton hysteresis parameters: Magnetic behavior under mechanical deformation. Physica B Condens **549**, 74–81 (2018)
5. Hergli, K., Marouani, H., Zidi, M., et al.: Magneto-mechanical behavior analysis using an extended Jiles-Atherton hysteresis model for a sheet metal blanking application. Appl. Comput. Electromagn. Soc. **35**, 727–734 (2020 a)
6. Hergli, K., Marouani, H., Zidi, M.: Extension of the Jiles–Atherton hysteresis model to characterize the magneto-mechanical behavior: experimental and numerical investigations for stator blanking. In: Lecture Notes in Mechanical Engineering, 2020, pp. 514–523, Springer (2020)
7. Da silva, L.G., Abdellehi, A., Demenjoud, M., et al.: An extension of the vector-play model to the case of magneto-elastic loadings. IEEE Access (2022)
8. Marouani, H., Hergli, K., Dhahri, H., et al.: Implementation and identification of preisach parameters: comparison between genetic algorithm, particle swarm optimization, and Levenberg–Marquardt Algorithm. Arab. J. Sci. Eng. **44**, 6941–6949 (2019)
9. Fang, X., Wang, Y., Wang, S., et al.: Nanomaterials assisted exosomes isolation and analysis towards liquid biopsy. Mater Today Bio **16**, 1–18 (2022)
10. Khan, R., Arshad, F., Hassan, I.U., et al.: Advances in nanomaterial-based immunosensors for prostate cancer screening. Biomed. Pharmacother. **155**, 1–12 (2022)
11. Zhenhui, M., Jeotikanta, M., Kecheng, W., et al.: Magnetic nanoparticles: synthesis, anisotropy and applications. Chem. Rev. **123**(7), 3904–3943 (2023)
12. Cleron, J., Baker, A., Nakotte, T., et al.: Exploring critical synthetic parameters for nanoscale ε-Fe 2 O 3 and their influence on magnetic behaviors. J. Phys. Chem. C **16**, 7256–7263 (2022)
13. Gich, M., Frontera, C., Roig, A., et al.: High- and low-temperature crystal and magnetic structures of ε-Fe2O3 and their correlation to its magnetic properties. Chem. Mater. **16**, 3889–3897 (2006)
14. Chirita, M., Grozescu, I..: Fe2O3—nanoparticles, physical properties and their photochemical and photoelectrochemical applications. Chemical Bulletin of Politehnica, University of Timisoara, Romania, 54 (2009)

15. Tucek, J., Zboril, R., Namai, A., et al.: ε-Fe2O3: an advanced nanomaterial exhibiting giant coercive field, millimeter-wave ferromagnetic resonance, and magnetoelectric coupling. Chem. Mater. **22**, 6483–6505 6483 (2010)
16. Jin, J., Ohkoshi, S., Hashimoto, K.: Giant coercive field of nanometer-sized iron oxide. Adv. Mater. **16**, 48–51 (2004)
17. Tučková, M., Harman, R., Tuček, P., et al.: Design of experiment for hysteresis loops measurement. J. Magn. Magn. Mater. **368**, 64–96 (2014)
18. Rayleigh, L.: On the behaviour of iron and steel under the operation of feeble magnetic forces. Phil. Mag. **142**, 225–245 (1887)
19. Fröhlich, O.: Investigations of dynamoelectric machines and electric power transmission and theoretical conclusions therefrom. Electrotech Z **2**, 134–141 (1881)
20. Shor, N.Z., Shabashova, L.P.: Solution of minimax problems by the method of generalized gradient descent with dilatation of the space. Cybern. Syst. Anal. **8**, 88–94 (1972)
21. Smeers, Y.: Generalized reduced gradient method as an extension of feasible direction methods. J. Optim. Theory Appl. **22**, 209–226 (1977)

Exploring the Effect of Sub-zero Treatment on the Performance of Steel Grades

E. A. Elsharkawi[1(✉)], Daniel Vincent[1], Mohamed H. Abdelaziz[2], D. MacNeil[1], and Hatem Mrad[3]

[1] Faculty of Science, Division of Engineering, Saint Mary's University, Halifax, NS, Canada
{Ehab.Elsharkawi,Don.macneil}@smu.ca
[2] Département PEC, Université Française d'Égypte, Ville Shorouk, Le Caire 4923116, Egypt
mohamed.abdelaziz@ufe.edu.eg
[3] École de Génie, Universite du Quebec en Abitibi-Temiscamingue, Québec, QC, Canada
hatem.mrad@uqat.ca

Abstract. Sub-zero treatment (SZT) is a low-temperature treatment used to improve the mechanical properties of ferrous and non-ferrous alloys. Such treatment is gaining interest on an industrial scale due to the attainable privileges, such as longer service life, better mechanical properties, and enhanced wear resistance. This study investigates the effect of deep cryogenic treatment (DCT) on the mechanical properties, hardness, and surface finish of different soft steel grades. The DCT process, conducted in this investigation, utilized a low-cost setup to carry out an 8-h ramp from ambient temperature down to -190 °C at a slow cooling rate, followed by 24 h at −190 °C, and then ramped up to ambient temperature at a slow heating rate. Four different grades of steel were used in the current investigation: cast iron, 304SS, CD4MCu, and 17-4PH stainless steel. For all grades of steel, results show an improvement in the surface bearing area of cryo-treated 304SS and 17-4PH samples compared to non-treated conditions. Moreover, enhanced strength and hardness values were reported for certain steels after applying that DCT protocol.

Keywords: Deep cryogenic treatment (DCT) · Mechanical properties · Ferrous alloys · surface bearing

1 Introduction

Steel of various grades is utilized for a range of applications and is often valued for its high strength and corrosive resistance. A typical method for improving steel material properties is standard thermal treatment. The thermal treatment process comprises annealing, normalizing, hardening, and tempering. A typical method for enhancing steel material properties is through standard thermal treatment. The thermal treatment process comprises annealing, normalizing, hardening, and tempering. The initial step in the hardening process involves heating the steel to the upper critical temperature to alter the iron crystal structure to a phase with an FCC structure, known as austenite (the soft phase), followed by rapid cooling to ambient temperature through quenching in a

M. Chouchane et al. (Eds.): CMSM 2023, LNME, pp. 52–59, 2024.
https://doi.org/10.1007/978-3-031-65007-9_6

suitable fluid such as oil or water, resulting in the formation of a very hard, brittle body-centered tetragonal (BCT) structure known as martensite. The tempering stage follows the hardening process to reduce brittleness and alleviate internal stresses developed during the quenching process [1, 2]. The hardness level of steels is mainly dependent on the amount of martensite formed in the microstructure after quenching. However, after conventional heat treatment, there is always some retained austenite that has not been transformed into martensite and will negatively affect the hardness values.

The thermal heat treatment process has a negative impact on the carbon footprint and energy consumption. However, an alternative low-cost and effective technique has attracted the attention of researchers in recent years, which is the Deep Cryogenic Treatment (DCT) [3]. Cryogenic treatment consists of cooling a material to a temperature below 0 °C (<273 K) to enhance the mechanical properties of steel alloys. There are three different types of Cryogenic Treatment reported, namely, Conventional Cold Treatment (CCT 223-193K), Shallow Cryogenic treatment (SCT-193-113K), and Deep Cryogenic Treatment (DCT- 113-77K) [4].

It is believed that using Sub-cold treatment after the quenching step transfers the ductile retained austenite phase into the hard martensite phase, which results in additional improvement in the wear resistance or hardness of the carbon steel alloys [3, 5]. The current work focused on deep cryogenic treatment (DCT), which starts with cooling the material to a defined sub-zero temperature, holding that temperature for a set period, and then gradually reheating the material back to room temperature. This type of sub-zero heat treatment has been implemented in the automotive, aerospace, and electronics industries as an effective process to improve component life and reduce manufacturing costs. Deep Cryogenic Treatment (DCT) minimizes the impact of greenhouse gas emissions by extending the life of items and lower maintenance [2, 3].

It has been reported immersion of certain types of steel alloys at −196 °C for 24 to 75 h can cause the grain structure to densify and release internal stresses, which enhances the wear characteristics [6]. Molinari et al. [7] reveals that deep cryogenic treatment of quenched and tempered high-speed steel tools results in a reduction in tool consumption. Most of the work done by researchers on steel alloys agrees that sub-zero treatment after quenching and before tempering results in converting the retained austenite to martensite.

There is inconsistency in the reported outcomes in the literature regarding the effect of DCT on altering the mechanical properties of steel alloys. Some results demonstrate positive outcomes, while other researchers have reported adverse outcomes for the same alloy steel. This contradiction may arise from the heat treatment conditions of the alloy steel used for the DCT. Some studies recommend applying the DCT on quenched steel, while others suggest using tempered steel. Consequently, there is a lack of reliability in this process, leading to the absence of certification and standardization. Most of the research has focused on applying the DCT on hard steels after quenching. However, there is limited data on the effect of DCT on soft steel alloys. Therefore, the objective of this study is to explore the extent to which DCT can significantly alter the properties of soft, non-heat-treated materials.

2 Materials and Methodology

In the current study, four different steel grades were used, and the composition of each alloy, based on the AISI standard, is presented in Table 1. In terms of hardness and surface roughness examinations, four 2.5 cm square samples were cut from each alloy plate and surface polished Fig. 1.

Table 1. Chemical composition (wt %) of the alloys used in the current study.

Alloy	Elements (wt%)							
	Cr	Ni	Mn	Mo	Cu	C	Si	Fe
Cast Iron	0.1	0.01	0.5	0	0	3.4	1.8	Bal
SS304	18	7	1.34	0.21	0	0.08	0.3	Bal
174-PH	15	3	0.2	0.01	3.5	0.07	0.2	Bal
CD4MCu	25	5	0.9	2	3.2	0.02	0.94	Bal

All samples were placed in a special liquid nitrogen tank designed by "Deep Cryogenics International" and slowly cooled from ambient temperature down to −190 °C at a slow cooling rate to avoid thermal cracking, followed by 24 h cold soaked at −190 °C then heated up to ambient temperature at a slow heating rate in a dry atmosphere before the annealing step at 200 °C using a low-cost and highly effective setup shown in Fig. 2.

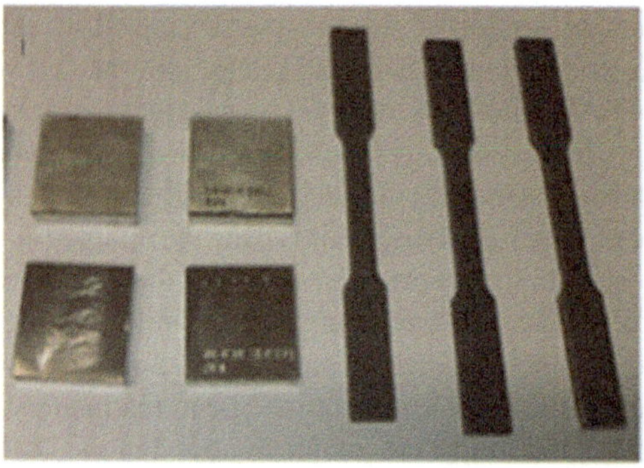

Fig. 1. Tensile and square steel samples.

Hardness measurements were performed using the DigiRock DR3 hardness tester. Five hardness measurements were taken in non-treated and cryo-treated samples. HRC hardness was taken for the 174 PH, CD4MCu, and HC materials. Measurements of the material's martensite were made using a Ferritscope FMP30C tester. A scanning White Light Interferometer was also used for measuring 3D surface roughness parameters Sa, S_{vk} and S_k. Tensile tests were also done on specific steel grades with an MTS Criterion model 45. All materials were subjected to DCT treatment protocol.

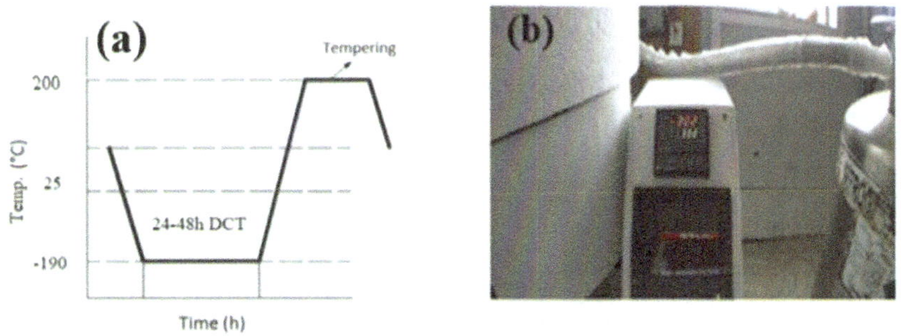

Fig. 2. (a) cooling cycle of DCT; (b) Cryogenic cooling setup [2]

3 Results and Discussion

3.1 Mechanical Testing

Figures 3(a) and (b) show the tensile test results of cryo-treated and non-treated cast iron, 304SS, and CD4MCu alloy samples. For cryo-treated cast iron samples, the tensile strength increased from 432 MPa to 519 MPa with a reduction in elongation of 5% compared to the non-treated samples. However, results show that a slight improvement in UTS was observed in cryo-treated samples of 304SS and CD4MCu compared to the non-treated condition.

The HRC and HRB hardness results of five different materials are shown in Fig. 4. The results show negligible amounts of change in the hardness values of 174-PH and Cd4MCu alloy samples. However, a slight improvement was observed in the hardness of 304 steel and cast-iron materials, which agrees with UTS results of the same alloys, Fig. 3(a). It has been reported in the literature that Cryogenic Treatment results in the transformation of ductile retained austenite to brittle martensite phase, which is responsible for improving the mechanical strength and hardness [8].

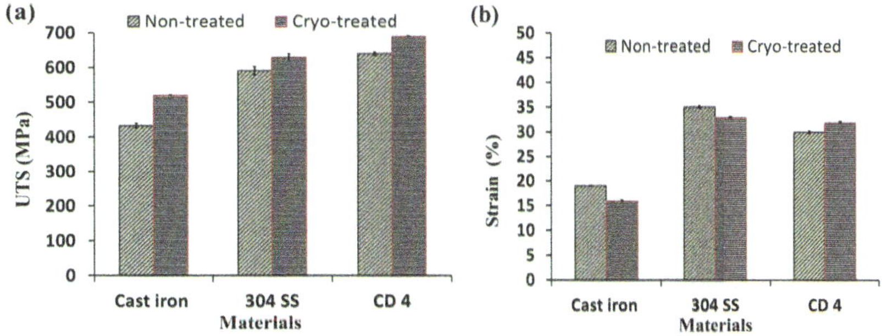

Fig. 3. UTS and Strain (%) values of cryo- treated and non-treated samples of different materials.

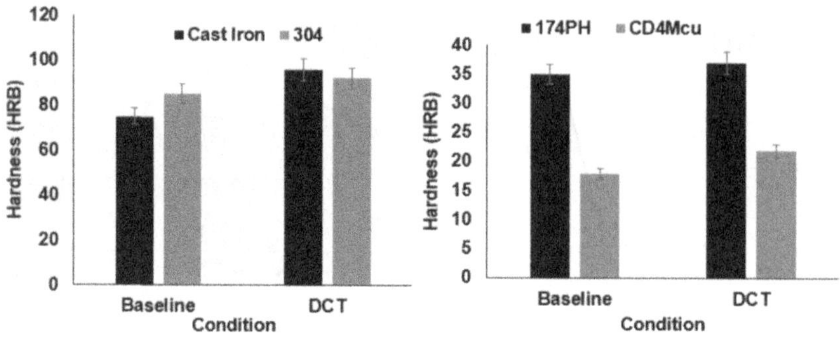

Fig. 4. Hardness test values of different types of materials before and after cryo-treatment

However, in the current study, no significant change in the amount of martensite was observed in cryo-treated samples compared to non-treated condition Fig. 5. This observation can be attributed to the fact that the samples used in the present study were soft, non-heat-treated materials. The improvement in the tensile strength of cast iron and 304SS materials may be attributed to grain size refinement and stress relief caused by the sub-zero treatment, as reported in the literature [9]. A significant improvement in hardness and wear resistance applied on hard steel was reported in the previous study [10].

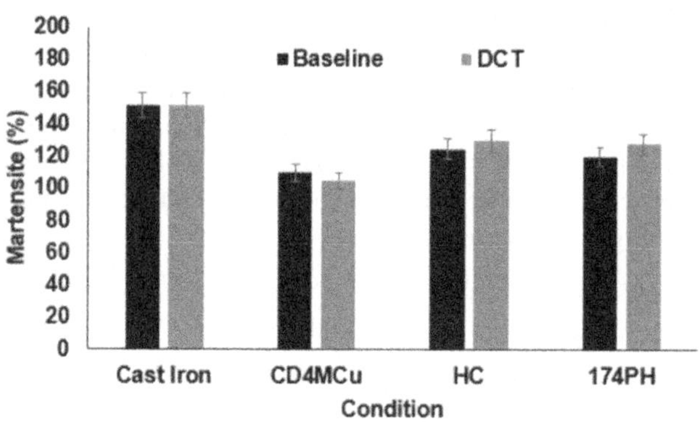

Fig. 5. Martensite measurement of different types of materials before and after cryo-treatment

3.2 Surface Roughness

The roughness of the metal surface plays an essential role in the performance of the mechanical system. Average surface roughness, R_a, is commonly used by researchers to evaluate the degree of smoothness or roughness of metal surfaces. The average roughness, R_a, is the average difference between peaks and valleys in a 2D surface profile graph [11]. However, the R_a does not provide all the required information about surface roughness, such as locations of the peaks and valleys along the length of the sample or if the peaks are clustered at specific areas. For example, a surface with a random profile can have the same R_a value as a surface with a well-ordered profile. Therefore, 3D texture profile was introduced by researchers using a 3D white light interferometer. The S_a parameter is equivalent to Ra and represents the average height of all measured points in the areal measurement. In some cases, S_a is not the proper parameter to be used to control the friction characteristics; for example, a brake rotor meets the required Sa value but still makes friction noise (squeaks). Therefore, other tribological parameters like S_k S_{pk} and S_{vk} may be used to assist in understanding the cause of the noise.

In the current study, S_a, S_k, and particle number per grain were measured for all alloy samples. S_k parameters are related to the surface texture valley depths, measuring the ability of the metal surface to hold lubricants. Therefore, it may be used to control the friction of lubricated moving parts like transmission clutch, bearings, and complex tidally contact mechanical systems. The graphical interpretation of the roughness parameters S_k, S_{vk}, and S_{pk} is illustrated in Fig. 6; where, "Smr1" denotes the areal material ratio that separates the reduced peaks "S_{pk}" from the core surface, and "Smr2" represents the areal material ratio that divides the reduced valleys "S_{vk}" from the core surface. S_{vk} reduced valleys are the pockets where the liquid can be stored on the surface resulting in improved lubricity [12].

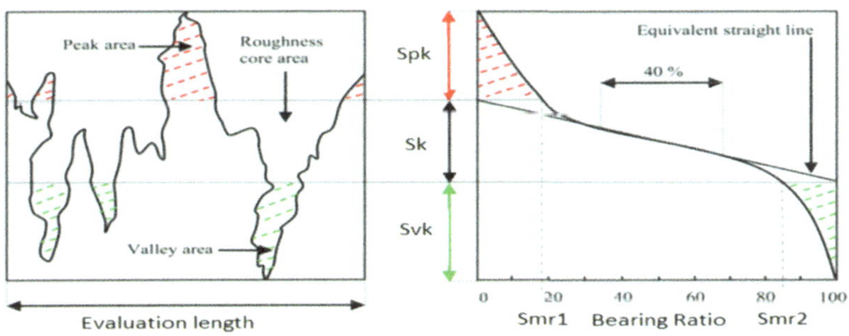

Fig. 6. Areal materials ratio curve showing the surface texture parameters S_{pk}, S_k, and S_{vk} [11]

Preliminary results of surface roughness reveal that there is no significant difference in the S_a parameter values of cryo-treated samples compared to non-treated conditions, which may be attributed to the softness of the materials used in the current study. However, a remarked improvement in surface bearing area parameter S_k by 27% was observed in the cryo-treated 304SS material Fig. 7(a) and (b). A lower value of Sk is preferred to achieve improved sliding contact between surfaces [13]. Results also show that the valley depth (S_{vk}) parameter increases from 17.9 μin for non-treated samples to 34.35 μin cryo-treated samples. No change in the S_{pk} parameter was observed in 304SS and CD4MCu alloy samples. It was reported that the S_{pk} and S_{vk} parameters are related to the wear volume of the lubricated mechanical system [14]. The particle number per grain was also increased by 30% for cryo-treated 174-PH material when compared to the non-treated condition, as shown in Figs. 8(a) and (b).

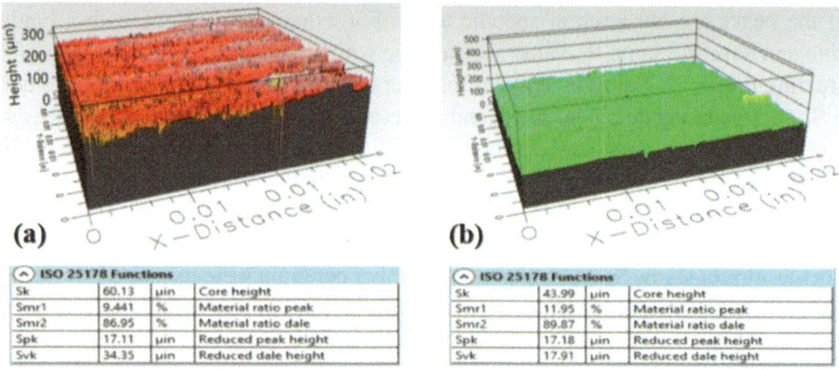

Fig. 7. 3D surface profile mapping; (a) non-treated 304SS and cryo-treated 304SS samples.

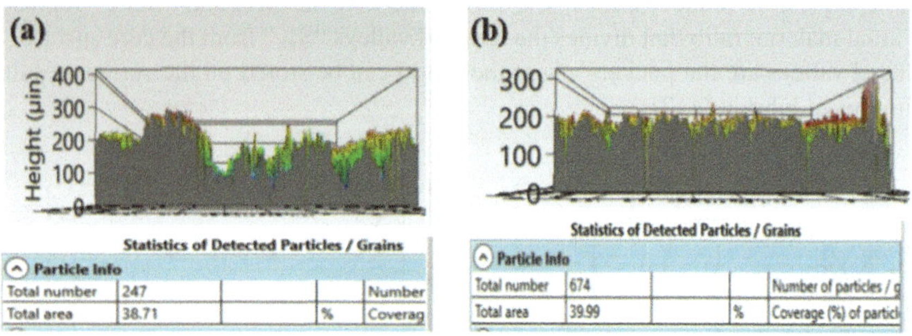

Fig. 8. 3D surface profile mapping: (a) non-treated 17-PH and (b) cryo-treated 174-PH samples.

4 Conclusion

The results of the current work explored the effect of cryogenic treatment on the mechanical properties and surface roughness of different types of soft steel materials. The results of cryogenic treated samples were compared with as-cast non-treated samples. The following conclusions can be drawn as follow:

1. A slight improvement was observed in the tensile properties and hardness of cryo-treated cast iron, 304SS, and CD4 materials. However, further investigation is needed for more detailed information about the effect of DCT on the tensile properties of hard steel.
2. It is evident that the cryogenic treatment improves the tribological parameters S_k and S_{pK} of 304SS and 17-PH materials. However, soaking time at −90 °C had no significant effect on the mechanical properties and surface roughness.
3. The analysis of the preliminary results recommends that DCT must be executed after the completion of quenching and before the tempering step.

Acknowledgments. The authors would like to recognize and appreciate the efforts and inputs of Mr. Jack Cohen, the founder and principal researcher at Deep Cryogenics International, AB, Canada, for providing technical support and access to lab facilities.

References

1. Kandpal, B.C., et al.: Effect of heat treatment on properties and microstructure of steels. Mater. Today: Proc. **44**(part 1), 199–205 (2021)
2. Singh, G.: A review on effect of heat treatment on the properties of mild steel. Mater. Today: Proc. **37**(Part 2), 2266–2268 (2021)
3. Jack, C.: Deep cryogenic treatment for marine and oil-and-gas applications. Ind. Heat. (2018)
4. Akincioğlu, S., Gökkaya, H., Uygur, İ: A review of cryogenic treatment on cutting tools. Int. J. Adv. Manuf. Technol. **78**, 1609–1627 (2015)
5. Meng, F., Tagashira, K., Azuma, R., Sohma, H.: Role of Eta-carbide precipitation in the wear resistance improvements of Fe–12Cr–Mo–V–1.4C tool steel by cryogenic treatment. ISIJ Int. **34**(2), 205–210 (2007)
6. Paolo, B., Delprete, C.: Deep cryogenic treatment: a bibliographic review. Open Mech. Eng. J. (2) (2008)
7. Molinari, A., Pellizzari, M., Gialanella, S., Straffelini, G., Stiasny, K.H.: Effect of deep cryogenic treatment on the mechanical properties of tool steels. J. Mater. Process. Technol. **118**(1–3), 350–355 (2001)
8. Jamali, A.R., Khan, W., Chandio, A.D., Anwer, Z., Jokhio, M.H.: Effect of cryogenic treatment on mechanical properties of AISI 4340 and AISI 4140 steel. Mehran Univ. Res. J. Eng. Technol. **38**, 755-766, (2019)
9. Serope, K., Schmid, S.: Manufacturing Engineering and Technology. 7th edition. Pearson Higher Education (2013)
10. https://www.researchgate.net/publication/344693324_Deep_Cryogenic_Microstructure_of_Cast_Iron
11. Ståhl, J.-E., Schultheiss, F., Hägglund, S.: Analytical and experimental determination of the ra surface roughness during turning. Procedia Eng. **19**, 349–356 (2011)
12. https://www.academia.edu/41463087/Introduction_to_surface_roughness_measurement
13. Das, J., Linke, B.: Evaluation and systematic selection of significant multi-scale surface roughness parameters (SRPs) as process monitoring index. J. Mater. Proces. Technol. **244**, 157–165 (2017)
14. Dzierwa, A., Pawlus, P., Zelasko, W.: The influence of disc surface topography after vapor blasting on wear of sliding pairs under dry sliding conditions. Coatings (10), 10 (2020)

Characterization of PEHD Films Using Uniaxial and Biaxial Testing

Fatma Kharrat(✉), Mohamed Khlif, and Chedly Bradai

Laboratory of Electromechanical Systems (LASEM), National Engineering School of Sfax, University of Sfax, 3038 Sfax, Tunisia
fatma.kharrat@hotmail.fr, fatma.kharrat@enis.tn

Abstract. The bulge test is an experimental technique used in materials science and engineering research to determine the mechanical properties and deformation behavior of thin films or membranes subjected to biaxial stress conditions. It involves subjecting the film to a controlled pressure, which causes it to bulge outward. In this present work, commercially available polyethylene high density (PEHD) films were used for characterization under uniaxial and biaxial loading conditions. The films underwent initial characterization through tensile testing along both the longitudinal direction (MD) and the transverse direction (TD) through tensile testing. Subsequently, they were further characterized using the bulge test. The contrast between the uniaxial and biaxial behavior of the film has highlighted the shortcomings of relying solely on uniaxial tests to evaluate the mechanical properties of the films under realistic operating conditions.

Keywords: PEHD films · mechanical characterization · bulge test · biaxial testing · tensile test

1 Introduction

The bulge test is a widely employed mechanical test specifically designed for evaluating the mechanical characteristics and functionality of thin films, membranes and coatings. These materials often pose challenges in characterization using conventional mechanical tests. The test involves subjecting a thin, circular sample to pressure, resulting in its deformation into a dome shape, as depicted in Fig. 1. By quantifying the resulting deformation, essential mechanical properties such as elastic modulus and tensile strength can be determined. A significant benefit of the bulge test lies in its capacity to mimic closely the real-world loading conditions experienced by these materials. This is due to the deformation mechanism that resembles the behavior observed in practical applications. The test is particularly useful for examining the response of materials under various loading conditions, including uniform or non-uniform pressure distributions. This versatility makes it highly valuable in sectors like microelectronics, where thin films are employed in the fabrication of electronic devices and in biomedical engineering, where understanding the properties of soft tissues and biomaterials is paramount. The bulge test presents several advantages, including its simplicity and minimal sample preparation

M. Chouchane et al. (Eds.): CMSM 2023, LNME, pp. 60–66, 2024.
https://doi.org/10.1007/978-3-031-65007-9_7

requirements, rendering it a cost-efficient method for assessing characterizing mechanical properties. Additionally, with a history of 70 years, this method has been widely employed for determining the elastic properties of thin films in microelectronics, nanodevices, microelectromechanical systems (MEMS), coatings and biomedical devices [1–3]. Numerous scientific research papers have focused on constructing and optimizing bulge test apparatus, with examples like Boyer et al. who introduced a hydraulic bulge test apparatus tailored for measuring sheet metal formability at high temperatures [4]. The apparatus was developed to simulate hot forming processes like stamping and forging, and the authors provided comprehensive details about its design, construction, and usage, along with experimental results showcasing its efficacy in measuring sheet metal formability under elevated temperatures. In another study, Vasilescu and Mario introduced a hydraulic bulge test setup to explore the work hardening behavior of sheet materials [5]. The authors elaborated on the apparatus design, including the hydraulic system and measurement instrumentation, and presented experimental findings that demonstrated its effectiveness in quantifying the work hardening behavior of various sheet materials. These instances underscore the extensive research devoted to developing and refining bulge test apparatuses, predominantly focusing on investigating the formability of metallic sheets. However, comparatively less attention has been directed towards utilizing this method to characterize polymeric films, with only a limited number of studies addressing this aspect [6–9].

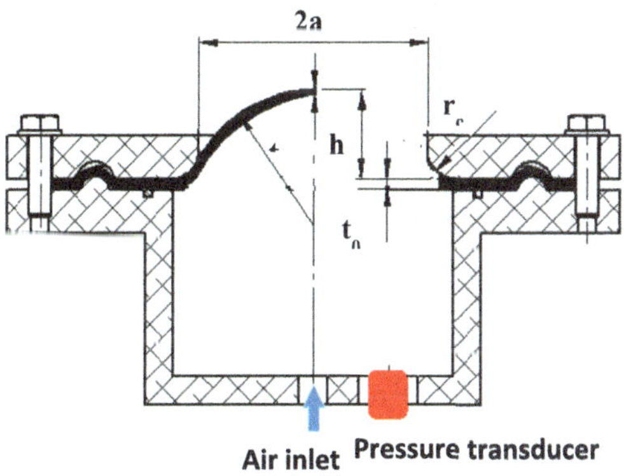

Fig. 1. Schematic depiction of the bulge test

2 Theoretical Approach of the Bulge Test

The stress analysis of the sheet or film subjected to bulge test can be conducted using the membrane theory, as documented in previous studies [10–12]. By assuming the semi-spherical form of the bulged sample, the equivalent stress, often referred to as biaxial stress, can be expressed as follows [13, 14].

$$\bar{\sigma} = \frac{P\rho}{2t}. \tag{1}$$

where P, represents the internal pressure applied to the film, ρ denotes the bulge radius and t stands for the instantaneous thickness at the apex of the film dome. See Fig. 1 representing all the geometrical parameters involved in the bulge test.

By considering the principle of volume conservation and assuming isotropic material deformation of, the von Mises effective strain is derived as follows;

$$\overline{\varepsilon} = -\varepsilon_z = ln\frac{t_0}{t} \tag{2}$$

where t_0 represents the initial thickness and t is the instantaneous thickness of the film at the apex of the dome.

The relationship between the load and the bulge height of a circular membrane, secured at its boundaries, can be described by a general equation proposed by Vlassak which reveals that the pressure applied to the membrane is a cubic function of the deflection of the film [1].

$$P = C_1\frac{\sigma_0 t}{a^2}h + C_2\frac{Et}{a^4(1-\vartheta)}h^3 \tag{3}$$

where P represents pressure, while C_1 and C_2 are constants influenced by the sample's geometry. The variables h, a, and t denote the bulge height, upper die radius, and film thickness, respectively. Additionally, σ_0 signifies the residual stress, E indicates the film's biaxial modulus, and ν represents the Poisson ratio.

"Several researchers have devised various analytical methods to calculate the bulge radius [15, 16] and thickness [11, 15, 17] using the aforementioned limiting assumptions. Tables 1 and 2 present the various techniques for determining thickness and height that have been documented in the literature.

Table 1. Bulge radius approaches

Approach	Bulge radius
Hill[15]	$\rho = \frac{a^2+h^2}{2h}$
Panknin[16]	$\rho = \frac{(a+r_c)^2+h^2-2r_c}{2h}$

Table 2. Dome thickness approaches

Approach	Thickness
Hill [15]	$t = t_0\left[\frac{1}{1+(h/a)^2}\right]^2$
Chakrabarty et al. [17]	$t = t_0\left[\frac{1}{1+(h/a)^2}\right]^{2-n}$
Kruglov[18]	$t = t_0\left[\frac{(a/\rho)}{\sin^{-1}(a/\rho)}\right]^2$

3 Materials and Methods

The material to be characterized consists on commercially available polyethylene high density (PEHD) 80 microns-thick films purchased from a Tunisian company working on film manufacturing. Test specimens are characterized through tensile and bulge testing. Tensile samples consist on rectangular film strips, measuring 20 × 10 mm, cut along both the machine direction (MD) and transverse direction (TD) of the film. Tensile tests are conducted using a universal tensile testing machine at a rate of 10 mm/min. To account for variations in film properties, a minimum of 5 specimens were tested.

Biaxial tests are conducted using a prototype bulge test apparatus constructed in the LASEM laboratory, following the setup commonly utilized in the literature [6, 8, 12, 19, 20]. The dimension of the tool mirror those employed by Chamekh [21]. Samples with a minimum diameter of 120 mm are secured between the upper die and the blank holder, and then deformed using air pressure supplied at a flow rate of 1 cm^3/min. The pressure is continuously monitored with a KOBOLD SEN-96 pressure sensor and recorded using "Labview" software via a" MyRio" acquisition card. A minimum of 5 repetitions are conducted to account for potential mechanical variations in the test specimens.

4 Results and Discussion

4.1 Tensile Tests

Figure 2 illustrates the behavior of the PLSF film for specimens taken in both the MD and TD, with the associated mechanical characteristics detailed in Table 3. An anisotropic behavior was noted, revealing notably higher stiffness in the TD compared to the MD and greater tensile strength in the MD. This underscores the intricate behavior of polymeric films under different loading conditions. Such anisotropic behavior suggests the non-uniform characteristics in the PEHD film which may be associated to a possible anisotropy in the crystalline structure of the film generated by the bubble stretching within both directions during the film blowing process. Comparison Contrasting with Fig. 3 reveals that the film demonstrates greater stiffness under biaxial tensile loading compared to uniaxial loading. Samples subjected to equi-biaxial showed a reduced elongation to failure of approximately 23%.

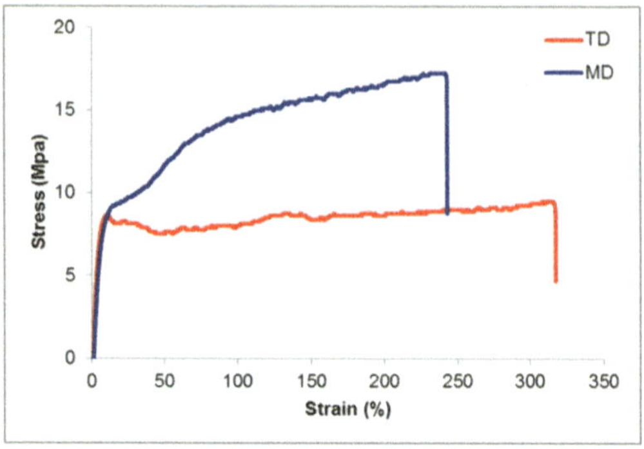

Fig. 2. Tensile behavior of PEHD films along the Machine Direction (MD) and Transverse Direction (TD)

Table 3. Tensile properties of PEHD film within the MD and TD

	Young modulus (MPa)	Tensile strength (MPa)	Strain at break (%)
MD	221.7 (±15.3)	18.1 (± 1.1)	265 (± 36)
TD	288.5 (±30.1)	9.7 (±0.7)	341 (± 22)
Bulge	485.9 (±13)	11.8 (±1)	22.7 (±4.3)

4.2 Bulge Test

The biaxial stress strain curve of the PEHD film was determined using Eq. (1) and (2). The instantaneous parameters (ρ,t) are estimated using Hill and Panknin assumption respectively (see Table 1 and 2). Figure 3 shows the biaxial behavior of the film and the corresponding properties are given in Table 3. The Young's modulus was determined by fitting the experimental data of pressure, denoted as P, against the dome height, labeled as h, with the general form of Eq. 3. The results show that the specimens subjected to biaxial tensile loading displayed a stiffer response compared to those under uniaxial loading. Moreover, Specimens undergoing equibiaxial loading exhibited significantly lower elongation to failure. Similar observations were made by Fatima in a study on thin composite film subjected to various loading conditions [22]. These findings highlight the limitations associated to uniaxial testing and emphasize the importance of using alternative characterization methods such as biaxial testing to enhance the comprehension of the film's behavior under real operating conditions.

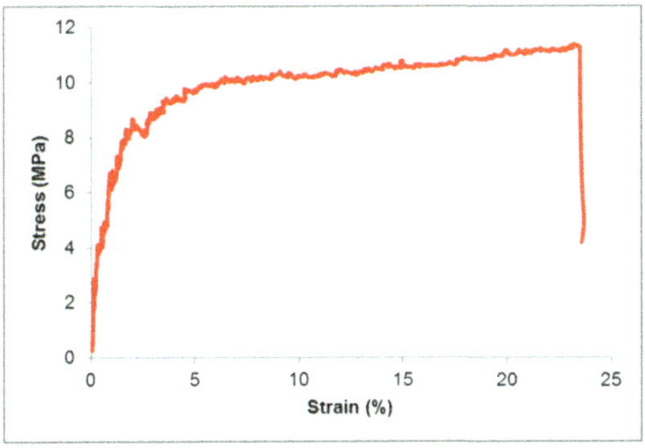

Fig. 3. Tensile behavior of PEHD films cut in the MD and TD

5 Conclusion

Reliable films characterization has received a crucial interest by many researchers in the last decades. In this work, PEHD films underwent characterization under both uniaxial and biaxial stress conditions. The results revealed higher stiffness with significantly reduced ductility under biaxial stress condition. This suggests that the film displays brittle-like behavior when subjected to biaxial loading, in contrast to its ductile-like response under uniaxial loading. These noteworthy findings underscore the limitations of uniaxial testing, emphasizing the necessity for the use of other characterization method like biaxial testing for a comprehensive understanding of the film behavior under real-world conditions. As a perspective for this work, it will be crucial to characterize the PEHD film under various bulge flow rates to examine the impact of the strain rate on the film behavior.

Acknowledgements. This research has received backing from the Tunisian Ministry of Higher Education and Scientific.

References

1. Vlassak, J.J., Nix, W.D.: A new bulge test technique for the determination of Young's modulus and Poisson's ratio of thin films. J. Mater. Res. **7**, 3242–3249 (1992). https://doi.org/10.1557/JMR.1992.3242
2. Berdova, M., et al.: Mechanical assessment of suspended ALD thin films by bulge and shaft-loading techniques. Acta Mater. **66**, 370–377 (2014). https://doi.org/10.1016/j.actamat.2013.11.024
3. Tinoco, H.A., Holzer, J., Pikálek, T., Fořt, T., Sobota, J., Buchta, Z., Lazar, J., Chlupová, A., Kruml, T., Hutař, P.: Numerical modeling approach for the assessment of elastic properties of bi-layer thin films measured by bulge test. In: Proceedings of the Tenth International Conference of Engineering Computational Technology, pp. 1–5 (2018). https://doi.org/10.31224/osf.io/qypw4
4. Boyer, A., Demazel, N., Coër, J., Carin, M., Laurent, H., Oliveira, M.: A novel hydraulic bulge test in hot forming conditions. J. Mater. Process. Technol. **316**, 117917 (2023). https://doi.org/10.1016/J.JMATPROTEC.2023.117917

5. Vasilescu, M.: Development of a Hydraulic Bulge Test to Determine the Work Hardening Behaviour of Sheet Materials (2016)
6. Zhang, J.J., Sun, Y.Y., Li, D. Sen, Cao, Y., Wang, Z., Ma, J., Zhao, G.Z.: Modeling the mechanics of graphene-based polymer composite film measured by the bulge test. J. Phys. D: Appl. Phys. **48**, 425302 (2015). https://doi.org/10.1088/0022-3727/48/42/425302
7. Savio, F. Lo, Rosa, G. La, Bonfanti, M.: A new theoretical-experimental model deriving from the contactless measurement of the thickness of bulge-tested elastomeric samples. Polym. Test. **87**, 106548 (2020). https://doi.org/10.1016/j.polymertesting.2020.106548
8. Çakmak, U.D., Major, Z.: Experimental thermomechanical analysis of elastomers under uni- and biaxial tensile stress state. Exp. Mech. **54**, 653–663 (2014). https://doi.org/10.1007/s11340-013-9820-8
9. Kojio, K., et al.: Advantages of bulge testing and rupture mechanism of glassy polymer films. Polymer **179**, 121632 (2019). https://doi.org/10.1016/J.POLYMER.2019.121632
10. Mulder, J., Vegter, H., Aretz, H., Keller, S., Boogaard, A.H.V., Den,: Accurate determination of flow curves using the bulge test with optical measuring systems. J. Mater. Process. Technol. (2015). https://doi.org/10.1016/j.jmatprotec.2015.06.034
11. Reis, L.C.D. dos: Constitutive Parameters Identification of Metal Sheets using circular bulge tests (2015)
12. Campos, H., Santos, A.D., Amaral, R.: Experimental and analytical evaluation of the stress-strain curves of AA5754-T4 and AA6061-T6 by hydraulic bulge test. Ciência & Tecnologia dos Materiais. **29**, 244–248 (2017). https://doi.org/10.1016/j.ctmat.2016.06.008
13. Lucian, L., Sorin, D., Banabic, D.: Determination of stress-strain curves of sheet metals by hydraulic bulge test. In: The 14th International ESAFORM Conference on Material Forming, vol. 1434, pp. 1429–1434 (2011). https://doi.org/10.1063/1.3589717
14. Kular, G.S., Hillier, M.J.: Re-interpretation of some simple tension and bulge test data for anisotropic metals. Int. J. Mech. Sci. **14**, 631–634 (1972). https://doi.org/10.1016/0020-7403(72)90020-3
15. Hill, R.: C. A theory of the plastic bulging of a metal diaphragm by lateral pressure. The London, Edinburgh, and Dublin. Philos. Mag. J. Sci. **4**, 1133–1142 (1950). https://doi.org/10.1080/14786445008561154
16. Pankin, W.: The hydraulic bulge test and the determination of the flow stress curves (1959)
17. Chakrabarty, J., Alexander, J.: Hydrostatic bulging of circular diaphragms. J. Strain Anal. **5**, 155–161 (1970)
18. Kruglov, A.A., Enikeev, F.U., Lutfullin, R.Y.: Superplastic forming of a spherical shell out a welded envelope. Mater. Sci. Eng. **323**, 416–426 (2002). https://doi.org/10.1016/S0921-5093(01)01376-4
19. Koç, M., Billur, E., Cora, Ö.N.: An experimental study on the comparative assessment of hydraulic bulge test analysis methods. Mater. Des. **32**, 272–281 (2011). https://doi.org/10.1016/j.matdes.2010.05.057
20. Lăzărescu, L., Comşa, D.S., Nicodim, I., Ciobanu, I., Banabic, D.: Characterization of plastic behaviour of sheet metals by hydraulic bulge test. Trans. Nonferrous Met. Soc. China **22**, 275–279 (2012). https://doi.org/10.1016/S1003-6326(12)61719-1
21. Chamekh, A., BelHadjSalah, H., Hambli, R., Gahbiche, A.: Inverse identification using the bulge test and artificial neural networks. J. Mater. Process. Technol. **177**, 307–310 (2006). https://doi.org/10.1016/j.jmatprotec.2006.03.214
22. Alabtah, F.G., Alkhouzaam, A., Khraisheh, M., Attia, H.: Characterization of surface properties of thin film composite (TFC) membranes under various loading conditions. CIRP Ann. **71**, 501–504 (2022). https://doi.org/10.1016/j.cirp.2022.04.069

Influence of Cold Rolling on Microstructure, Mechanical Properties and Fracture Behavior of CuZn30 Brass Alloy

Wafa Taktak(✉) and Riadh Elleuch

Laboratory of Systems Electro-Mechanical (LASEM), National School of Engineers of Sfax, University of Sfax, 599-3018 Sfax, BP, Tunisia
riadh.elleuch@gnet.tn

Abstract. In this research work, the CuZn30 brass alloy sheets were cold rolled in various cold rolling amount (25%, 50% and75%) at ambient temperature to study the effect of cold rolling on microstructural, mechanical properties and fracture toughness behavior of this alloy. Microstructure evolution of specimens was examined through optical microscope before and after cold rolling. These analyses indicated that cold rolling triggers the increase of shear bands areas and the orientation of grains along rolling direction. The mechanical properties of the unrolled and rolled CuZn30 brass alloy were investigated by tensile tests and micro-Vickers hardness tests. These analyses revealed the enhance of the cold rolling amount leads to increase in strength and hardness and a decrease in ductility. Fracture behavior of the unrolled and rolled specimens is studied through ASTME 1820 standard fracture tests methods for determining resistance curves (J-Δa) and fracture toughness parameter ($j_{0.2}$) using center-cracked panels (CCP). The results suggested a significant loss in the fracture toughness parameter of the material (the tenacity at crack initiation J0,2) when the cold rolling amount increases. These finding results are ascribed to the noticeable increase in the number of dislocation and the considerable stretching of equiaxed grains in the rolling direction, due to the hardening created during cold rolling.

Keywords: CuZn30 brass alloy · Cold rolling · Microstructure · Mechanical properties · Fracture toughness behavior

1 Introduction

The 70% copper and 30% zinc alloy, commonly known as 70–30 brass (CuZn30 brass) is a single- phase (alpha phase) solid solution substituted for Zinc in the FCC structure of copper (Mahardika et al.2020, Igelegbai et al. 2016). This alloy is extensively carried out in technology, and subsequent to light metals they are among the most widespread used alloys with the non-ferrous metals institution they are used in various fields of industry, including civil engineering, the aeronautic industry, automobile production, the arms industry, the electrical and several other fields, as well as in the production of musical instruments (Kabash 2015). The impressive increase in the use of this alloy is attributable

M. Chouchane et al. (Eds.): CMSM 2023, LNME, pp. 67–76, 2024.
https://doi.org/10.1007/978-3-031-65007-9_8

to the higher ductility, good mechanical properties, high corrosion resistance, excellent machinability and very cold forming properties (Mahardika et al. 2020, Igelegbai et al. 2016, Rahman et al.2018, Yilmaz et al. 2020). The CuZn30 brass alloy is a single-phase alloy. This implies that the mechanical properties of this alloy cannot be enhanced by heat treatment, but can be enhanced its properties by cold forming. As we all know, cold deformation is a plastic deformation process carried out at low temperature to obtain strain hardening (Taktak and Elleuch 2022). Cold deformation of alpha brass alloys not only produces this sheet, it also improves strength of material through strain hardening phenomenon. However, the plasticity or ductility of cold deformed was decreased. Several studies have been investigated the effect of cold rolling on microstructure and mechanical properties of CuZn30 brass alloy (Darmawan et al. 2013, Bai et al. 2018, Darmawan et al.2018). It was indicated that the grains become elongated and deformed along rolling direction with increase the cold rolling amount. Due to strain hardening the tensile strength (tensile and yield strength) was improved with increasing the percentage of cold rolling. However, the plasticity (ductility) was decreased. So far, numerous investigators have mostly concentrated on the effect of cold rolling on mechanical properties and microstructure. Whereas the effect of cold rolling on fracture toughness behavior of cold rolled CuZn30 alloy have been little investigated. The aim of the following study is to evaluate the effect of cold rolling on the microstructure, mechanical and fracture toughness behavior of CuZn30 brass alloy sheets. In this study, we investigate this effect in more depth by comparing the rolled and unrolled materials. The results revealed that the cold rolling could notably impact the fracture toughness of the CuZn30brass alloy. Furthermore, by increasing the cold rolling amount, the grains became denser and more extended in the rolling direction which creates microstructure oriented and predominant by shear bands and finally results in reduced the ductility and increased the hardness and strength tensile properties.

2 Material and Methods

The material chosen in this research is a α-brass CuZn30 alloy plates of 3 mm thickness. Its nominal chemical composition was 30% Zn, 0.3% Ni, 0.05%Fe, 0.1%Sn, 0.02%Al and the balance of Cu (Wt %). The as received α-brass CuZn30 alloy sheets with the dimensions of 250 mm long, 100 mm wide and 3 mm thick were cold-rolled using laboratory rolling mill at ambient temperature. The cold rolling of CuZn30 plates enabled a reduction in thickness from 3 mm to 2.25 mm, 1.5 mm and 0.75 mm, at cold rolling reduction rates of 25, 50 and 75%.

To evaluate the microstructure of as received and cold rolled specimens, optical microscopy (OM) observation was used. The specimens were coated using resin and catalyst, and prepared according to a conventional metallographic process involving grinding using SiC papers of different sizes, surface polishing using polished metal and etching using a solution consisting of 10 cm^3 hydrochloric acid (HCl), 3 g ferric chloride (FeCl3) and 90 cm^3 ethanol (C2H5 OH) for 5–10 s.

Mechanical properties of unrolled and rolled specimen were identified by hardness and tensile tests. Hardness measurements were carried out by the Vickers microhardness machine under of 50 g and a loading length of 15 s at room temperature. Five measurements were performed for each condition. The elongation and strength properties were measured by tensile test utilizing the universal testing machine LLOYD of capacity 50 kN at speed of 10mm/min. The value of strain was measured by an extensimeter with a 25 mm length installed on the gauge section of specimen. All tensile specimens were machined a long rolling direction in accordance with the NF EN 10002-1 standard as mentioned in Fig. 1. The number of samples for tensile test was three for each case.

Where t corresponding to the thickness of the specimens which is variable. it is a function of cold rolling amount. The values of thickness t were 3mm for unrolled alloy, 2.25mm (25% reduction), 1.5 mm (50% reduction) and 0.75mm (75% reduction).

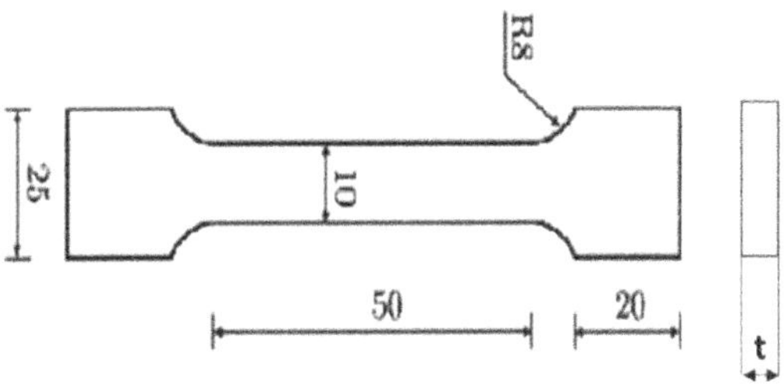

Fig. 1. A schematic of the tensile specimens.

In this study, the crack growth resistance, curve (J-Δa), was employed to evaluate the fracture toughness parameter ($J_{0.2}$) of the unrolled and rolled α-brass CuZn30 alloy was studied. The centre-cracked panel (CCP) specimens were used in the fracture toughness tests. All these specimens were machined a long rolling direction. Before the fracture tests, fatigue pre-cracks were developed in the specimen up to a/w equal to 0.36 (where a and W are the initial half crack length and half width of the specimen respectively). The J-R curves were produced using the single specimen technique, in accordance with the instructions and products defined in the ASTM E1820 standards (Imad et al. 2003). The crack propagation length was measured with the images captured by the high-resolution video camera located in front of the CCP specimen. Calculation of the J integral is in based on the estimation of the charge-displacement registration of the fracture toughness test with the following formula (Schwalbe et al. 1995):

$$J = \frac{K^2}{E} + \frac{A_{pl}}{B(W - a_0)} \tag{1}$$

With W: specimen half width, B: specimen thickness, *a0* specimen half initial crack length, E: Young's Modulus and K: stress intensity factor donated by (Lei and Neale 1997):

$$K = \frac{P}{B\sqrt{2W}} f\left(\frac{a}{w}\right) \tag{2}$$

For CCP type specimen the $f\left(\frac{a}{w}\right)$ is donated by:

$$f\left(\frac{a}{w}\right) = \left[1 - 0.025\left(\frac{a}{w}\right)^2 + 0.06\left(\frac{a}{w}\right)^4 \sqrt{\frac{\pi\,(a/_w)}{cos(\pi\,a/_w)}}\right] \tag{3}$$

A_{pl} represents the plastic strain, obtained from load-displacement curve record. Fracture tests were carried out using a 50 kN capacity LLOYD universal testing machine at a rate of 1 mm/min. For ductile material according to ASTME1820, the critical fracture toughness ($j_{0.2}$) was defined by intersecting the 0.2 mm offset line with the J-Δa curve.

3 Resultants and Discussion

3.1 Microstructure Analysis of Cold Rolled CuZn30 Brass Alloy

Figure 2 (a-d) introduces the experimental optical microstructures observation of the studied CuZn30 brass alloy before and after cold rolling at different cold rolling amount (25%, 50% and 75%). Figure 2.a shows coarser equiaxed grains with no preferred orientation in the as received CuZn30 brass alloy. The Fig. 2.b clearly shows the forming of twining with wide bands but in small density in the in the case of CuZn30 alloy rolled at the cold rolling amount of 25%. The twinning is the effect of identical movements of atoms in a multiple number of rows parallel to a twin plane in the initial lattice. For the cold rolling amount 50% (Fig. 2.c), twinning is most dominant with the onset of a twinning boundary in the CuZn30 brass alloy. When the cold rolling amount become 75% (Fig. 2.d), the mechanism changes radically and is now dominated by the shear band. Extremely elongated grains have been observed at this amount. As cold rolling amount increase, shear deformation bands become elongated and dislocation accumulates. On other hand, the width of the shear bans decreases, leading to difficulty in distinguishing a single grain as depicted in the Fig. 3 (c,d). This is in agreement with studies reported by other researchers (Sofyan and Basori 2016, Basori et al 2017, Horizon et al.2019, Sun and Li 2020, Darmawan et al. 2018).

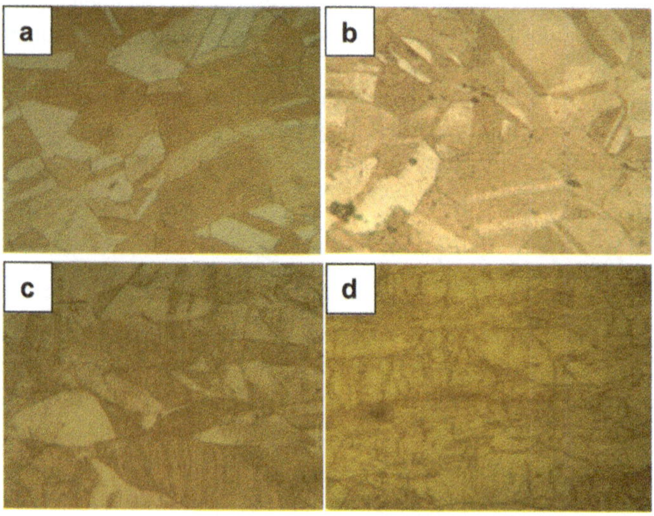

Fig. 2. The microstructure of unrolled and rolled CuZn30 brass alloy specimens. (a) unrolled brass, (b) cold rolled 25%, (c) cold rolled 50% and (d)cold rolled 75%.

3.2 Mechanical Properties of Cold-Rolled CuZn30 Brass

According to Fig. 3, the cold rolling amount is graphed against the microvikers hardness. The initial steps of cold rolling, show a rapid increase in hardness with a 60 Hv0.05 number when the cold rolling amount increases from 0% to 25%. After that, the rate at which microhardness increase slows. There is a 25 Hv0.05 observed with the increase of cold rolling amount from 25% to 75%. It is clear that the level of hardness greatly increased as a result of the cold rolling. This implies that there is a strain hardening during the process of rolling. The CuZn30 brass alloy is more susceptible to hardening in the initial steps of cold rolling after the amount of 25%, the rate at which the hardness increases decrease significantly. This can be attributed to firstly, the increase of the nucleation and movement of dislocation which enhances the density of the dislocation. Secondly, the creation of the extended and oriented structure due to cold rolling has a significant role in enhancing the level of hardness. These results have been mentioned by previous studies (Sofyan and Basori 2016, Klein and Indacochea 2016).

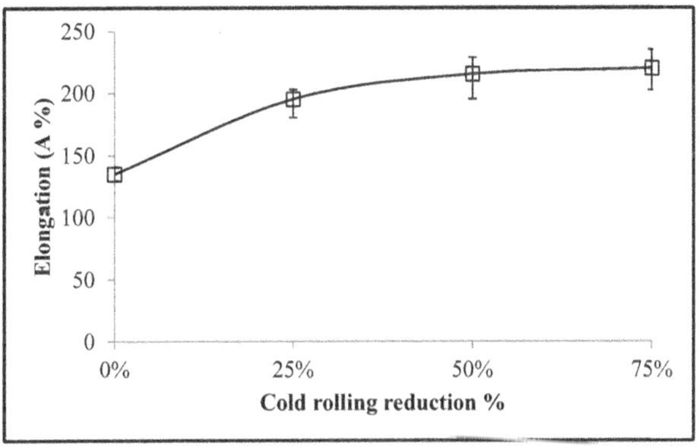

Fig. 3. The microhardness of unrolled and rolled CuZn30 brass alloy specimens.

As displayed in Fig. 4, the cold rolling amount is shown against the percentage of elongation (%A). Percentage of elongation is defined as the strain at moment of rupture throughout the tensile test. It is obvious observed that the linear decrease in percentage of elongation with increasing the cold rolling amount. It is clear that the CuZn30 brass alloy lose its ductility with the increase of cold rolling amount.

As depicted in Fig. 5 presents the yield strength (σ_Y)and the ultimate tensile strength (σ_{UTS}) as function of cold rolling amount on the similar axes. Same as the hardness results the strength tensile increase rapidly in the initial step (from 0% to 25%) and the rate of increase becomes slower. The rise in yield strength and ultimate tensile strength is 328 MPa and 216MPa respectively when the cold rolling amount increases from 0% to 25%. This rise of yield strength and ultimate tensile strength slowed, mounting to a nearly equal 141MPa and 72MPa respectively with the increase of cold rolling amount from 25% to 75%. This implicated that the cold deformation improved the strength of CuZn30 brass alloy. It is obviously proved that the strength and ductility are inversely linked. It can attribute that to the appearance of the work hardening phenomenon due to the plastic deformation with increase the mobility of this dislocation. This following result is in agreement with the results reported by some investigations (Taktak and Elleuch 2022, Horizon et al.2019, Chao et al 2011).

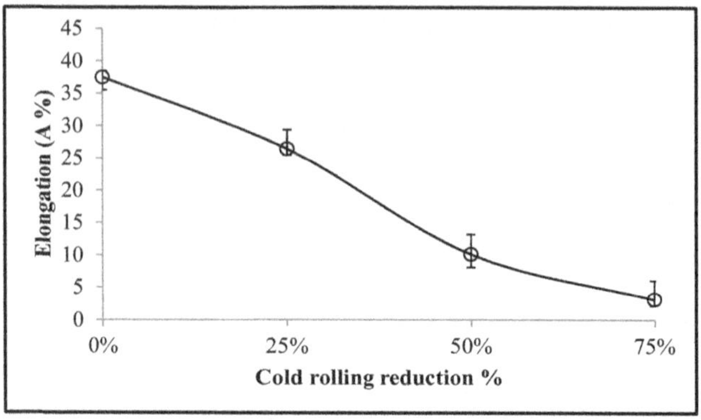

Fig. 4. The Elongation values of unrolled and rolled CuZn30 brass alloy specimens.

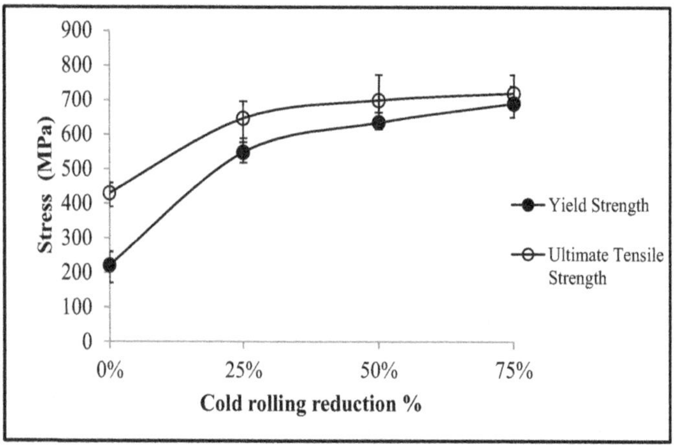

Fig. 5. The tensile strength properties (σ_u and σ_{UTS}) values of unrolled and rolled CuZn30 brass alloy specimens.

3.3 Fracture Toughness Behavior of Cold-Rolled CuZn30 Brass

In Fig. 6 the load-displacement curves of the unrolled and rolled CuZn30 brass alloy are shown. All curves contain a completely linear part, plastic deformation to maximum force and the decrease of the force to failure. The linear part presents the elastic region when the crack propagation does not occur. After linear part, plastic regions found where the crack propagation does not take but the mouth of the crack is deformed. Thus, down to the maximum load point, crack don't develop and the material resist to crack growth.

When the force reaches its maximum, crack growth starts to dissipate the CCP specimen. Finally, the maximum force can be compared with the crack growth resistance for alloy. It can be seen that the maximum load decreased when the cold rolling amount increase. The hard work obtained occur the cold rolling presents the fundamental factor in the reduction of the maximum force of brass alloy after cold rolling.

The fracture resistance (J-Δa) curves for the unrolled and rolled CuZn30 brass alloy are illustrated in Fig. 7. All the specimens show stable growth behavior, and the J-integral enhances in a monotonical way with Δa. But with the increase in cold rolling amount, the slope of the J-R curves decreases. Indicating reduction of the resistance to stable crack extension for specimens cold rolled at higher amount. The $J_{0.2}$ value of unrolled specimens are displayed in Fig. 8. it can be seen that the $J_{0.2}$ values of CuZn30 brass alloy decreased from 47.24 to 14.42 kJ/m^2 (by 69,5%) with the increase of cold rolling. The results approved that the fracture toughness parameter ($J_{0.2}$) of CuZn30 brass alloy reduces when the cold rolling amount increase. it may be confirmed that the cold rolling process reduces the fracture toughness behavior of the CuZn30 brass alloy tested (Madi et al.2020). This marked decrease in the crack initiation resistance ($J_{0.2}$) of rolled specimens can be attributed to grain extension along rolling direction, the increase in the number of dislocations and the occurrence of the microcracks. These findings results were observed in various FCC metals, notably steel alloy (Madi et al.2020) and aluminum alloy (Tajally et al.2010).

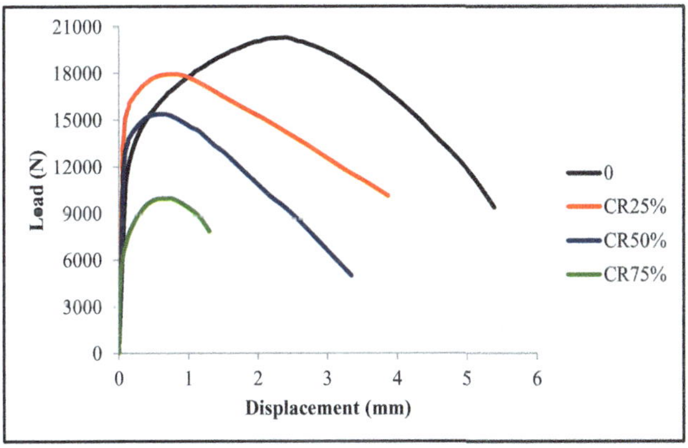

Fig. 6. Load-displacement curves of unrolled and rolled CuZn30 brass alloy specimens.

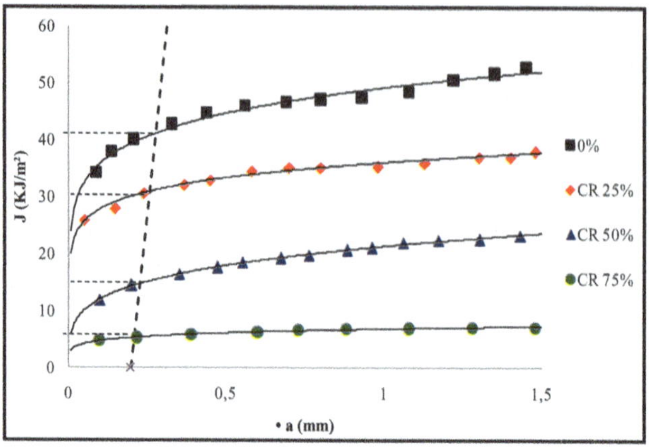

Fig. 7. J-Δa curves of unrolled and rolled CuZn30 brass alloy specimens.

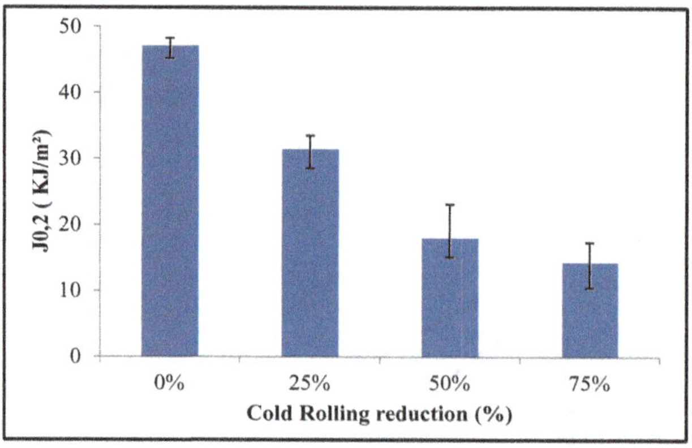

Fig. 8. The resistance at crack initiation J0.2 of unrolled and rolled CuZn30 brass alloy specimens.

4 Conclusion

This study gives a better insight into the influence of cold rolling on the development of the microstructure, on the mechanical characteristics and on the fracture toughness of the CuZn30 brass alloy. The following conclusions were reached from the investigation:

1. The formation of a slip mechanism is more favored than twining at 25% cold rolling amount. At the 50% cold rolling amount, the mechanism is strongly dominated by twining, with the occurrence of certain shear bands. At the cold rolling amount of 75%, the mechanism was predominated by the shear band, associated with a modified grain structure, from equiaxed to extended.
2. The cold rolling procedure has a considerable influence on the mechanical characteristics of this alloy. A noticeable rise in yield strength, tensile strength and hardness and strongly decrease of elongation with the increase in the cold-rolling amount. Cold rolling leads to strengthening and fragility of the α-brass CuZn30 alloy.

3. The fracture toughness parameter ($J_{0.2}$) is considerably decrease with the enhance of the cold rolling amount. The cold rolling procedure causes a reduction in the fracture toughness behavior of CuZn30 brass alloy.

References

Mahardika, A.S., Setiawan, I.A., Sofyan, B.T.: Effect of cold rolling and annealing temperature on the characteristics of Cu-28Zn-5.5 Al alloy produced by gravity casting for bullet case application. Key Eng. Mat. **833**, 3–7 (2020)

Igelegbai, E.E., Alo, O.A., Adeodu, A.O., Daniyan, I.A.: Evaluation of mechanical and microstructural properties of α-brass alloy produced from scrap copper and zinc metal through sand casting process. J. Mine Mater. Charac. Eng. **5**(1), 18–28 (2016)

Kabash, Z.A.: Effect of annealing on mechanical properties of brass alloy type C38500. Diy. J. Eng. Sci. 16–26 (2015)

Rahman, M.M., Sufian, M.A., Shahid, R., Kaiser, M.S., Ahmed, S.R.: Effect of hot and cold rolling on electromechanical properties of a commercial high conductive metallic material. In: The 11th International Conference on Marine Technology, MARTEC (2018)

Atay, H.Y., Uslu, G., Kahmaz, Y., Atay, Ö.: Investigations of microstructure and mechanical properties of brass alloys produced by sand casting method at different casting temperatures. In: IOP Conference Series: Materials Science and Engineering, vol. 726(1), 012018 (2020)

Taktak, W., Elleuch, R.: Fracture toughness resistance and mechanical tensile properties of cold rolled CuZn30 brass alloy. In: Advances in Materials, Mechanics and Manufacturing II: Proceedings of the Third International Conference on Advanced Materials, Mechanics and Manufacturing (A3M'2021), March 25–27, 2021, pp. 231–239 (2022)

Darmawan, A.S., Siswanto, W.A., Sujitno, T.: Comparison of commercially pure titanium surface hardness improvement by plasma nitrocarburizing and ion implantation. Adv. Mater. Res. **789**, 347–351 (2013)

Bai, Y., He, T., Liu, Y.: Effects of Sn microalloying on cold rolling and recrystallization textures and microstructure of a ferritic stainless steel. Mater. Charac. **13**, 142–150 (2018)

Darmawan, A.S., Febriantoko, B.W., Anggono, A.D., Riyadi, T.W.B., Hamid, A.: Effect of thickness reduction on cold rolling process to microstructure and brass hardness. In: MATEC Web of Conferences, 248 (2018)

Imad, A., Wilsius, J., Abdelaziz, M.N., Mesmacque, G.: Experiments and numerical approaches to ductile tearing in an 2024–T351 aluminium alloy. Int. J. Mech. Sci. **45**(11), 1849–1861 (2003)

Schwalbe,K.H., Neale, B.: A procedure for determining the fracture behaviour of materials—the unified fracture mechanics test method EFAM GTP 94. Fat Frac. Eng. Matet. Struct. **18**(4), 413–424 (1995)

Lei, Y., Neale, B.K.: The fracture behavior of a centre cracked tensile specimen. Fat Fract. Eng. Mater. Struct. **20**, 201–216 (1997)

Sofyan, B.T., Basori, I.: Effects of deformation and annealing temperature on the microstructures and mechanical properties of Cu-32% Zn Brass ARPN JEAS **11**(4), 2741–2745 (2016)

Basori, I., Gadhu, R., Sofyan, B.T.: Effects of deformation and annealing temperature on the microstructures and hardness of Cu-29Zn-0.6 Bi brass. Key Eng. Mater. **748**, 218–223 (2017)

Horizon, W., Nisa, S.K., Sofyan, B.T.: Effect of cold rolling and annealing temperature on the characteristics of Cu-28Zn-3.2 Mn Alloy. In: IOP Conference Series: Materials Science and Engineering 517(1), 012002 (2019, April)

Sun, S., Li, T.Q.: Effect of cold rolling process on microstructure and properties of T1 Copper sheet. J. Phys.: Conf. Ser. **1605**(1), 012132 (2020)

Klein J, Indacochea J (2016) Recrystallization Behavior of 70/30 Brass hardness
Chao, H.Y., Sun, H.F., Wang, E.D.: Working hardening behaviors of severely cold deformed and fine-grained AZ31 Mg alloys at room temperature. Trans. Nonfer. Met. Soc. China **21**, s235–s241 (2011)
Hajizadeh, K., Tajally, M., Emadoddin, E., Borhani, E.: Study of texture, anisotropy and formability of cartridge brass sheets. J. Allo. Comp. **588**, 690–696 (2014)
Ohlsson, R.: Variations in hardness and microstructure in cartridge cases at annealing (2018)
Tajally, M., Huda, Z., Masjuki, H.H.: A comparative analysis of tensile and impact-toughness behavior of cold-worked and annealed 7075 aluminum alloy. Int. J. Imp. Eng. **37**(4), 425–432 (2010)
Madi, Y., Shinohara, Y., Besson, J.: Effect of prestrain on ductility and toughness in a high-strength line pipe steel. Int. J. Fract. **224**, 15–29 (2020)

Identification and Identifiability Analysis of Material Parameters Using Hybrid Optimization Algorithms and Optimized Artificial Neural Networks

R. Ben Hmida[1(✉)], R. Bahloul[1], F. Richard[2], and S. Thibaud[3]

[1] National Engineering School of Monastir (ENIM), Mechanical Engineering Laboratory (LGM), University of Monastir, Avenue Ibn Eljazzar, 5019 Monastir, Tunisia
ramzi.benhmida@issatkr.u-kairouan.tn

[2] Université de Franche-Comté, CNRS, Institut FEMTO-ST, 25000 Besançon, France
fabrice.richard@univ-fcomte.fr

[3] SUPMICROTECH-ENSMM, CNRS, Institut FEMTO-ST, 25000 Besançon, France
sebastien.thibaud@univ-fcomte.fr

Abstract. Accurate data sets for material behavior model simulation are necessary for realistic numerical simulation. The true material parameter's identification is still a difficult task. When the optimization process is combined with a finite element simulation, this is typically constrained by the computational time. In this study, a procedure for the identification of material parameters based on a hybrid approach is described. This methodology suggests a strategy for decreasing the computing cost by substituting the FEM simulations by an artificial neural network (ANN) model in the optimization loop. For this reason, a parametric study of the FE simulation is carried out to generate an ANN training database. A high-predictive-performance ANN model is also created by optimizing the hyper-parameters. To quantify the conditioning of the inverse problem and to justify the replacement of the FE model with an ANN model, an identifiability analysis based on an identifiability indicator (I-index) is also proposed. The classical characterization tensile test is used to apply this optimization approach. Finally, numerical and experimental stress-strain tensile curves are compared to evaluate the effectiveness of this methodology.

Keywords: Numerical simulation · Artificial Neural Networks · Parametric identification · Identifiability

1 Introduction

Material parameter identification plays a crucial role in the field of engineering and science, allowing to analyze and predict the behavior of materials under various loading conditions, which in turn leads to improved designs. Additionally, the material parameters identification can present various challenges that may affect accuracy and efficacy. Therefore, the correct parameter identification is critical for producing reliable numerical simulations that can impact real-life applications and decision-making.

M. Chouchane et al. (Eds.): CMSM 2023, LNME, pp. 77–85, 2024.
https://doi.org/10.1007/978-3-031-65007-9_9

Finite element model updating (FEMU) is one of the most important techniques for improving and refining numerical models based on experimental data [1, 2]. Although FEMU has proven its effectiveness in various engineering applications, some limitations exist, particularly regarding the computational cost for material parameter identification [3].

Artificial neural networks (ANNs) have revolutionized the field of material science, opening up new possibilities for material parameter identification thanks to their ability to model nonlinear relationships [4–6]. This innovative approach presents a significant reduction in computational cost compared to traditional methods. ANNs have demonstrated remarkable capabilities for solving complex problems and provide efficient solutions by mimicking the processes within biological neural systems. ANNs can then replace computationally costly Finite Element Method (FEM) simulations in optimization loops.

In the current study, the parameters of plastic behavior are determined from uniaxial tensile test using inverse analysis based on a hybrid optimization algorithm, combining genetic algorithms and LM algorithms. In the optimization loops, the ANN model is employed instead of finite element computations. To reflect the stability of the inverse problem solution and to justify the replacement of the FE model with an ANN model, an identifiability analysis is then carried out.

Finally, the experimental stress-strain curve is compared with the numerical counterpart as well as to the ANN model prediction in order to validate the identification methodology.

2 Experimental Setup

Uniaxial tensile tests were carried out on flat copper alloy specimens annealed at 450 °C for 30min. The tests were conducted on an MTS testing machine with a crosshead speed of 0.03 mm/s. The measurement of the specimen's elongation ΔL was performed by a laser extensometer and the reaction force F is measured by the way of load cell of 1 kN. The used specimens have a rectangular Sect. (0.21 mm × 5 mm) with an initial gage length equal to 15 mm. The specimens were elongated up to fracture and the true stress–true strain curves were obtained.

3 Numerical Modelling of Tensile Test

A finite element parametric model, programmed in MATLAB language, has been developed to numerically simulate the tensile tests. The LS-DYNA software with the explicit integration method is used to the numerical simulation. The specimen is meshed by 4-nodes quadrilateral shell element to decrease the computational time. For this test, the left boundary is fixed while the right boundary is moved in the x-direction, as shown in Fig. 1. The total reaction force is calculated by summing all nodal forces in X-direction at the fixed end.

Fig. 1. Finite element modeling of the tensile test.

The mechanical behavior of the copper alloy is considered through an elastic–plastic law with isotropic hardening. The generalized Hooke's law is used to describe the elastic behavior of the material. Here, the isotropic yield criterion of von Mises is used, and the plastic behavior is modelled by the isotropic hardening law (Voce's law), as follows:

$$\dot{R} = b(Q - R)\dot{p} \tag{1}$$

R and p indicate respectively the isotropic hardening variable and the equivalent plastic strain. The parameters Q and b designate the asymptotic value and the exponent of the isotropic Voce's law, respectively [7].

For this material, the elastic parameters were obtained from ultrasonic characterization. Finally, three parameters characterizing the plasticity mechanism should be identified.

4 Artificial Neural Network Model

In this section, the artificial neural network (ANN) is used to predict the relationship between the material parameters and the true stress-strain curve. In particular, the inputs are the material parameters, and the outputs are the true stress-strain curves associated with these values. To achieved that, a set of numerical tensile tests obtained from the FE parametric modeling is performed to generate an ANN training database. The feed forward neural network model (FFNN), which belongs to the class of multi-layer perceptron (MLP), is employed in this study. In the FFNN model, the neurons are completely interconnected and arranged in successive layers, mainly input, hidden and output layers. In this ANN model, the sets of the weights and biases are adjusted by a scaled conjugate gradient backpropagation algorithm, where the error between the output values and the target responses is minimized during training phase.

To obtain an ANN model with high predictive performance, it is necessary to optimize some hyperparameters defining the architecture design of the model, namely the number of hidden layers, the number of neurons per hidden layer and the learning algorithm. Hence, the Bayesian optimization method (BOA) was used to tune the hyperparameters, which saves time and enhances the training model's performance [8]. Hyperparameters are extracted by minimising the objective function, which is the difference between the predicted and trained responses. Convergence occurs when the objective function reaches a predetermined value or reaches its maximum number of iterations. The flowchart of optimization hyperparameters is shown in Fig. 2.

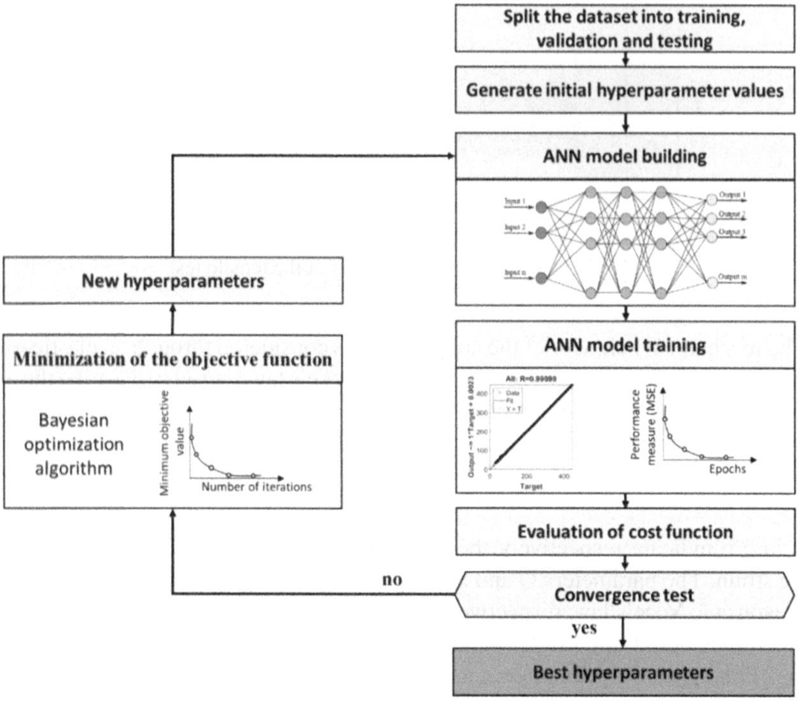

Fig. 2. Hyperparameters optimization procedure.

To validate the deep learning approach used to accurately predict the true stress strain tensile curves, a set of 324 simulations is performed to generate the database for ANN training according to the Taguchi orthogonal array approach (L = 18), as described in [9].

Note that ANN performance analysis is evaluated using the correlation coefficient R and the mean square error (MSE). During the training period (Epoch), the MSE of training, validation, and testing data decrease continuously until a value of 0.08 without any significant traces of overfitting, as shown in Fig. 2. A good correlation, with a regression coefficient R close to one (R≈1), between target and predicted stress strain-curves is demonstrated, as shown in Fig. 4. The results show that the ANN model is well-generalized, indicating that it can be utilized as a substitute method for accurately predicting true tensile stresses (Fig. 3).

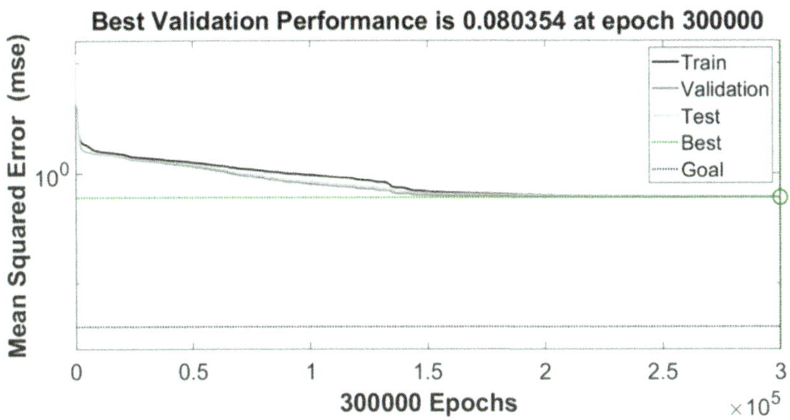

Fig. 3. Performance plot of the ANN model.

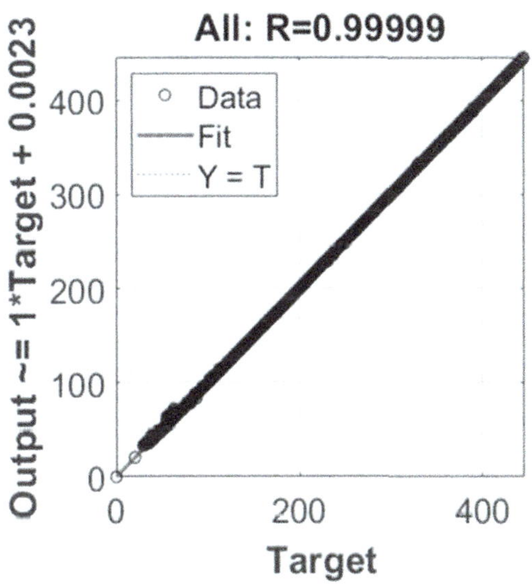

Fig. 4. ANN model regression plot.

5 Identification Methodology and Identifiability Analysis

In this section, the identification methodology and identifiability analysis of the material parameters will be presented. An inverse problem is used for the parameter's identification. The purpose of this inverse formulation is to find the material parameters that minimize the gap between the numerical and experimental stress-strain curves of the tensile test. During the identification process, the material parameters are adjusted using a hybrid optimization algorithm that combines genetic and Levenberg-Marquardt (LM) optimization methods. The LM algorithm starts by using the best set of parameters provided by the genetic algorithm in order to achieve an optimal result. The numerical stress-strain curves used in the present identification method are predicted using the ANN model mentioned above since the ANN model response is almost instantaneous, which considerably reduces the computational cost.

In this study, the genetic algorithm uses 100 successive generations to find an initial solution, requiring 3030 simulations. This computation will take approximately 757.5 h, if the FE model is applied using a massively parallel processing (MPP) version of LS-DYNA with 8 processors. By using the ANN model to predict the numerical stress-strain curves, the online computation time is achieved in only 30 mm. Furthermore, the offline computation time, corresponding to the dataset preparation, takes 81 h.

The proposed methodology was implemented in MIC2M developed by Richard [10]. The current identification procedure is presented in the flowchart given in Fig. 5.

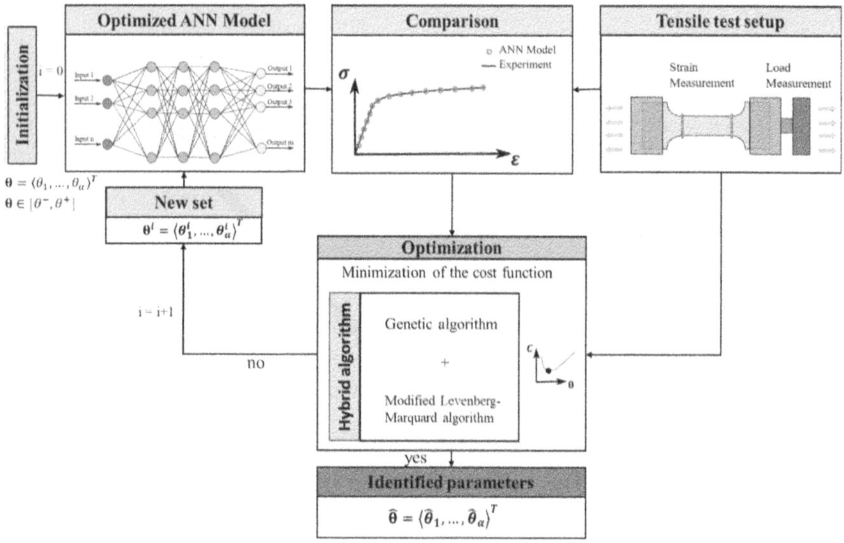

Fig. 5. Flowchart of the identification procedure.

The numerical stress predicted by FE and ANN models, using the identified plastic parameters, is compared in order to evaluate the ANN model robustness and validate the proposed identification approach, as given in Fig. 6.

The maximum error is around 2.1% at the beginning of the tensile test then it decreases when the plastic strain accrues. Overall, the mean absolute percentage error (MAPE) is equal to 0.56%. Therefore, the optimized ANN model accurately forecasts the numerical stress, which confirms the ANN model's replacement of the FE model.

The identifiability analysis based on an indicator is applied to quantify the conditioning of the inverse problem and to justify the substitution of the FEM by an ANN model for true-stress curve prediction. The present analysis reflects the stability of the inverse problem solution. As reported by Richard et al. [11], the measure of multicollinearity of the sensitivity functions of a set of parameters is a function of the ratio $\lambda_{max}/\lambda_{min}$. λ_{max} and λ_{min} are, respectively, the largest and smallest eigenvalue of the pseudo-Hessian

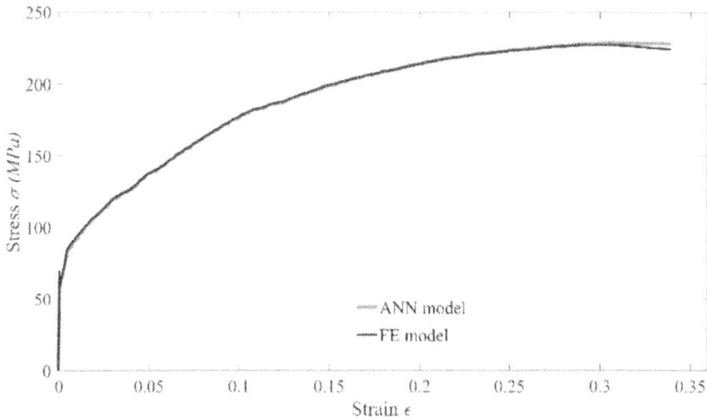

Fig. 6. Comparison of true stress-strain curves obtained by ANN and FE models.

matrix $\boldsymbol{H}$, defined from the sensitivity matrix. More details of the identifiability indicator can be found in [12]. An identifiability index of set of k parameters can be written as follows:

$$I_k = \log_{10}\left(\frac{\lambda_{max}}{\lambda_{min}}\right) \tag{2}$$

Figure 7 shows the measurement of identifiability index with respect to the tensile test strain using FE and ANN models. The identifiability indicator decreases to a value less than 3 when the tensile strain increases, revealing that the problem is well-posed. After the tensile test, the identifiability index stabilizes at a value less than 3 for FE and ANN models, confirming the ability to identify the set of parameters using the tensile test.

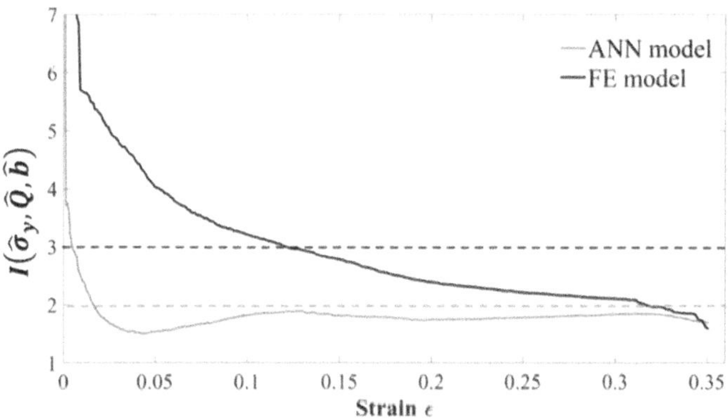

Fig. 7. Identifiability index measurement during the tensile test.

Figure 8 shows a comparison between the experimental and the numerical true stress-strain curves. A good agreement is observed between the stress-strain identified and the experimental one.

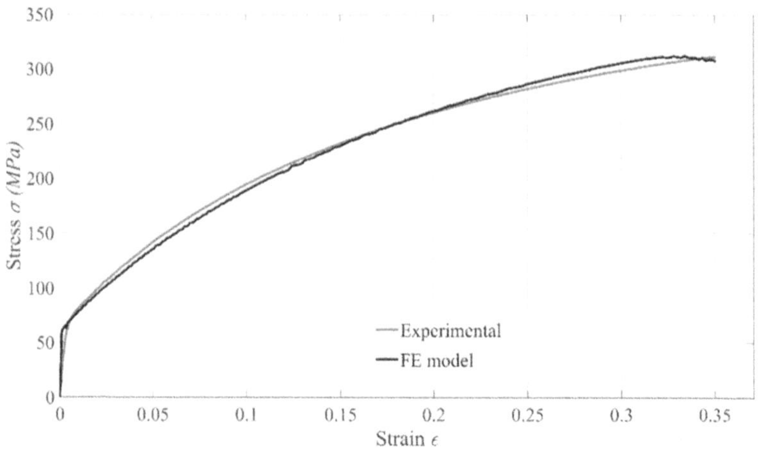

Fig. 8. Comparison of true stress-strain curves between EF modeling and experimentation.

6 Conclusion

In this work, a methodology for identifying material parameters based on a hybrid optimization algorithm, combining the genetic and the Levenberg-Marquardt algorithms, and an ANN model is developed.

The identifiability analysis, based on an indicator of multicollinearity, was conducted to estimate the reliability of the identified material parameters and to validate the conditioning of the inverse problem. This confirms that the artificial neural networks can then be used as an alternative way for stress-strain curve prediction with good accuracy. To validate the proposed identification strategy, the numerical simulation finding of the tensile test using the identified parameters and given in terms of stress-strain curve is compared with the same results predicted by ANN model and measured experimentally. The parameters identified by inverse analysis showed good agreement compared with the experimental measurements. The proposed methodology allows to reduce the computational cost of the optimization procedure. Future studies will focus on the extension of this method to try the identification of more complex constitutive models from more complex experimental tests.

References

1. Kavanagh, K.T., Clough, R.W.: Finite element applications in the characterization of elastic solids. Int. J. Solids Struct. **7**(1), 11–23 (1971). https://doi.org/10.1016/0020-7683(71)90015-1
2. Kavanagh, K.T.: Extension of classical experimental techniques for characterizing composite-material behavior: the experimental-analytical method described in this paper is shown to yield material descriptions from specimen shapes previously considered intractable. Exp. Mech. **12**(1), 50–56 (1972). https://doi.org/10.1007/BF02320791
3. Chaparro, B.M., Thuillier, S., Menezes, L.F., Manach. P.Y., Fernandes, J.V.: Material parameters identification: Gradient-based, genetic and hybrid optimization algorithms. Comput. Mater. Sci. **44**(2), 339–346 (2008). https://doi.org/10.1016/j.commatsci.2008.03.028

4. Aguir, H., BelHadjSalah, H., Hambli, R.: Parameter identification of an elasto-plastic behaviour using artificial neural networks–genetic algorithm method. Mater. Des. **32**(1), 48–53 (2011). https://doi.org/10.1016/j.matdes.2010.06.039
5. Abbassi, F., Belhadj, T., Mistou, S., Zghal, A.: Parameter identification of a mechanical ductile damage using artificial neural networks in sheet metal forming. Mater. Des. **45**, 605–615 (2013). https://doi.org/10.1016/j.matdes.2012.09.032
6. Zhang, J.P., Gao, P.F., Fang, F.: An ATPSO-BP neural network modeling and its application in mechanical property prediction. Comput. Mater. Sci. **163**, 262–266 (2019). https://doi.org/10.1016/j.commatsci.2019.03.037
7. Voce, E.: A practical strain hardening function. Metallurgia **51**(307), 219–226 (1955)
8. Noè, U., Husmeier, D.: On a new improvement-based acquisition function for bayesian Optimization, arXiv preprint arXiv: 1808.06918 (2018). https://doi.org/10.48550/arXiv.1808.06918
9. Cimbala, J.M.: Taguchi Orthogonal Arrays, Pennsylvania State University (2014)
10. Richard, F.: MIC2M: Modélisation et Identification du Comportement Mécanique des Matériaux (1999). http://mic2m.univ-fcomte.fr
11. Richard, F., Villars, M., Thibaud, S.: Viscoelastic modeling and quantitative experimental characterization of normal and osteoarthritic human articular cartilage using indentation. J. Mech. Behav. Biomed. Mater. **24**, 41–52 (2013). https://doi.org/10.1016/j.jmbbm.2013.04.012
12. Fauvel, V., Gaillard, Y., Guillemet, G., Garabédian, P., Richard, F.: Simultaneous extraction of the elastic-plastic properties of a thin film and its substrate from an identifiability-based design of depth-sensing nanoindentation testing. J. Mech. Phys. Solids **163**, 104860 (2022). https://doi.org/10.1016/j.jmps.2022.104860

Physico-chemical Characterization of the Phosphate Sludge of the Metlaoui Plant

Meryam Zahra[1,2], Soufien Essahbi[1], Taktak Mohamed[1](✉), and Mohamed Haddar[1]

[1] Laboratory of Mechanics, Modeling and Production (LA2MP), National Engineering School of Sfax ENIS, University of Sfax, BP.1173, 3038 Sfax, Tunisia
mohamed.taktak@fss.usf.tn

[2] National Engineering School of Gabes ENIG, University of Gabes, Gabes, Tunisia

Abstract. Tunisia is a very important producer of phosphate, which is mainly used in the chemical industries. However, the extraction and processing of phosphate generates a significant amount of waste. Their accumulation can lead to adverse effects on human health, biodiversity, and aquatic ecosystems. The Gafsa Phosphate Company (CPG) is a Tunisian company specializing in the exploitation and production of phosphate. It operates mainly in the mining region of Gafsa-Metlaoui. Our work is located in the mining region of Metlaoui which is located in the southwest of Tunisia, Governate of Gafsa. The discharge of Metlaoui phosphate waste after sludge treatment is a major environmental problem. To limit this industrial pollution, it is essential to find valuable materials in this waste, which forms a complex mixture.

The objective of this work is to perform physicochemical research on ore waste, sludge from Metlaoui phosphate laundries, as well as looking for main materials used in the manufacture of cement and plasters.

Keywords: Phosphate washing sludge · Sludge composition · FX Analysis · DRX Analysis

1 Introduction

In order to process the raw phosphate ore extracted at Metlaoui, phosphate is separated from sterile material. The waste that is generated is called sludge (Johnson C, 1987), (Gnandi K. 1998), (Bodjok K. 2003). This sludge was decanted to obtain the solid material that was the subject of this study. Phosphate (P) is an essential element in our lives. It is used in various fields such as fertilizers, feed, corrosion protection, detergents, cosmetics, and pharmaceutical products (Oliveira MS et al., 2011), (Alsafasfeh A et al., 2017).

Adequate processes and techniques should be used to separate coarse grains (quartz, sand), remove mineral particles, and concentrate apatite. This treatment indeed leads to a considerable accumulation of phosphate sludge. The presence of these elements allows us to think about the use of phosphate sludge in the construction field. Focus on phosphate sludge, also called phosphate washing sludge (Bachouâ H et al., 2014,

M. Chouchane et al. (Eds.): CMSM 2023, LNME, pp. 86–93, 2024.
https://doi.org/10.1007/978-3-031-65007-9_10

Dabbebi R et al., 2018, 2019, Loutou M et al., 2013, 2016) this sludge is produced with an amount equal to 1 tonne per tonne of phosphate (Jellali S et al., 2010, 2011) that is released into rivers, the ocean, or free areas causing air pollution, water, and land. To date, several studies have been done about the use of these sludge in another field. The mineralogical composition of this mud is similar to that of phosphate ore, with the presence of francolite, fluorapatite, carbonates, quartz, and traces of natural clays and zeolites.

2 Materials and Methods

2.1 Raw Material

Our base material is phosphate slurry, also known by phosphate waste, it is a by-product of the phosphate mining industry that is generated during the extraction and processing of phosphate ore. As a rule, the process begins with the extraction of phosphate ore from underground deposits or quarries. This process generates large amounts of waste in the form of mud, often composed of fine particles of rock, sand, silt and phosphate that cannot be recovered.

2.2 The Raw Material Preparation

Mud formation begins with mining, once extracted, the phosphate rocks are crushed and crushed, subsequently, flotation is found which is used to separate the phosphate minerals from other undesirable minerals present in the rock. In this process (Fig. 1.), the crushed rock is mixed with water and chemical reagents and then stirred in tanks. Once the flotation is complete, the phosphate foam is recovered and separated from the rest of the solution (the sludge) as it's shown in (a) this step is the solid-liquid separation. The phosphate slurry is then subjected to a dehydration process to remove excess water. After dehydration, the phosphate sludge was dried in a 100 °C oven (Fig. 2.) from 24 to 48 h. Once dried (b), the mud is crushed (c) and transformed into powder (d) to apply our physico-chemical research.

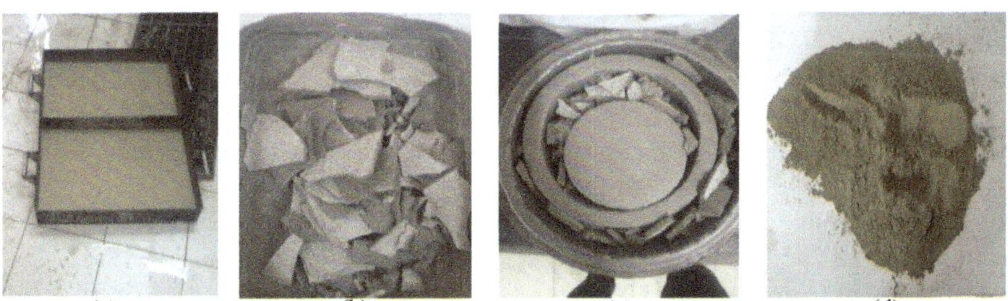

(a) (b) (c) (d)

Fig. 1. The raw material preparation process

Fig. 2. The oven used to dehydrate the phosphate sludge

2.3 Characterization

2.3.1 Chemical Composition (FX)

X-fluorescence is an analytical technique (Hasna Ait Bouh, 2020) used in various fields to determine the composition of samples such as the analysis of geological materials such as rocks, soils, and minerals, in our case mud, in addition, it is used for quality control in industries. To recap, X-ray fluorescence is a versatile method that allows rapid, non-destructive and quantitative analysis of the elemental composition of samples, making it valuable in many scientific and industrial fields.

2.3.2 Spectroscopic Analysis (DRX)

X-ray diffraction is used to identify minerals and can also be used to quantify them, which is useful for evaluating mineral composition (Semën Gorfman et al. 2021). Similarly, this method provides detailed information on the crystal structure including unit cell dimensions, bond angles and interatomic distances, that is, XRD can detect crystalline phases such as transformation phases or crystallographic impurities. Using software like DIFFRAC EVA, TOPAS, High Score, and databases like ICDD PDF-4+ and COD, one can determine which phases are present, their textures, and the residual stresses through measurement interpretation. To obtain diffractionograms, the intensity of the diffracted beam must be recorded as a function of the 2ϴ deviation angle of the incident beam.

3 Results

3.1 Chemical Composition (FX)

The mixture was subsequently ground and placed in tubes of 30g weight after sampling, the elemental composition of a sample can be determined using the X-ray fluorescence (XRF) analysis technique. The principle is that atoms excited in the sample (Fig. 3.) by X-rays produce fluorescence radiation when they return to their ground state. This method is commonly employed in various disciplines like archaeology, geology, metallurgy, materials science, chemistry, physics, biology, forensics, and many others (Table 1).

Fig. 3. The sampling used for the X-ray fluorescence machine

Table 1. Chemical characteristics of sludge

Element	Value(%)
Calcium	16.36
Sulfur	7.8
Phosphorus	4.13
Iron	1.42
Potassium	0.43
Magnesium	0.25
Titanium	0.22

The typical composition of phosphate sludge can include Phosphate (often in the form of calcium phosphate) - main constituent, Silica, Clay, Calcium, Iron, Aluminium, Fluorine, Sulfate, Magnesium, Carbonate, and Organic matter.

The presence of these components may vary depending on the source and the processing of the phosphate sludge. The exact composition details may vary depending on a specific production plant or deposit. Phosphate sludge can also contain contaminants, such as heavy metals, depending on the source of the phosphate and the associated production processes.

The composition of the phosphate sludge can vary also depending on various factors such as the source of the phosphate, the other minerals present in the deposit, as well as the processing techniques used. Generally, phosphate sludge is a by-product of the phosphate mining industry and generally contains a combination of minerals and chemicals. Different studies have shown that heavy metals such as «Cd, Cr, Mn, Mo, Ni, Pb, U, V and Zn» can be found in phosphate sludge and the possibility of their recovery using various techniques and adjusting separate parameters (El Ouardia E, 2010), (Elgharbi S et al., 2015).

3.2 Spectroscopic Analysis (DRX)

To analyze the crystalline structure of materials, the X-ray diffraction technique is often used. It is applied to samples such as phosphate waste, it can be used to evaluate the crystalline composition of these materials. This can be useful for identifying the mineral phases present in phosphate waste, which can be important for various applications such as waste management, resource recovery, or other industrial processes.

XRD is commonly used to locate minerals in unknown samples. Each mineral has a unique crystal structure that generates characteristic diffraction curves when exposed to X-rays. By comparing the curvatures of an unknown sample to a database of known mineral diffraction patterns, we can determine what minerals are present. All samples needed a PANalytical X'Pert Pro with a Cu Kα radiation source (Fig. 4.) to measure their crystallinity. The measurements were taken at a room temperature of around 25 °C. The 2θ angle was calibrated to be between 20 and 45 degrees.

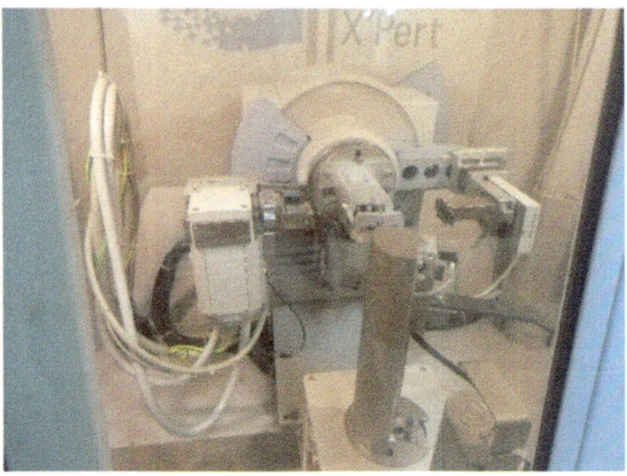

Fig. 4. X-Ray Diffraction machine

Using the Highscore Plus software and his databases, we can analyze the curve below and identify the different phases present, according to the figure. The peaks we find in our mud play an important role in our research because they are indispensable components in the field of construction. The curve shows the presence of Calcite, iron, sodium, and sulfide in their oxide forms and Gypsum in these hydrated and anhydrite forms in their crystallized form. Crystalline materials generate sharp and intense peaks on the X-ray diffraction diagram due to the atomic order present in their crystalline structure, as shown in Fig. 5.

Amorphous materials do not have sharp peaks on the diagram. On the other hand, they have a diffuse diffusion, which indicates that diffraction occurs at different angles than in specific directions, for example for potassium, magnesium and titanium in their oxidized form (Table 2).

Table 2. The chemical formula of the elements presents in the sludge

Element	Chemical Formula
Calcium Oxide	CO
Sulfur Oxide	SO_2
Phosphorus Oxide	P_2O_5
Iron Oxide	Fe_2O_3
Potassium Oxide	K_2O
Sodium Oxide	Na_2O
Titanium Oxide	Ti_2O
Magnesium Oxide	MgO
Calcium Carbonate	CaCO3
Gypsum	CaSO4

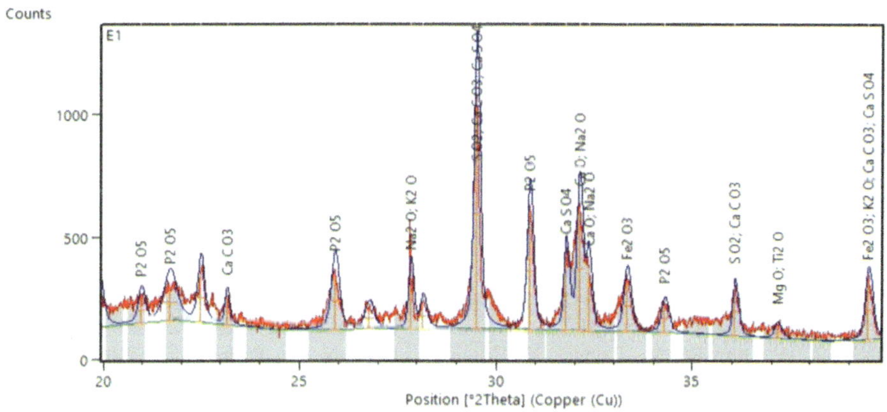

Fig. 5. XRD curvzes of the sludge samples

This curve concerns chemical elements that have an intensity such as SO_2, $CaCO_3$, $CaSO_4$, CaO, and Na_2O.

SO_2 is a key element in the cement industry because of its role in clinker formation. In addition, it has a positive influence on the quality of cement produced because of its ability and potential to reduce air emissions. $CaCO_3$ calcium carbonate plays a crucial role in the manufacture of cement by contributing to its physical, chemical, and economic properties. $CaSO_4$ gypsum is a very important element in the construction sector because it is important given its recovery time control, hydration heat, expansion, and improvement of the workability of the construction material such as cement. Its presence in the mud helps to obtain a better quality building material and reduces the risk of defects in the concrete structure, for example. For the CaO is a crucial component in the manufacture of cement because it reacts with other components to form the key phases of cement, influences the hardening and resistance of the final product, acts as a reactivity regulator, and allows precise control of cement properties, to avoid unwanted effects on performance, it is common to regulate the presence of Na_2O.

4 Conclusion

As a result of this work, the sludge is rich in essential elements such as CaO, Fe_2O_3, Na_2O, K_2O, SO_2, P_2O_5, $CaCO_3$ and $CaSO_4$. This means that they should be considered as by-products and not as waste. The presence of these elements allows us to do more research and even make the addition of new materials to improve the sludge and exploit it in new areas. By studying the chemical behavior of by-products, their use as raw materials for the manufacture of high-end ceramics and environmentally friendly building materials can be considered. In order to improve our results, we will seek to recover other industrial waste such as wood waste, paper waste and of course gypsum. Waste has a major impact on phosphate sludge. And as a research reference we will make a comparison between our results and the results of these experiments on cement, it will help us to clearly specify the quantities of each element. Our second step is to improve our sample mechanically for our next research will be based on mechanical strength tests and acoustic tests which are the objectives of our work.

References

Johnson, C.: Le bassin côtier à phosphate du Togo, Thèse, Université de Bourgogne, France (1987)

Oliveira, M.S., Santana, R.C., Ataíde, C.H., Barrozo, M.A.S.: Recovery of apatite from fotation tailings. Sep. Purif. Technol. **79**, 79–84 (2011)

Alsafasfeh, A., Alagha, L.: Recovery of phosphate minerals from plant tailings using direct froth fotation. Minerals. **7**, 145 (2017)

Bachouâ, H., Othmani, M., Coppel, Y., Fatteh, N., Debbabi, M., Badraoui, B.: Structural and thermal investigations of a Tunisian natural phosphate rock. **8** (2014)

Dabbebi, R., Barroso de Aguiar, J.L., Camões, A., Samet, B., Baklouti, S.: Effect of the calcinations temperatures of phosphate washing waste on the structural and mechanical properties of geopolymeric mortar. Constr Build Mater. **185**, 489–498 (2018)

Dabbebi, R., de Aguiar, J.L.B., Samet, B., Baklouti, S.: Mineralogical and chemical investigation of Tunisian phosphate washing waste during calcination. J. Therm. Anal. Calorim. **137**, 1827–1840 (2019)

Loutou, M., Hajjaji, M., Mansori, M., Favotto, C., Hakkou, R.: Phosphate sludge: thermal transformation and use as lightweight aggregate material. J Environ Manag. **130**, 354–360 (2013)

Loutou, M., Hajjaji, M., Mansori, M., Favotto, C., Hakkou, R.: Heated blends of clay and phosphate sludge: microstructure and physical properties. J Asian Ceram Soc. **4**, 11–18 (2016)

Jellali, S., Wahab, M.A., Hassine, R.B., Hamzaoui, A.H., Bousselmi, L.: Adsorption characteristics of phosphorus from aqueous solutions onto phosphate mine wastes. Chem. Eng. J. **169**, 157–165 (2011)

Jellali, S., Wahab, M.A., Anane, M., Riahi, K., Bousselmi, L.: Phosphate mine wastes reuse for phosphorus removal from aqueous solutions under dynamic conditions. J. Hazard. Mater. **184**, 226–233 (2010)

Semën, G., David, S., Netanela, C., Peter, S.: Multipurpose diffractometer for in situ X-ray crystallography of functional materials (2021)
El Ouardia, E.: Étude de la calcination du phosphate clair de youssoufa (Maroc). Afrique Science: Revue Internationale des Sciences et Technologie (2010)
Elgharbi, S., Horchani-Naifer, K., Férid, M.: Investigation of the structural and mineralogical changes of Tunisian phosphorite during calcination. J. Therm. Anal. Calorim. **119**, 265–271 (2015)

Book

Gnandi, K.: Cadmium et autres polluants inorganiques dans les sols et les sédiments de la région côtière du Togo : une étude géochimique, Thèse, Université Friedrich Alexander, Erlangen-Nuremberg, République Fédérale d'Allemagne (1998)
Bodjok, K.: Etude chimique de l'impact sur l'environnement du rejet des effluents du traitement des phosphates de Hahotoé-Kpogamé (Togo) dans la mer. Université de Lomé (Togo), Mémoire d'ingénieur de travaux en gestion de l'eau et de l'environnement (2003)
Hasna, A.B.: X-Ray fluorescence Technique Analysis (Principles and instrumentations), National Center for Energy Sciences and Nuclear Techniques, Rabat, Morocco (2020)

Online document (no DOI available)

Gypse et anhydrite. http://www.rncan.gc.ca/mms/c,y/contenu/2002/34.pdf
Le plâtre. http://www.perso.orange.fr/smart2000/le%20platre.htm

Composite and Bio-Materials

UV Aging of Poly (Butylene) Succinate/Date Palm Fibers Composites

Rania Chaari[1(✉)], Mohamed Khlif[1], Chedly Bradai[1], Catherine Lacoste[2], and Philippe Dony[2]

[1] Electromechanical Systems Laboratory LASEM, University of Sfax, Soukra Km 4, 3038 Sfax, Tunisia
rania.chaarigem@gmail.com

[2] Laboratory of ITheMM (EA7548), Institute of Thermics, Mechanics and Materials, UFR Exact and Natural Sciences, Campus du Moulin de la Housse, BP 1039, 51687 Reims, Cedex 2, France

Abstract. Natural fibers, when used as reinforcement, offer a wide range of attractive characteristics and provide an alternative solution for waste materials. To investigate the performance of natural fiber-reinforced composites in outdoor applications, it is essential to analyze their environmental degradation properties, particularly under UV conditions. In this study, biocomposites based on date palm fibers were fabricated using the injection molding process. An enzymatic hydrolysis process, employing a synergistic activities effect of xylanase and pectinase enzymes, was applied to treat date palm fibers and compare the resulting composites. Tensile testing and SEM analysis were served for this evaluation. Additionally, the impact of accelerated UV environmental aging on the mechanical properties of all the fabricated composites was assessed. The results obtained revealed that UV aging caused a certain reduction in the tensile properties of composites loaded with raw fibers. This reduction was attributed to weak interfacial bonding, as evidenced by SEM images, which may have been further degraded after artificial aging. In contrast, composites filled with treated fibers exhibited improved preservation of their tensile properties, suggesting a more robust interfacial interaction between the treated fibers and the matrix material.

Keywords: UV conditions · biocomposites · date palm fibers · enzymatic hydrolysis process · interfacial bonding

1 Introduction

Polymers have emerged as substitutes for metals and ceramics in various applications due to their advantageous properties, including lightweight and mechanical strength, as well as for economic reasons related to their low cost. However, the growing need for sustainable materials has recently spurred the use of bio-based materials for producing biocomposites, in which natural fibers serve as suitable reinforcement components [1–3]. These fiber-reinforced composites have gained widespread popularity in packaging, as well as in sectors such as automotive, construction, and maritime industries, due to their technical attributes [4–6].

M. Chouchane et al. (Eds.): CMSM 2023, LNME, pp. 97–105, 2024.
https://doi.org/10.1007/978-3-031-65007-9_11

In order to improve the expected composite performance, various treatments have been applied to ligno-cellulosic fibers, resulting in improved interfacial adhesion with the matrix. While physical and even chemical treatments have been common, recent focus has shifted to enzymatic treatment due to its eco-friendly extraction method. In our previous studies, various enzymatic treatments were considered for treating date palm trunk fibers [7–9]. These treatments utilize xylanase and/or pectinase enzymes, known for their reaction specificity and their abilities to target the hydrophilic components of the fiber, degrading at low temperatures and thus affecting the quality of the resulting composites [10, 11]. The contribution of these researches lies in the validation that the treatment technique combining these two enzymes proved to be the most effective in preventing fiber aggregation during processing and enhancing the composite's moisture resistance by addressing interfacial compatibility issues. As a result, mechanical characteristics of the composite materials are achieved.

Despite numerous studies on natural fibers, there remains a lack of comprehensive understanding regarding the effects of utilizing readily available date palm fibers on the mechanical properties of polymer-based composites materials. The polymer matrix, like other polymeric systems, is susceptible to degradation when exposed to natural weathering (UV radiation, temperature, moisture and environmental pollutants). Chain scission and crosslinking competing mechanisms occur during accelerated weathering, which induce polymer emrittlement [12]. Furthermore, the durability of plant fiber-based composites is significantly influenced by several factors including the composite components, fiber content, fiber extraction method, and interfacial adhesion [13, 14]. However, numerous of their impacts on biocomposites are not yet fully understood. Matuana et al. found that an increase in fiber concentration could increase the oxidation rate in the composite, thereby leading to accelerated degradation. With the addition of a compatibilizer, there was an improvement of the interfacial interaction that led to a decrease in the composite's degradation rate [15]. The objective of this study was to assess the influence of modifying the PBS/date palm fibers interface, through the use of a blend of xylanase and pectinase enzymes, on the composite's durability. To evaluate the long-term behavior, an accelerated aging procedure involving UV radiation was employed. Subsequently, SEM analysis and mechanical property testing were employed for composites characterization.

2 Materials and Methods

2.1 Materials

The palm fiber utilized in this study was extracted from date palm tree waste, specifically from leaves and petioles, in southern Tunisia (Kebili, Tunisia). The fibers were screened to remove any unwanted debris, using mesh sizes of 160 and 630 μm. As for the PBS matrix, it was procured from Nature Plast in France.

2.2 Fibers Modification

The enzymatic treatment of palm fibers was realized in an incubator, employing a combination of xylanase (with an activity of 72 U/g) and pectinase (with an activity of 5100 U/g) enzymes. The xylanase juice was generated using the AX4 strain of Talaromyces thermophilus, while the pectinase juice was obtained from the CT1 mutant of Penicillium occitanis. These enzymes were manufactured within the Laboratory of Molecular Biotechnology of Eukaryotes (LBME) at the Center of Biotechnology of Sfax (CBS). The enzymatic treatment process was conducted at 50 °C with continuous agitation for a duration of 8h. The ratio of fiber matter (g) to liquid (ml) was (1:130). Prior to the enzymatic treatment, a mild NaOH solution (0.4 M) was used as a pretreatment for the fibers to enhance the enzyme effectiveness. The fibers were immersed in an Erlenmeyer flask with a ratio of fibers (g) to NaOH solution (ml) of 1:6.25, to ensure complete wetting. Subsequently, they were autoclaved at 120 °C and 1 bar pressure for 20 min. The collected fibers, following this treatment, were thoroughly rinsed with distilled water until reaching a pH of 7. Then, they were dried for 48h at 50 °C in an oven. This pretreatment facilitated the subsequent enzymatic breakdown of non-cellulosic components in the fibers, thereby enhancing fiber properties for reinforcing the composite material.

2.3 Manufacturing of PBS-DPF Composites

Palm fibers and PBS matrix underwent a drying process at 85 °C for 4h before being introduced into a Clextral BC21 extruder, where the materials were blended to prepare the compounded composite. The resulting mixture was further dried for 48h at 50 °C and subsequently pelletized for use in the injection molding process. Standard samples with dimensions of 75 mm × 10 mm × 4 mm were obtained. For raw fibers, various mass contents (0%, 10%, and 20%) were considered, while a fixed mass ratio of 20% was used for the enzymatically treated fibers.

2.4 Particle Size Analysis

The analysis of the fiber size distribution was carried out using a Fiber Quality Analyzer of FQA Kajaani FS300 Metso model at the University of Quebec in Abitibi-Temiscamingue, Canada (Laboratory of UQAT E214 Biomaterials). After acquiring the length and width measurements, statistical analyses were performed to determine the aspect ratio (Length/Diameter) of the fibers.

2.5 SEM Analysis

To assess the interfacial fiber/matrix interaction, scanning electron microscopy images of the biocomposite tensile fracture surfaces were captured. An FEI Q250/EDAX Thermo Fisher machine type was used with an acceleration voltage of 20 kV.

2.6 Mechanical Properties

Prior to conducting mechanical tensile characterization, the injected samples were subjected to a one-week conditioning period at 50% relative humidity and 23 °C. Subsequently, tensile testing was performed utilizing the Instron 33R4204, operating at a displacement rate of 20 mm/min.

2.7 Accelerated Weathering Test

In a Q-Panel climate chamber, the samples were suspended and exposed to UVB radiation emitted by 8 fluorescent lamps (UVB 313). The biocomposites underwent an artificial aging process for a total duration of 700h, with the chamber temperature maintained at 40 °C. This controlled aging procedure enabled the assessment of the biocomposites' performance and durability under accelerated UV exposure, simulating harsh environmental conditions over an extended period.

3 Results and Discussion

3.1 Particle Size Analysis

Particle size analysis of fibers holds significant importance in the manufacturing of biocomposites, as it directly impacts their mechanical properties. Figure 1 presents the morphological distribution of diameter and the mass distribution of aspect ratio (L/D) of crude palm fibers. The results reveal that the fibers have a diameter range of up to 80 μm, with the most significant population fraction (23.5%) falling within the range of 31 to 40 μm (Fig. 1 a)). Moreover, approximately 76.6% of the palm fibers have a diameter below 40 μm, indicating the prevalence of finer fibers in the sample. As for the aspect ratio L/D, the Fig. 1 b) demonstrates a significant mass distribution, with approximately 92.5% of fibers falling between 1 and 5. Numerous researchers have emphasized the crucial impact of reducing diameter and enhancing the (L/D) ratio to improve the final composite properties [16, 17]. Therefore, enzymatic treatment of palm fibers holds promise for enhancing composite properties. This enzymatic approach targets non-cellulosic (adhesive) materials, facilitating their removal and resulting in the acquisition of finer and rougher fibers. These improvements in fiber characteristics have the potential to positively influence the overall performance of the composites.

a)

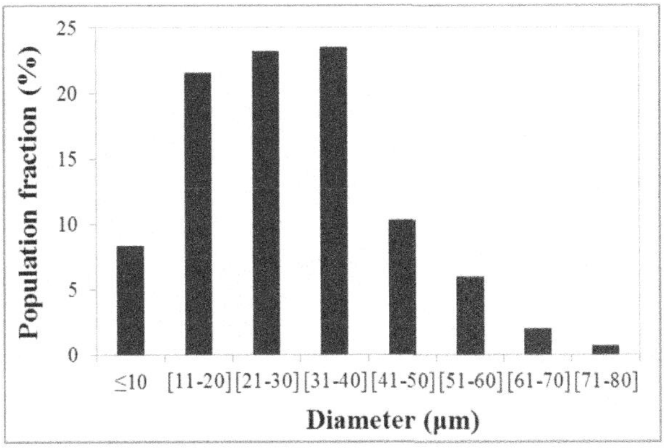

b)

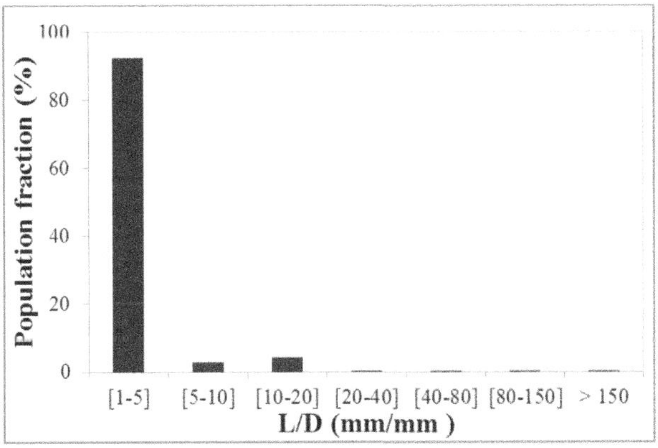

Fig. 1. Distribution of a) the diameter and b) the aspect ratio (L/D) for raw palm fibers.

3.2 SEM Analysis

SEM micrographs depicting the tensile fractured surfaces of composites loaded with 20% native and modified fibers are presented in Fig. 2. As observed from Fig. 2a, the palm fibers are loosely embedded in the polymer matrix, with indications of potential fiber detachment. The presence of gaps already indicates pulled-out raw fibers resulting from mechanical testing. This was attributed to the inadequate wettability of the fibers within the PBS, which causes poor interaction at the interface and may lead to reduced mechanical properties of the composite [18]. Conversely, the structure appears more refined when using enzymatically treated fibers, as shown in Fig. 2b. It is evident that the combined action of xylanase and pectinase enzymes has noticeably reduced the presence of adhesive components, serving as a matrix within the fibers. This fact has improved the compatibility between the PBS polymer and palm fibers and promoted a more effective interfacial interaction. These findings are consistent with mechanical test results, wherein composites with treated fibers exhibited higher tensile strength.

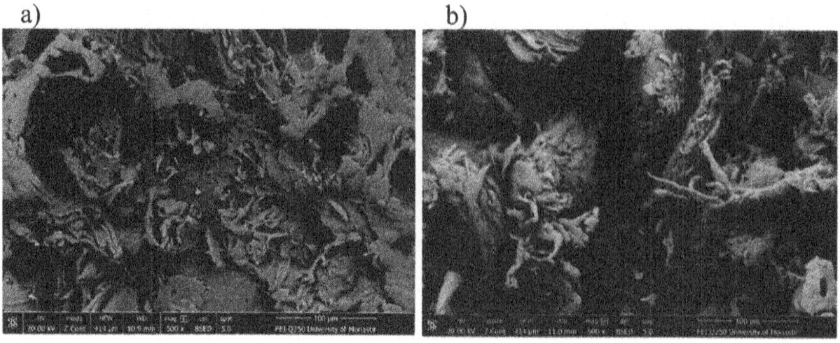

Fig. 2. SEM micrographs of the composites reinforced with a) 20% raw fibers b) 20% treated fibers.

3.3 Mechanical Characterization

The results of tensile testing, before and after aging, for composites with different formulations, are presented in Table 1. It is evident from the table that the Young's modulus increased with higher fiber content in the PBS matrix. The highest Young's modulus is achieved by composite samples containing 20% treated fibers, with a value of 1595 MPa. This figure represents an approximate 62.5% increase compared to unreinforced composites and a 30% increase compared to composites reinforced with 20% raw fibers. This improvement was linked to the removal of amorphous components after fiber treatment, as evidenced by analysis, which results in a decrease in interfacial tension between the polymer molecular chains and the modified fibers [19]. Quite interestingly, even after undergoing the accelerated UV process, composites reinforced with treated fibers maintained the highest stiffness modulus value. Moreover, it is noticeable that composites loaded with raw fibers displayed lower tensile strength than unreinforced composites. The reduction amounts to approximately 29% at a 20% raw fiber mass content. However, when enzymatically treated fibers are used, a slight improvement in this property is observed in comparison to the case of raw fibers (at 20% loading). According to the aforementioned, this fact is a result of enhanced interfacial adhesion achieved through enzymatic treatment, leading to better stress transfer from the polymer to the cellulosic fibers. After 700h of aging, there was a slight decrease in the composite's tensile strength, although this change was almost negligible in the case of modified fibers.

Regarding the strain at break property, the aging cycle significantly reduced the flexibility of PBS by almost 94%. This can be attributed to the breaking of tie molecules in the polymer, which are responsible for maintaining its mechanical characteristics [20–23]. Radical-based oxidation processes induced by ultraviolet (UV) rays are responsible for altering the physicochemical properties of PBS. Photo-oxidation leads to the degradation of the polymer matrix through chain scission via Norrish type I and II reactions [24]. However, the incorporation of date palm fibers resulted in a moderate decrease in this property after UV exposure compared to pure PBS. This finding is consistent with a previous study by P.V. Joseph et al., where they reported that the addition of sisal fibers to a polymeric matrix system decreased the loss in tensile properties following aging [22]. This can be explained by the presence of lignin, a component of the fiber that may contribute to protecting the polymer matrix from UV degradation by absorbing ultraviolet light [25]. Particularly, the treated fibers exhibited notably better retention of the composite's flexibility.

Table 1. Tensile characteristics of unaged and aged composites.

Composite's state	Fiber's content in the composite	Fiber State	E (MPa)	σ (MPa)	Ɛ (%)
Unaged	0%	-	600 ± 6,18	40 ± 1,12	187 ± 6,1
	10%	Untreated	744 ± 3,86	31,5 ± 0,2	10,7 ± 0,34
	20%	Untreated	1131 ± 5,31	28,3 ± 0,34	6,8 ± 1,02
	20%	treated	1595 ± 22,73	30 ± 0,27	4,27 ± 0,09
Aged	0%	-	593 ± 67,88	38 ± 0,22	11,5 ± 0,64
	10%	Untreated	720 ± 15,5	29 ± 0,55	7,3 ± 0,31
	20%	Untreated	970 ± 16,82	27,3 ± 0,24	4,98 ± 0,07
	20%	treated	1356 ± 21,2	29,3 ± 0,19	3,62 ± 0,04

4 Conclusion

Prolonged UV light exposure resulted in a loss of stiffness modulus and tensile strength in pure PBS, accompanied by a dramatic decrease in its strain at break value. Biocomposites containing raw fibers also experienced a minor reduction in tensile properties compared to virgin PBS under long-term UV radiation exposure. In contract, the addition of palm fibers treated with enzymes enhanced the tensile strength and modulus of the unaged composites and demonstrated higher preservation of tensile properties than the composites with untreated fibers (at equivalent loading) after the aging process. This fact can be attributed to the observed strong physical attachment between the treated fibers and PBS, as demonstrated by SEM analysis. Hence, enzymatic fiber treatment emerges as a highly promising and competitive substitute for typically modifying agents. Enzymes are cost-effective, recyclable, and operate under gentle conditions, making them an eco-friendly and sustainable option for fabricating composites with desired high mechanical properties and excellent durability. The envisaged perspectives for this work include the evaluation of the potential applications of these biocomposites in various industries, such as automotive or construction.

References

1. Whelan, T., Kronthal-Sacco, R.: Actually, consumers do buy sustainable products. Harvard Bus. Rev. 1–4 (2019)

2. Väisänen, T., Das, O., Tomppo, L.: A review on new bio-based constituents for natural fiber-polymer composites. J. Clean. Prod. **149**, 582–596 (2017). https://doi.org/10.1016/j.jclepro.2017.02.132
3. Chang, B.P., Mohanty, A.K., Misra, M.: Studies on durability of sustainable biobased composites: a review. RSC Adv. **10**(31), 17955–17999 (2020). https://doi.org/10.1039/C9RA09554C
4. Oushabi, A., Sair, S., Hassani, F.O., Abboud, Y., Tanane, O., El Bouari, A.: The effect of alkali treatment on mechanical, morphological and thermal properties of date palm fibers (DPFs): study of the interface of DPF–polyurethane composite. S. Afr. J. Chem. Eng. **23**, 116–123 (2017). https://doi.org/10.1016/j.sajce.2017.04.005
5. AL-Oqla, F.M., Sapuan, S.M.: Polymer selection approach for commonly and uncommonly used natural fibers under uncertainty environments. JOM **67**(10), 2450–2463 (2015). https://doi.org/10.1007/s11837-015-1548-8
6. Dhakal, H., et al.: Mechanical properties of leaf sheath date palm fibre waste biomass reinforced polycaprolactone (PCL) biocomposites. Ind. Crops Prod. **126**, 394–402 (2018). https://doi.org/10.1016/j.indcrop.2018.10.044
7. Chaari, R., et al.: Enzymatic treatments effect on the poly (butylene succinate)/date palm fibers properties for bio-composite applications. Ind. Crops Prod. **148**, 112270 (2020). https://doi.org/10.1016/j.indcrop.2020.112270
8. Khlif, M., Chaari, R., Bradai, C.: Physico-mechanical characterization of poly (butylene succinate) and date palm fiber-based biodegradable composites. Polym. Polym. Compos. **30**, 09673911221080157 (2022). https://doi.org/10.1177/09673911221080157
9. Chaari, R., Kharrat, F., Khlif, M., Bradai, C., Lacoste, C., Dony, P.: Effect of processing conditions on the rheological and mechanical properties of composites based on a PBS matrix and enzymatically treated date palm fibers. Int. Polym. Proc. **38**(1), 19–29 (2023). https://doi.org/10.1515/ipp-2022-4251
10. George, M., Mussone, P.G., Alemaskin, K., Chae, M., Wolodko, J., Bressler, D.C.: Enzymatically treated natural fibres as reinforcing agents for biocomposite material: mechanical, thermal, and moisture absorption characterization. J. Mater. Sci. **51**, 2677–2686 (2016). https://doi.org/10.1007/s10853-015-9582-z
11. Chaari, R., Khlif, M., Bradai, C., Lacoste, C., Dony, P.: A comparative study on the influence of fiber enzymatic treatment in thermal, morphological and mechanical properties of PBS matrix reinforced with different date palm fibers: under accelerated UV radiation exposure. J. Compos. Mater. **58**(3), 287–298 (2024). https://doi.org/10.1177/00219983231223821
12. Butylina, S., Hyvärinen, M., Kärki, T.: A study of surface changes of wood-polypropylene composites as the result of exterior weathering. Polym. Degrad. Stab. **97**(3), 337–345 (2012). https://doi.org/10.1016/j.polymdegradstab.2011.12.014
13. Chang, S.T., Chou, P.L.: Photodiscoloration inhibition of wood coated with UV-curable acrylic clear coatings and its elucidation. Polym. Degrad. Stab. **69**(3), 355–360 (2000). https://doi.org/10.1016/S0141-3910(00)00082-3
14. Seldén, R., Nyström, B., Långström, R.: UV aging of poly (propylene)/wood-fiber composites. Polym. Compos. **25**(5), 543–553 (2004). https://doi.org/10.1002/pc.20048
15. Matuana, L.M., Jin, S., Stark, N.M.: Ultraviolet weathering of HDPE/wood-flour composites coextruded with a clear HDPE cap layer. Polym. Degrad. Stab. **96**(1), 97–106 (2011). https://doi.org/10.1016/j.polymdegradstab.2010.10.003
16. Iannace, S., Ali, R., Nicolais, L.: Effect of processing conditions on dimensions of sisal fibers in thermoplastic biodegradable composites. J. Appl. Polym. Sci. **79**(6), 1084–1091 (2001)
17. Hakeem, K.R., Jawaid, M., Rashid, U.: Biomass and Bioenergy Processing and Properties. 2014th edn. Springer, New York, USA (2014). https://doi.org/10.1007/978-3-319-07578-5

18. Chaari, R., Kharrat, F., Khlif, M., Bradai, C., Lacoste, C., Dony, P.: Investigating the influence of the aging process on the rheological and mechanical properties of composites based on PBS matrix and enzymatically treated date palm fibers. J. Compos. Mater. **57**(20), 3189–3199 (2023). https://doi.org/10.1177/00219983231184631
19. Mulinari, D.R., Voorwald, H.J., Cioffi, M.O., da Silva, M.L.: Cellulose fiber-reinforced high-density polyethylene composites—mechanical and thermal properties. J. Compos. Mater. **51**(13), 1807–1815 (2017). https://doi.org/10.1177/0021998316665241
20. Azwa, Z.N., Yousif, B.F., Manalo, A.C., Karunasena, W.: A review on the degradability of polymeric composites based on natural fibres. Mater. Des. **47**, 424–442 (2013). https://doi.org/10.1016/j.matdes.2012.11.025
21. Wei, L., McDonald, A.G.: Accelerated weathering studies on the bioplastic, poly (3-hydroxybutyrate-co-3-hydroxyvalerate). Polym. Degrad. Stab. **126**, 93–100 (2016). 10.1016/j.polymdegradstab.2016.01.023
22. Joseph, P.V., Rabello, M.S., Mattoso, L.H.C., Joseph, K., Thomas, S.: Environmental effects on the degradation behaviour of sisal fibre reinforced polypropylene composites. Compos. Sci. Technol. **62**(10–11), 1357–1372 (2002). https://doi.org/10.1016/S0266-3538(02)00080-5
23. Matuana, L.M., Kamdem, D.P.: Accelerated ultraviolet weathering of PVC/wood-flour composites. Polym. Eng. Sci. **42**(8), 1657–1666 (2002). https://doi.org/10.1002/pen.11060
24. Jabarin, S.A., Lofgren, E.A.: Photooxidative effects on properties and structure of high-density polyethylene. J. Appl. Polym. Sci. **53**(4), 411–423 (1994). https://doi.org/10.1002/app.1994.070530404
25. Cogulet, A., Blanchet, P., Landry, V.: Wood degradation under UV irradiation: a lignin characterization. J. Photochem. Photobiol. B **158**, 184–191 (2016). https://doi.org/10.1016/j.jphotobiol.2016.02.030

Hydrothermal Aging Behavior and Tensile Properties of Olive Stone Flour-Reinforced High Density Polyethylene Bio-Composites

Saoussen Belghith[1(✉)], Hend Ben Hadj Salah[1], and Hamza Habbachi[2]

[1] Laboratory of Mechanical Engineering (LGM), University of Monastir, National School of Engineers of Monastir (ENIM), Monastir, Tunisia
belghith.saoussen@gmail.com

[2] University of Sousse, Higher Institute of Applied Science and Technology of Sousse, Sousse, Tunisia

Abstract. For environmental causes, bio-composites are gradually increasing. The main difficulty with these materials is their resistance to moisture absorption which limits their longevity. This study investigated the behavior of High Density Polyethylene /Olive Stone Flour (HDPE/OSF) bio-composites immersed in distilled water and sea water at three different temperatures then compared to neat HDPE. The absorption rate was increased for the charged composites compared to the neat material. Nevertheless, the rigidity of the aged composites remains slightly higher than the aged neat HDPE.

Keywords: bio-composites · hydrothermal aging · mechanical properties

1 Introduction

The used olive pomace has been the majority of nations in the Mediterranean region cultivate olive trees for economic purposes, with approximately 75% of the world's total olive oil production originating from this area. For instance, and according to the Tunisian Oil Office in 2019, Tunisia is the second-largest global producer of olive oil, following the European Union, producing about 340 million tonnes annually. However, the olive oil extraction process generates significant liquid and solid waste, which can lead to environmental contamination, with one hundred kilograms of olives producing 35 kg of pomace, 100 L of margine, 20 L of oil, and 25 kg of leaves, as noted by (Nefzaoui 1990). Environmental preservation is becoming a global priority, prompting the development of sustainable and eco-friendly materials with properties like low density, good thermal insulation, mechanical strength, affordability, and easy disposal.

Natural fiber and polymer composites face challenges related to microstructure heterogeneity and sensitivity to environmental conditions such as moisture, temperature, UV light, and mechanical stress. Despite the substantial solid waste produced by olive oil extraction, only few studies have explored the use of olive stones as fillers in polymer-based composite materials. Different studies focused on various types of polymers, such

M. Chouchane et al. (Eds.): CMSM 2023, LNME, pp. 106–112, 2024.
https://doi.org/10.1007/978-3-031-65007-9_12

as PP. In various studies (Boukerrou et al 2011, Tasdemir 2017), the introduction of OSF negatively affected the mechanical properties of PP composites. Tensile strength, elongation at break, and impact resistance generally decreased as OSF content increased. The main reason for this was the poor dispersion of the flour in the polymer matrix, leading to the formation of aggregates and a weaker composite. The use of chemical treatments and coupling agents, such as maleic anhydride-grafted polypropylene (MAPP), was found to improve the compatibility between OSF and PP. This improved compatibility led to enhanced mechanical properties, better thermal stability, and reduced moisture absorption in the composites (Khemakhem and Jaziri 2016, Naghmouchi et al. 2015, Gümüş et al. 2019, Contreras et al. 2022).

On the other hand, OSF has been studied in other polymers like PS, PVC, PLLA, and HDPE, leading to changes in various properties, including mechanical, thermal, and water uptake behavior. (Boussehel et al. 2011) studied PS/OSF compo-sites, revealing poor interfacial adhesion, lower thermal stability, and increased moisture uptake. The introduction of a silane coupling agent 3-(Trimethoxysilyl)propyl methacrylate, TMSPMA, improved tensile strength, Young's modulus, and thermal stability, enhancing filler-matrix interaction. ((Naghmouchi et al. 2014) studied PVC composites with up to 50% OSF content. These composites were stiffer but had reduced mechanical strength and impact resistance, with a 30% decrease in tensile strength and an 80% decrease in impact strength at 40% filler loading. (Ilhamouchen et al. 2017), focused on enhancing high density polyethylene (HDPE) with olive stone flour (OSF). They used maleic anhydride-grafted polyethylene (MAPE) as coupling agent and vinyl triacetoxy silane (VTAS) OSF treatment to improve compatibility, resulting in better mechanical and thermal properties, reduced moisture uptake, and increased modulus. The inclusion of compatibilizers also improved mechanical and thermal characteristics, with varying effects on tensile strength, crystallinity, and thermal behavior.

(Koutsomitopoulou et al. 2014), combined OSF with PLA for biodegradable materials. They assessed different grinding methods for OSF and found that particle size and morphology varied. The addition of fillers improved crystallinity but reduced transition temperatures. Tensile modulus increased with filler content, while tensile strength slightly decreased due to poor interfacial bonding. (Perınoviç et al. 2010), studied PLLA with varying OSF content, observing its impact on crystallinity and thermal stability. Higher OSF concentrations hindered crystallization. Cooling rates affected crystallite size.

In summary, the inclusion of OSF in various polymer matrices can affect properties such as mechanical strength, thermal stability, and water absorption, with the size and treatment of the OSF, the choice of polymer, and the presence of coupling agents playing crucial roles in these effects. Environmental factors also need to be considered. It is the aim of this study to assess the effects of accelerated hydrothermal aging on the mechanical and structural properties of high density polyethylene (HDPE) reinforced with olive stone flour (OSF). HDPE/OSF samples, with a fixed 30% OSF content, were immersed in distilled and sea waters at three different temperatures (25 °C, 50 °C, and 80 °C). The study focused on analyzing water absorption behavior, changes in mechanical properties via tensile testing.

2 Materials and Methods

2.1 Materials

The used olive pomace has been collected from local olive producer in the region of Monastir-Tunisia. It was obtained using three phase centrifugal process. Then submitted to a solvent extraction and fractionated to get an exhausted olive stone. The main components of this by-product residue are cellulose, hemicellulose, lignin, fat and protein. Exhausted olive stones were dried in air temperature for two weeks then steamed in an oven at 60 °C until crumbly mass was obtained. Besides, they were grinded then sifted through electrical sieve with a square mesh of 300μm to obtain Olive Stone Flour (OSF).

A polyethylene is among the most produced thermoplastics worldwide (Khelif et al 2007), in particularly, High Density Polyethylene (HDPE). Due to the high consumption of this variety, the HDPE is chosen as polymer matrix of the composite material. The used HDPE has a density of 0.95 g/cm^2, a melting temperature of 128–135 °C, a transition temperature Tg = −100 °C and a melt flow index 1.51 g/10 min at 190 °C.

2.2 Preparation of the Composite

HDPE was mixed with 30% weight of OSF oven-dried. The HDPE and OSF were processed using twin screw extruder. The temperature of different zones used during the extrusion was 175 °C/180 °C/185 °C/190 °C/200 °C and the screw speed was 30 rpm. The products obtained were subsequently injection-molded to give standard dogbone specimens (Ihamouchen C. 2017).

2.3 Hydrothermal Procedure

Natural sea water and distilled water was used for aging test. Specimens having dimensions of 168x10x4 mm^3 was prepared for hydrothermal testing. Before absorption, all specimens were dried in a vacuum oven at 55 °C until a steady weight is reached. After that, they were sealed with an isolating tape to prevent absorption into thickness dimension. Following this phase, specimens were immersed into distillated water and sea water at three different temperatures (T_1 = 25 °C, T_2 = 50 °C and T_3 = 85 °C). The total duration of immersion test is 960 h. Every 24 h, specimens were drawn out, soaked with tissue and then weight using precision balance.

The evolution of the moisture weight is calculated as (Petchwattana N. 2018).

$$W_t = \frac{W - Wd}{Wd} \tag{1}$$

where W is the weight of the moist material under testing and W_d is the weight of the dry material.

2.4 Mechanical Test

Mechanical tests have been conducted for filled and neat specimens using tensile testing machine LLOYD EZ-20. Testing parameters have been chosen according to ASTM D538-03.

3 Results and Analysis

3.1 Distilled Water Immersion

Figure 1 shows the experimental absorption curves in distilled water at different temperature for neat HDPE and HDPE/OSF composite versus the square root of the immersion time normalized with respect of the sample thickness ($\sqrt{t}/h$).

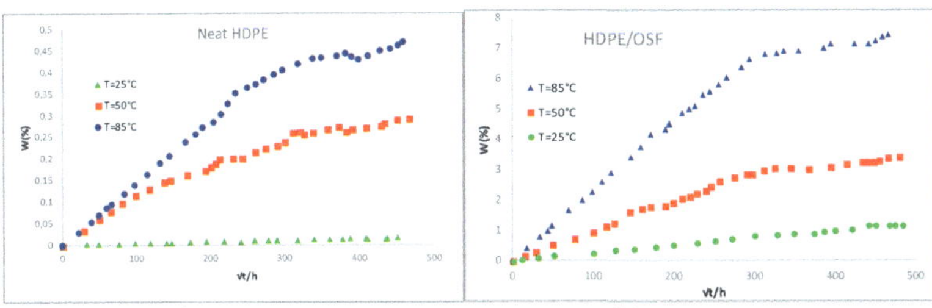

(a) *(b)*

Fig. 1. The weight changes as a function of $t^{1/2}$/h in distilled water at different temperature for: (a) Neat HDPE and (b) charged bio-composites.

Each point of the curves represents the mean over three specimens, for each the results agreed within about ±5%. Results show that the moisture content is increased with the addition of OSF given that OSF has a hydrophilic nature. Furthermore, for both material as the temperature increases the rate of moisture content and the maximum moisture content raise. It's clear that the water temperature reduces de saturation time, and at a given temperature, saturation occurs faster for the composites than for neat HDPE. This tendency agrees with the result obtained by Lin Q. et al. (2002) when studying the hydrothermal aging behavior of polypropylene filled with wood flour bio composites at different temperatures.

3.2 Sea Water Immersion

Figure 2 shows the experimental sea water absorption curves at different temperature for charged HDPE versus the square root of the immersion time normalized with respect of the sample thickness ($\sqrt{t}$/h).

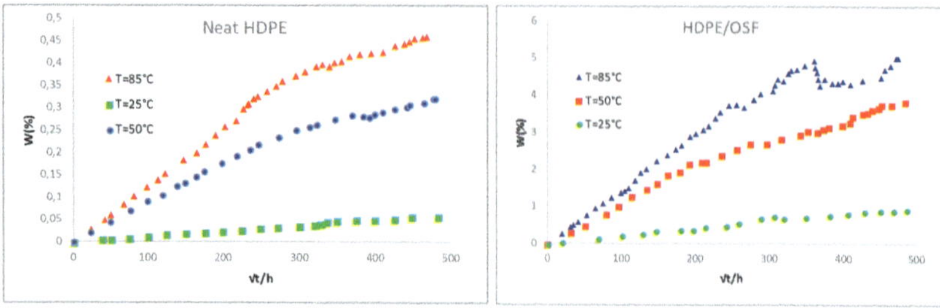

(a) *(b)*

Fig. 2. The weight changes as a function of $t^{1/2}$/h in natural sea water at different temperature for: (a) Neat HDPE and (b) charged bio-composites.

Results reveals a similar behavior to immersion in distilled water. However, a slow-down of water absorption is noted for both neat and charged HDPE. In fact, the electrolytes present in sea water probably slower the absorption process. Similar results has been reported in the literature (Ben Daly et al. 2007). It has been proposed that the presence of dissolved salt may be responsible of the ddegradation of wtaer diffusion process.

3.3 Mechanical Tests

Tensile tests were performed on neat HDPE and HDPE/OSF composite before and after aging test. Figure 3 shows the evolution of the modulus and mechanical resistance of both materials before aging test.

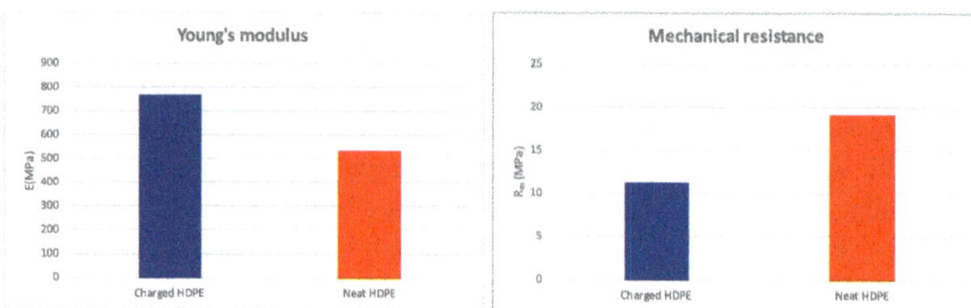

Fig. 3. Young's Modulus and Mechanical resistance of tested materials before aging.

The results for dry specimens confirmed the role of olive stone flower as reinforcer of the HDPE matrix. They show an increase by 43% of the rigidity, nevertheless, the reinforcer limits the HDPE ductility (Fig. 3 (b)) by about 41%. These results were also demonstrated by (Hamida B. 2017) for le Polystyrene PS reinforced with OSF et (Naghmouchi et al 2015) for Polyvinyl chloride reinforced with OSF.

Figure 4 presents the evolution of the rigidity of neat HDPE (Fig. 4(a) and (b)) and reinforced HDPE (Fig. 4 (c) and (d)) after hydrothermal aging process respectively in distilled water and sea water for different temperatures.

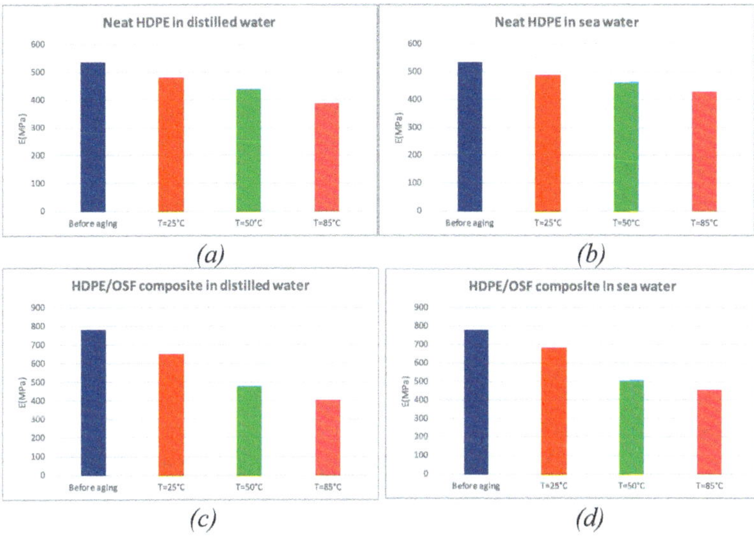

Fig. 4. Young's Modulus for tested materials after aging in distilled (a &c)) and sea water (b &d).

The absorbed water reduces the rigidity of neat HDPE by respectively by 5, 13 and 37% in distilled water and by 12.7, 13 and 49% in sea water for respectively 25°, 50 °C and 85 °C.

The moisture also the rigidity of reinforced HDPE by respectively 13, 51.5 and 48.9% in distilled water and by 27, 53 and 48.5% for respectively 25, 50 and 85 °C.

4 Conclusion

The presence of OSF in HDPE matrix facilitates water absorption and increases the maximum amount absorbed water, probably as a consequence of hydrophilic nature of OSF. Temperature growth from 25 °C to 85 °C enhances the maximum water absorption and the rate of water absorption. This growth is more pronounced for distilled water aging environment compared to sea water.

The addition of OSF in HDPE matrix increased de Young's modulus and decreased the mechanical resistance. Both neat HDPE and HDPE/OSF composites were less stiff with the rise of aging temperature respectively in distilled and sea water. The decline is more pronounced in distilled water.

The absorbed moisture decreases de Young's modulus due to degradation of HDPE – OSF interface area and plasticization of the matrix.

In future work, it will be interesting to conduct a more tribological characterization before and after exposure to hydrothermal aging.

References

Nefzaoui, A.: Valorisation des sous-produits de l'olivier. In: Tisserand J.-L., Alibés, X. (eds.) Fourrages et sous-produits méditerranéens, pp. 101–108. CIHEAM, Zaragoza (1990)

Boukerrou, A., Salem, K., Djidjelli, H., Ihamouchen, C., Martinez, J.J.: Study and characterization of composites materials based on polypropylene loaded with olive husk flour. J. Appl. Polym. Sci. **122**(2), 1382–1394 (2011)

Naghmouchi, I., Mutjé, P., Boufi, S.: Olive stones flour as reinforcement in polypropylene composites: a step forward in the valorization of the solid waste from the olive oil industry. Ind. Crops Prod. **72**, 183–191 (2015)

Khemakhem, M., Jaziri, M.: Composites based on (ethylene-propylene) cop lymer and olive solid waste: Rheological, thermal, mechanical, and morphological behaviors. Polym. Eng. Sci. **56**, 27–35 (2016). https://doi.org/10.1002/pen.24188

Tasdemir, M.: Polypropylene/olive pit & almond shell polymer composites: wear and friction. IOP Conf. Ser.: Mater. Sci. Eng. **204**, 012015 (2017). https://doi.org/10.1088/1757-899X/204/1/012015

Gümüş, B., Yağci, Ö., Ceylan Erdoğan, D., Tasdemir, M.: Dynamical mechanical properties of polypropylene composites filled with olive pit particles. J. Test. Eval. **47**, 20180198 (2019). https://doi.org/10.1520/JTE20180198

Jurado-Contreras, S., Navas-Martos, F.J., Rodríguez-Liébana, J.A., et al.: Manufacture and characterization of recycled polypropylene and olive pits biocomposites. Polymers **14**, 4206 (2022). https://doi.org/10.3390/polym14194206

Hamida, B., Meghezzi, A., Nadia, N.: Polystyrene/olive stone flour composites: effects of filler contents on morphological, thermal, and water absorption properties. Courrier du Savoir **23**, 125–128 (2017)

Naghmouchi, I., Mutjé, P., Boufi, S.: Polyvinyl chloride composites filled with olive stone flour: mechanical, thermal, and water absorption properties. J. Appl. Polym. Sci. **131**, 41083 (2014). https://doi.org/10.1002/app.41083

Ihamouchen, C., Hocine, C., Amar, B., Fenouillot, F., Barrès, C.: Effect of compatibilizing agents on the mechanical properties of high-density polyethylene/olive husk flour composites. J. Appl. Polym. Sci. **128** (2012). https://doi.org/10.1002/app.38434

Koutsomitopoulou, A.F., Bénézet, J.C., Bergeret, A., Papanicolaou, G.C.: Preparation and characterization of olive pit powder as a filler to PLA-matrix biocomposites. Powder Technol. **255**, 10–16 (2014). ISSN 0032–5910. https://doi.org/10.1016/j.powtec.2013.10.047

Perinovic, S., Andričić, B., Erceg, M.: Thermal properties of poly(L-lactide)/olive stone flour composites. Thermochim. Acta **510**, 97–10 (2010)

Khelif, R., Chateauneuf, A., Chaoui, K.: Reliability-based assessment of polyethylene pipe creep lifetime. Int. J. Press. Vessels Pip. **84**(12), 697–707 (2007)

Ihamouchen, C., Djidjelli, H., Boukerrou, A., Martinez Vega, J.J.: Number of processing cycle effect on physical mechanical properties of composites reinforced by olive husk flour. Appl. Res. J. **3**(12), 365–375 (2017)

Petchwattana, N.: Wood-plastic biocomposites prepared from recycled high density polyethylene bottles and wood flour: a comparative study with virgin high density polyethylene/wood flour composites. ICBBB '18 (2018). https://doi.org/10.1145/3180382.3180393

Lin, Q., Zhou, X., Dai, G.: Effect of hydrothermal environment on moisture absorption and mechanical properties of wood flour-filled polypropylene composites. J. Appl. Polym. Sci. **85**, 2824–2832 (2002)

Ben Daly, H., Ben Hadj Salah, H., Cole, K., Denault, J., Dogui, A.: Effect of moisture absorption on the properties of clay-reinforced polypropylene nanocomposites (2007)

Development and Characterization of Polylactique Acid-Based Bio-Composites Reinforced with Posidonia Oceanica Fibers

Ines Ghanmi[1], Faouzi Slimani[2(✉)], Samir Ghanmi[3], and Mohamed Guedri[4,5]

[1] Laboratory of Physics, Mathematics, Quantum Modeling and Mechanical Design, IPEIN Nabeul, Carthage University, Carthage, Tunisia

[2] LMPE/ LR18ES01, National Higher Engineering School of Tunis, University of Tunis, Tunis, Tunisia
slimani.faouzi@yahoo.fr

[3] Research Unit UR 17/ES/19, FS Bizerte. Department of Technology IPEIN Nabeul, Carthage University, Carthage, Tunisia
samir.ghanmi@ipein.rnu.tn

[4] Laboratory/LR18ES45 - Physics, Mathematics, Quantum Modeling and Mechanical Design, IPEIN Nabeul, Nabeul, Tunisia
mohamed.guedri@isetn.rnu.tn

[5] National Higher Engineering School of Tunis, University of Tunis, Tunis, Tunisia

Abstract. This study involves the characterization of a novel thermoplastic material, a biocomposite derived from Posidonia Oceanica (PO) fibers, which is inherently natural and does not undergo any physical or chemical treatment. The new material is formulated from commercial-grade poly-lactic acid (PLA), intended for 3D printing spools in additive manufacturing, and is fortified with Posidonia Oceanica fibers. These fibers, collected from the southern Mediterranean Sea beaches, are in the form of balls of varying volumes. The utilization of Posidonia Oceanica, or sea balls, is due to the fact that this bio-natural material is not suitable for industrial production and the exploitation of marine residues. The biocomposite underwent hot compression tests in a sandwich form under specific conditions. A sequence of uni-axial tensile tests was conducted at ambient temperature to establish the mechanical characteristics of the novel material. Identification tests revealed an enhancement in the mechanical characteristics of the new material with a low percentage of Posidonia Oceanica.

Keywords: PLA · Posidonia Oceanica · characterizing · biocomposite

1 Introduction

Biocomposite materials, fortified with natural fibers and employing biodegradable polymers as their matrix, are witnessing an increasing trend in the industrial sector. The objective of this research is to discover materials that can substitute metallic materials currently used in various industrial applications, while maintaining appropriate mechanical properties. These materials find their use in a broad spectrum of industrial applications, such as

M. Chouchane et al. (Eds.): CMSM 2023, LNME, pp. 113–121, 2024.
https://doi.org/10.1007/978-3-031-65007-9_13

automotive, transportation, and aeronautics. The fabrication of a biocomposite requires the appropriate choice of fibers and compatible matrices to achieve a desired mechanical attribute. Achieving satisfactory adhesion between fibers and matrix is a common goal among researchers and industrialists. Numerous researchers have concentrated on natural fibers, especially cellulose-based fibers like kenaf, flax, hemp, and banana [1–4], or fibers extracted from the date palm tree [5]. These fibers can be employed with different types of matrices, such as thermoplastics or thermosets [6, 7]. This study aids in the creation and analysis of a novel bio-composite substance, fortified with Posidonia Oceanica fibers, employing poly-lactic acid (PLA) as a foundational matrix. Posidonia Oceanica is a marine plant found along the coast (Fig. 1). The selection of fibers derived from sea balls is justified by their abundance on our beaches and their biodegradable nature. The selection of PLA is also rationalized by its market accessibility at a fair cost and its capacity to biodegrade. This biocomposite is fabricated through a process of hot compression. To achieve this, we have engineered and produced an electromechanical system, primarily composed of a die and a digitally regulated punch, which can be affixed to a conventional traction-compression apparatus. To develop this biocomposite, a factorial experimental design was implemented. The production parameters of the novel biodegradable substance encompass the fiber content, the thermal conditions, the duration of punch retention in the matrix, and the pressure exerted between the matrix and punch. A sequence of tensile tests were conducted on standard specimens, precision-cut by laser, derived from the biocomposite. The results indicate that this new biocomposite possesses promising mechanical properties.

Fig. 1. Representation of the Posidonia Oceanica plant [8]

2 Materials and Methods

2.1 Materials

In order to produce this bio-composite, which employs PLA as the foundational matrix and Posidonia Oceanica (PO) as the strengthening element, an experimental device was designed and assembled. This apparatus facilitates the production of a biodegradable material under high temperature and pressure conditions (Fig. 2).

The mechanism was composed of the following components:

- A die and punch.
- Warming components to heat the die and the punch.

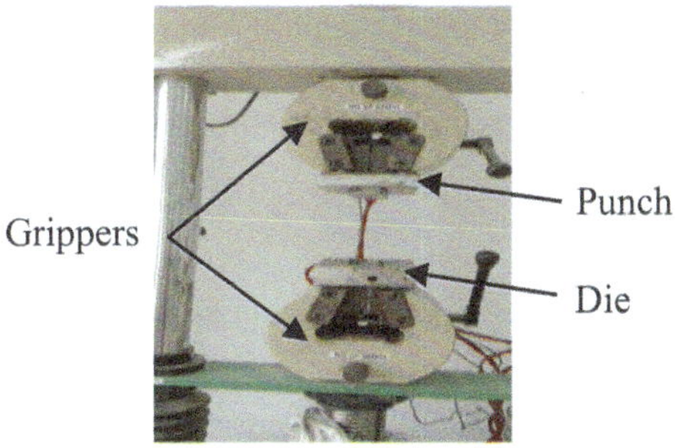

Fig. 2. Experimental set-up for producing the new material

- Two Type J sensors are incorporated, one positioned within the die and the other within the punch.
- A thermal regulator.

The die, designed to house the layers of bio-composite, is attached to the lower part of the tensile testing apparatus, while the punch is linked to the upper part. The die, punch, and the composition of the material are illustrated in Fig. 3. The pair of holes indicate the positions of the heating resistors and thermocouples, which are connected to the thermal regulator. The thermomechanical setup was mounted on a pulling machine of the HM-S 200 KN series 04M4210.

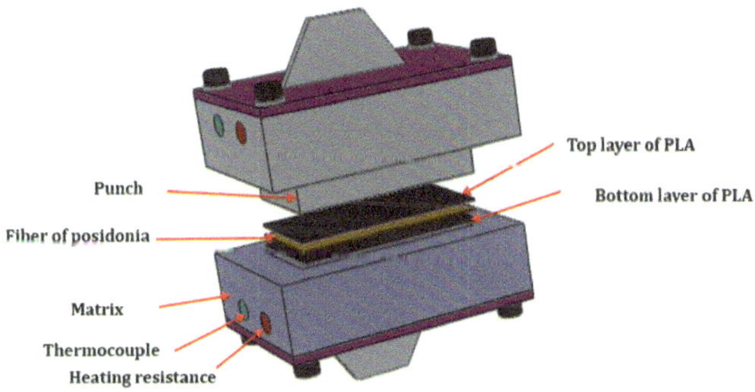

Fig. 3. 3D illustration of the experimental mechanism.

2.2 Methods

The process of characterizing our novel bio-composite can be broadly categorized into three primary parts:

- Part 1:
    - Preparation of PO fibers to match the dimensions of the intended bio-composite plate (includes cleaning and detangling of fibers) (Fig. 4).

  - Preparation of PLA: this involves cutting into small pieces of 0.5 to 1 cm and weighing.
  - New bio-composite sheets are produced through hot pressing.

- Part 2:

  - Laser cutting of the biocomposite specimens into dumbbell shapes (as shown in Fig. 5).
  - Executing a series of tensile tests.

- Part 3: Aggregation of findings (ascertaining the physical attributes of the bio-composite).

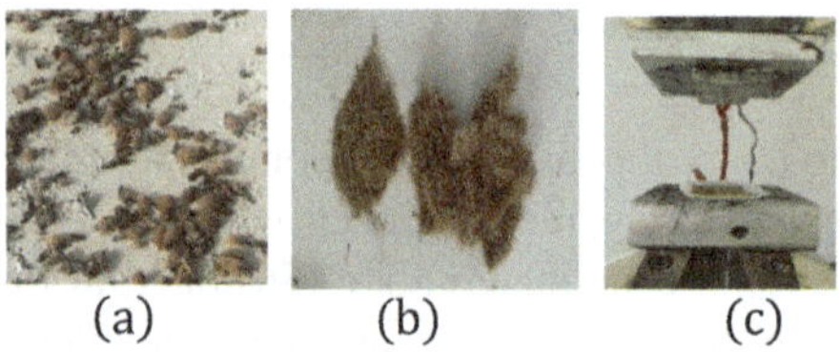

Fig. 4. Process of Preparing Posidonia Oceanica as Reinforcement: (a): Posidonia Oceanica prior to processing. (b): Separation of fibers; (c): Material Before Being Compressed.

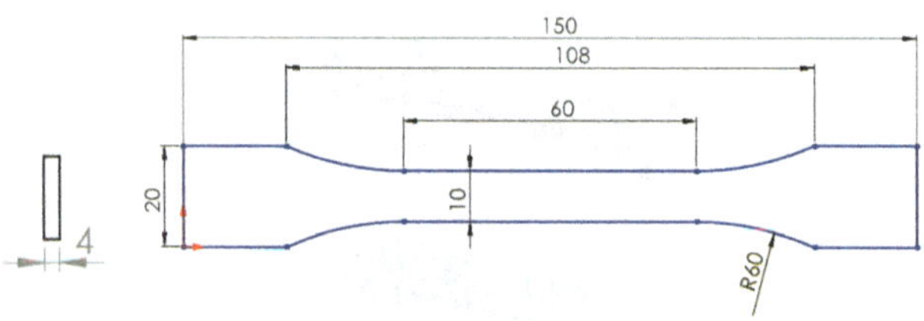

Fig. 5. Tensile Test Samples Compliant with NF ISO 527–2 Standard [9]

3 Experimental Protocol

In the production of the biocomposite material, we exploited the experimental mechanism. This research considers four design factors for the fabrication of the novel substance:

- The temperature for heating (T°C),
- The percentage of fiber (% PO),
- The period of time during which pressure is exerted (t(min)),
- The applied pressure force (F (10^3.daN)).

All parameters, at their minimum and maximum levels, are depicted in Table 1. Eight bio-composite plates were fabricated as per the specifications in Table 2 (Fig. 6).

Table 1. Parameters for the production of the new material.

Parameters	Factor Levels	
Matrix temperature (T (°C))	170	180
% of fiber	5	10
Holding time (t (min))	10	20
Pressure force (F (10^3.daN))	10	14

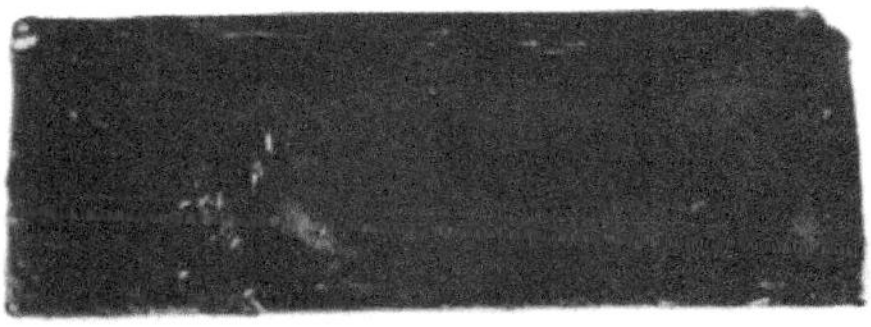

Fig. 6. Illustration of a PLA/PO plate

4 Results and Analysis

In order to assess our novel biocomposite material, we performed tensile tests on samples taken from the 8 plates, following the ISO 527 standards [10]. The plates were produced by hot pressing using the thermomechanical system as per the specifications in Table 2. Each plate was then laser-cut into 3 specimens.

Table 2. Experimental parameters for the plates

N° of plate	T° (C)	PO (%)	t (min)	F (10^3daN)
1	170	5	20	10
2	170	5	20	14
3	180	5	20	10
4	180	5	20	14
5	170	10	10	10
6	170	10	10	14
7	180	10	10	10
8	180	10	10	14

The specimens underwent tensile tests under optimal conditions, and the resulting stress-strain curves are illustrated in Figs. 7 and 8. The Fig. 7 depicts the characteristics of the novel biocomposite with 5% fibers (PLA/PO-5), whereas Fig. 8 shows the characteristics of the biocomposite with 10% fibers (PLA/PO-10). The performance of all samples can be segmented into three main phases:

- The initial phase represents the linear increase in stress with strain, covering the deformation zone [0.05–0.25%] in accordance with ISO 527–4 [10]. Within this elastic part of the stress-strain graph, the elasticity modulus can be calculated employing Hooke's law as per Eq. (1) [1]:

$$E = \frac{\Delta\sigma}{\Delta\varepsilon} \quad (1)$$

where E: Young's Modulus (MPa), $\Delta\sigma$: stress variation (MPa), and $\Delta\varepsilon$: strain variation.
- During the second stage, the initial part is inclined, succeeded by a linear rise to the maximum stress value.
- In the final stage, the tensile specimen experiences breaking after the stresses have been released.

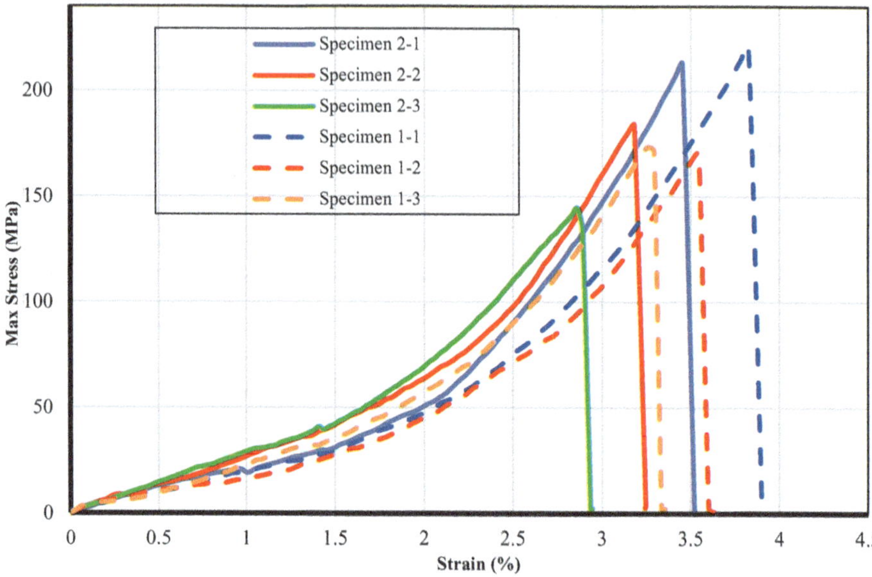

Fig. 7. Tensile Graphs for 5% PO Samples (Plate 1-- and Plate 2 -)

The Fig. 7 displays the stress-strain curves for Plates 1 and 2, while Fig. 8 illustrates the same for Plates 5 and 6. In both figures, the stress-strain curves for Plates 2 and 5 reach a peak before descending, suggesting a failure point. On the other hand, Plates 1 and 6 show a consistent rise in stress, achieving a higher peak at an increased strain. It's important to recognize that the amount of Posidonia Oceanica fibers in the PLA used for these plates can vary, as can the dispersion of these fibers. These fluctuations can influence the mechanical characteristics of the material, including the peak stress and strain. Such changes in fiber content can have a significant impact on the material's overall performance. Hence, it's vital to take into account the intended use and the ideal fiber content during the design and manufacturing processes of PLA-based materials.

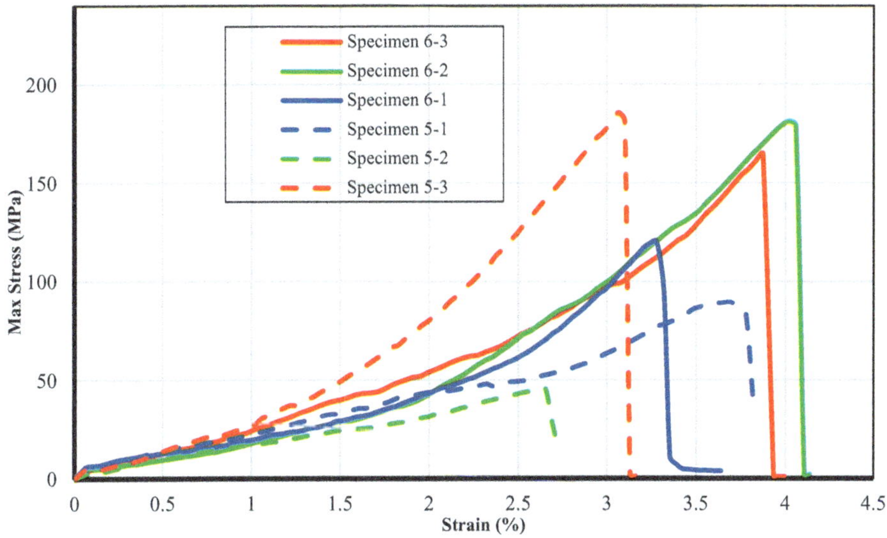

Fig. 8. Tensile Graphs for 10% PO Samples (Plate 5-- and Plate 6 -)

Tables 3, 4, 5 and 6 provides a synopsis of the results from the tensile tests conducted on the two biocomposites, PLA/PO-5 and PLA/PO-10, as well as for the standalone PLA. This includes the average values and standard deviations of their mechanical properties such as density, maximum stress, strain, and elasticity modulus under tensile stress, in comparison to PLA alone. The data indicates a significant rise in the peak stress, strain, and density for both PLA/PO-5 and PLA/PO-10 biocomposites in comparison to pure PLA. Conversely, there is a decrease in the Young's modulus for PLA/PO-5. For PLA/PO-5, the maximum stress, strain, and density values for plates 1 and 2 are 221.4 MPa, 3.9, and 950 kg/m^3 respectively. For PLA/PO-10, these values for plates 5 and 6 are 186.11 MPa, 4.14, and 940 kg/m^3 respectively.

Table 3. Mean density of PLA/PO bio-composites [6]

Material	ρ(kg/m^3)
PLA	850 ±12
PLA-PO-5	945 ±57
PLA-PO-10	922 ±23

Table 4. Mean maximum stress of PLA/PO bio-composites [6]

Material	σ_{Max}(MPa)
PLA	46 ±14
PLA-PO-5	167.87 ±43.46
PLA-PO-10	137.74 ±24.32

Table 5. Mean strain of PLA/PO bio-composites [6]

Material	ε(%)
PLA	1.26 ±0.43
PLA-PO-5	3.50 ±0.39
PLA-PO-10	3.9 ±0.30

Table 6. Mean elastic modulus of PLA/PO bio-composites [6]

Material	E(GPa)
PLA	3.32 ± 1.77
PLA-PO-5	2.82 ± 0.11
PLA-PO-10	3.29 ± 0.34

5 Conclusion

This study contributed to the production and characterization of a new thermoplastic, biocomposite, and biodegradable material. This material is a blend of polylactic acid, serving as a matrix, reinforced with Posidonia Oceanica. Commercial polylactic acid, which is not typically recommended for additive manufacturing, and unused sea balls collected from beaches, form this biocomposite material. We engineered and fabricated a thermo-mechanical device to produce this new bio-composite using hot pressure. Uni-directional tensile tests were conducted on co2 laser-cut samples taken from the plates to characterize the new material. These plates were produced considering four experimental parameters: heating temperature (T°C), reinforcement percentage (% of PO), the duration the bio-composite was kept under pressure (t(min)), and the pressure force (F (10^3.daN)). The results demonstrated a noticeable increase in the maximum stress and strain of the new PLA/PO-10 and PLA/PO-5 bio-composites compared to PLA alone, albeit at the expense of a reduction in the Young's modulus of PLA/PO-5.

References

1. Huda, M.S., Drzal, L.T., Mohanty, A.K., Misra, M.: Effect of fiber surface-treatments on the properties of laminated biocomposites from poly-lactic acid (PLA) and kenaf fibers. Compos. Sci. Technol. **68**, 424–432 (2008)
2. Poilâne, C., Cherif, E., Richard, F., Vivet, A., Ben Doudou, B., Chen, J.: Polymer reinforced by flax fibres as a viscoelastoplastic material. Compos. Struct. **112**, 100–112 (2014)
3. Pietruszka, B., Gołębiewski, M., Lisowski, P.: Characterization of hemp-lime bio-composite. IOP Conf. Ser. Earth Environ. Sci. **290**, 012027 (2019)
4. Sivaranjana, P., Arumugaprabu, V.: A brief review on mechanical and thermal properties of banana fiber based hybrid composites. SN Appl. Sci. **3**, 176 (2021)
5. Kumar, S.S., Raja, V.M.: Processing and determination of mechanical prop-erties of prosopis juDPFslora bark, banana and coconut fiber reinforced hybrid bio composites for an engineering field. Compos. Sci. Technol. **208**(26), 108695 (2021)
6. Slimani, F., Ghanmi, I., Ghanmi, S., Guedri, M.: Production of thermo-plastic composites reinforced with posidonia oceanica fibers. Eng. Technol. Appl. Sci. Res. **14**(2), 13243–13247 (2024)
7. Ghanmi, I., Slimani, F., Ghanmi, S., Guedri, M.: Development and charac-terization of a PLA biocomposite reinforced with date palm fibers. Eng. Tech-nol. Appl. Sci. Res. **14**(2), 13631–13636 (2024)
8. Trivedi, A.K., Gupta, M.K., Singh, H.: PLA based biocomposites for sus-tainable products: a review. Adv. Ind. Eng. Polym. Resea. **6**(4), 382 (2023). https://doi.org/10.1016/j.aiepr.2023.02.002
9. ISO 527–2: 2012, Plastiques-Détermination des propriétés en traction-Partie 2: conditions d'essai des plastiques pour moulage et extrusion (2012)
10. ISO 527–4:2021, Plastiques—détermination des propriétés en traction—partie 4: conditions d'essai pour les composites plastiques renforcés de fibres isotropes et orthotropes (2021)

Numerical Investigation of Shearing Processes on the Deformation Behavior of Elasto-Plastic Fe-TiB_2 Composite

Hanen Jrad[1,3](✉), Abir Bouhamed[1,2], Manel Dammak[1], Mondher Wali[1], and Fakhreddine Dammak[1]

[1] LEE Laboratory, National Engineering School of Sfax, University of Sfax, Sfax, Tunisia
hanen.jrad@essths.u-sousse.tn

[2] Institut Supérieur des Sciences Appliquées et de Technologie de Gabès, University of Gabes, Rue Omar ibn el Khattab, 6072 Gabes, Tunisia

[3] École Supérieure des Sciences et de la Technologie de Hammam Sousse, University of Sousse, Rue Lamine Abassi, 4 11 Hammam Sousse, Tunisia

Abstract. The progress in sheet metal forming owes a lot to cutting-edge tools, especially in the vast field of finite element simulations. These simulations are a game-changer for delving into numerical investigations. To ensure the reliability and quality of results, it's crucial to accurately model material behavior throughout the forming process. This involves characterizing and determining the material properties under scrutiny and translating them into numerical material models. In the business of creating a ferritic steel matrix reinforced with brittle ceramic particles like TiB2, the need to know how these materials behave under different loads is necessary. This understanding is gold when it comes to modeling their mechanical properties and predicting how they'll shape up. This study describes new ground by applying a 12% TiB_2 content Fe-TiB_2 composite to the shearing process. A user interface material subroutine (VUMAT) is performed. The focal point of this investigation resides in the anticipation of how specific pivotal process parameters, such as the thickness of the blank and the applied holding load, influence the quality attributes of the final product. Additionally, the objective is to show the shearing deformation capabilities of Fe-TiB_2 composite and to represent its performance.

Keywords: Iron-Titanium Diboride composite · shearing process · User Material subroutine (UMAT) · elastoplastic framework

1 Introduction

Nowadays, metal matrix materials have garnered significant attention due to their potential for reducing structural weight, achieved by enhancing specific modulus while maintaining a favorable strength-to-ductility ratio, as evidenced in prior research, [1–3]. Recent advancements in composites, featuring a ferritic steel matrix reinforced with brittle ceramic particles, underscore the urgent need to understand their response to

M. Chouchane et al. (Eds.): CMSM 2023, LNME, pp. 122–129, 2024.
https://doi.org/10.1007/978-3-031-65007-9_14

varied loading conditions, particularly given their intended use in automotive shaping processes. Simple shear tests conducted by Dammak and colleagues [4–6] revealed substantial rupturing of primary and eutectic particles as the primary damage mechanism in Fe-TiB2 composites, with minimal particle debonding observed. Conversely, Lartigue et al. [7] found significant plastic deformation in TiB2 particles induced by hot rolling, potentially contributing to lower damage levels in the hot-rolled composite. Plastic strains induced by thermal and forming processes have the potential to alter the properties of these materials, necessitating assessment, especially regarding the ductility of TiB2 and the ferritic matrix, to optimize processing capabilities. Previous numerical analyses of Fe-TiB2 composites have focused on damage development and subsequent stiffness loss under various mechanical loading conditions. In a recent study, Dorhmi et al. [8, 9] delved deeper into investigating the effective properties of the composite following hot and cold rolling processes.

In recent years, the field of mechanical engineering has witnessed a remarkable evolution, catalyzed by the advent of cutting-edge tools such as finite element analysis (FEA). FEA has emerged as a pivotal tool in addressing intricate engineering challenges, particularly in the optimization of manufacturing processes. By leveraging computational simulations, FEA enables engineers to model and analyze the behavior of complex systems under various loading conditions, offering insights that are often impractical or impossible to attain through traditional experimental methods alone, [10–14]. Further, the proliferation of advanced tools, such as finite element simulations, has significantly enhanced the optimization of manufacturing processes, particularly those integrating innovative techniques. The past two decades have witnessed notable progress in the sheet shearing process, which has emerged as a focal area in engineering applications, driven by increasing demands across diverse industries including automotive, aerospace, and consumer electronics. Numerous researchers have dedicated efforts to understanding the influence of process parameters aimed at improving the quality and integrity of final products. The effectiveness of manufacturing processes relies heavily on factors such as equipment, tooling, and operator proficiency. For example, various studies have investigated the effects of punch-die clearance, tooling geometry, friction, blanking velocity, and indenter blank holder on the mechanical behavior of sheet materials, as evidenced by works such as those by [15, 16]. It has been conclusively demonstrated that these factors wield a significant effect not only on product accuracy but also on the lifespan of the tooling. This study presents into a numerical investigation of the mechanical behavior of Fe-TiB_2 composite during shearing processes under various operational conditions. It has been conclusively demonstrated that these factors wield a significant effect not only on product accuracy but also on the lifespan of the tooling. This study presents into a numerical investigation of the mechanical behavior of Fe-TiB_2 composite during shearing processes under various operational conditions.

2 Constitutive Model

This section aims to present an elastoplastic constitutive model, implemented within the VUMAT subroutine, to characterize the behavior of a Fe-TiB_2 blank. This blank consists of a ferritic steel matrix composite containing 12% titanium diboride (TiB2) particles, undergoes deformation through the shearing process.

Box 1. Elasto-plastic constitutive equations

-Deformation tensor	$\varepsilon = \varepsilon^e + \varepsilon^p$
-Hook's law	$\sigma = D(d) : \varepsilon^e$
-Elasticity function	$f = \frac{\varphi_f(\sigma)}{\sqrt{g_e(d)}} - \frac{R}{\sqrt{g_p(d)}} - \sigma_Y$
-Elasticity criterion	$\varphi_f(\sigma) = \lVert\sigma\rVert_H = \sqrt{\sigma^t H \sigma}$
-Function describing the degradation of elasticity	$g_e(d) = (1-d)^{2\delta_e}$
-Function describing the degradation of plasticity	$g_p(d) = (1-d)^{2\delta_p}$
-Isotropic hardening (double exponential rule)	$R(r) = Q_1(1-e^{-\beta_1 r}) + Q_2(1-e^{-\beta_2 r})$
-J2 criterion tensor	$H = \begin{bmatrix} 1 & -0.5 & -0.5 & 0 & 0 & 0 \\ & 1 & -0.5 & 0 & 0 & 0 \\ & & 1 & 0 & 0 & 0 \\ & & & 3 & 0 & 0 \\ & Sym & & & 3 & 0 \\ & & & & & 3 \end{bmatrix}$
-Ductile damage model (Lemaître. 1992)	$F_d = \frac{S}{(s+1)(1-d)^\beta} \left\langle \frac{Y - Y_0}{S} \right\rangle^{s+1}$
-Flow rule	$\dot{\varepsilon}^p = \dot{\gamma} \frac{\partial F}{\partial \sigma} = \frac{\dot{\gamma}}{\sqrt{g_e}} n, \quad n = \frac{1}{\varphi} H\sigma$

The detailed elasto-plastic constitutive equations are outlined below in Box1, and for a deeper understanding, more comprehensive explanations can be found in the publications by [17, 18].

s, S, β and Y_0 denote the ductile damage variables, Q_1, Q_2, β_1 and β_2 correspond to the isotropic hardening variables, while $\dot{\gamma}$ is the plastic multiplier represents the elastic tensile stress and **D** signifies the elastic stiffness tensor.

3 Shearing Process Simulation

The investigated composite material is an industrially produced Fe-TiB2 composite, synthesized via an in situ precipitation technique followed by subsequent hot rolling. This composite comprises a TiB_2 particle content constituting 12% of its volume and exhibits distinctive crystallographic and morphological characteristics. The metal matrix composite comprises an elastoplastic matrix derived from ferritic steel, reinforced with ceramic particulates (TiB_2). Table 1 provides an overview of the mechanical properties of the Fe-TiB_2 composite.

The blanking procedure involving sheet metal entails substantial shear distortion caused by the punch, leading to the separation of the sheet, as depicted in Fig. 1. Consequently, there is notable plastic indentation accompanied by damage, primarily localized between the plunging punch and the die. Within this framework, a two-dimensional (2D) axisymmetric assumption is utilized to simulate the shearing operation using the finite element software ABAQUS/Explicit. The established model is then interfaced with user-defined subroutines (VUMAT) to integrate the mechanical response of the Fe-TiB2 composite. Regarding the processed blank, a 2D model is discretized using a 4-node bilinear axisymmetric quadrilateral element (CAX4R). A refined mesh is implemented within the fracture-prone region, as illustrated in Fig. 2.Two cases are simulated and discussed by varying: (a) blank thickness (2mm, 3mm); (b) holding load (3000N, 4000N).

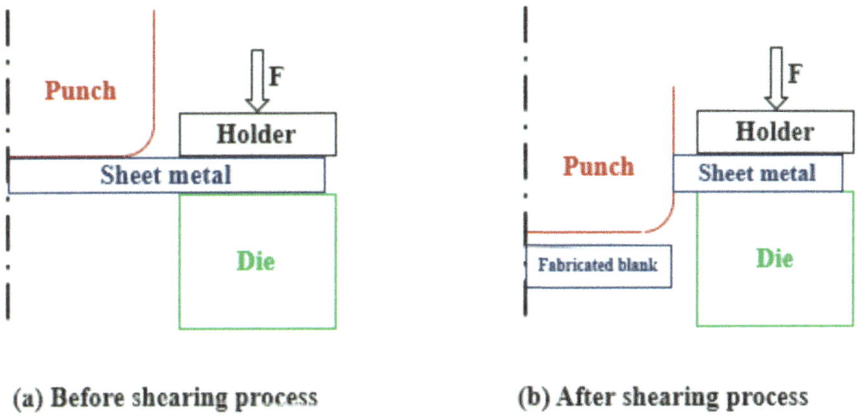

Fig. 1. Representation of shearing process.

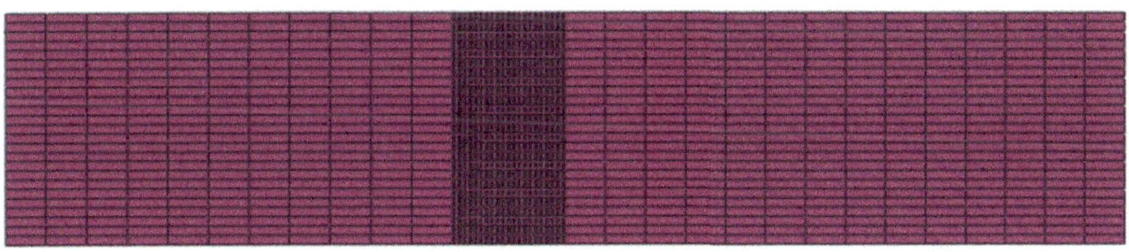

Fig. 2. Blank mech.

Table 1. Mechanical properties of Fe-TiB_2 composite.

E (MPa)	235557					
υ	0.283					
Double exponential Hardening (MPa)	$\sigma(r) = 250 + 249\left(1 - e^{-14.3r}\right) + 79.7\left(1 - e^{-98.5r}\right)$					
Damage parameters	β	s	S	Y_0	δ_e	δ_p
	49	0.64	510	0	0.5	0.5

4 Parametric Study: Results and Discussion

In this pioneering study, an exhaustive examination of the deformation mechanism inherent in the shearing operation is undertaken. The primary aim of this computational inquiry is to scrutinize the influence of distinct process variables, specifically the initial thickness of the blank and the applied holding force, on diverse facets of the shearing process, encompassing the forming load and the resultant quality of the produced component. A systematic parametric investigation is carried out within this computational framework.

Two shearing process simulations is adopted with distinct initial thickness values for the blank (2mm and 3mm) with fixed holding load (4000N), demonstrating the behavior of the Fe-TiB2 during process. Figure 3 depicts the evolution of punch load and punch displacement throughout the shearing process for two distinct initial blank thicknesses (2mm and 3mm). This visualization highlights the impact of the initial blank thickness on the geometry of the produced blank. It is apparent from the graphical representation that as the initial blank thickness increases, the extent of deformation zones expands, and the irregularities along the sheared surface edges become more conspicuous. Furthermore, Table 2 underscores the necessity for greater shear force with decreasing thickness. This finding underscores the direct correlation between the initial blank thickness parameter and the deformation of the workpiece.

Furthermore, two other shearing process simulations are incorporated to present the effect of holding load (3000N and 4000N) on the response of Fe-TiB_2 composite during shearing processing. Specifically, the maximum punch load decreases from 12.18 to 12.06 when the holding load changes from 3000 N to 4000 N, as indicated in Table 3.

Table 2. Maximum punch load as function of initial blank thickness.

Initial blank thickness (mm)	2	3
Maximum punch load (kN)	12.06	16.33

Table 3. Maximum punch load as function of holding load.

Holding load (N)	3000	4000
Maximum punch load (kN)	12.18	12.06

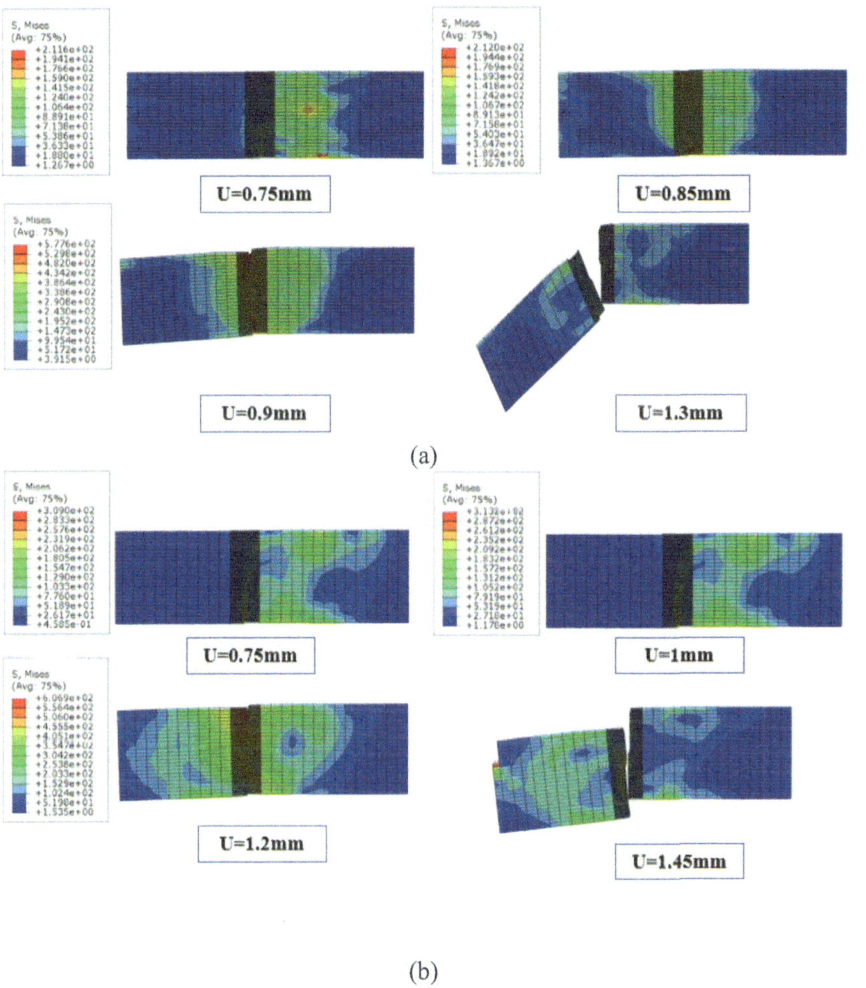

Fig. 3. Analyzing Von Mises stress distribution and numerically predicting punch displacement in shearing operations at Two Thickness Values: h = 2mm (a) and h = 3mm (b) for 4000N holding load.

5 Conclusion

This study offers a methodology for simulating the sheet shearing process employing an advanced elastoplastic finite element model, which is comprehensively linked with ductile damage criteria. The formulated model is incorporated into the Abaqus software through a custom user-defined subroutine (VUMAT) to predict the load-displacement characteristics of the punch and assess the quality of components manufactured from the $Fe\text{-}TiB_2$ composite. In pursuit of optimal process performance, the impact of shearing process parameters, such as initial blank thickness and the holding load, on the quality of the processed product is thoroughly investigated. A parametric investigation has been conducted, providing valuable insights and guidance for improving the quality of products produced through the shearing process and efficiently shaping the $Fe\text{-}TiB_2$ blank.

References

1. Cha, L., Lartigue-Korinek, S., Walls, M., Mazerolles, L.: Interface structure and chem- istry in a novel stccl-bascd composite Fc-TiB2 obtained by eutcctic solidification. Acta Mater. **60**(18), 6382–6389 (2012)
2. Chawla, N., Chawla, K.K.: Metal-matrix composites in ground transportation. JOM **58**(11), 67–70 (2006)
3. Kaczmar, J.W., Pictrzak, K., Wlosinski, W.: The production and application of metal matrix composite materials. J. Mater. Process. Technol. **106**(1), 58–67 (2000)
4. Dammak, M., Barbier, D., Gasperini, M.: Kinematic strain hardening of Fe-TiB2 composite: experimental analysis and phenomenological modeling. J. Mater. Eng. Perform. **33**(9), 4359 (2023)
5. Gasperini, M., Dammak, M., Franciosi, P.: Stress estimates for particle damage in Fe-TiB2 metal matrix composites from experimental data and simulation. Eur. J. Mcch. A Solids. **64**, 85–98 (2017)
6. Dammak, M., Bouhamed, A., Jrad, H., Dammak, F.: In situ experimental characterization and numerical investigation of Fe-TiB2 Steel matrix composite behavior considering fully coupled damage model: simulation during incremental forming process. Mater. Today Commun. **38**, 107741 (2024). https://doi.org/10.1016/j.mtcomm.2023.107741
7. Lartigue-Korinek, S., Walls, M., Haneche, N., Cha, L., Mazerolles, L., Bonnet, F.: Interfaces and defects in a successfully hot-rolled steel-based composite Fe-TiB2. Acta Mater. **98**, 297–305 (2015)
8. Dorhmi, K., Derrien, K., Hadjem-Hamouche, Z., Morin, L., Bonnet, F., Chevalier, J.P.: Experimental study and micromechanical modelling of the effective elastic properties of $Fe—TiB_2$ composites. Compos. Struct. **272**, 114–122 (2021)
9. Dorhmi, K., Morin, L., Derrien, K., Hadjem-Hamouche, Z., Chevalier, J.P.: A homogenization-based damage model for stiffness loss in ductile metal-matrix composites. J. Mech. Phys. Solids **137**, 103812 (2020)
10. Elleuch, S., Jrad, H., Wali, M., Dammak, F.: Finite element analysis of the effect of porosity on biomechanical behaviour of functionally graded dental implant. Proc. Inst. Mechan. Eng. Part E J. Process Mechan. Eng. **3**, 095440892 (2023)
11. Elleuch, S., Jrad, H., Wali, M., Dammak, F.: Mandibular bone remodeling around osseointegrated functionally graded biomaterial implant using three dimensional. Int. J. Numer. Methods Biomed. Eng. **39**(9), c3750 (2023)

12. Elleuch, S., Jrad, H., Wali, M., Dammak, F.: Agglomeration effect on biome-chanical performance of CNT-reinforced dental implant using micromechanics-based approach. J. Mechan. Behav. Biomed. Mater. **145**, 106023 (2023)
13. Ellouz, H., Jrad, H., Wali, M., Dammak, F.: Large deflection analysis of FGM/magneto-electro-elastic smart shells with porosities under multi-physics loading. Mechan. Adv. Mater. Struct. (2023). https://doi.org/10.1080/15376494.2023.2243938
14. Ellouz, H., Jrad, H., Wali, M., Dammak, F.: Numerical modeling of geometrically nonlinear responses of smart magncto-clcctro-clastic functionally graded double curved shallow shells based on improved FSDT. Comput. Math. Appl. **151**, 271–287 (2023)
15. Krinninger, M., Feistle, M., Golle, R., Volk, W.: Notch shear cutting of aluminum alloys. Procedia Eng. **183**, 53–58 (2017)
16. Sachnik, P., Hoque, S.E., Volk, W.: Burr-free cutting edges by notch-shear cutting. J. Mater. Process. Technol. **249**, 229–245 (2017)
17. Bouhamed, A., et al.: Identification of fully coupled non-associatcd-Ductilc damage constitutive equations for thin sheet metal applications: numerical feasibility and experimental validation. Thin-Walled Struct **176**, 109365 (2022)
18. Bouhamed, A., Mars, J., Jrad, H., Wali, M., Dammak, F.: Experimental and numerical methodology to characterize 5083-aluminium behavior considering non-associated plasticity model coupled with isotropic ductile damage. Int. J. Solids Struct. **229**, 111139 (2021)

Numerical Study of the Impact of Procedural Parameters on the Deformation of Fe-TiB_2 Composite Throughout Deep-Drawing Operation

Abir Bouhamed[1,2](✉), Hanen Jrad[1,3], Mondher Wali[1], and Fakhreddine Dammak[1]

[1] LEE Laboratory, National Engineering School of Sfax, University of Sfax, Sfax, Tunisia
abir.bouhamed@enis.tn

[2] Institut Supérieur des Sciences Appliquées et de Technologie de Gabès, University of Gabes, Rue Omar ibn el Khattab, 6072 Gabes, Tunisia

[3] École Supérieure des Sciences et de la Technologie de Hammam Sousse, University of Sousse, Rue Lamine Abassi, 4011 Hammam Sousse, Tunisia

Abstract. The advancement of metal forming techniques and procedures has been notably impacted by the integration of state-of-the-art tools, specifically within the extensive realm of FE simulations for numerical analysis. Obtaining precise numerical representations of material behavior throughout the forming process is crucial to ensure the reliability and excellence of the resultant outcomes. It is imperative to precisely characterize and determine the material attributes under examination and subsequently translate them into numerical material models. The development of steel matrix composite with TiB2 reinforcement requires a thorough insight of how they respond to loading conditions variation. This comprehension is essential for their mechanical characteristics modeling and their malleability prediction. This work unveils the inaugural application of an elasto-plastic material with a 12% ceramic reinforcement volume ratio, undergoing a Deep-drawing operation. The numerical simulation is conducted using Abaqus/Standard through the UMAT subroutine. The aim is to estimate the influence of specific procedural variables, namely the space between the die and the punch, the initial blank thickness, and the friction coefficient between the workpiece and the forming tool, to comprehend the deformation mechanics and the consequent influence of procedural variables. This analysis offers valuable insights into how these three procedural variables can be manipulated to improve the produced component quality and achieve enhanced malleability throughout the Deep-drawing operation.

Keywords: UMAT subroutine · Deep-drawing operation · Fe-TiB_2 steel matrix composite · elasto-plastic model

1 Introduction

Composites comprising a steel matrix with ceramic reinforcement particles have garnered considerable interest for their potential in decreasing structural weight, provided that enhancements in their particular modulus are characterized by a favorable ratio of

M. Chouchane et al. (Eds.): CMSM 2023, LNME, pp. 130–140, 2024.
https://doi.org/10.1007/978-3-031-65007-9_15

strength to ductility, as evidenced in prior investigations (Kaczmar et al., 2000; Chawla and Chawla, 2006; Cha et al., 2012). The recent and encouraging advancements in these composites highlights the necessity to comprehend their reaction to diverse conditions of loading. This comprehension is essential not only for simulating their mechanical characteristics but also for forecasting their malleability, particularly concerning their intended utilization in shaping applications within the automotive sector.

Recent investigations conducted by Dammak and colleagues (2024; 2023) showcased, via comprehensive shear assessments encompassing a broad spectrum of plastic deformations, that the primary mechanism of deterioration in the Fe-TiB2 composite requires considerable rupture of both sizable primary and eutectic particles, with reduced occurrences of particle detachment during experimentation. In contrast, findings by Lartigue et al. (2015), as elucidated through metallurgical scrutiny, unveiled a remarkable plastic distortion in the ceramic particles caused by the hot rolling procedure, potentially contributing to the observed diminished levels of impairment in the hot-rolled composite. The plastic deformations caused by diverse thermal and shaping procedures possess the capability to modify the heightened properties of these substances, thereby constraining their pragmatic applications. Consequently, the evaluation of these attributes, encompassing the malleability of TiB2 and the ferritic substrate, becomes crucial to enhance processing capacities and trajectories.

Previous research on Fe-TiB2 alloys have primarily concentrated on understanding the process of degradation, as depicted in Gasperini et al. study (2017), and the reduction in stiffness, as highlighted by Dorhmi et al. (2020). Both investigations primarily scrutinized the reaction of the composite material under diverse conditions of mechanical loading. In addition, Dorhmi et al. (2021) investigated the performance characteristics of the composite subsequent to cold and hot rolling procedures. Across the past two decades, the Deep-drawing procedure has amassed notable attention and witnessed burgeoning enthusiasm in the fabrication of sheet metal commodities, propelling extensive scholarly and industrial exploration. This procedure is poised to sustain its enduring significance and persist in exerting a significant influence on various sectors (Chen et al., 2022; Atul and Babu, 2019; Kumar and Mohamed, 2016). The Deep-drawing examination encompasses highly intricate distortions influenced by myriad process variables, including the shape of the forming tool and die, initial part layout, space amid the forming tool and die, imposed force, lubrication, and material attributes of the blank. Given the complex interconnections dictating procedural outcomes, particularly regarding material traits, devising an appropriate material paradigm for the Deep-drawing procedure represents a noteworthy endeavor. Recently, the academic and investigative circles, as exemplified by Bergweiler et al. (2020) and Marcin et al. (2020) have demonstrated a burgeoning interest in exploring novel materials within this manufacturing methodology and comprehending their dynamics.

In summary, this research fills a significant void in the existing body of literature by conducting the inaugural examination of the Fe-TiB2 composite behavior during the Deep-drawing procedure. We replicate the conduct of Fe-TiB2 blanks containing a 12% TiB2 volume ratio subjected to the Deep-drawing process. Our integration aims to elucidate the impact of specific procedural variables, including initial blank thickness, space between the die and the punch, and the friction coefficient between the part and the punch.

Grasping the mechanics of deformation and the ramifications of procedural variables is vital for adeptly devising Deep-drawing operations. Consequently, this inquiry provides valuable insights into how these variables can be adjusted to augment product excellence and attain enhanced malleability during Deep-drawing sheet metal forming activities, thereby making significant strides in the progression of manufacturing methodologies and materials science.

2 Constitutive Equations of Elastoplastic Material

The objective of this segment is to introduce an elastoplastic constitutive framework, integrated within the UMAT subroutine, for delineating the conduct of a Fe-TiB2 blank as it undergoes the Deep-drawing procedure to yield a cylindrical cup comprised of a composite matrix of ferritic steel embedded with 12% TiB2 particles. The precise elastoplastic constitutive formulations are delineated in Table 1, with more extensive elucidations provided in the studies by Bouhamed et al., (2022, 2021) and Ben Said et al., (2021).

In Table 1, the initial yield stress is depicted by σ_Y, the stress tensor is denoted by σ, and the plastic strain rate tensor is indicated as $\dot{\varepsilon}^p$. Moreover, the isotropic hardening variables are depicted as Q_1, Q_2, β_1 and β_2, which are described by a double exponential law, while the plastic multiplier is presented by $\dot{\gamma}$. Additionally, a fourth-order tensor following the conventional J2 plasticity yield criterion is symbolized by **P**.

Table 1. Constitutive equations of elastoplastic material.

a/Yield function	**c/Isotropic hardening**
$f = \varphi_f(\sigma) - \sigma_p \leq 0$ (1)	$R = R(r) = Q_1(1 - e^{-\beta_1 r}) + Q_2(1 - e^{-\beta_2 r})$ (5)
$\varphi_f(\sigma) = \lVert\sigma\rVert_P = \sqrt{\sigma^t P \sigma}$ (2)	**d/J2 plasticity yield criterion tensor**
$\sigma_p = \sigma_Y + R(r)$ (3)	$P = \begin{bmatrix} 1 & -0.5 & -0.5 & 0 & 0 & 0 \\ & 1 & -0.5 & 0 & 0 & 0 \\ & & 1 & 0 & 0 & 0 \\ & & & 3 & 0 & 0 \\ & Sym & & & 3 & 0 \\ & & & & & 3 \end{bmatrix}$ (6)
b/Plastic strain tensor	
$\dot{\varepsilon}^p = \dot{\gamma} \frac{\partial f}{\partial \sigma} = \dot{\gamma} n \quad ; \quad n = \frac{1}{\varphi} P \sigma$ (4)	

3 Homogenization Procedure

The composite under scrutiny is an industrially manufactured Fe-TiB2 amalgam, produced via an in-situ precipitation technique, succeeded by subsequent hot rolling. This amalgam encompasses a 12% volume ratio of TiB2 particles, exhibiting distinctive crystallographic and morphological patterns. It comprises an elastoplastic framework comprising ferritic steel (F1), fortified with TiB2 reinforcement particles. In Table 2, elaborate mechanical attributes of the distinct material phases are delineated. The Mori-Tanaka model furnishes an analytical methodology for approximating both the Poisson's fraction of this innovative composite and also their Young's modulus under investigation (Table 3).

Table 2. Mechanical properties of the Constituent's material phases.

Materials	Young modulus (GPa)	Poisson's ratio
F_1	210	0.3
TiB_2	565	0.108

Table 3. Effective elastic properties of Fe-TiB2 composite.

Material	Young modulus (MPa)	Poisson's ratio
Fe-TiB_2	235557	0.283

The isotropic hardening parameters delineated in Eq. (5) are adjusted to guarantee a robust association between the analytical curve embodying the Double exponential principle and the empirical curve witnessed in the work presented in (Dammak et al., 2023), as illustrated in Fig. 1. In Table 4, the enhanced isotropic hardening variables are specified.

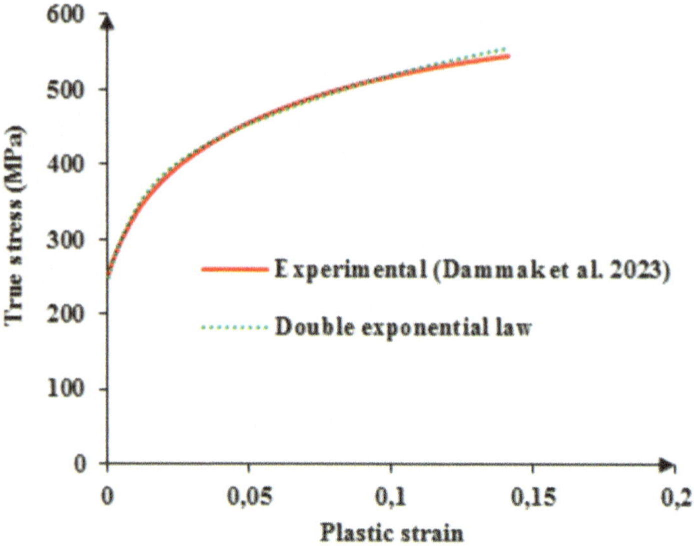

Fig. 1. Double exponential law for the elastoplastic Fe-TiB2 composite: Experimental and analytical confrontation.

Table 4. Identified Double exponential parameters of Fe-TiB_2 composite.

Double exponential parameters	σ_Y (MPa)	Q_1 (MPa)	β_1	Q_2 (MPa)	β_2
	250	249	14.3	79.7	98.5

4 FE Model

A thorough scrutiny of the deformation procedure linked with the deep drawing of a cylindrical cup is performed. In Fig. 2, the particular dimensions of the geometry procedure are shown. The elastoplastic material under examination is a Fe-TiB2 part, featuring 12% ceramic volume ratio. An FE model is created utilized Abaqus/Standard to simulate the deep-drawing sequence. The aforementioned material paradigm is incorporated into the simulation via the User Material Subroutine UMAT, facilitating efficient control of contact interactions between the workpiece surface and the forming tool. Symmetry plane is imposed as boundary conditions. Employing C3D8 kind of mesh, quarter Fe-TiB2 part is divided into 29388 elements. A 5500 N holding force is imposed throughout the simulation, with an assumed friction coefficient of 0.1 between the forming tools (punch and die) and the blank. Within this inquiry, three procedural variables are recognized in the Deep-drawing sequence: and the friction coefficient between the forming tool and the blank, the clearance between the die and punch and the initial blank thickness. The purpose of investigating these various parameters, detailed in Table 5, is to assess how they influence both Von Mises stress distribution and manufactured product surface quality.

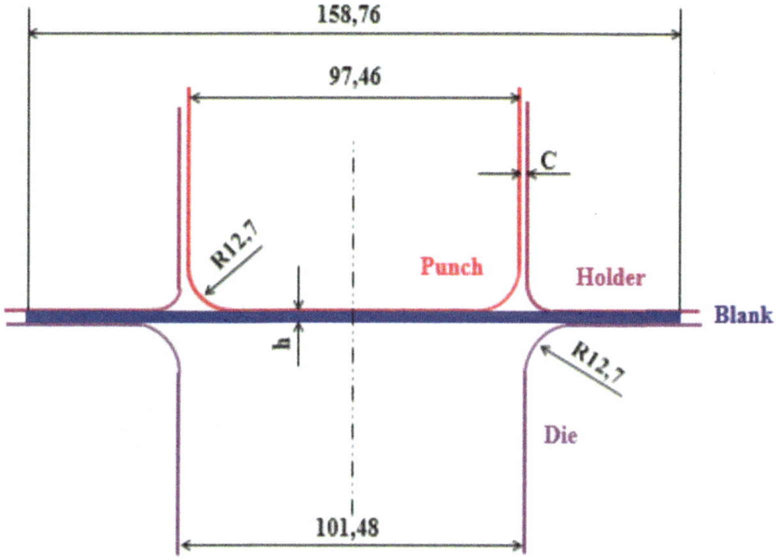

Fig. 2. Geometry of Deep-drawing process.

Table 5. Values of Deep-drawing process parameters.

Parameter	Value		
Clearance (C) (mm)	1.5	2	2.5
Blank thickness (h) (mm)	2	3	4
Friction coefficient (f)	0.05	0.1	0.15

5 FE Simulation: Results and Discussions

The objective of this computational analysis is to explore the impact of specific procedural variables on various facets of the Deep-drawing process, encompassing Von Mises stress dispersion, malleability, and the final fabricated quality product. An evaluation is performed among three fabricated components with distinct spacings (C) between the die and the punch. The numerical FLD representation, depicted in Fig. 3, elucidates the malleability througout the deep-drawing operation for varying spacings. Moreover, this figure illustrates that the important deformation values are attained when the space between the die and the punch is minimal. Therefore, reducing the spacing emerges as a critical design variable, giving a more uniform biaxial deformation dispersion. Figure 4 presents the numerical FLD depiction, delineating the malleability throughout the deep-drawing operation for distinct initial part thicknesses (h). It is evident that an increase in part thickness leads to higher deformation values.

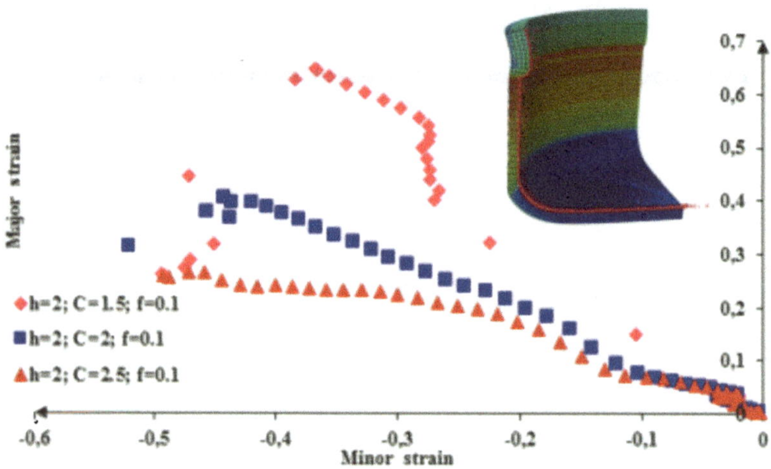

Fig. 3. Representation od numerical FLD for various space (C) between the forming tool and the die.

Additionally, Fig. 5 illustrates the distribution of Von Mises stress for varying friction coefficients. It is apparent that an escalation in the friction coefficient profoundly affects the Von Mises stress dispersion of the fabricated component. Taken together, Figs. 3–5 that alterations in the first part thickness, the gap between the die and the punch, and the friction coefficient between the part and the forming tool wield a substantial impact on the malleability of the Fe-TiB2 workpiece throughout the Deep-drawing operation. Table 6 showcases the optimal design values for the parameters of the deep-drawing process. These values represent the ideal selection for ensuring enhanced formability throughout the deep-drawing operation.

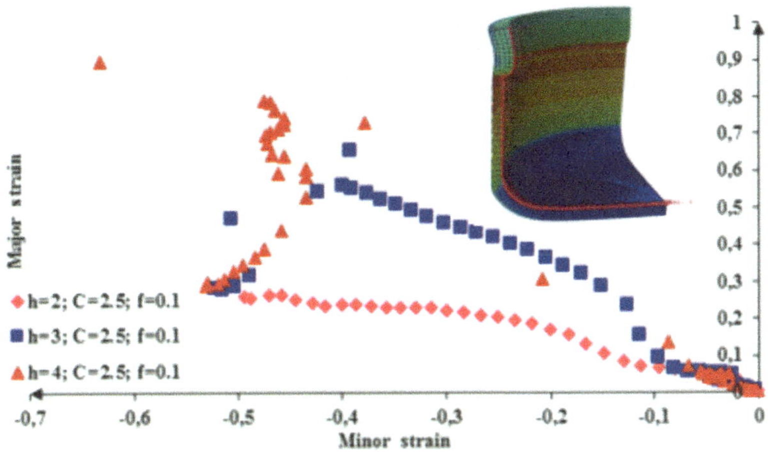

Fig. 4. Numerical FLD representation for various initial blank thickness.

Table 6. Values of optimal design Deep-drawing process parameters.

Parameter	Optimal design
Clearance (C)	1.5 mm
Blank thickness (h)	4 mm
Friction coefficient (f)	0.05

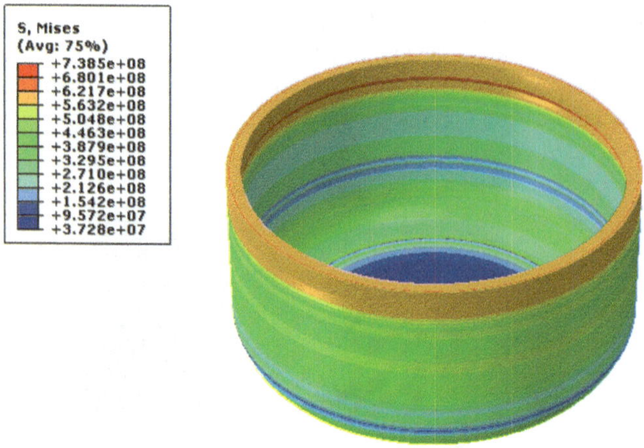

(a) h=3 ; C=2.5 ; f=0.05

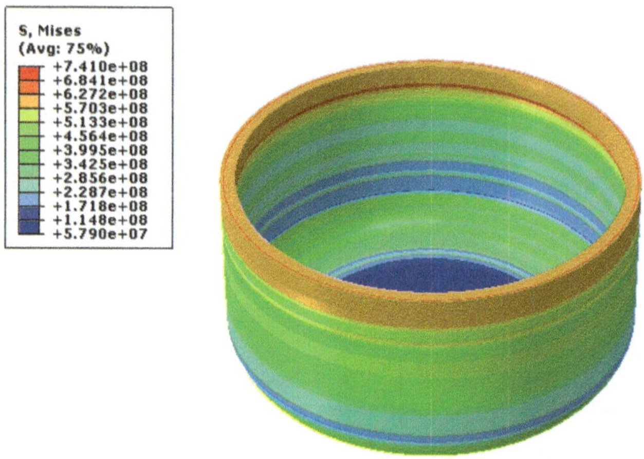

(b) h=3 ; C=2.5 ; f=0.1

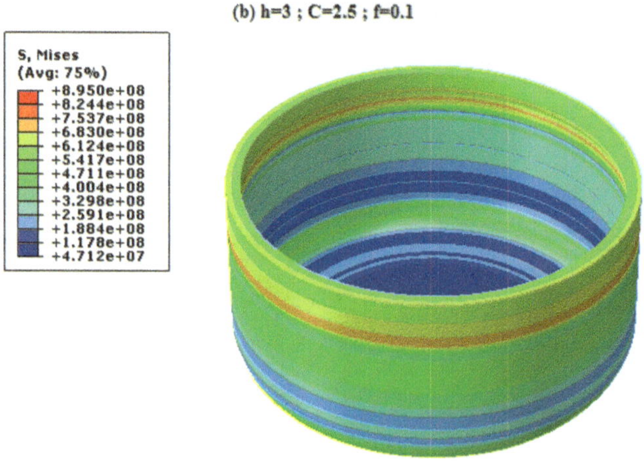

(c) h=3 ; C=2.5 ; f=0.15

Fig. 5. Dispersion of Von Mises stress for various friction coefficient between the blank and the forming tool (punch).

6 Conclusion

In this investigation, an FE model incorporating a custom material subroutine (UMAT) was utilized to simulate the Deep-drawing operation for cylindrical cups fabricated from steel matrix composite with 12% ceramic reinforcement. The numerical findings illuminate the impact of specific procedural variables, such as the initial part thickness, the gap between the die and the punch, and the friction coefficient between the blank and the punch, on the Von Mises stress dispersion and malleability during the Deep-drawing procedure. This examination underscores that reducing the gap and increasing the initial blank thickness enhance malleability. Furthermore, augmenting the friction coefficient seems to markedly influence the Von Mises stress dispersion for the produced component.

Furthermore, this study opens avenues for future research directions, such as exploring additional process parameters or investigating the effects of varying material compositions within the composite. Additionally, the insights gained from this parametric investigation offer valuable guidance for enhancing the quality of products produced through the Deep-drawing process and effectively shaping Fe-TiB2 blanks.

References

Atul, S.T., Babu, M.L.: A review on effect of thinning wrinkling and spring-back on deep drawing process. Proc. Inst. Mech. Eng. Par B J. Eng. Manuf. **233**(4), 1011–1036 (2019)

Ben Said, L., et al.: SPIF Manufacture of a dome part made of AA1060-H14 alummum alloy using CNC lathe machine: numerical and experimental investigations. Arab. J. Sci. Eng. **46**(12), 12207–12220 (2021)

Bergweiler, G., Fiedler, F., Shaukat, A., Loffler, B.: Expenmental investigation of dimensional precision of deep drawn cups using direct polymer additive tooling. J. Manuf. Mater. Proc. **5**(1), 3 (2020)

Bouhamed, A., et al.: Identification of fully coupled rion-associated-Ductile damage constimtive equations for thin sheet metal applications: Numerical feasibility and experimental validation. Thin-Walled Struct. **176**, 109365 (2022)

Bouhamed, A., Mars, J., Jrad, H., Wali, M., Dammak, F.: Experimental and numerical methodology to characterize 50S3-aluminhmi behavior considering non-associated plasticity model coupled with isotropic ductile damage. Int. J. Solids Struct. **229**, 111139 (2021)

Cha, L., Lartigue-Korinek, S., Walls, M., Mazerolles, L.: Interface structure and chemistry in a novel steel-based composite Fe-TiB2 obtained by eutectic solidification. Acta Mater. **60**(18), 6382–9 (2012)

Chawla, V., Chawla, K.K.: Metal-matrix composites in ground transportation. JOM **5S**(11), 67–70 (2006)

Chen, K., Carter, A.J., Korkolis, Y.P.: Flange wrinkling in deep-drawing: Experiments, simulations and a reduced-order model. J. Manuf. Mater. Proc. **6**(4), 76 (2022)

Dammak, M., Barbier, D., Gasperini, M.: Kinematic strain hardening of Fe-TiB2 composite: experimental analysis and pheriomeuological modeling. J. Mater. Eng. Perform. **33**(9), 4359–4366 (2023)

Dammak, M., Bouhamed, A., Jrad, H., Dammak, F.: In situ experimental characterization and numerical investigation of Fe-TiB2 steel matrix composite behavior considering fully coupled damage model: Simulation during incremental forming process. Mater. Today Commun. **3**, 107741 (2024). 10.1016vj.nitcamm.2023.107741

Gasperini, M., et al.: Expenmental study and micromechanical modelling of the effective elastic properties of Fe-TiB2 composites. Compos. Struct. **272**, 114–122 (2021)
Dorhmi, K., Morin, T., Derrien, K., Hadjem-Hamouche, Z., Chevalier, J.P.: A homogeniza-tion-based damage model for stiffness loss in ductile metal-matrix composites. J. Mech. Phys. Solids **137**, 103S12 (2020)
Gasperini, M., Dammak, M., Franciosi, P.: Stress estimates for particle damage in Fe-TiB2 metal matrix composites from experimental data and simulation. Eur. J. Mech. A Solids **64**, 5–98 (2017)
Kaczmar, J.W., Pietrzak, K., Wlosiiiski, W.: The production and application of metal matrix composite materials. J. Mater. Process. Technol. **106**(1), 58–67 (2000)
Kumar, A.P., Mohamed, M.N.: A novel automated design calculating system for axisymmet-ric thin w·alled structures by deep drawing process. J. Adv. Eng. Res. **3**(2), 128–133 (2016)
Lartigue-Korinek, S., Walls, M., Haneche, N., Cha, L., Mazerolles, L., Bonnet, F.: Interfaces and defects in a successfully hot-rolled steel-based composite Fe—TiB2. Acta Mater. **98**, 297–305 (2015)
Marcin, K., Łukasz, L., Paulina, L.-G., Krzysztof, M., Janusz, M.: Analysis of the deep drawing process of three-layered explosive welded composite. Proc. Manuf. **50**, 153–158 (2020). https://doi.org/10.1016/j.promfg.2020.08.029

A Comparative Investigation on the Impact of Reprocessing on Reinforced Thermoplastic Composites

Nihel Ketata[1,2](✉), Noamen Guermazi[1], Yves Grohens[2], and Bastien Seantier[2]

[1] LGME Lab, National Engineering School of Sfax, University of Sfax, Sfax, Tunisia
nihel.ketata@enis.tn

[2] UMR CNRS 6027, IRDL, Université de Bretagne Sud, BP 92116, 56321 Lorient Cedex, France

Abstract. Because they are biodegradable, natural fiber-reinforced biopolymers, also referred to as bio-composites, are a practical substitute for petrochemical polymers reinforced with glass fiber. This research aims to compare the impact of reprocessing PLA-PBS matrix-reinforced flax fibers with glass fiber composites on their morphological and tensile properties. Reprocessing does not completely achieve the dispersion of single fibers, according to morphology studies. However, the second extrusion has a slight effect on fiber dispersion. Uniaxial tensile tests revealed that the mechanical properties of all the materials were not significantly damaged during the second reprocessing. In return, for the GF/PLA-PBS, significant higher stress at break is noted, with 53% more than the FF/PLA-PBS. This finding suggests that the mechanical recycling of FF/PLA-PBS and GF/PLA-PBS composites for two cycles of compounding-injection molding is technically feasible without significantly affecting the mechanical performance.

Keywords: PLA/PBS matrix · Flax fibers · Glass fibers · thermoplastic composites · Extrusion cycles · Injection molding · Mechanical properties

1 Introduction

Requirements have changed with technological developments. Sometimes traditional materials are insufficient to meet these exigencies. Particularly in marine applications and spacecraft, high strength in addition to low weight is required. Due to their low density and high strength, fiber-reinforced composites are used in many engineering applications, including aerospace engineering and military engineering. However, Over 95% of the reinforcements used in the composites industry are made of glass fibers. (Alagarraja et al., 2014). The vast majority of plastics processors adhere to ISO 14001 environmental management system requirements by recycling the waste plastics generated during manufacturing. This lowers costs and lessens the process's impact on the environment. Besides, a novel class of fiber-reinforced materials, known as bio-composites, has emerged and continues to evolve. These bio-composites involve the integration of natural

M. Chouchane et al. (Eds.): CMSM 2023, LNME, pp. 141–149, 2024.
https://doi.org/10.1007/978-3-031-65007-9_16

reinforcing fibers, such as flax, hemp, and ramie, within a biopolymeric matrix (Alvarez et al., 2004; Keller, 2003; Ketata et al., 2022a). These vegetable fibers are renewable, nonabrasive, can be thermally recycled, and show less concern for safety and health. Natural fiber-reinforced thermoplastic composites provide high mechanical performance as well as good recycling properties (Bourmaud et al., 2011; Bourmaud and Baley, 2009, 2007; Le Duigou et al., 2008). Nevertheless, these composites' mechanical performance is still inferior than thermoplastic composites reinforced with glass fibers. (Ausias et al., 2013). The present investigation focuses on the effect of reprocessing on the morphology and uniaxial tensile properties of the thermoplastic blend of poly (lactic acid) (PLA) and poly (butylene- succinate) (PBS) reinforced with flax fibers. The results are compared with glass fiber composites GF/PLA-PBS elaborated with the PLA-PBS matrix. We have selected material with a fiber concentration of 20%. Both injection molding and compound extrusion reprocessing were highlighted to understand the fiber dispersion and fiber-matrix adherence using the injection-molded surface's tensile cracked SEM micrographs. The tensile properties exhibited by reinforced thermoplastics at two compounding-injection cycles are also discussed and compared. The results indicated that composites were not significantly damaged during the second reprocessing.

2 Materials and Methods

2.1 Material

The poly (L-lactic acid) grade IngeoTM 3D850 PLA, provided by Natureworks (Minnetonka, MN, USA), and the grade Bio PBS FZ91PM PBS, provided by Mitsubishi Chemical Europe GmbH, form the basis of the matrix of the material under study. The supplier of the flax fibers is Procotex SA Corporation. Gazechim Composites France is the supplier of glass fiber grade CS 272-10P. As indicated in Table 1, two series of composites were created, each with 20% of their weight made up of fibers. These composites will be referred to as FF/PLA-PBS composites and GF/PLA-PBS composites.

Table 1. Concentrations by weight of the studied materials

	% PLA	% PBS	% Flax fibers	% Glass fibers
PLA-PBS	80	20	0	0
FF/PLA-PBS	64	16	20	0
GF/PLA-PBS	64	16	0	20

2.2 Mixing Conditions and Samples Preparation

In this study, the Labstation Plasticorder The single-screw Brabender extruder was chosen over the twin-screw model keeping the integrity of the fibers preserved and reduce the shear rate (Ketata et al., 2022b). PLA and PBS pellets were subjected to a drying process under vacuum at 60 °C for 48 h to eliminate any remaining moisture before extrusion. After that, fibers were extruded with the matrix (Table 1). The rotating speed remained at around 25 rpm. Extrusion was performed at a blending temperature of 180 °C. In order to extrude the materials two times and after being allowed to cool in the air, the films were ground into flakes using a commercial plastics grinder. The flakes were then dried and injected into the molds. Without extrusion or further processing, Cycle 1 is thought to represent the compound's initial qualities.

2.3 Injection Molding Parameters

Compound pellets were also vacuum dried for 48 h at 60 °C. Injection molding was then carried out on a Battenfeld injection press machine. A barrel temperature of 180 °C was used. A mold intended to create normalized tensile specimens was filled with materials. Because the molten compound appeared to have a higher viscosity, the injection pressure was maintained constant at 1600 bar to get appropriate samples. A set injection time of 0.25 s was used. A temperature of 25 °C was maintained for the mold. Two injection cycles were used for the recycling experiments. Thus, we obtained randomly dispersed short-fiber composites.

2.4 Scanning Electron Microscopy (SEM)

A scanning electron microscope Jeol JSM IT500HR was used to analyze the tensile fracture surfaces. A thin layer of gold was sputtered on the samples in an Edwards Sputter Coater. This microscopic characterization allow to understand the fibers' shape, structure, and adhesion to the matrix at the microscopic level, as well as the impact of each extrusion or injection operation on the dispersion of flax and glass.

2.5 Testing and Evaluation

In a lab setting, static tensile tests were conducted on ISO-527–2 normalized specimens (Fig. 1). Tensile specimens were tested using an MTS Synergy 1000 RT apparatus equipped with a 10KN cell. The loading speed was 10 mm/min. To obtain accurate values of the tensile modulus (E) (at little strain values), an axial extensometer was coupled to the tested samples, and the corresponding values of the tensile modulus were measured and averaged. The tensile strength (σ) and the elongation at break (ε %) were obtained from the corresponding stress-strain diagrams. At least five different specimens were carried out for each formulation.

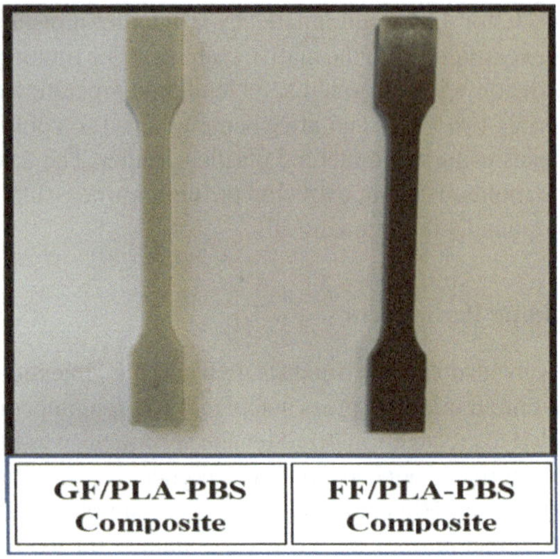

Fig. 1. Injected composite samples (GF: Glass Fibers and FF: Flax Fibers).

3 Results and Discussions

3.1 Morphological Analysis

A morphological analysis of the reinforced composites gave details on the adherence and bonding of the fibers to the matrix. The tensile fractured injection-molded surfaces of FF/PLA-PBS are presented in the SEM micrographs. (Fig. 2) and GF/PLA-PBS (Fig. 3). These observations give qualitative information about fiber dispersion and orientation. Fibers are known to be damaged during the production of composites. (Dickson et al., 2014).

It is crucial that the polymer blend matrix adheres to the reinforcing elements, particularly the fibers. This is a result of stress passes via the interface from the fibers to the matrix. After the first compounding-injection cycle, many bundles of flax fibers (Fig. 2a) demonstrated a lack of homogeneity (Ketata et al., 2024). The interaction between matrix and flax fibers can also be discussed. Fibres had no PLA/PBS matrix coating traces sticking on their surfaces, which confirm insufficient wetting, poor interface adhesion, and mediocre interactions between the flax fiber surface and the matrix, (Dickson et al., 2014; Ketata et al., 2022b). These findings can be explained by the fibers' tendency to agglomerate together (Chuai et al., 2001), which causes them to become irregularly distributed throughout the matrix (Dickson et al., 2014). This is because the flax fiber bundles and the matrix have very different characteristics. Interactions between glass fibers and PLA-PBS matrix are very limited because glass fibers have a polar surface (Roux et al., 2000; Valente et al., 2011).

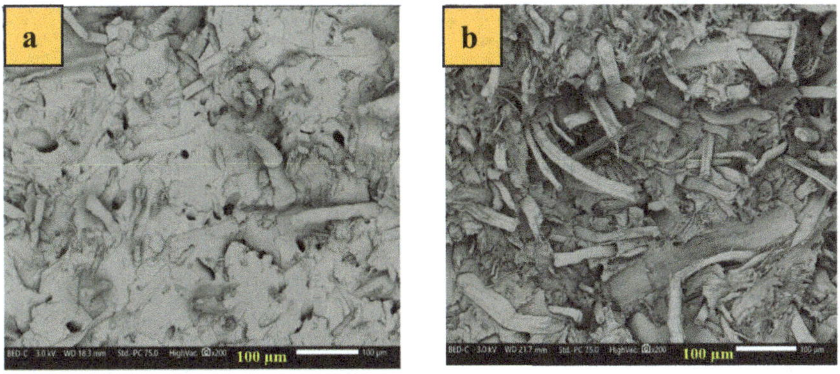

Fig. 2. Fracture surfaces of FF/PLA-PBS composite: (a): after one compounding- injection cycle and (b): after two compounding- injection cycles

Figure 3.a. shows a GF/PLA-PBS SEM micrograph. The observation of a large number of holes on the fracture surface of composites with pulled-out long fiber length and clean fibers with nearly no residual matrix adhesion on the fiber surfaces confirms the poor interfacial adhesion between glass fiber and the matrix (Valente et al., 2011). This behavior agrees with results reported by Hornsby et al. (HORNSBY et al., 1997) and Arbelaiz et al. (Arbelaiz et al., 2005) in theirs studies on similar materials. To solve similar problems, some additives can be added to the composites to improve the adhesion between the fibers and the matrix. As an example, Roux et al. (Roux et al., 2000) analyzed the fracture surfaces of glass fiber/PP composites containing MAPP (maleic anhydride-polypropylene copolymer) and showed a large amount of matrix adhered to glass fibers, indicating improved fiber-matrix adhesion. Even while some flax bundles are always left behind and the dispersion of individual fibers is not completely achieved after the second compounding-injection cycle, this indicates that the second extrusion has a small impact on fiber dispersion (Fig. 2b). This confirms some bundle division during recycling. In the GF/PLA-PBS SEM micrograph, similar results have been observed (Fig. 3b). There were also a few gaps between the fiber and the matrix, suggesting a small improvement in adhesion (Valente et al., 2011). The fracture surfaces also showed a mixture of clean fibers and fiber pull-out. Fiber dispersion is an important factor in achieving good mechanical properties. This will be the focus of the following section.

3.2 Composites Tensile Properties– One Extrusion Cycle

This study shows that the tensile modulus of FF/PLA-PBS and GF/PLA-PBS is 48% and 63% higher, respectively, than the properties of unreinforced PLA-PBS (Ketata et al., 2022a). (Fig. 4). FF/PLA-PBS biocomposite has been shown to be promising. But as expected, the tensile modulus and strength values obtained on the FF/PLA-PBS were 28% lower than those of the GF/PLA-PBS. The addition of glass fibers results in a noticeable rise in the matrix's modulus. This might result from the matrix's increased rigidity after fiber reinforcement, which allowed a greater degree of stress transfer at the interface (Joseph et al., 2003). As a result, the mechanical properties of the

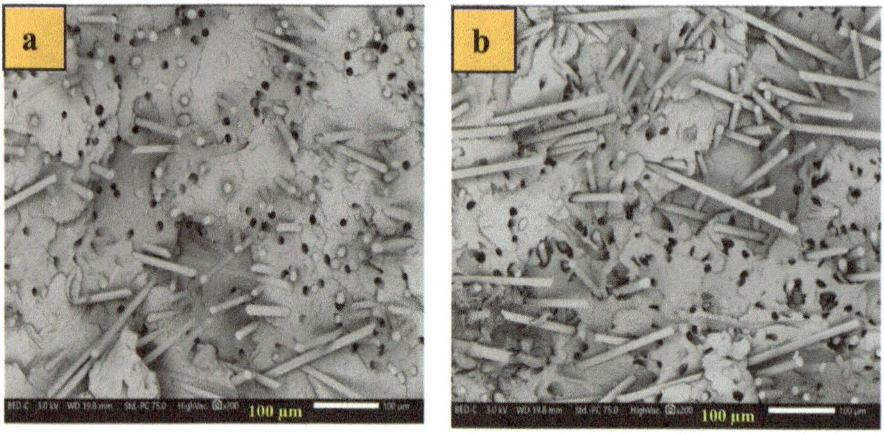

Fig. 3. Fracture surfaces of GF/PLA-PBS composite: (a): after one compounding- injection cycle and (b): after two compounding- injection cycles

composites are mainly governed by the mechanical properties of the fibers. The FF/PLA-PBS matrix exhibits strains at break of 1.7%, compared with 4.5% for the PLA/PBS matrix. For a variety of reasons, including flax fiber anisotropy, injected samples may not exhibit improved properties at break even though fibers are added (Baley et al., 2006), bundles or elementary fiber arrangement, strain concentrations caused by fibers, and poor interface properties (Le Duigou et al., 2008). The microscopy of fractured injection-molded specimens supported this finding.

3.3 Recycling Investigation of Composites

The histograms illustrate the evolution of injected composite tensile properties during recycling on tensile modulus (Fig. 4.a), tensile strength (Fig. 4.b), and strain at break (Fig. 4.c). The results indicate that the mechanical properties of all the materials were not significantly damaged during the second reprocessing. With the second cycle, the slight decrease in PLA-PBS tensile modulus can be a consequence of the reduction in molecular weight induced by the shear rate, and crystallization occurs during cooling (Bourmaud et al., 2011). Our results are in good agreement with the study shown by Pillin et al. (Pillin et al., 2008). They found a small reduction of PLLA modulus during recycling. Interestingly, FF/PLA-PBS properties globally remain stable. Tensile characterization revealed good strength conservation. This result may refer to the effect of vegetal fiber mechanical degradation after processing (Arbelaiz et al., 2005). Our results are in accordance with other studies, which showed only minor changes with reprocessing. **Le Duigou et al.** (Le Duigou et al., 2008) **for the** flax/PLA composites, **Bourmaud et al.** (Bourmaud et al., 2011) **for the** hemp fiber in recycled PP**, Ausias et al.** for PP-flax composites (Ausias et al., 2013), **and Bourmaud and Baley** (Bourmaud and Baley, 2007) **for** hemp and sisal/PP composites. These findings lend credence to the widely held belief that natural fiber composites maintain their stiffness and strength after recycling. (Dickson et al., 2014). After two processing cycles, tensile modulus stability for GF/PLA-PBS is established. In return, significantly higher tensile strength is noted, with 53% more than FF/PLA-PBS. Moreover, the behavior of tensile modulus and tensile strength have revealed the importance of fiber dispersion in obtaining mechanical properties. Compared to values reported in the literature at similar loadings, while our small tensile modulus variation of FF/PLA-PBS and GF/PLA-PBS was not statistically significant, the reductions founded by **Dickson et al.** (Dickson et al., 2014) were highly founded. Consequently, it's possible that recycling has a big influence on their outcomes. Similar some changes in reprocessed biopolymer composites stiffness were seen (Bodros et al., 2007), but they were accompanied by significant reductions in strength performance with reprocessing. The high reductions in strength are contrary to our results. However, our results did show only 4% strength loss. In contrast, GF/PLA-PBS showed a 19% expansion, which can be explained by improved glass fiber dispersion in the PLA-PBS matrix. It is crucial to keep in mind that a biocomposite made entirely of recycled materials would not be used in an industrial setting (recycled material is typically combined with virgin material).

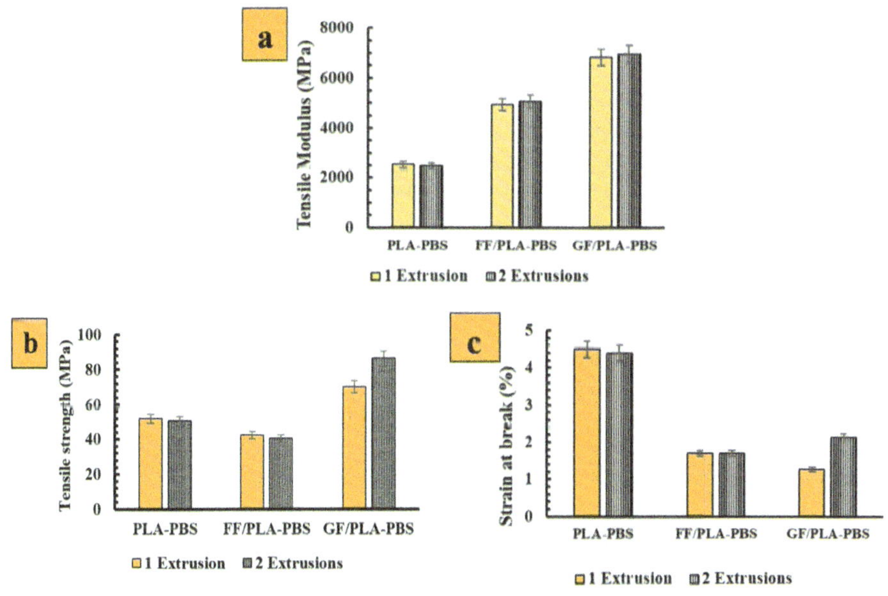

Fig. 4. Effect of compounding-injection cycles on: tensile modulus (a), tensile strength (b) and the strain at break (c).

4 Conclusion

A thermoplastic blend of PLA-PBS biopolymer was reinforced with 20 wt% of flax fibers (FF) to investigate the effect of reprocessing on reinforced natural fibers compared to synthetic glass fibers (GF). Composites were then subjected to two compounding-extrusion cycles for the purpose of studying the degradation of FF/PLA-PBS and GF/PLA-PBS composites. The reprocessing impact on mechanical and morphological properties has been investigated. First, it has been shown that the mechanical properties of the two composite materials are not significantly damaged during the second reprocessing. The tensile modulus remains quite unchanged for all the composites. For the GF/PLA-PBS, significant higher tensile strength and strain at break are noted. These results can be explained by the fact that reprocessing leads to improved glass fiber dispersion in the matrix. Good mechanical qualities are mostly dependent on fiber dispersion, which is achieved during the mixing and shaping procedures. We need to learn more about how the fibers in molten thermoplastic disperse and break down under flow.

Acknowledgements. The authors declare that they have no known competing financial interests or personal relationships that could have appeared to influence the work reported in this paper.

References

Alagarraja, K., et al.: Fabrication and testing of fibre reinforced polymer composites material. IOSR J. Mechan. Civ. Eng. **2**, 18 (2014)

Alvarez, V.A., Fraga, A.N., Vázquez, A.: Effects of the moisture and fiber content on the mechanical properties of biodegradable polymer–sisal fiber biocomposites. J. Appl. Polym. Sci. **91**, 4007–4016 (2004). https://doi.org/10.1002/app.13561

Arbelaiz, A., Fernández, B., Cantero, G., Llano-Ponte, R., Valea, A., Mondragon, I.: Mechanical properties of flax fibre/polypropylene composites. Influence of fibre/matrix modification and glass fibre hybridization. Compos. A Appl. Sci. Manuf. **36**, 1637–1644 (2005). https://doi.org/10.1016/j.compositesa.2005.03.021

Ausias, G., Bourmaud, A., Coroller, G., Baley, C.: Study of the fibre morphology stability in polypropylene-flax composites. Polym. Degrad. Stab. **98**, 1216–1224 (2013). https://doi.org/10.1016/j.polymdegradstab.2013.03.006

Baley, C., Perrot, Y., Busnel, F., Guezenoc, H., Davies, P.: Transverse tensile behaviour of unidirectional plies reinforced with flax fibres. Mater. Lett. **60**, 2984–2987 (2006). https://doi.org/10.1016/j.matlet.2006.02.028

Bodros, E., Pillin, I., Montrelay, N., Baley, C.: Could biopolymers reinforced by randomly scattered flax fibre be used in structural applications? Compos. Sci. Technol. **67**, 462–470 (2007). https://doi.org/10.1016/j.compscitech.2006.08.024

Bourmaud, A., Baley, C.: Rigidity analysis of polypropylene/vegetal fibre composites after recycling. Polym. Degrad. Stab. **94**, 297–305 (2009). https://doi.org/10.1016/j.polymdegradstab.2008.12.010

Bourmaud, A., Baley, C.: Investigations on the recycling of hemp and sisal fibre reinforced polypropylene composites. Polym. Degrad. Stab. **92**, 1034–1045 (2007). https://doi.org/10.1016/j.polymdegradstab.2007.02.018

Bourmaud, A., Le Duigou, A., Baley, C.: What is the technical and environmental interest in reusing a recycled polypropylene–hemp fibre composite? Polym. Degrad. Stab. **96**, 1732–1739 (2011). https://doi.org/10.1016/j.polymdegradstab.2011.08.003

Chuai, C., Almdal, K., Poulsen, L., Plackett, D.: Conifer fibers as reinforcing materials for polypropylene-based composites. J. Appl. Polym. Sci. **80**, 2833–2841 (2001). https://doi.org/10.1002/app.1400

Dickson, A.R., Even, D., Warnes, J.M., Fernyhough, A.: The effect of reprocessing on the mechanical properties of polypropylene reinforced with wood pulp, flax or glass fibre. Compos. A Appl. Sci. Manuf. **61**, 258–267 (2014). https://doi.org/10.1016/j.compositesa.2014.03.010

Hornsby, P., Hinrichsen, E., Tarverdi, K.: Preparation and properties of polypropylene composites reinforced with wheat and flax straw fibres: part I Fibre characterization. J. Mater. Sci. **32**, 443–449 (1997). https://doi.org/10.1023/A:1018521920738

Joseph, P.V., Mathew, G., Joseph, K., Groeninckx, G., Thomas, S.: Dynamic mechanical properties of short sisal fibre reinforced polypropylene composites. Compos. A Appl. Sci. Manuf. **34**, 275–290 (2003). https://doi.org/10.1016/S1359-835X(02)00020-9

Keller, A.: Compounding and mechanical properties of biodegradable hemp fibre composites. Compos. Sci. Technol. Eco-Compos. **63**, 1307–1316 (2003). https://doi.org/10.1016/S0266-3538(03)00102-7

Ketata, N., Guermazi, N., Grohens, Y., Seantier, B.: Effect of bleaching treatment on the mechanical and thermal behavior of flax fibers reinforced poly-(lactic acid) (PLA) and poly-(butylene succinate) (PBS) biocomposites. In: Amar, M.B., Souf, M.A.B., Beyaoui, M., Trabelsi, H., Ghorbel, E., Tounsi, D., El Mahi, A., Chaari, F., Haddar, M. (eds.) Advances in Materials, Mechanics and Manufacturing III: Proceedings of The Fourth International Conference on Advanced Materials, Mechanics and Manufacturing (A3M'2023), March 20-21, 2023, Sousse, Tunisia, pp. 132–140. Springer, Cham (2024). https://doi.org/10.1007/978-3-031-57324-8_15

Ketata, N., Seantier, B., Guermazi, N., Grohens, Y.: On the development of a green composites based on poly (lactic acid)/poly (butylene succinate) blend matrix reinforced by long flax fibers. Materials Today: Proceedings **52**, 95–103 (2022)

Ketata, N., Seantier, B., Guermazi, N., Grohens, Y.: Processing and properties of flax fibers reinforced PLA/PBS biocomposites. Mater. Today Proc. Polym. Mediterranean Fiber Int. **53**, 228–236 (2022). https://doi.org/10.1016/j.matpr.2022.01.047

Le Duigou, A., Pillin, I., Bourmaud, A., Davies, P., Baley, C.: Effect of recycling on mechanical behaviour of biocompostable flax/poly(l-lactide) composites. Compos. A Appl. Sci. Manuf. **39**, 1471–1478 (2008). https://doi.org/10.1016/j.compositesa.2008.05.008

Pillin, I., Montrelay, N., Bourmaud, A., Grohens, Y.: Effect of thermo-mechanical cycles on the physico-chemical properties of poly(lactic acid). Polym. Degrad. Stab. **93**, 321–328 (2008). https://doi.org/10.1016/j.polymdegradstab.2007.12.005

Roux, C., Denault, J., Champagne, M.F.: Parameters regulating interfacial and mechanical properties of short glass fiber reinforced polypropylene. J. Appl. Polym. Sci. **78**, 2047–2060 (2000)

Valente, M., Sarasini, F., Marra, F., Tirillò, J., Pulci, G.: Hybrid recycled glass fiber/wood flour thermoplastic composites: manufacturing and mechanical characterization. Compos. A Appl. Sci. Manuf. **42**, 649–657 (2011). https://doi.org/10.1016/j.compositesa.2011.02.004

Experimental Investigation into the Influence of Stacking Sequence on Fatigue Damage in a Glass/Epoxy Composite

Hana Driss[1,2](✉), Abderrahim ElMahi[1], Mourad Bentahar[1], Moez Beyaoui[2], and Mohamed Haddar[2]

[1] LAUM UMR CNRS 6613, University of Maine, Av. O. Messiaen, 72085 Le Mans Cedex 9, France
hana.driss@enis.tn

[2] Laboratory of Mechanic Modelling and Production (LA2MP), National School of Engineering of Sfax (ENIS), Sfax University, Route de Soukra, 3038 Sfax, Tunisia

Abstract. In this study, we conduct static tensile and fatigue tests on laminated composites to evaluate their mechanical properties. The experimental assessments aim to gauge the durability and energy-dissipating capabilities of the material. Initially, static tests are performed to analyze the tensile behavior of the beams, determine fracture parameters, and establish characteristics for subsequent fatigue tests. Subsequently, fatigue analyses are carried out to assess the fatigue resistance of the structures. The impact of stacking sequence on stiffness, hysteresis loop, energy absorption, and loss factor is investigated for two different laminates, and a comparative analysis of composite behaviors is presented. The findings reveal that the composite with a $[0_2,90_2]$ s sequence demonstrates superior energy dissipation, resulting in higher loss factors. These identified properties serve as sensitive indicators of damage in composites throughout their service life. The primary objective of this research is to characterize the static and fatigue properties of the material, thereby exploring its potential applications in various industries.

Keywords: Damage · Composite · Stacking sequence · Fatigue

1 Introduction

Glass/epoxy composites find increasing applications in diverse structural roles within the automotive and aerospace sectors, owing to their superior mechanical performance and lower specific weight compared to alternative materials. Engineers must carefully consider fatigue loading, a critical factor that can lead to laminated composite failure. Consequently, extensive studies have explored the mechanical properties of glass/epoxy composites, focusing on microstructural aspects like anisotropy, heterogeneity, and properties of the fiber/matrix interface.

Numerous studies (El Mahi et al. 1995; Kashtalyan and Soutis, 2000; Zhang et al. 2008; Rebière and Gamby, 2004) underscore the pivotal role of stiffness loss during fatigue cycles in shaping the fatigue behavior. This loss unfolds across three distinct

M. Chouchane et al. (Eds.): CMSM 2023, LNME, pp. 150–157, 2024.
https://doi.org/10.1007/978-3-031-65007-9_17

phases: an initial rapid decrease in stiffness, followed by a gradual and mostly linear degradation up to 90% of the total life, culminating in stiffness degradation associated with the rapid development of various fatigue damage types, including fiber/matrix interface failure leading to ultimate laminated composite failure.

Several parameters, such as the number of plies, stacking sequence, laminate thickness, loading mode, and fiber density, among others, exert influence on the fatigue properties of composite structures. Brillaud and El Mahi, (1991) emphasize the substantial impact of stacking sequence on transverse cracking in composite laminates subjected to cyclic fatigue tests. Johannesson, (1993), through experimental investigations, explores the fatigue stress limit of glass/epoxy composites, seeking to understand the influence of enhanced ductility in the matrix material on stress limits.

This study specifically delves into the fatigue behavior of glass/epoxy composites under different bend orientations. Cyclic tensile testing is employed to elucidate the impact of stacking sequences ($[0_2,90_2]s$; $[90_2,0_2]s$) on both static and fatigue properties. Experimental static analysis is conducted to identify the parameters of each configuration from the stress-strain diagram. The outcomes of this investigation provide valuable insights for selecting the optimal stacking sequence, establishing boundary conditions, and achieving the maximum fatigue life for the laminate.

2 Materials and Methods

2.1 Preparation of Samples

The manufacturing process employed for fabricating laminated composite plates is based on the wet lay-up method. In this approach, fibers undergo impregnation with a catalyzed resin characterized by a certain level of fluidity. Subsequently, the fibers are arranged onto the mold, which has been previously waxed or coated with a release agent. The resin is then manually applied using either a roller or a brush. Through this meticulous process, a uniform impregnation of the fibers by the resin is achieved, thereby enhancing the bond between the reinforcement and the matrix. This method not only ensures operational flexibility but also empowers the operator to meticulously control the process, guaranteeing an even distribution of the resin over the fibers. The resulting plates are then cut using a diamond saw, with dimensions measuring 200*20.

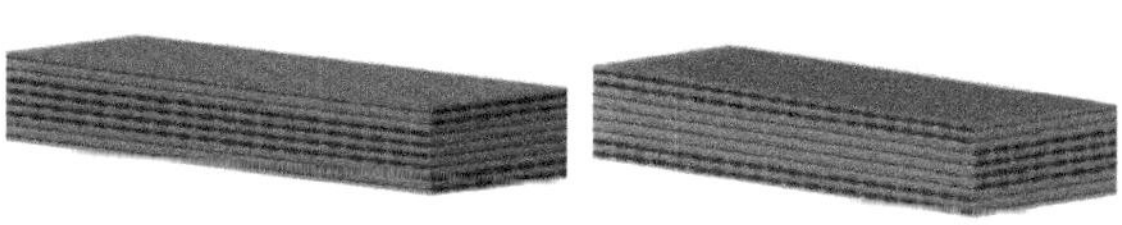

(a) (b)

Fig. 1. The two configurations (a) $[0_2,90_2]s$; (b) $[90_2,0_2]s$

2.2 Experimental Procedure

Static and fatigue tests were conducted using an INSTRON 8801 hydraulic universal testing machine with a 10 kN capacity. The load was measured using a force cell, and computer control and data acquisition were employed. To ensure result validity, five specimens were tested for each configuration. Static tests involved applying a displacement signal at a testing speed of 1 mm/min. Fatigue tests utilized a sinusoidal waveform with a charge signal frequency of 10 Hz. The specimens were subjected to a specified displacement level (r_d). This comprehensive testing approach aimed to assess both static and fatigue behaviors, providing a robust evaluation of the material's mechanical performance under various loading conditions.

During testing, specimens are subjected to an inter-jaw length of 100 mm. The displacement ratio, noted R_d and defined by Eq. (1), is set at 0.65:

$$R_d = \frac{d_{min}}{d_{max}} \tag{1}$$

where, d_{min} and d_{max} represent the minimum and maximum displacements applied during the cycle, respectively. Following this, adjustments are made to both the average displacement, denoted as d_{mean}, and the amplitude of displacement d_{amp} to introduce diversity into the displacement level 'r_d' as specified by Eq. (2):

$$r_d = \frac{d_{max}}{d_{rup}} \tag{2}$$

where d_{rup} stands for the failure displacement observed in quasi-static tensile tests. To account for potential variability in experimental results, at least two specimens per loading configuration are tested. One load level is explored, namely $r = 0.65$.

3 Results

3.1 Static Results

Figure 2 illustrates the stress-strain characteristic curves derived from laminated samples featuring various stacking sequences. Each configuration reveals three distinct zones, offering a nuanced understanding of the material's performance. Initially, an identifiable linear segment emerges, representing the linear response of the material. Notably, laminates whose layers have a 0° angle at the ends exhibit higher stress levels than those whose layers have a 90° angle at the corresponding points of this linear interval.

The second region delimits the non-linear behavior of laminates, significantly influenced by stacking properties. This phase is characterized by non-linear responses attributed to tension, and damage initiation is distinctly detected in this region. Finally, the third region marks the ultimate failure of laminates. As the load increases, damage accumulates steadily until the samples reach failure.

Strain analysis at the point of failure reveals consistent deformation patterns for all laminates. Of particular interest is the fact that laminates with outer layers at an angle of 0° exhibit the highest load forces and stiffness among the configurations examined. This in-depth examination offers valuable insights into the distinct phases of load-deformation

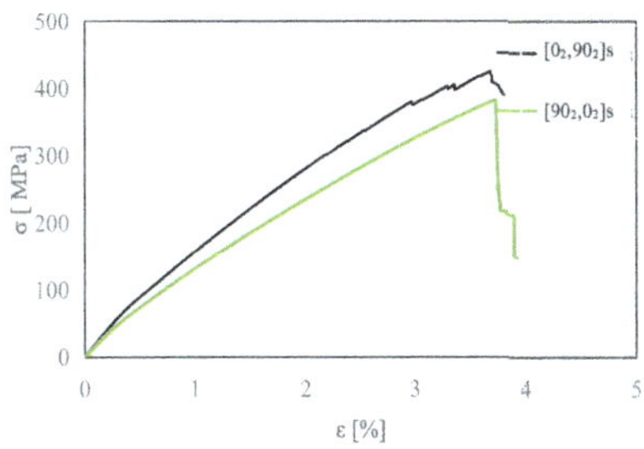

Fig. 2. Tensile characteristics of beams with diverse stacking sequences

behavior exhibited by the laminate samples, elucidating their mechanical responses under varying stacking conditions.

Recently explored is the tensile behavior of laminated composites, unveiling the influential role of stacking topology in shaping stress-strain curves. In fatigue testing with displacement control, the determination of the load level necessitates identifying the displacement at failure. An average displacement, denoted as d_{mean}, is computed, and cyclic displacement with a damping amplitude is subsequently applied around this central value. To ensure a distinct loading application in both linear and non-linear domains, the ultimate displacement is pinpointed at the boundary of the linear domain. Tabulated data in Table 1 presents the values of d_{rup} (breaking displacement) and F_{rup} (breaking load) for each laminate configuration.

Table 1. Parameters for the design of laminates

Parameters	$[0_2,90_2]s$	$[90_2,0_2]s$
Failure displacement d_{rup} (mm)	4	4
Failure load F_{rup} (N)	17140	16510

3.2 Fatigue Loading

The fatigue-tension response of a sandwich structure featuring diverse stacking sequences is examined, focusing on the variations in stiffness evolution during cyclic loading and its correlation with the number of cycles. This methodology stands as a widely adopted approach for investigating damage progression in fatigue studies. Figure 3 illustrates the progression of the maximum applied load (F_{max}) with respect to the number of cycles, normalized by the initial maximum load (F_{max0}), using a semi-logarithmic scale to study stiffness loss (F_{max}/F_{max0}) at a displacement level of (r_d = 65%). Results across configurations reveal a distinct three-phase pattern in stiffness loss leading to failure. The initial cycles witness a reduction in stiffness, signifying the initial degradation of the structure's mechanical properties and the onset of damage with the multiplication of cracks in the specimens.

The subsequent phase is characterized by a gradual and sustained reduction in stiffness, encompassing nearly the entire service life of the beams. Notably, the $[90_2,0_2]$s laminate experiences a relatively rapid initiation of cracking during this phase. Lastly, the third phase manifests an abrupt decline in stiffness, marking the continued progression of the damage mechanism with an increasing number of cycles, culminating in the total failure of the samples. Specifically, for the $[90_2,0_2]$s laminate, a higher degradation rate and earlier failure are observed compared to the $[0_2,90_2]$s configuration.

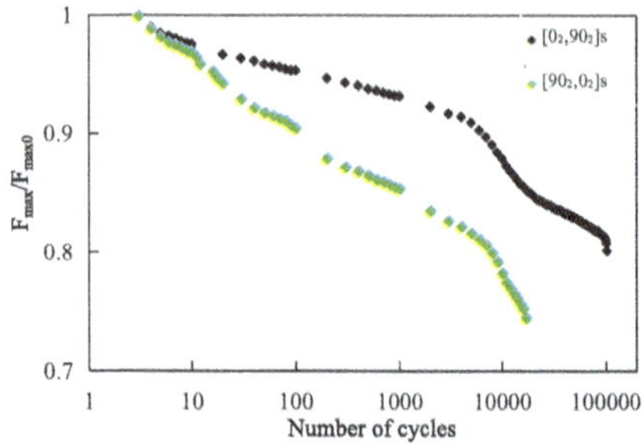

Fig. 3. Variation in stiffness loss with respect to the number of cycles across different stacking sequences

3.3 Hysteresis Curves

For the analysis of fatigue tests in different configurations, we examine the energy characteristics by studying the evolution of hysteresis curves. The load level is consistently set at 80%, and each cycle is defined by 200 experimental points. Figure 4 displays load-displacement curves for loading and unloading cycles in laminates with various stacking sequences. These curves, marked by congestion and unclear patterns, are adjusted along the displacement axis in each cycle for better visibility. Despite differences in stacking sequences, the behavior of the hysteresis curves remains comparable for both configurations. Notably, the $[0_2,90_2]$s laminate exhibits a higher maximum load, and as the number of cycles increases, the area under the hysteresis loops decreases. Initial cycles also show higher maximum loads compared to later cycles.

3.4 Energy Analyses

Experimental investigations into fatigue aim to quantify the energy dissipation capacity of composites with diverse stacking sequences. These structures exhibit notable prowess in absorbing energy. The study focuses on cyclic fatigue tension testing of specimens. Illustrated in Fig. 5, the hysteresis cycle outlines the area encompassed by the loops, representing the dissipated energy (E_d) for each cycle. The load-displacement characteristics of laminates are contingent on deformation energy, where the stored potential energy (E_p) is determined by the area beneath the top (loading segment) of the hysteresis loop. Utilizing trapezoidal summation, depicted in Fig. 5, facilitates the determination of both dissipated energy (E_d) and maximum potential energy (E_p).

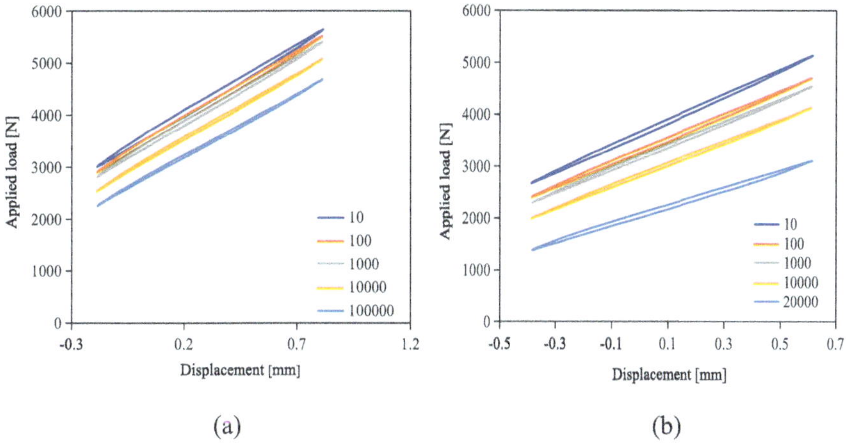

Fig. 4. Hysteresis curves corresponding to various stacking sequences and cycle numbers, (a) $[0_2,90_2]$s; (b) $[90_2,0_2]$s

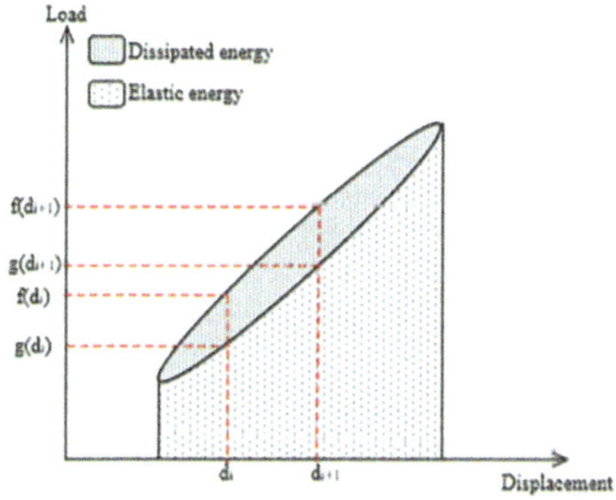

Fig. 5. Depiction of a hysteresis cycle.

To mitigate the computed surface error, the hysteresis curves undergo division by a substantial number denoted as 'n.' The expressions for dissipated energy and potential energy are provided by Eqs. (3) and (4) (Alomarah et al. 2020):

$$E_p = \frac{1}{2}\sum_{i=1}^{n}(d_{i+1} - d_i)\left[f(d_{i+1}) + f(d_i)\right] \tag{3}$$

$$E_d = \frac{1}{2}\sum_{i=1}^{n}(d_{i+1} - d_i)\left\{\left[f(d_{i+1}) + f(d_i)\right] - \left[g(d_{i+1}) + g(d_i)\right]\right\} \tag{4}$$

The variation in dissipated energy over the number of cycles for distinct configurations is illustrated in Fig. 6. These graphs offer a comparative analysis of dissipated energy across composites featuring different stacking sequences. The curves depict the amount of energy dissipated by the structures during their operational cycles, providing insights into the material's lifespan. Specifically, $[0_2,90_2]$s laminates exhibit a higher dissipation of energy compared to $[90_2,0_2]$s laminates as the cycle count rises.

In the initial cycles, there is a considerable amount of dissipated energy, followed by a subsequent decrease with a higher number of cycles. Notably, laminates with the applied loading level exhibit significant disparities in energy loss per unit volume. The fatigue response reveals substantial variations from the early stages to the end of the fatigue life, indicative of increased damage in the composites during fatigue testing. Furthermore, the stacking sequence of layers emerges as a crucial factor in energy dissipation throughout fatigue. Loading and unloading cycles result in substantial energy dissipation through layer collapse, accounting for the heightened energy dissipation in the $[0_2,90_2]$s laminate, characterized by 0° layers.

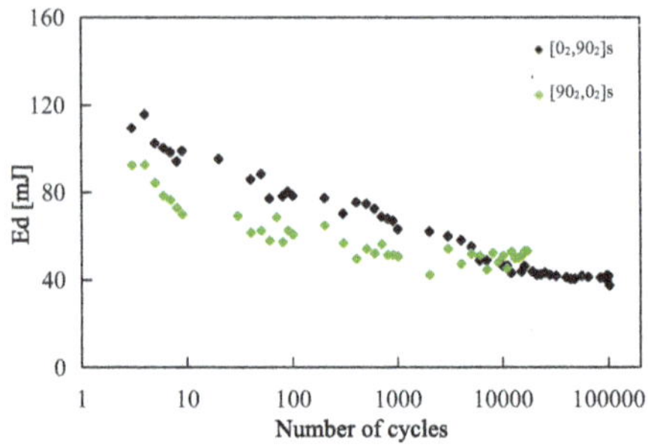

Fig. 6. Energy dissipation trends across different stacking sequences as a function of the number of cycles.

3.5 Loss Factor

The dissipation of energy in laminated composites is influenced by several factors, including the stacking sequence and the viscoelastic properties of the material constituting the structure. In cyclic fatigue tests, the loss factor, denoted as η, represents the ratio of energy dissipated per cycle (E_d) to the maximum potential energy per cycle (E_p). This relationship is expressed by Eq. (5) (Monti, 2016; Allagui, 2021):

$$\eta = \frac{E_d}{2\pi E_p} \tag{5}$$

Figure 7 illustrates the progression of the fatigue loss factor η for structures tested with a span length of 100 mm, employing a semi-logarithmic scale. The loss factor experiences a decline as the number of cycles increases. A notable rapid decrease is observed in the initial cycles for both configurations, attributable to the initial damage mechanisms induced by the initial load applied to the structure. Subsequently, an increase is noted over the initial thousand cycles. Eventually, the evolution of the loss factor stabilizes across all sandwich configurations. Towards the end of the tests, an increase in the loss factor becomes evident.

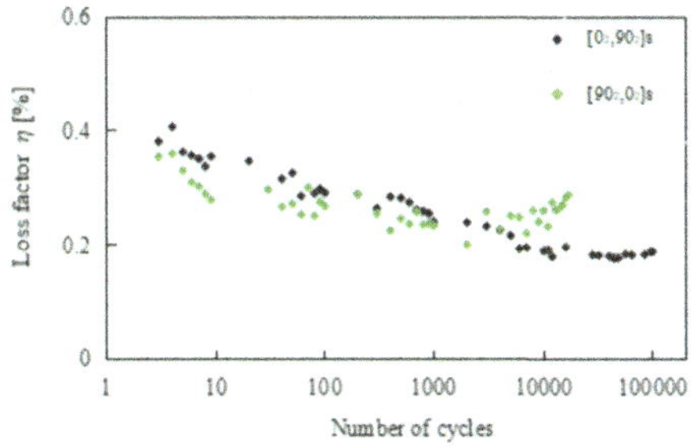

Fig. 7. Variation in the loss factor relative to the number of cycles across various stacking sequences

4 Conclusion

The paper addresses the evaluation of static and dynamic tensile properties in laminated composites. The findings highlight a significant influence of stacking sequence on the structures' stiffness, energy dissipation, and loss factor. The observed changes in stiffness serve as a reliable indicator of specimen condition, and the use of the loss factor is recommended for assessing damage propagation in sandwich materials, proving more substantial than stiffness alone. Notably, the static and cyclic characteristics of the laminate, representing layers at a 90° angle, exhibit suboptimal performance in this orientation. Integrating experimental tests, including a non-destructive acoustic emission technique for damage monitoring, could enhance the precision of monitoring damage initiation and its correlation with damping ratio evolution.

References

Allagui, S.: Développement et caractérisation d'un composite biosourcé instrumenté et recyclable (Doctoral dissertation, Le Mans Université; University of Sfax) (2021)

Alomarah, A., Masood, S.H., Sbarski, I., Faisal, B., Gao, Z., Ruan, D.: Compressive properties of 3D printed auxetic structures: experimental and numerical studies. Virt. Phys. Protot. **15**(1), 1–21 (2020)

Monti, A.: Élaboration et caractérisation mécanique d'une structure composite sandwiche à base de constituants naturels (Doctoral dissertation, Université du Maine) (2016)

Brillaud, J., El Mahi, A.: Numerical simulation of the influence of stacking sequence on transverse ply cracking in composite laminates. Compos. Struct. **17**(1), 23–35 (1991)

El Mahi, A., Berthelot, J.M., Brillaud, J.: Stiffness reduction and energy release rate of cross-ply laminates during fatigue tests. Compos. Struct. **30**(2), 123–130 (1995)

Johannesson, T.: Fatigue strength of glass fiber/epoxy angle-ply laminates with different matrix ductility. Mech. Compos. Mater. **28**(5), 439–444 (1993)

Kashtalyan, M., Soutis, C.: Stiffness degradation in cross-ply laminates damaged by transverse cracking and splitting. Compos. A Appl. Sci. Manuf. **31**(4), 335–351 (2000)

Rebière, J.L., Gamby, D.: A criterion for modelling initiation and propagation of matrix cracking and delamination in cross-ply laminates. Compos. Sci. Technol. **64**(13–14), 2239–2250 (2004)

Zhang, Y., Vassilopoulos, A.P., Keller, T.: Stiffness degradation and fatigue life prediction of adhesively-bonded joints for fiber-reinforced polymer composites. Int. J. Fatigue **30**(10–11), 1813–1820 (2008)

Numerical Simulation of Elastic Wave Propagation in Composite Structure

Raslen Nemer[1] (✉), Henia Arfa[2], Faker Bouchoucha[3], and Mohamed Ichchou[4]

[1] Physics, Mathematics, Modeling and Mechanical Design Research Laboratory (LR-18ES45), National Engineering School of Sousse (ENISo), University of Sousse, BP 264 Sousse Erriadh, 4023 Sousse, Tunisia
raslennemer97@gmail.com

[2] Laboratory of Mechanical Engineering (LGM), National Engineering School of Monastir (ENIM), University of Monastir, Avenue Ibn Eljazzar, 5019 Monastir, Tunisia

[3] Physics, Mathematics, Modeling and Mechanical Design Research Laboratory (LR-18ES45), Nabeul Preparatory Institute for Engineering Studies (IPEIN), Carthage University, University Campus, Merazka, 8000 Nabeul, Tunisia

[4] Laboratory of Tribology and Dynamics of Systems (LTDS), Ecole Centrale de Lyon, 36 Avenue Guy de Collongues, 69130 Ecully, France
Mohamed.Ichchou@ec-lyon.fr

Abstract. This paper delves into the concept of multimodal waveguide structures and their modeling across a relatively wide frequency spectrum encompassing low and medium frequencies. The chosen methodology combines wave propagation with finite element modeling. The proposed wave/finite element (WFE) technique involves a post-processing step applied to the mass and stiffness matrices, which are initially obtained through finite element methods, for a specific substructure. To establish connections between degrees of freedom and node forces, periodicity conditions are implemented. Consequently, real, imaginary, or complex wave numbers and frequencies are derived from distinct eigenproblems that arise. The exact form of these eigenproblems hinges on the desired solution's nature and can take the shape of a linear, quadratic, or polynomial eigenproblem. The primary focus here is on determining guided multimodal dispersion curves. This propagative approach, grounded in a finite element model (WFE), offers an efficient means to compute dispersion curves for both straightforward and intricate guided structures. The paper also derives certain properties of guided waves through this formulation and provides a comprehensive account of its numerical implementation. Furthermore, the paper showcases the practicality of the Wave Finite Element (WFE) method in predicting wave propagation within single- and multilayer structures across various frequency ranges.

Keywords: Wave finite element method · Dispersion curves · Wave propagation · Numerical simulation

M. Chouchane et al. (Eds.): CMSM 2023, LNME, pp. 158–167, 2024.
https://doi.org/10.1007/978-3-031-65007-9_18

1 Introduction

Research on guided waves remains extensive due to their prevalence in various engineering domains. As a result, a substantial portion of the existing literature is dedicated to investigate the characteristics and applications of guided waves. The Wave Finite Element Method (WFEM) represents a versatile approach that combines periodic structure theory with widely-used industry-standard finite element software. The extensive body of WFEM literature (Duhamel et al. 2022) covers numerous applications, including damage detection, met materials, vibro-acoustics, and inverse identification. (Fan et al. 2017) has developed an enhanced version of WFEM with the objective of investigating the one-dimensional wave propagation characteristics in periodic structures. For the purpose of determining wave amplitudes generated by mobile loads applied to a periodic railway track, (Hoang et al. 2018) presents a technique that employs WFEM in the frequency domain. Additionally, using WFEM, (Miranda and Santos 2018) conducts an examination of the forced response of bending waves propagating within a one-dimensional beam. Moreover, WFEM can be effectively employed for analyzing the properties of 2D wave propagation within periodic structures. Also, (Zergoune et al. 2017) has introduced a modeling approach based on WFEM to study the vibro-acoustic bending behavior of sandwich panels. He has derived an expression for the acoustic transmission attenuation of these panels. To illustrate the band gap characteristics of reinforced shell structures, Hong (Hong et al. 2018) has presented an improved iteration of WFEM. Mencik (Mencik. 2018) has developed a numerical method grounded in WFEM for calculating the 2D wave modes of cyclically symmetric periodic structures.

On the other hand, over the last decade there has been intensive research into numerical methods for analyzing the vibratory behavior of complex composite or periodic structures over a wide range of frequencies, as described by Reddy (Reddy. 1997), who investigated the analysis of wave dispersion in symmetrical sandwich panels. Renji (Renji et al. 1996) introduced a new analytical formulation to obtain a central stiffness matrix for transverse shear and bending in both directions in orthotropic sandwich panels. Recently, (Guillaumie.2015) proposed an analytical solution for the eigenmodes and modal densities of symmetrical sandwich structures with composite skins. (Florence and Renji 2015) have also developed explicit solutions for the modal density of composite cylinders.

Thus, the wave finite element method (WFEM) has been successfully applied to the analysis of waves in complex periodic waveguides to predict the radiation efficiency of many thick-film structures (Mencik and Ichchou. 2005). Although the dispersion characteristics of sandwich panels have recently attracted much attention (Alkhader et al. 2015).

In this document, we focus on the field of linear elastodynamics, with an emphasis on the vibrational behavior of such structures, which can be studied by determining the characteristics of the waves propagating within them. More specifically, we study the propagation and scattering of guided waves in simple and complex elastic structures.

2 Overview of the WFE 1D

2.1 Formulation of 1D-Waveguides

In this section, we present a formulation of the WFE method for one-dimensional free waves propagation of in a straight, this formulation corresponds to elastic waveguide. The structure is composed by N identical subsystems with d length. Figure 1 illustrates the wave structure along the x-direction.

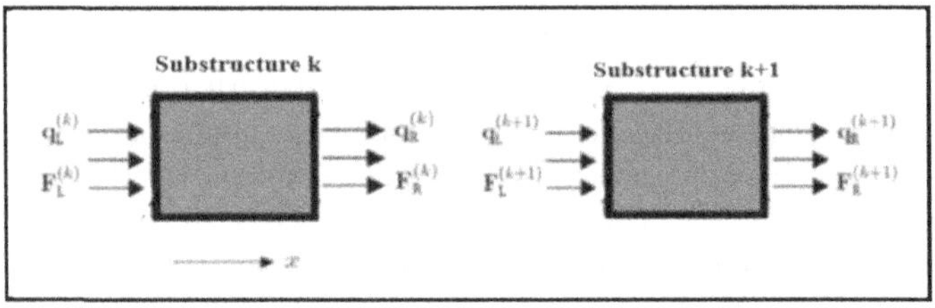

Fig. 1. Illustration of a waveguide along the x-direction.

Displacements and forces are denoted by q and f, respectively and subscripts 'L' and 'R' refer to the left and right edges of a cell. Both edges have the same number n of degrees of freedom. Mesh compatibility are applicated among the N subsystems. The discrete dynamic equation of a cell is formulated as follows:

$$\left(K + i\omega C - \omega^2 M\right)q = F \tag{1}$$

where M, C, and K represent the mass, damping, and stiffness matrices, respectively. In the case of periodic structures, it is necessary to condense the internal degrees of freedom at the left and right cross-sections. By introducing the condensed dynamic stiffness operator D.

where: $D = K + i\omega C - \omega^2 M$

In the case where there are no external forces on the inner nodes, the equation of motion can be partitioned into:

$$\begin{pmatrix} D_{LL} & D_{LI} & D_{LR} \\ D_{IL} & D_{II} & D_{IR} \\ D_{RL} & D_{RI} & D_{RR} \end{pmatrix} \begin{pmatrix} q_L \\ q_I \\ q_R \end{pmatrix} = \begin{pmatrix} F_L \\ 0 \\ F_R \end{pmatrix} \tag{2}$$

In some cases of finite element analysis, there may be no internal degrees of freedom, which can then be eliminated from Eq. (2), this leads to:

$$\begin{bmatrix} D^*_{LL} & D^*_{LR} \\ D^*_{RL} & D^*_{RR} \end{bmatrix} \begin{pmatrix} q_L \\ q_R \end{pmatrix} = \begin{pmatrix} F_L \\ F_R \end{pmatrix} \tag{3}$$

where the partitions of the dynamic stiffness matrix D^* are:

$$\begin{aligned} D^*_{LL} &= D_{LL} - D_{LI}D_{II}^{-1}D_{IL}, D^*_{LR} = D_{LR} - D_{LI}D_{II}^{-1}D_{IR}D^*_{RL} \\ &= D_{RL} - D_{RI}D_{II}^{-1}D_{IL}, D^*_{RR} = D_{RR} - D_{RI}D_{II}^{-1}D_{IR} \end{aligned} \tag{4}$$

2.2 Resolution of the Eigenvalue Problem

The propagation constant describing wave propagation over the cell length λ_i, where $\lambda_i = e^{-iK_i d}$ withe d and K_i are respectively the cell length and the wave number. In the base of force equilibrium in a cell:

$$\lambda F_L + F_R = 0 \tag{5}$$

And applying Bloch's theorem (Droz et al.2014):

$$q_R = \lambda q_L \tag{6}$$

By introducing Eqs. (4) and (5) into the system (3), we obtain:

$$\begin{bmatrix} D_{LL}^* & -D_{LR}^* \\ D_{RL}^* & D_{RR}^* \end{bmatrix} \begin{pmatrix} q_L \\ \lambda q_L \end{pmatrix} = \begin{pmatrix} F_L \\ \lambda F_L \end{pmatrix} \tag{7}$$

Expanding the system (6) and simplifying the injected forces F_L we find:

$$\left[D_{RL}^* + \lambda\left(D_{RR}^* + D_{LL}^*\right) + \lambda^2 D_{LR}^*\right] q_L = 0 \tag{8}$$

This quadratic system presents a second-order eigenvalue problem, with 2n pairs of eigenvalues λ_i and eigenvectors q_i, each pair corresponding to a wave.

The corresponding eigenvalue problem can be expressed using an appropriate state vector $\varnothing_i = \left[(\varnothing^q)^t, \left(\varnothing^F\right)^t\right]^t$, related to a symplectic transfer matrix S.

$$S\begin{Bmatrix} \varnothing^q \\ \varnothing^F \end{Bmatrix} = \lambda_i \begin{Bmatrix} \varnothing^q \\ \varnothing^F \end{Bmatrix} \tag{9}$$

With

$$S = \begin{bmatrix} D_{LR}{}^{-1} D_{LL} & D_{LR}{}^{-1} \\ D_{RL} - D_{RR} D_{LR}{}^{-1} D_{LL} & -D_{RR} D_{LR}{}^{-1} \end{bmatrix} \tag{10}$$

The eigenvectors $\varnothing_i = \begin{Bmatrix} \varnothing_i^q \\ \varnothing_i^F \end{Bmatrix}_{i=1\ldots 2n}$ indicate the shape of the motion on the transverse cell, and n is the number of nodal degrees of freedom on each side of the cell. The eigenvector $\varnothing_i = \begin{pmatrix} q_L \\ F_L \end{pmatrix}$ can be partitioned into 2n vectors associated with nodal degrees of freedom and nodal forces.

Therefore, the properties of free wave propagation are determined by solving this eigenvalue problem:

$$\begin{cases} S\varnothing_i = \lambda_i \varnothing_i \\ |S - \lambda_i I_{2n}| = 0 \end{cases} \tag{11}$$

According to the previous analysis, the solutions $\{\lambda_i, \varnothing_i\}_{i=1\ldots 2n}$ of eigenvalue problems are called the wave modes of the periodic structure.

3 Numerical Modeling by Using WFEM

The Waveguide Finite Element Method (WFEM), necessitates computer-based implementation for the development of programs that demonstrate enhanced simplicity, robustness, and speed. Indeed, the WFEM, based on the finite element method, which includes the generation and division of mass and stiffness matrices, is executed through the use the ABAQUS software combined with MATLAB.

Two test structures are presented to illustrate the developed method. The first example corresponds to isotropic cylindrical pipe. While the second example focuses on a composite cylindrical pipe with the same characteristics.

3.1 Cylindrical Isotropic Waveguide

A cylindrical pipe is a simple form of 1D periodic structure that can be experimentally tested. Table.1 shows the geometrical characteristics of the cylindrical waveguide cell studied.

Table 1. Geometrical characteristics of the cylindrical waveguide 1

	Unit (mm)
Radius, R	100
Thickness, E	1
Length, L	20

The used material is the steel AISI 1095 (table.2). The meshing of the substructure using shell element (S4R) is given in the Fig. 2, with 36 nodes per section; each node is characterized by a set of 6 degrees of freedom (including three translational degrees and three rotational degrees) and a total of 18 elements.

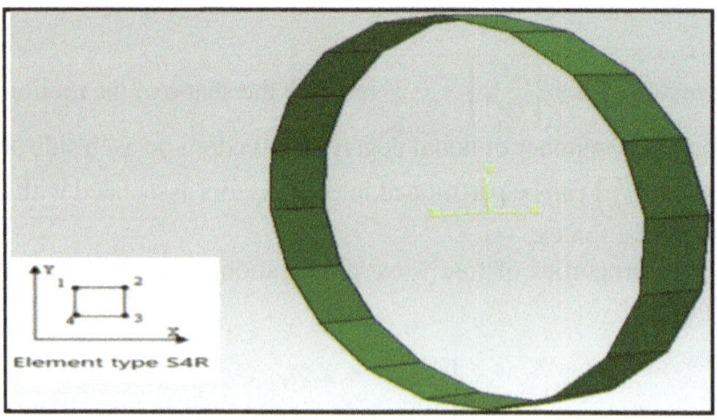

Fig. 2. The finite element Meshing of the cylindrical waveguide.

3.2 Composite Cylindrical Waveguide

The structure of the second example is a multilayer cylindrical tube, considered as a symmetrical layers sandwich waveguide composed by steel AISI 1095 and aluminum-1050. This second cell of the cylindrical waveguide studied (Fig. 3) has the following characteristics: have the following characteristics: Radius = 50 mm, Thickness = 1 mm, Length = 30 mm.

Consider two uniform waveguides: each layer constructed from 1 mm-thick.

The arrangement investigated in this study is: Steel-Aluminum (Steel/Alum), as shown in Fig. 3.

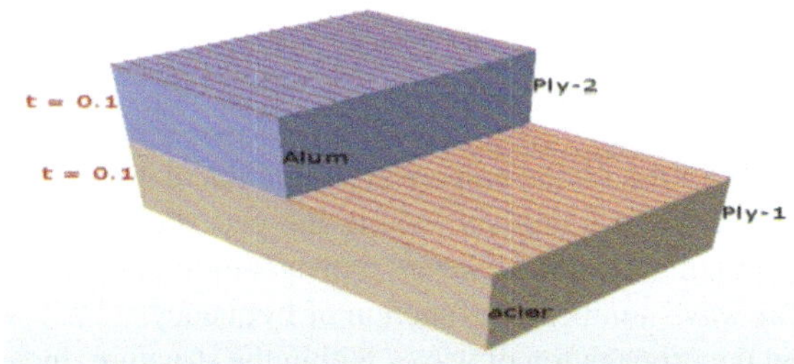

Fig. 3. Arrangement Steel-Aluminum

The meshing of the substructure is performed similarly to what was studied in example 1, using the surface element (S4R), with number of nodes = 65 per section, this is equivalent to a total of 390 degrees of freedom (6 degrees of freedom per node), and with a total number of elements equals 52.

Table 2 summarizes the mechanical properties of both materials (Steel and aluminum).

Table 2. Material properties of waveguides

	Steel AISI 1095	aluminum-1050
Density (kg/m^3)	7800	2700
Young's modulus, E (GPa)	210	71
Poisson's ratio, ν	0.3	0.28

Figure 4 shows the geometry of the model studied and its mesh using an S4R element.

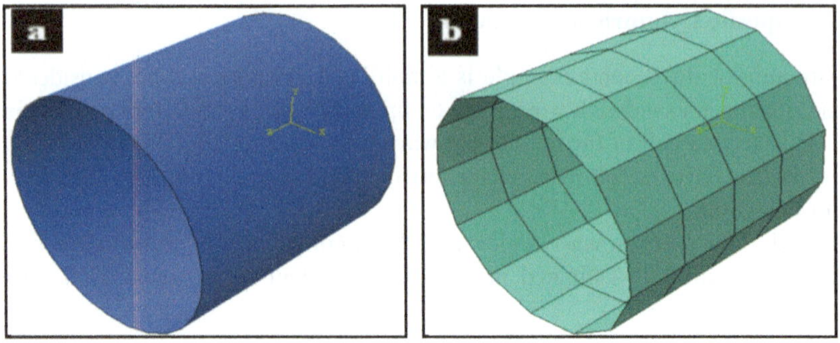

Fig. 4. (a) Geometrical model of waveguide2 (b) Meshing of the Composite waveguide.

3.3 Numerical Results and Discussion

The WFEM approach allows for the numerical simulation of dispersion curves by studying the dynamic behavior of a unit cell within the periodic waveguide. This analysis is based on characterizing of guided waves propagation parameters. These dispersion curves illustrate the wave number as a function of frequency.

Thus reflecting the propagation of waves within the structure. Incident modes propagating in the positive x-direction and reflected modes (symmetric to the incidents) propagating in the negative x-direction are observed. This result is confirming by the formulation of elastic wave propagation in waveguides. As the frequency increases, other modes within the structure are gradually attenuated.

Notably, the emergence of the longitudinal tension-compression mode, torsional mode, and flexural mode are observed. These modes are not sensitive to defect detection. At relatively high frequencies, section modes appear, influencing the cross-sectional properties of the waveguide and subsequently serving for defect detection. The dispersion curves allow different velocities for each wave mode reflecting the behavior of progressive waves within the structure.

3.3.1 Dispersion Curves of Isotropic Waveguide

The WFEM is used to characterize wave propagation parameters in periodic structures. Solving the eigenvalue problem is employed to extract the corresponding eigenmode and determine the dispersion curves. These dispersion curves provide information on wave numbers. The wave numbers reflect the behavior of guided modes within the structure. Specifically, incident and reflected modes (where $\lambda_{n+i} = \lambda_i^{-1}$ for i $\in \{1\ldots n\}$) are highlighted. These incidence and reflected modes, which are symmetric, are used to classify the eigenmodes. Additionally, displacement modes are important for energy transport to detect defaults (Bouchoucha et al. 2010).

Moreover, Fig. 5 also presents the evolution of displacement modes with frequencies, including torsional, flexural and cross-sectional modes at higher frequencies. These displacement modes constitute the reduced modal basis used for modal decomposition of kinematic variables. The article by (Bouchoucha et al. 2010) also discusses the validation of these dispersion curves by comparing them to analytical results found in literature.

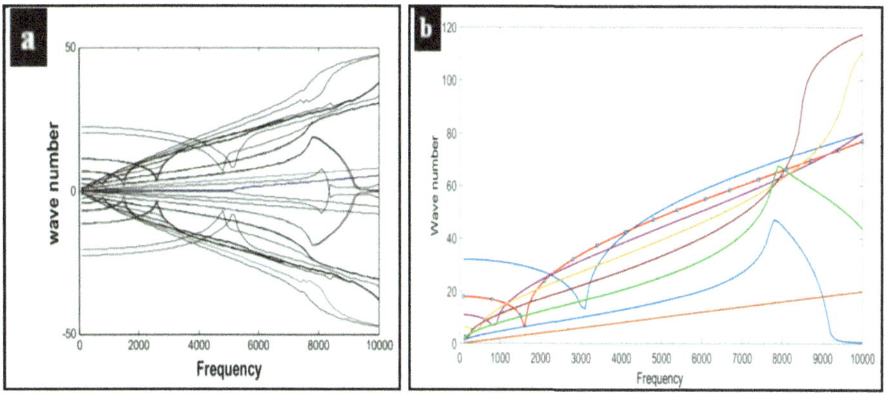

Fig. 5. (a) Dispersion curves Bouchoucha et al (2012). (b) Dispersion curves (isotropic waveguide) obtained by WFEM.

3.3.2 Dispersion Curves of Composite Waveguide

For this multilayer waveguide, dispersion curves can be obtained over a frequency range of [0 Hz, 20000 Hz] (see Fig. 6). In this frequency range, the most significant propagation modes, both incident and reflected, appear, including the "classical" modes corresponding to torsional and bending wave movements (as described in reference (Graff, 1991)). In addition, an additional mode, called the cross-section mode, can be observed. The results obtained are clearly validated at low frequencies (LF) compared to linear elasticity theory. As frequency increases, solutions to the eigenvalue problem vary.

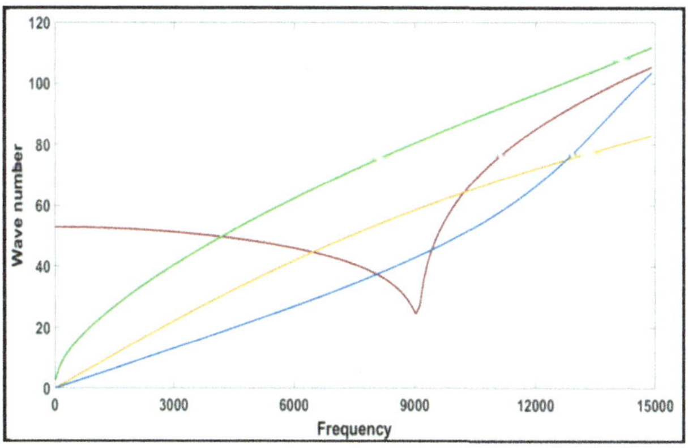

Fig. 6. Dispersion curves (Composite waveguide) obtained by WFEM.

4 Conclusions

In this contribution, we have developed a WFE propagative approach for wave-based characterization of guided structures. The methodology consists in developing this WFE approach coupled with finite element modeling for the numerical calculation of propagation parameters. The WFE method was applied to one-dimensional structures, not only isotropic guided waves, but also composites. An eigenvalue problem was formulated to determine dispersion curves numerically. These dispersion curves for simple and complex guided structures have been obtained for a wide range of frequencies. The formulation presented is general and can be adapted to various types of guided structures. In addition, this procedure has been implemented in a finite element software package, providing an efficient tool for the analysis of free waves under realistic conditions. The results of this study provide valuable insights into wave propagation within the structures considered, whether single- or multi-layered.

References

Duhamel, D., Mencik, J.-M.: Time response analysis of periodic structures via wave-based absorbing boundary conditions. Europ. J. Mechan. A/Solids **91**, 104417 (2022)

Fan, Y., Collet, M., Ichchou, M., Li, L., Bareille, O., Dimitrijevic, Z.: Enhanced wave and finite element method for wave propagation and forced response prediction in periodic piezoelectric structures. Chin. J. Aeronaut. **30**(1), 75–87 (2017)

Hoang, T., Duhamel, D., Foret, G., Pochet, J.-L., Sabatier, F.: Wave finite element method ˆ and moving loads for the dynamic analysis of railway tracks. In: 13th World Congress on Computational Mechanics (WCCM XIII), pp. 25–28 (2018)

de Miranda, E.J.P., Santos, J.M.C.D.: Flexural wave band gaps in phononic crystal Euler Bernoulli beams using wave finite element and plane wave expansion methods. Mater. Res. **20**, 729–742 (2018)

Zergoune, Z., Ichchou, M., Bareille, O., Harras, B., Benamar, R., Troclet, B.: Assessments of shear core effects on sound transmission loss through sandwich panels using a two-scale approach. Comput. Struct. **182**, 227–237 (2017)

Hong, J., He, X., Zhang, D., Zhang, B., Ma, Y.: Vibration isolation design for periodically stiffened shells by the wave finite element method. J. Sound Vib. **419**, 90–102 (2018)

Mencik, J.-M.: A wave finite element approach for the analysis of periodic structures with cyclic symmetry in dynamic sub structuring. J. Sound Vib. **431**, 441–457 (2018)

Reddy, J.N.: Mechanics of laminated composite plates - Theory and analysis. CRC Press, Boca Raton, FL (1997)

Renji, K., Nair, P., Narayanan, S.: Modal density of composite honeycomb sandwich panels. J. Sound Vib. **195**(5), 687–699 (1996)

Guillaumie, L.: Vibroacoustic flexural properties of symmetric honeycomb sandwich panels with composite faces. J. Sound Vib. **343**, 71–103 (2015)

Florence, S.J.K., Renj, K.: Modal density of thin composite cylindrical shells. J. Sound Vib. (2015). https://doi.org/10.1016/j.jsv.2015.11.030i

Mencik, J.M., Ichchou, M.N.: Multi-mode propagation and diffusion in structures through finite elements. Eur J Mech A-Solid **24**(5), 877–898 (2005)

Alkhader, M., Iyer, S., Shi, W., Venkatesh, T.: Low frequency acoustic characteristics of periodic honeycomb cellular cores: the effect of relative density and strain fields. Compos. Struct. **133**, 77–84 (2015)

Duhamel, D., Mace, B.R., Brennan, M.J.: Finite element analysis of the vibration of waveguides and periodic structures. J. Sound Vib. **294**, 205–220 (2006)

Droz, C., Lainé, J.-P., Ichchou, M.N., Inquiété, G.: A reduced formulation for the free-wave propagation analysis in composite structures. J. Sound Vib. Compos. Struct. **113**, 134–144 (2014)

Bouchoucha, F., Akrout, M., Fakhfakh, T., Ichchou, M.N., Haddar, M.: "Detection of the defects in cylindrical pipes: application to the wave finite element method", international journal of modeling. Identif. Control (IJMIC) **9**(4), 336–349 (2010)

Bouchoucha, F., Akrout, M., Fakhfakh, T., Ichchou, M.N., Haddar, M.: Damage detection in cylindrical pipe through diffusion matrix in wave finite element method. Adv. Struct. Eng. **15**(3), 435–445 (2012). https://doi.org/10.1260/1369-4332.15.3.435

Damage Parameters Identification Based on Experimental Data for Advanced Steel Matrix Composite

Manel Dammak[1(✉)], Abir Bouhamed[1,2], Hanen Jrad[1,3], and Fakhreddine Dammak[1]

[1] LEE laboratory, National Engineering School of Sfax, University of Sfax, Sfax, Tunisia
manele.dammak@gmail.com

[2] Institut Supérieur des Sciences Appliquées et de Technologie de Gabès, University of Gabes, Rue Omar ibn el Khattab, 6072 Gabes, Tunisia

[3] École superieure des sciences et de la technologie de Hammam Sousse, University of Sousse, Rue Lamine Abassi, 4011 Hammam Sousse, Tunisia

Abstract. In metal forming, improving the performance of the final products involves mainly understanding, reducing and predicting damage during forming. Damage models and mechanical behavior are coupled to simulate the material's progressive softening as a result of damage increase. However, identifying damage parameters is still challenging researchers. In this paper, main damage mechanisms are presented for a Fe-12% TiB_2 Steel Matrix Composite SMC under monotonic simple shear tests. A sensitivity analysis approach is setup identify ductile damage parameters, based on experimental results for both the considered SMC and a ferritic steel representative of the composite matrix. Optimization algorithm enables the identification of a set of damage parameters that provides a suitable correlation between numerical and experimental stress-strain for both Fe-TiB_2 SMC and the ferritic unreinforced steel.

Keywords: Steel matrix composite · simple shear · damage · Parameters identification

1 Introduction

The study of the mechanical behavior of materials has always been, from an industrial point of view, necessary for the dimensioning of mechanical parts. For a long time, the study of the only reversible (and linear) behavior of a certain class of materials was the focus of the main research communities, through continuum mechanics. However, industrials rapidly became much more demanding due to increasingly complex mechanical stresses, the diversity of materials employed, the drive to lower production costs and the growing obligation to quality and safety. These naturally led researchers to take an interest in irreversible, non-linear physical mechanisms such as plasticity and strain hardening. However, these methodologies proved insufficient to describe the ultimate evolution of ductile materials, modelled as elastoplastic. Thus, a physical mechanism leading macroscopically to a decrease in material stiffness and describing

M. Chouchane et al. (Eds.): CMSM 2023, LNME, pp. 168–175, 2024.
https://doi.org/10.1007/978-3-031-65007-9_19

the loss of internal cohesion of the material was then introduced, consequently was developed the continuous damage mechanics (CDM). Damage is defined as the progressive deterioration of materials under solicitation, which leads to the volume element rupture. Damage modelling can be carried out using two different approaches: one is based on a micro-physical approach, often heavy and time consuming (Gurson 1977; Monchiet et al. 2008; Keralavarma and Benzerga 2010; Benzerga & Leblond 2010; Fritzen et al. 2012; Mohr & Marcadet 2015; Pineau et al. 2016; Chen and Dong 2009; Gaspérini et al. 2017; Park et al. 2020; He et al. 2021; Wu et al. 2022), while the other uses a macro-phenomenological approach. The latter is based on the introduction of state variables associated with the various phenomena revealed by experimentation and based on CDM models suggested by (Lemaitre and Chaboche, 1978). The damage, which is represented by an internal variable, enables adjustment of the evolution of plasticity and decay of the material's mechanical properties. The damage law serves in determining the material's remaining strength and undamaged amount. The identification of damage parameters is crucial for obtaining reliable and accurate results. The effectiveness of the optimization method chosen will determine how well the damage parameters can be assessed. Some researchers have determined the damage parameters based on the inverse method (Yue et al. 2013; Bouchard et al. 2011; Yin et al. 2015) which present a major drawback related to the difficulty in parameters identification. Another more commonly used approach, consists in a sensitivity study that describes the effect of each damage parameter on the change of the stress-strain curve, as depicted in the works of (Cherouat et al. 2018; Barriere et al. 2021; Kadhim et al. 2022). (Bouhamed et al. 2022, 2021) present a novel optimization algorithm, which is used to analytically determine Lemaître damage parameters. This method is highlighting its ability to predict the damage parameters of Lemaitre ductile damage model. For this reason, this approach is adopted in the current study. As regards materials development for automotive sector, increasingly severe European legislation regarding co_2 emissions has led to the emergence of a new generation of steels, presenting an improved specific stiffness while adding low-density and hard ceramic reinforcements in the steel matrix (Arcelor 2008). In this context, an in-situ industrially cast Steel Matrix Composite (SMC) reinforced by a 12% volume fraction of TiB_2 particle is considered. In a first part, experimental investigation involving microstructural features and mechanical behaviour under shear tests is presented for both the SMC and the unreinforced ferritic Steel. In the second part, the damage parameters identification procedure is used to determine the damage parameters for both materials and adequate set of parameters is assessed.

2 Experimental Procedure

2.1 Material and Mechanical Behavior

Two industrially cast and hot rolled materials are considered in this study: a Fe-TiB_2 SMC composite and an unreinforced ferritic steel, hereafter namely F1, with a composition that best matches that of the composite matrix, and which is particularly designed to serve as reference for the composite analysis. The materials were provided by ArcelorMittal in rolled and annealed sheet form. Microstructure of both as-received materials is shown in Fig. 1 through (RD, TD) sheet plane EBSD micrographs, with RD: Rolling direction and TD: transverse direction.

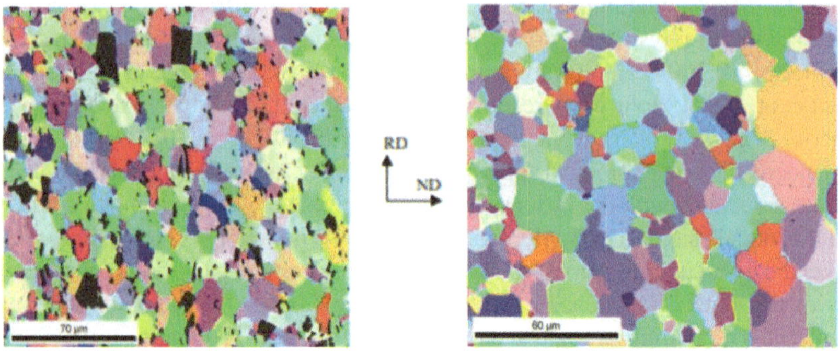

Fig. 1. Inverse pole figure map of α-Fe (IPF//RD) for (a) Fe-12% TiB_2. TiB_2 particles appear in black and (b) F1 unreinforced ferritic steel.

As the composite initial microstructure shows a clear preferred orientation of TiB_2 Particles along the rolling Direction (Fig. 1a), attention is to be paid regarding the relative global orientation of particles and mechanical loading direction. In the current study, simple monotonic shear tests are performed along the transverse direction TD, as illustrated in Fig. 2. Shear tests and microscopy conditions are detailed in (Dammak et al. 2024, 2024a, 2024b).

The TiB_2 reinforcing effect on both steel strength and strain hardening is depicted in Fig. 2 through the plot of the evolution of shear stress with plastic slip γ_p obtained under simple monotonic shear tests up to specimen macroscopic failure. This representation requires numerical processing of the (τ, γ) raw curves using MATLAB routines. The initial plastic flow stress is determined with a 0.1% Von Mises equivalent plastic strain offset.

2.2 Shear Damage Investigations

The samples used for the SEM investigations were cut from the center of the useful zone of each deformed sample for the sake of homogeneity of strain, taking care to avoid the specimen's free edges, as illustrated in Fig. 3a. A shear reference frame (SD, NS, ND) is defined (Fig. 3a), with SD: Shear Direction, NS: Normal to Shear plane, ND: Normal Direction to the sheet plane. For each deformed sample, two distinct specimens were taken (Fig. 3a) so as to examine two perpendicular sections (SD, NS) and (SD, ND), on which the morphological structure of particles are very different. Three different amount of global shear deformation are considered for SEM investigation: $\gamma = 5\%$, $\gamma = 17\%$ and $\gamma = 49\%$, as indicated on Fig. 3b.

SEM examination of Fe-TiB_2 composite microstructure after different monotonic simple shear tests shows that after $\gamma = 5\%$ (Fig. 4a), the microstructure is very similar

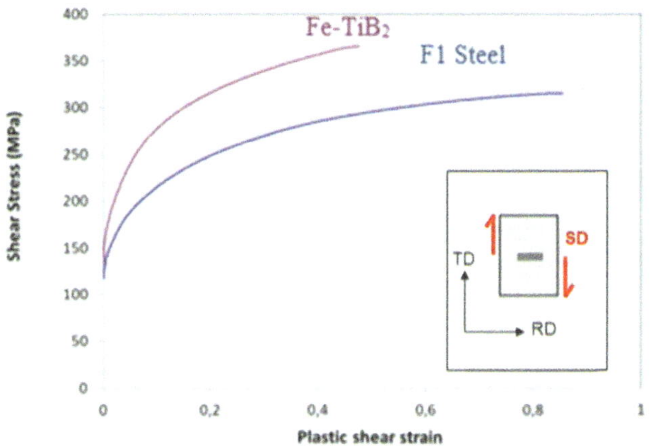

Fig. 2. Shear stress vs plastic shear strain of TD monotonic simple shear, with schematic illustration of a volume of composite material showing the Shear Direction SD with respect to the RD. The dark rectangle shows the preferred orientation of TiB2 particles.

to that of the initial state, with no significant damage marks. Significant damage manifests for $\gamma = 17\%$ (Fig. 4b) by particle ruptures, and this mechanism remains globally unchanged i.e. by rupture of particles even aflter $\gamma = 49\%$ on shear planes (Fig. 4c) or the transverse direction (Fig. 4d), without signifiant deohesions consistently with MET analysis of (Cha et al. 2012) which evidenced a very strong and clean matrix/ TiB2 interface quality.

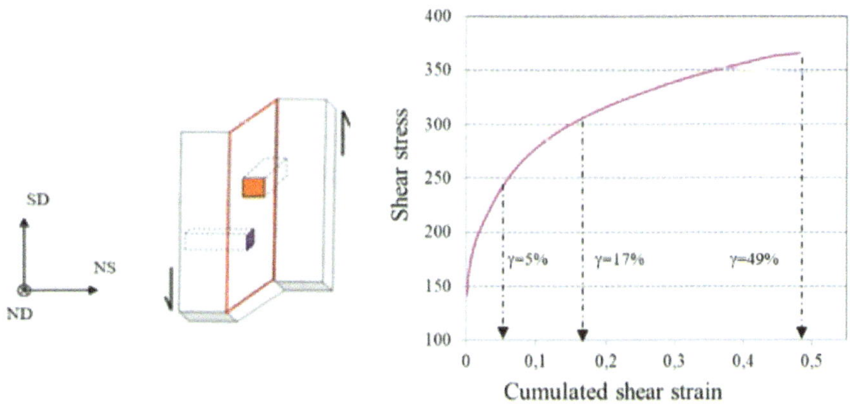

(a) (b)

Fig. 3. (a) Schematic drawing of samples taken from specimens deformed in simple shear for SEM investigations. Arrows indicate shear direction, (b) illustration on the composite overall shear curve of the deformation amount involved in macroscopic observations.

Fig. 4. Damage observation within the composite for (a) $\gamma = 5\%$, (b) $\gamma = 17\%$ and $\gamma = 49\%$ cumulated shear strain on (c) (SD, NS) sheet plane and (d) (SD, ND) sheet plane

3 Damage Parameters Identification of $FeTiB_2$ Composite

Reducing the propensity for damage to occur is a key focus of structural safety. To this end, one approach is to identify damage parameters. The majority of research reported in the literature used sensitivity analysis to determine the damage parameters by explaining how each damage parameter affected the change of the curve up to the rupture. In this particular study, an objective function, presented in the work of (Bouhamed et al. 2022, 2021), is adopted to determine the damage material parameters (S, s, Y0, β) of (Lemaitre and Chaboche, 1978) ductile damage model, which is presented in Eq. (1) as below:

$$\mathrm{Fd} = \frac{S}{(s+1)(1-d)^{\beta}} < \frac{Y - Y0}{S} >^{s+1} \tag{1}$$

d is the scalar damage variable, Y illustrate the damage force while s, S, Y0 and β depict the ductile damage parameters.

The optimization algorithm proposed in the study by (Bouhamed et al. 2021) focuses on minimizing an objective function. This objective function is responsible for computing the area enclosed by curves generated from both experimental measurements and numerical simulations. Achieving the lowest possible value of this objective function signifies the optimal set of damage parameters.

This approach generates the following parameters for both unreinforced F1 steel and Fe-TiB_2 composite, listed in the Table 1.

Table 1. Estimated damage parameters of Ferritic steel (F1) and FeTiB2 composite.

	β	s	S(MPa)	Y0 (MPa)
Ferritic steel (F1)	65	0.922	690	0
Fe-TiB_2 composite	49	0.64	510	0

Comparison between the experimental and numerical (FE) equivalent stress–equivalent plastic strain curves of the Fe-TiB_2 composite and F_1 behavior, presented in Fig. 5, highlights an appropriate match, which were established through the actual optimization process using the identified set of damage parameters given above.

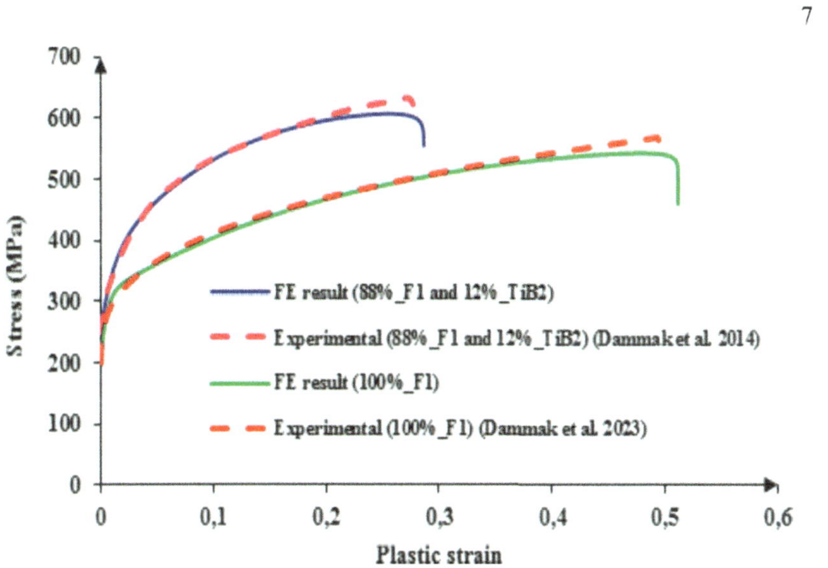

Fig. 5. Numerical and experimental equivalent Stress – Equivalent plastic strain curves of Ferritic steel (F1) and FeTiB2 composite (88%_F1 and 12%_TiB2) with the estimated damage parameters.

4 Conclusion

The study of material ductility during a forming operation remains the most important industrial problem. During these operations, the material is submitted to large plastic strain that may eventually cause ductile damage to develop and lead to final fracture. This study highlighted the statistically representative damage characteristics of an advanced industrial hot rolled Fe-12% TiB_2 SMC submitted to monotonic simple shear. This test enables to attend large plastic strains and highlight the strengthening effect of TiB_2 particles when comparing the experimental mechanical behavior of the SMC to the unreinforced steel. Furthermore, damage parameters identification using the proposed optimization algorithm is achieved and gives rise to satisfactory numerical experimental confrontations.

Acknowledgements. The authors gratefully acknowledge LSPM-CNRS laboratory for the experimental investigations and ArcelorMittal Maizière les Metz Company for providing materials.

References

ARCELOR Research Group: Patent EP 1 897 963 A1. Bulletin 2008/11, 20 p. (2008)

Barriere, T., Cherouat, A., Gabrion, X., Holopainen, S.: Short-to long-term deformation behavior, failure, and service life of amorphous polymers under cyclic torsional and multiaxial loadings. Int. J. Plast. **147**, 103106 (2021)

Benzerga, A.A., Leblond, J.B.: Ductile fracture by void growth to coalescence. Adv. Appl. Mech. **44**, 169–305 (2010)

Bouchard, P.O., Gachet, J.M., Roux, E.: Ductile damage parameters identification for cold metal forming applications. In: AIP Conference Proceedings, vol. 1353, No. 1, pp. 47–52. American Institute of Physics (2011)

Bouhamed, A., et al.: Identification of fully coupled non-associated-Ductile damage constitutive equations for thin sheet metal applications: numerical feasibility and experimental validation. Thin-Walled Struct. **176**, 109365 (2022)

Bouhamed, A., Mars, J., Jrad, H., Wali, M., Dammak, F.: Experimental and numerical methodology to characterize 5083-aluminium behavior considering non-associated plasticity model coupled with isotropic ductile damage. Int. J. Solids Struct. **229**, 111139 (2021)

Cha, L., Lartigue-Korinek, S., Walls, M., Mazerolles, L.: Interface structure and chemistry in a novel steel-based composite Fe–TiB2 obtained by eutectic solidification. Acta Mater. **60**, 6382–6389 (2012)

Chen, Z., Dong, X.: The GTN damage model based on Hill'48 anisotropic yield criterion and its application in sheet metal forming. Comput. Mater. Sci. **44**(3), 1013–1021 (2009)

Cherouat, A., Borouchaki, H., Zhang, J.: Simulation of sheet metal forming processes using a fully rheological-damage constitutive model coupling and a specific 3D remeshing method. Metals **8**(12), 991 (2018)

Dammak, M., Barbier, D., Gaspérini, M.: Kinematic strain hardening of Fe-TiB2 composite: experimental analysis and phenomenological modeling. J. Mater. Eng. Perform. **33**(9), 4359–4366 (2024)

Dammak, M., Bouhamed, A., Jrad, H., Dammak, F.: In situ experimental characterization and numerical investigation of Fe-TiB2 Steel Matrix Composite behavior considering fully coupled damage model: simulation during incremental forming process. Mater. Today Commun. **38**, 107741 (2024). https://doi.org/10.1016/j.mtcomm.2023.107741

Dammak, M., Bouhamed, A., Jrad, H.: Experimental analysis and modeling of mixed hardening of Steel Matrix Composites under monotonic and reverse simple shear loading: application to SPIF process. Mech. Adv. Mater. Struct. 1–13 (2024b). https://doi.org/10.1080/15376494.2024.2354935

Fritzen, F., Forest, S., Böhlke, T., Kondo, D., Kanit, T.: Computational homogenization of elasto-plastic porous metals. Int. J. Plast. **29**, 102–119 (2012)

Gaspérini, M., Dammak, M., Franciosi, P.: Stress estimates for particle damage in Fe-TiB2 metal matrix composites from experimental data and simulation. Eur. J. Mech. A/Solids **64**, 85–98 (2017)

Gurson, A.L.: Continuum theory of ductile rupture by void nucleation and growth: part I—yield criteria and flow rules for porous ductile media (1977)

He, Z., Zhu, H., Hu, Y.: An improved shear modified GTN model for ductile fracture of aluminium alloys under different stress states and its parameters identification. Int. J. Mech. Sci. **192**, 106081 (2021)

Ibrahim, I., Mohamed, F., Lavernia, E.: Particulate reinforced metal matrix composites—a review. J. Mater. Sci. **26**, 1137–1156 (1991). Int. J. Fatigue **48**, 39–47 (2013)

Kadhim, K.J., Jaber, J.A., Ibrihim, H.R.: Implementation of finite element analysis for solving the constraints in forming process of large steel parts. East.-Eur. J. Enterp. Technol. **118**(1) (2022)

Keralavarma, S.M., Benzerga, A.A.: A constitutive model for plastically anisotropic solids with non-spherical voids. J. Mech. Phys. Solids **58**(6), 874–901 (2010)

Lemaitre, J., Chaboche, J.L.: Aspect phénoménologique de la rupture par endommagement (in French). J. de Mecanique Appliquée **2**(3), 317–365 (1978)

Mohr, D., Marcadet, S.J.: Micromechanically-motivated phenomenological Hosford-Coulomb model for predicting ductile fracture initiation at low stress triaxialities. Int. J. Solids Struct. **67**, 40–55 (2015)

Monchiet, V., Cazacu, O., Charkaluk, E., Kondo, D.: Macroscopic yield criteria for plastic anisotropic materials containing spheroidal voids. Int. J. Plast. **24**, 1158–1189 (2008)

Park, N., Stoughton, T.B., Yoon, J.W.: A new approach for fracture prediction considering general anisotropy of metal sheets. Int. J. Plast. **124**, 199–225 (2020)

Pineau, A., Benzerga, A.A., Pardoen, T.: Failure of metals I: Brittle and ductile fracture. Acta Mater. **107**, 424–483 (2016)

Wu, H., Zhuang, X., Zhang, W., Zhao, Z.: Anisotropic ductile fracture: experiments, modeling, and numerical simulations. J. Mater. Res. Technol. **20**, 833–856 (2022)

Yin, Q., Soyarslan, C., Isik, K., Tekkaya, A.E.: A grooved in-plane torsion test for the investigation of shear fracture in sheet materials. Int. J. Solids Struct. **66**, 121–132 (2015)

Yue, Z.M., Soyarslan, C., Badreddine, H., Saanouni, K., Tekkaya, A.E.: Inverse identification of CDM model parameters for DP1000 steel sheets using a hybrid experimental-numerical methodology spanning various stress triaxiality ratios. Key Eng. Mater. **554**, 2103–2110 (2013)

Tribological Characterization of Aged Mollusc Shells-UHMWPE Biocomposites Manufactured by Hot Compression Molding Process

Besma Sidia(✉) and Walid Bensalah

Mechanical Engineering Laboratory, National Engineering School of Monastir, University of Monastir, LR99ES32 Monastir, Tunisia

sidia.besma@gmail.com

Abstract. This study examined the impact of hygrothermal aging on the tribological properties of biocomposites fabricated by hot compression molding. Mollusc shell particles (MS) are incorporated, at variable weight percentage, into a UHMWPE matrix to create biocomposites. These biocomposites then undergo accelerated hygrothermal aging by immersion in a Ringer's solution at 80 °C. The friction and wear behavior of the biocomposites was evaluated using a pin-on-disk tribometer under dry and ambient conditions. The results indicate that hygrothermal aging leads to an improvement in the tribological behavior of the biocomposites and particularly their wear resistance, which is influenced by the amount of MS particles incorporated. The study investigated the wear behavior of the aged biocomposite and proposed a corresponding wear mechanism.

Keywords: Hot compression molding · aging · MS-UHMWPE · wear resistance · Ringer's solution

1 Introduction

Ultra-high molecular weight polyethylene (UHMWPE) is a remarkable polymer prized for its outstanding combination of mechanical and tribological properties. It boasts exceptional wear and impact resistance, exhibits self-lubricating qualities, resists chemical attack, and demonstrates biocompatibility with the human body [1, 2]. UHMWPE is therefore frequently used in commercial and medical uses [3]. For the replacement of the hip joint, it is planned to use a total hip prosthesis made of a UHMWPE polymer as an acetabular cup articulated against a stainless-steel metal femoral head [2]. Despite a history of relative success, the UHMWPE component degrades due to wear in total hip joint replacements [2]. The constant motion within a joint subjects UHMWPE to abrasive, fatigue, and adhesive wear mechanisms, leading to the generation of wear particles [4, 5]. For that, numerous studies are focusing on the fillers added to the UHMWPE matrix to further improve the wear behavior of UHMWPE. For instance, to improve various properties of polymers, researchers have incorporated diverse fillers, including natural materials like natural coral [3], hydroxyapatite [4], zeolite [6], graphene oxide [6], carbon fiber [7], and mollusk shell (MS) [7], among others.

M. Chouchane et al. (Eds.): CMSM 2023, LNME, pp. 176–185, 2024.
https://doi.org/10.1007/978-3-031-65007-9_20

However, aging is another major cause of deterioration of the UHMWPE prosthetic component in the human body. In order to simulate natural aging and "in vivo" circumstances, the tribological behavior of biocomposites subjected to rapid aging can be studied. Hygrothermal aging experiments on graphene oxide-UHMWPE composites were conducted by Belotti et al. [8] in purified water at room temperature and below 80 °C for 20 days. From the results, compounds that cured at room temperature absorbed less than 0.1% of water. The limited solubility and affinity of UHMWPE for water was considered the cause of this low uptake. Drying at 80 °C appears to induce a long-term water uptake effect in the composites [8]. Fouad et al. [9] reported a notable decline in the mechanical properties of UHMWPE, including elastic modulus, yield strength, fracture stress, and strain at break, attributed to natural aging. Mohamed et al. [10] also observed a decrease in the nanomechanical properties of UHMWPE, such as elastic modulus and hardness, as a result of thermal and serum aging, and an increase in these properties with UV irradiation. These findings collectively underscore the need for careful consideration of the aging process in the design and use of UHMWPE-based materials.

This study explored how hygrothermal aging affects the tribological properties of biocomposites made from UHMWPE reinforced with varying weight fractions of mollusc shell (MS) particles. The biocomposites were fabricated using a heated compression molding technique. A Ringer's solution was used to age the MS-UHMWPE biocomposite samples at 80 °C. Wear measurements were then performed using a pin-on-disk tribometer. In this study, the friction coefficient and wear behavior of aged and unaged MS-UHMWPE biocomposites were evaluated as they rubbed against stainless steel pins. Morphological studies were performed on the aged MS-UHMWPE biocomposite disk. The wear morphology of aged biocomposites will be addressed.

2 Experimental and Test Details

2.1 Material Preparation

To perform the wear tests, a hemispherical M30NW pin was used that imitates the femoral head of a prosthesis. The pin was supplied in a mirror state by C2T implant in Tunisia, with a diameter of 10 mm, as shown in Fig. 1.a.

To reinforce the UHMWPE matrix, mollusc shells (MS) were collected directly from a Tunisian Mediterranean beach. Prior to use, the MS underwent a cleaning process involving rinsing with deionized water and immersion in alcohol for one hour to eliminate any soluble organics and adherent proteins. Following the cleaning process, the MS was dried in an oven at 50 °C for four hours. It was then pulverized into a fine powder using a blade mill, achieving an average particle size of 51 μm.

The resulting MS particles were chosen as the reinforcing material for the UHMWPE matrix, where GUR 4113 from Celanese (Germany) was used as the UHMWPE with an average molecular weight of 3.9 Mio g/mol (as mentioned in the data sheet of Celanese's GUR® 4113 UHMWPE).

2.2 Process for Preparing MS-UHMWPE Composites

The hot compression molding process was used to fabricate MS-UHMWPE composites (Fig. 2).

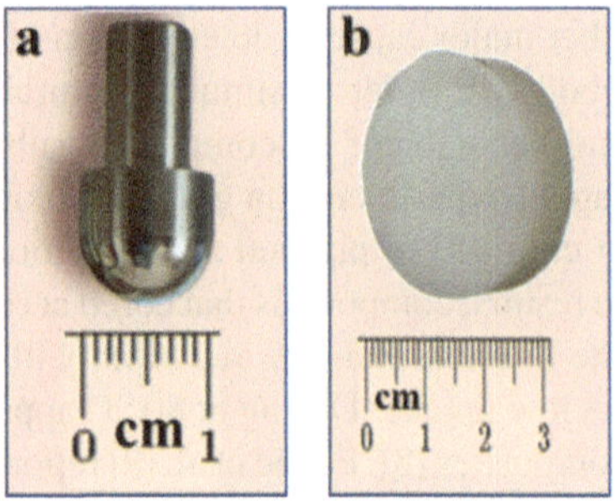

Fig. 1. (a) M30NW pin, (b) Biocomposite disk.

For this aim, MS and UHMWPE powders were prepared with suitable mixing ratios. Different weight percentage of MS were used namely: 0, 5, 10, 15 and 25 wt. %. Then, to achieve a uniform distribution of MS particles throughout the UHMWPE matrix, the powders were ball-milled at a speed of 120 m per minute for 20 min. The mixed powder is then placed layer by layer packed into a mold with cylindrical cavities to give the desired shape for the final part.

Then, the mold containing the powder mixture is heated to a temperature above the melting temperature of the UHMWPE (200 °C), but below the decomposition temperature of the MS. This temperature is reached using a JOOS type hot press.

Molding lasts 4 h at an average pressure of 18 MPa. During the compression, the UHMWPE particles melt and bond with the MS particles, forming a solid, dense part.

An optimization of the compression molding parameters was performed in order to obtain a solid final part without molding defects, as explained in our previous works [11, 12]. After compression molding, the resulting composite disc is cooled slowly in the mold to avoid deformation due to rapid cooling. The cylindrical disks had a radius of 30 mm and a height of 8 mm, which simulates the acetabular cup of the prosthesis used in this study. The biocomposite surfaces were polished with abrasive papers ranging from P180 to P4000 prior to wear testing. The biocomposite surfaces exhibited a final average roughness (Ra) of 0.18 μm, with a standard deviation of 0.04 μm.

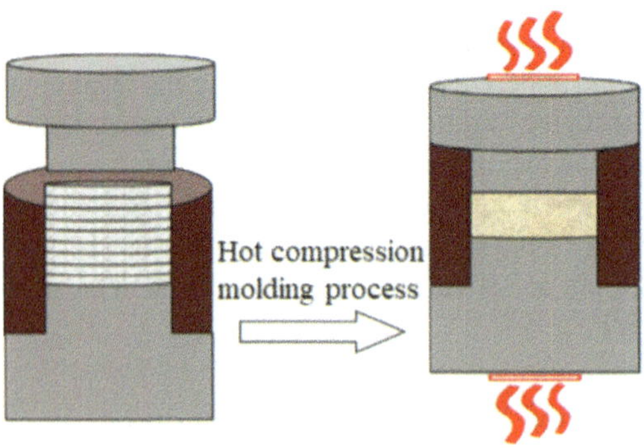

Fig. 2. Hot compression molding process.

2.3 Hygrothermal Aging Process

Accelerated hygrothermal aging was conducted on both unfilled UHMWPE and MS-UHMWPE samples. The aging process followed the guidelines of F2003-02 standard [13], where the discs underwent a 21-day immersion in Ringer's solution at 80 °C the chemical composition of the Ringer's solution is shown in Table 1.

Table 1. Chemical composition of Ringer's solution [8]

Chemical composition [g.L-1]	KCl	NaHCO3	NaCl	CaCl2
	4.30	0.20	9	0.24

2.4 Wear Test

Unaged and aged biocomposites were tested for wear using a reciprocating pin-on-disk tribometer (Fig. 3). The tribometer setup utilized an M30NW pin sliding linearly against a disc-shaped MS-UHMWPE biocomposite sample. The sliding conditions mimicked those of a total hip prosthesis, with a stroke length of 30 mm per cycle (corresponding to a sliding speed of 30 mm/s) and a frequency of 1 Hz. This matches the average walking frequency of a human (1 Hz), while the sliding velocity falls within the range experienced by a total hip prosthesis (0–50 mm/s) [9, 10]. To assess wear behavior, the biocomposite samples underwent 15,000 sliding cycles, corresponding to a cumulative wear distance of 450 m. After tribological testing, a 2D profilometer, the Taylor Hobson Surtronic 116, was used to quantify the wear track profiles on both aged and unaged discs. The data acquisition process employed the Gold Talyprofile software. The specific wear rate was then calculated using the following equation [10]:

$$K = \frac{\Delta V}{F_N \times L} \tag{1}$$

The equation uses the following symbols: F_N (N) for normal load, L (m) for total sliding distance, and ΔV (mm^3) for average volumetric wear.

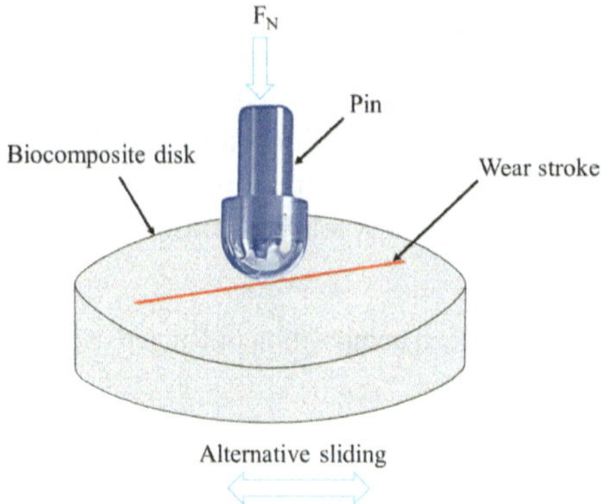

Fig. 3. Pin-on-disk tribometer.

3 Results and Discussion

3.1 Friction Coefficient Behavior

The influence of hygrothermal aging on the frictional behavior of MS-UHMWPE biocomposites is illustrated in Fig. 4. The graph reveals that, in general, the coefficient of friction trajectories of both aged and unaged biocomposites are similar. For unfilled UHMWPE, hygrothermal aging in Ringer's solution has no influence on the coefficient of friction, which remains close to 0.13 regardless of whether the material is aged or unaged. These findings align with those of previous studies conducted by other researchers [7, 13, 14]. In contrast, the coefficient of friction of the various biocomposites in Ringer's solution decreased following hygrothermal aging, as demonstrated in Fig. 4. For example, the aged 5 wt. % of MS composite exhibited a coefficient of friction of 0.07, which was half that of its unaged counterpart (0.13). Indeed, the release of water molecules from the Ringer's solution during friction would be responsible for this phenomenon, as the water molecules act as a lubricant and are discharged at the sliding-interface due to the pressure exerted by the pin. This outcome aligns with findings in existing literature, specifically in the context of reinforcing UHMWPE with short carbon fiber [8].

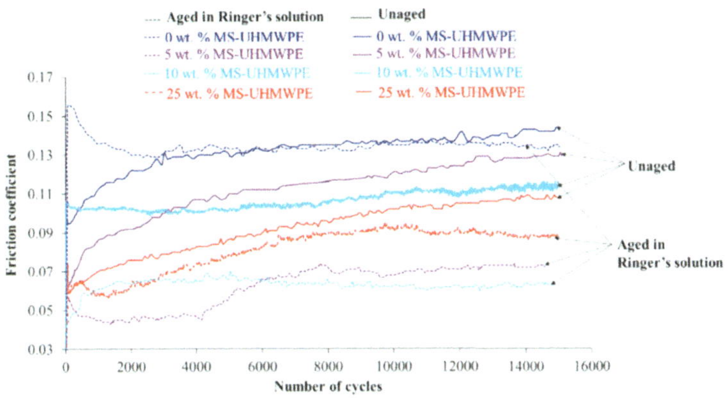

Fig. 4. Variation of the coefficient of friction over the number of cycles for MS-UHMWPE biocomposites rubbing against dry stainless steel, with and without hygrothermal aging.

3.2 Wear Measurement

The influence of biocomposite composition (aged and unaged with varying MS content) on specific wear rate is illustrated in Fig. 5.

Hygrothermal aging in Ringer's solution significantly reduced the specific wear rate of unfilled UHMWPE by 20% compared to pure unaged UHMWPE, as shown in Fig. 5. This observation aligns with previous research findings [11–14]. Furthermore, Fig. 5 also demonstrates that incorporating mollusc shell (MS) particles into UHMWPE further improves the wear behavior after hygrothermal aging. This improvement increases with the weight percentage of MS. Indeed, for 25 wt. % of MS, the specific wear rate of the unaged composite is 25 x 10–6 mm3/N.m and decreases with hygrothermal aging to reach 10 x 10–6 mm3/N.m corresponding to 60% reduction. Moreover, 5 wt.% MS presents the lowest specific wear rate.

The observed results may be related to the lubricating effect of water molecules released from Ringer's solution in the contact. Moreover, the waterlogged MS particles become softer and will be compacted easily by the pin in the wear trace [15]. The generated debris will play, thereafter, the role of lubricant.

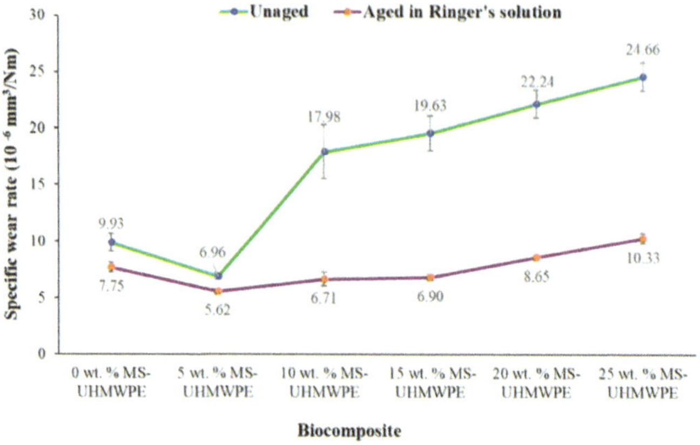

Fig. 5. Effect of MS content (wt.%) on the specific wear rate of aged and unaged biocomposites.

4 Wear Morphology

Figure 6 illustrates wear morphological of the wear stroke of aged unfilled UHMWPE disk after 15000 cycles of sliding against stainless-steel pin in a dry state. As shown in Fig. 6, unfilled UHMWPE wear (0 wt.% MS-UHMWPE), parallel plowing with scratches and grooves was observed in the sliding direction.

Moreover, examination of the wear track revealed detached UHMWPE debris in the form of small crushed. This phenomenon can be attributed to the mechanical interaction between the asperities (rough peaks) of the harder stainless steel counterface and the softer UHMWPE surface, leading to the breakage and fragmentation of UHMWPE polymer chains.

As an instance, Fig. 7 illustrates wear stroke morphology of aged 5 wt.% of MS-UHMWPE biocomposite.

The analysis of the wear morphology on the aged 5wt. % MS-UHMWPE (as depicted in the Fig. 7 demonstrates the existence of porosity and small pores within the worn surface. It is likely that the formation of these micropores occurred as a result of the separation of the charge-matrix interface. This observation aligns with the findings reported by Alothman et al. [16], who also noted a similar occurrence of aged hydroxyapatite-high density polyethylene composite.

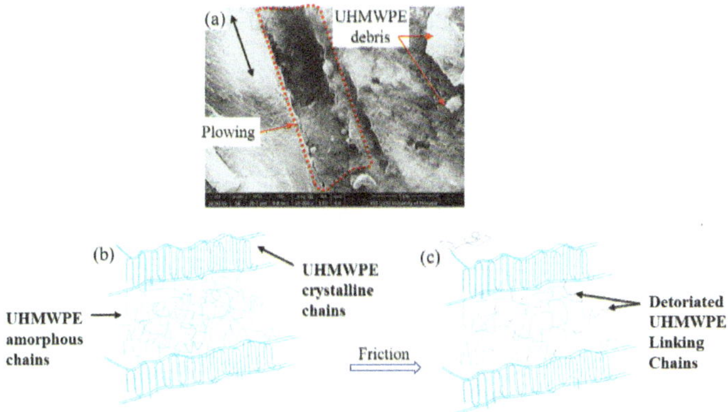

Fig. 6. Wear stroke morphology of aged unfilled UHMWPE.

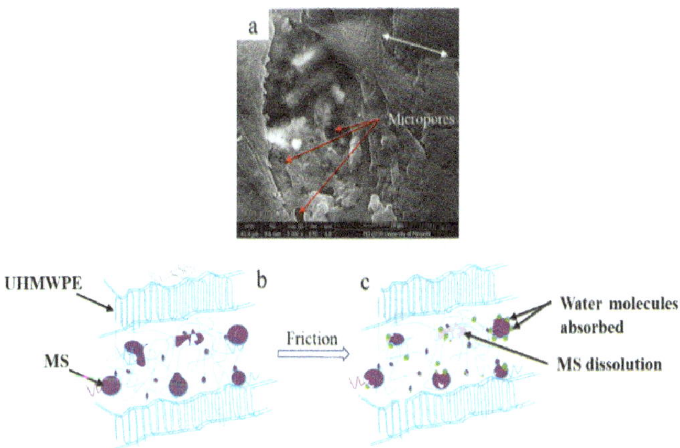

Fig. 7. Wear stroke morphology of aged 5 wt.% MS-UHMWPE biocomposite.

5 Conclusion

This study explored the effects of hygrothermal aging in Ringer's solution at 80 °C on the wear and friction characteristics (tribological behavior) of MS-UHMWPE biocomposites. With different weight percentages, the composites were manufactured using hot compression molding. The results of hygrothermal aging experiments indicate that aging influenced the tribological properties of the biocomposites. Hygrothermal aging resulted in beneficial changes to the biocomposites, lowering their coefficient of friction and specific wear rate. Indeed, a 60% reduction in specific wear rate was observed with the addition of 25 wt. % of MS-UHMWPE. This improvement can be attributed to the effect of water absorption at the filler-matrix interface in the biocomposite. In conclusion, the studied composites showed interesting tribological performances, friction and wear behaviors, under hygrothermal aging. The findings suggested that the optimal content of MS was around 5 wt. %. These biocomposites exhibit promising potential for *in vivo* applications.

Acknowledgment. The authors gratefully acknowledge the support of the Ministry of Higher Education and Scientific Research of Tunisia and the University of Monastir (LGM: LAB-MA-05). We would also like to extend special thanks to Mr. Didier Voillemin, manager of C2T Implants, for his invaluable cooperation throughout this study.

References

1. Moreno, S.S.: Carbon reinforced UHMWPE composites for orthopaedic applications: characterization and biological response to wear particles. Luleal University of Technology, Sweden (2013). http://pure.ltu.se/portal/files/41870673/Silvia_Su_er.Komplett.pdf
2. Shanmugam, L., Kazemi, M.E., Qiu, C., Rui, M., Yang, L., Yang, J.: Influence of UHMWPE fiber and Ti6Al4V metal surface treatments on the low-velocity impact behavior of thermoplastic fiber metal laminates. Adv. Compos. Hybrid Mater. **3**, 508–521 (2020). https://doi.org/10.1007/s42114-020-00189-7
3. Diabb Zavala, J.M., et al.: Manufacture and mechanical properties of knee implants using SWCNTs/UHMWPE composites. J. Mech. Behav. Biomed. Mater. **120**, 104554 (2021). https://doi.org/10.1016/j.jmbbm.2021.104554
4. Ming Song, S.T., Ashkanfar, A., English, R., Rothwell, G.: Computational method for bearing surface wear prediction in total hip replacements. J. Mech. Behav. Biomed. Mater. **119**, 104507 (2021). https://doi.org/10.1016/j.jmbbm.2021.104507
5. Longhofer, L.K., Chong, A., Strong, N.M., Wooley, P.H., Yang, S.Y.: Specific material effects of wear-particle-induced inflammation and osteolysis at the bone–implant interface: a rat model. J. Orthop. Transl. **8**, 5–11 (2017). https://doi.org/10.1016/j.jot.2016.06.026
6. Liu, C.Y., Ishigami, A., Kurose, T., Ito, H.: Wear resistance of graphene reinforced ultra-high molecular weight polyethylene nanocomposites prepared by octa-screw extrusion process. Compos. Part B Eng. **215**, 108810 (2021). https://doi.org/10.1016/j.compositesb.2021.108810
7. Chen, X., Zhang, S., Zhang, L., Zhu, P., Zhang, G.: Design and characterization of the surface porous uhmwpe composite reinforced by graphene oxide. Polymers (Basel) **13**, 1–14 (2021). https://doi.org/10.3390/polym13040482
8. Belotti, L.P., Vadivel, H.S., Emami, N.: Tribological performance of hygrothermally aged UHMWPE hybrid composites. Tribol. Int. **138**, 150–156 (2019). https://doi.org/10.1016/j.triboint.2019.05.034
9. Fouad, H., Elleithy, R., Alothman, O.Y.: Thermo-mechanical, wear and fracture behavior of high-density polyethylene/hydroxyapatite nano composite for biomedical applications: effect of accelerated ageing. J. Mater. Sci. Technol. **29**, 573–581 (2013). https://doi.org/10.1016/j.jmst.2013.03.020
10. Mohamed, H.F., Mourad, A.H.I., Barton, D.C.: UV irradiation and aging effects on nanoscale mechanical properties of ultra high molecular weight polyethylene for biomedical implants. Plast. Rubber Compos. **37**, 346–352 (2008). https://doi.org/10.1179/174328908X314370
11. Sidia, B., Bensalah, W.: Tribological properties of High Density Polyethylene based composite: the effect of mollusc shell particles under dry condition. J. Compos. Mater. **55**, 1119–1130 (2021). https://doi.org/10.1177/0021998320965655
12. Sidia, B., Bensalah, W.: Tribological and mechanical behaviors of mollusk shell-ultrahigh molecular weight polyethylene bio-composite. Polym. Compos. **42**, 3206–3215 (2021). https://doi.org/10.1002/pc.26051

13. F2003: Standard practice for accelerated aging of ultra-high molecular weight polyethylene after gamma irradiation in air. **02**, 2008–2011 (2010). https://doi.org/10.1520/F2003-02R15. Copyright
14. Bárány, T., Földes, E., Czigány, T.: Effect of thermal and hygrothermal aging on the plane stress fracture toughness of poly(ethylene terephthalate) sheets. Express Polym Lett **1**, 180–187 (2007). https://doi.org/10.3144/expresspolymlett.2007.28
15. Alothman, O.Y., Almajhdi, F.N., Fouad, H.: Effect of gamma radiation and accelerated aging on the mechanical and thermal behavior of HDPE/HA nano-composites for bone tissue regeneration. Biomed. Eng. Online **12**, 1–15 (2013). https://doi.org/10.1186/1475-925X-12-95

Investigating Mechanical Attributes of Biomaterials for Bone Tissue Engineering Using Numerical Methodology

Monia Trimeche[1(✉)], Hichem Smaoui[2], and Ridha Ben Cheikh[1]

[1] Laboratory of Materials, Optimization, and Energy for Sustainability (LAMOED), Department of Industrial Engineering, Tunis El Manar University, National School of Engineers of Tunis, Tunis, Tunisia
monia_trimeche@yahoo.fr

[2] College of Engineering Prince Sattam bin Abdulaziz University, Al-Kharj, Kingdom of Saudi Arabia

Abstract. The individual use of polymers and ceramics has not successfully reproduced the intricate properties of bone. Yet, their combination within a composite material exhibits the potential for enhanced mechanical and biological performance, yielding a more faithful emulation of bone attributes. This research endeavors to appraise the advantageous mechanical traits inherent in a degradable bone implant composite, comprising β-TCP/DCPD and PHBV biomaterials. The assessment employs a numerical approach to scrutinize the composite's mechanical prowess.

The numerical investigation of the (β-TCP/DCPD)-PHBV composite reveals a mechanical response within the matrix/reinforcement interface region strikingly reminiscent of genuine bone behavior. This likeness is substantiated by the distribution of Von Mises stress, strain, and Young's modulus.

Keywords: Bone implant · calcium phosphate ceramic-biopolymer · mechanical properties · compressive strength · Young's Modulus · 3D model · numerical simulation

1 Introduction

In the realm of healthcare, a diverse array of bioresorbable and biodegradable bone substitutes is essential to address the pressing demand for effective solutions to bone trauma. To enhance the capabilities of existing materials, rigorous research endeavors are being undertaken to explore their versatility and rejuvenate their efficacy. The application of biodegradable biomaterials is steadily on the rise for the repair of severe bone defects, employing scaffolds crafted from these materials that offer a temporary structure for the growth of new bone tissue, subsequently breaking down and assimilating into the organism [1, 2].

A focal point of extensive investigation lies in the realm of "ceramic-biopolymer" composite biomaterials, intended for bone repair. Additionally, there is a burgeoning

M. Chouchane et al. (Eds.): CMSM 2023, LNME, pp. 186–194, 2024.
https://doi.org/10.1007/978-3-031-65007-9_21

interest in the development of multi-component scaffolds that can be tailored to exhibit specific properties [3–6]. Surgical implantation of temporary bone substitutes often entails significant mechanical stress. As a result, the mechanical attributes of such substitutes must be sufficiently robust to overcome any hindrances that could impede their effectiveness [7–9].

The biobased polymer PHBV holds great promise for tissue engineering applications due to its remarkable qualities, including its natural origin and biocompatibility. Diverging from more common biodegradable polymers such as PGA, PLGA, or PLA, PHBV boasts a degradation rate conducive to bone growth and maintains a stable local pH during degradation [10]. Nevertheless, PHBV does exhibit comparatively lower strength and stiffness when contrasted with other biodegradable materials that possess superior mechanical properties, a characteristic that presents both an advantage and a limitation [11]. Notably, the incorporation of PHBV has demonstrated favorable effects on the ductility, dimensional stability, and cellular proliferation rate in the majority of biocomposites it is added to [12]. However, its use is constrained by its limited strength and, particularly, its absence of bioactivity [13]. Its clinical applications span cardiovascular stents, absorbable surgical sutures, medical packaging, and porous scaffolds for treating certain bone defects arising from disease or trauma where conventional therapies fall short [14–16]. In the context of potential applications, numerous studies have showcased that introducing bioceramics into the PHBV polymer matrix enhances the resulting biomaterial's bioactivity, degradation rate, and mechanical performance [17].

In the past two decades, the utilization of bioresorbable ceramics has gained substantial traction in the field of osseous tissue repair. Hydroxyapatite (HA), tricalcium phosphate, dicalcium phosphates, ceramic phosphocalcium powder (CPC), and bioglass have found extensive biomedical applications due to their biocompatibility, corrosion resistance, and biological activity [13]. Nonetheless, these materials are not without limitations, featuring pronounced rigidity, fragility, and susceptibility to rupture.

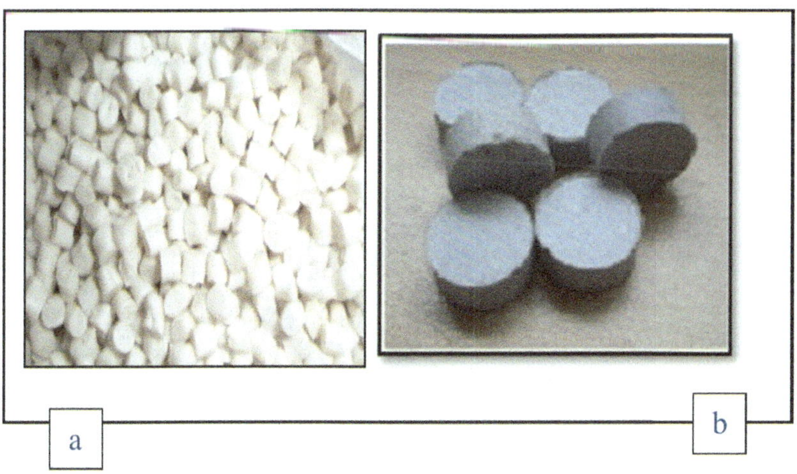

Fig. 1. (a) PHBV Granulates; (b) (β-TCP/DCPD)-PHBV biocomposite

The biocomposite (β-TCP/DCPD-PHBV), depicted in Fig. 1(b), was developed at the Laboratory of Materials, Optimization, and Energy for Durability (LAMOED), comprising a blend of calcium phosphate ceramic (β-TCP/DCPD) and polyhydroxybutyrate-co-hydroxyvalerate (PHBV) copolymer, as visualized in Fig. 1(a). This specialized bone substitute is meticulously engineered to possess pivotal mechanical properties, aptly guiding the tissue regeneration process while offering ample resistance to support neoformed bone.

This study aims to investigate the mechanical properties of the (β-TCP/DCPD)-PHBV composite, including modulus of elasticity, yield strength, and compressive strength, which could be measured by numerical testing.

2 Materials and Methods

2.1 Chosen Model

To create a geometric model that most closely represents the biomaterial, the researchers firstly consider the PHBV matrix as a parallelized (Fig. 2, a, b), in which the β-TCP/DCPD reinforcement was incorporated in a hexagonal shape of different volumes (Fig. 2, c). The β-TCP/DCPD granules are randomly dispersed in the PHBV matrix.

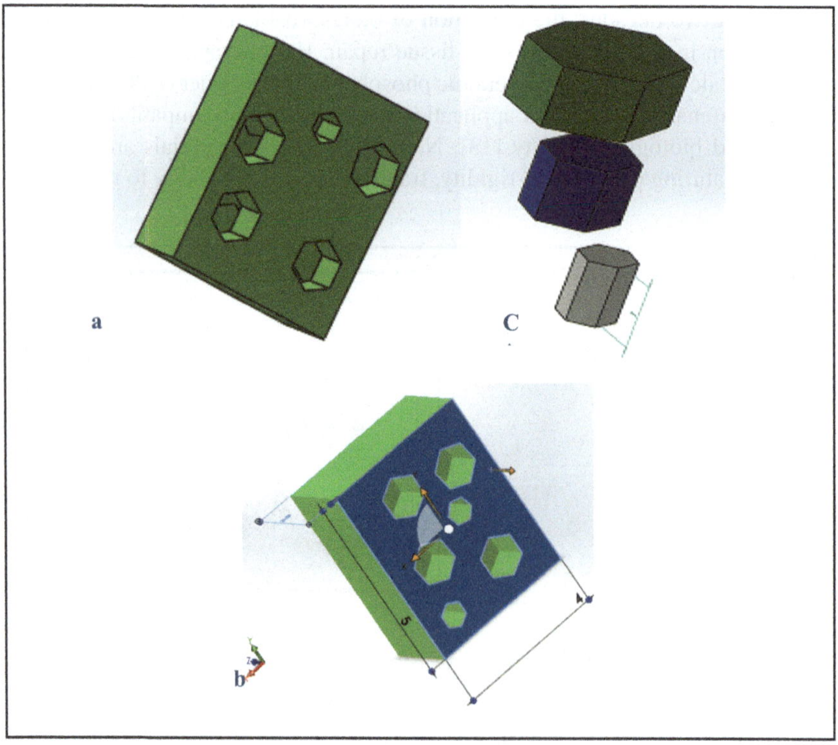

Fig. 2. (a and b) Matrix faces of β-TCP/DCPD-PHBV biocomposite, (c) Ceramic particles

The weight fractions of the composite conditioned the number and size of the granules. Table 1 groups the different dimensional specificities of the biocomposite, while, Table 2 presents the mechanical characteristics of PHBV and β-TCP/DCPD required by this numerical analysis.

Table 1. Dimensional characteristics of the designed model

Component	Shape	Size (mm)	Thickness (mm)
Matrix	Parallelepiped	3 X 4	2
Reinforcement	Helical	D = 0.6;1; 1.6	1

Table 2. Physical properties of β -TCP / DCPD and PHBV [17]

	Poisson's ratio	Young's modulus (MPa)	Density (g/cm^3)
PHBV	0.27–0.3	971	1.2
β-TCP/ DCPD	0.3	3337	2.689

Overall, tetrahedral meshes are easier to generate and remesh than hexahedral meshes. Given the relatively simple nature of the model under consideration, it is preferable to use a tetrahedral mesh. To strike a balance between computational efficiency and the accuracy of the results, a mesh with a reasonable number of elements was generated (Fig. 4, a). An initial evaluation of the model was then carried out. If the results did not meet expectations, the model was remeshed and re-evaluated. Finally, efforts were made to ensure that the results achieved satisfactory convergence.

We deliberately opted for full integration of the PHBV matrix with the β-TCP/DCPD reinforcement. This is supported by the obvious close contact between the PHBV matrix and the β-TCP/DCPD reinforcement observed in the composite material, as illustrated in Fig. 3. The mechanical characteristics analyzed by the simulation are the following: Von Mises Stresses, displacements, strains and Young's modulus. These mechanical quantities are then put in perspective with those obtained through experimental tests.

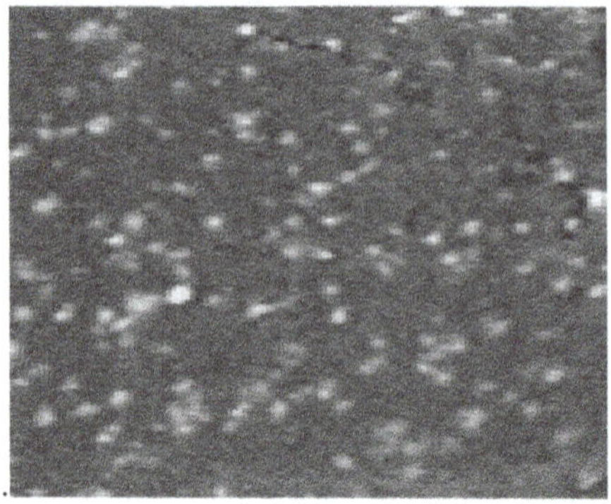

Fig. 3. SEM observation showed the dispersion and the inclusion of β -TCP/DCPD granule in the PHBV matrix, (100X).

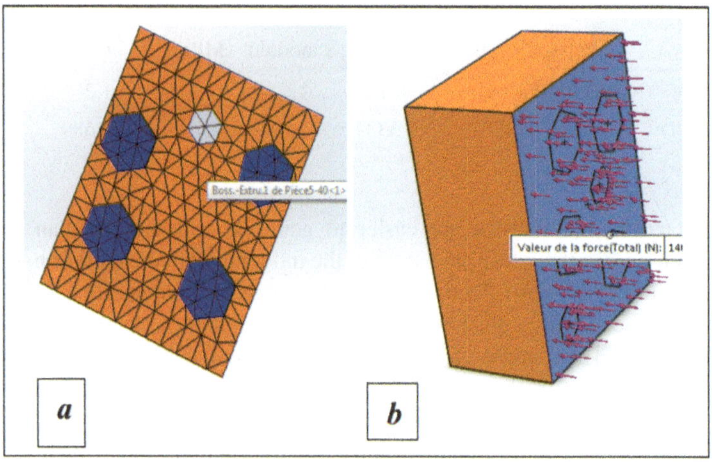

Fig. 4. (a) Composite mesh, (b) Force applied normally to the superior base (pink arrows), model (40/60)

2.2 Conditions in the Limits

The upper base of the biomaterial (Fig. 4a, indicated by pink arrows) is subjected to a vertical load, while the lower part is fixed to an immobile platform.

3 Results

3.1 Constraint of Von Mises

The results of the static analysis of the compressive mechanical behavior indicate that the maximum stress observed is 484 MPa, as shown in Fig. 5a. To avoid any potential damage or failure, the stress is limited to a maximum value of 11.7 MPa. Additionally, it was observed that the stress experienced at the interface between the matrix and the reinforcement is lower compared to that observed at the β-TCP/DCPD ceramic.

3.2 Displacement and Deformation

From Fig. (5b), it is evident that the simulated block boundary undergoes the greatest displacement when subjected to compression, while the β-TCP/DCPD granules show a comparatively lower response. This is consistent with the observation of minimal deformation at these boundaries, as shown in Fig. (5c).

3.3 Young's Modulus (MPa)

Based on the simulation results, Young's modulus can be categorized into three distinct levels, with the blue line representing PHBV, the green line representing the matrix-reinforcement interface, and the red line representing the β-TCP/DCPD biomaterial, as illustrated in Fig. (5a).

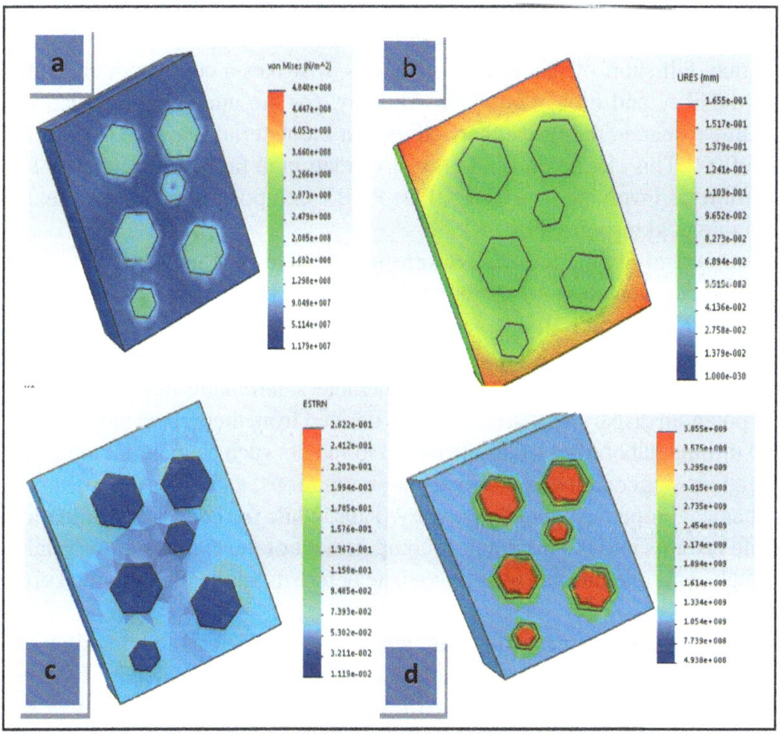

Fig. 5. (a) Von Mises constraint and its distribution in the composite (40/60), (b): Displacement in 40/60, (c) Deformation (40/60), (d) Young's Modulus 40/60 (N/m^2)

4 Discussion

Calcium phosphate (CaP) bioceramics have gained popularity as the preferred choice for bone substitutes due to their attractive attributes, including bioactivity, osteoconductivity, injectability, hardening induced by a low-temperature setting reaction and the ability to be shaped. However, CaP has brittleness, limited compressive strength and limited degradability, requiring it to be combined with certain materials to produce substitutes with suitable characteristics. As a result, a multitude of research projects have been undertaken to formulate bioresorbable polymer-ceramic composites aimed at increasing the effectiveness of the resulting substitute [18].

Poly (3-hydroxybutyrate-co-3-hydroxyvalerate), or PHBV, a polymer, has been combined with diverse ceramics to yield bio-composites with enhanced mechanical attributes. For instance, M. Zhou integrated bioglass (BG) into PHBV scaffolds (PHBV scaffolds with 10% BG (w/w)), resulting in heightened hydrophilicity and mechanical robustness, as indicated in their research [20].

Within this study, the fusion of β-TCP/DCPD (phosphocalcic cement) with PHBV polymer was orchestrated to amplify its mechanical performance [17, 18]. Following the comprehensive assessment of mechanical and biological properties across varied β-TCP/DCPD composite ratios, the researchers identified the 40% β-TCP/DCPD composite as the most optimal convergence of both constituents [17]. While its mechanical strength may fall short of that of pure ceramics, it strikes a commendable equilibrium between ductility and bioactivity. Consequently, for the numerical simulation undertaken in this research, the researchers adopted a biomaterial featuring a β-TCP/DCPD fraction of 0.4. This choice finds additional validation in favorable outcomes from animal experiments involving (β-TCP/DCPD)-PHBV composites, which have exhibited a positive biological response [18, 19].

The numerical simulation undertaken herein yields a significant advantage over laboratory experimentation by enabling the creation of a virtual prototype of the β-TCP/DCPD-PHBV material, amenable to various numerical analyses. As depicted in Fig. 5.a, the maximum Von Mises stress approximates 484 MPa, closely aligning with the experimental value of 494 MPa [17]. Nonetheless, it remains imperative to acknowledge the potential disparities between results derived from theoretical methods and those obtained through laboratory trials, attributed to factors such as material microstructure, simplifications in geometric reinforcement assumptions, and other parameters impacting mechanical properties, notably porosity. Thus, while the numerical simulation offers invaluable insights into the mechanical comportment of the composite material, validation through experimental data is a prerequisite before its application in actual orthopedic contexts.

These numerical findings affirm the researchers' selected model and attest to the relatively modest interface between reinforcement and matrix, evidenced by lower maximum Von Mises stress values (ranging from 125 MPa to 248 MPa) in Fig. 5.a, akin to values observed in human cortical bone (ranging from 90 MPa to 230 MPa). Further, Fig. 5.d unveils that these same regions exhibit a lower Young's modulus in contrast to β-TCP/DCPD, indicating diminished rigidity. Conversely, Figs. 5.b and 5.c illustrate more pronounced displacement and deformation within the PHBV matrix, underscoring the necessity for composite homogenization.

By reducing the particle size of β-TCP/DCPD, the specific surface area can be increased, facilitating the interaction between the biomaterial and the biological environment. This, in turn, could lead to increased bioactivity, cell adhesion and ultimately advanced tissue regeneration. In addition, reducing particle size could also enhance mechanical properties, through more homogeneous integration into the matrix, promoting better load transfer and mitigating stress concentrations. However, it is important to recognise that reducing particle size can lead to higher production costs as additional processing steps are required, which must be duly taken into account.

5 Conclusion

The numerical simulation further confirmed the mechanical similarity of the composite to bone, particularly at the interface between the matrix and reinforcement. However, future simulations should consider additional parameters such as the porosity and dimensional characteristics of the β-TCP/DCPD to improve the convergence between the developed biomaterial and the simulation model. These findings have promising implications for the development of biomaterials with improved mechanical and biological properties for bone substitutes.

References

1. O'brien, F.J.: Biomaterials & scaffolds for tissue engineering. Mater. Today **14**(3), 88–95 (2011)
2. Wei, S., Ma, J.X., Xu, L., et al.: Biodegradable materials for bone defect repair. Military Med. Res. **7**, 54 (2020). https://doi.org/10.1186/s40779-020-00280-65
3. Shuai, C., et al.: NMgO-incorporated PLLA bone scaffolds: enhanced crystallinity and neutralized acidic products. Mater. Des. **174**, 107801 (2019). https://doi.org/10.1016/j.matdes.2019.107801
4. Feng, P., et al.: Characterizations and interfacial reinforcement mechanisms of the multicomponent biopolymer-based scaffold. Mater. Sci. Eng. C **100**, 809–825 (2019). https://doi.org/10.1016/j.msec.2019.03.030
5. Shuai, C., et al.: Antibacterial polymer scaffold based on mesoporous bioactive glass loaded with in situ grown silver. Chem. Eng. J. **374**, 304–315 (2019). https://doi.org/10.1016/j.cej.2019.03.273
6. Feng, P., et al.: A multi-material scaffold with tunable properties: toward bone tissue repair. Adv. Sci. **5**, 1700817 (2018). https://doi.org/10.1002/advs.201700817
7. Donate, R., Monzón, M., Alemán-Domínguez, M.E.: Additive manufacturing of PLA-based scaffolds intended for bone regeneration and strategies to improve their biological properties. e-Polymers **20**(1), 571–599 (2020). https://doi.org/10.1515/epoly-2020-0046
8. Morgan, E.F., Unnikrisnan, G.U., Hussein, A.I.: Bone mechanical properties in healthy and diseased states. Annu. Rev. Biomed. Eng. **20**(1), 119–143 (2018). https://doi.org/10.1146/annurev-bioeng-062117-121139
9. Chanlalit, C., Shukla, D.R., Fitzsimmons, J.S., An, K.N., O'Driscoll, S.W.: Stress shielding around radial head prostheses. J. Hand Surg. **37**(10), 2118–2125 (2012)
10. Reddy, C.S.K., Ghai, R., Kalia, V.: Polyhydroxyalkanoates: an overview. Bioresour. Technol. **87**(2), 137–146 (2003)
11. Goonoo, N., Bhaw-Luximon, A., Passanha, P., Esteves, S.R., Jhurry, D.: Third generation poly (hydroxyacid) composite scaffolds for tissue engineering. J. Biomed. Mater. Res. B Appl. Biomater. **105**(6), 1667–1684 (2017)
12. Trimeche, M.: Biomaterials for bone regeneration: an overview. Biomater. Tissue Technol. **1**, 1–5 (2017). https://doi.org/10.15761/BTT.1000107

13. Ray, S., Patel, S.K.S., Singh, M., Singh, G.P., Kalia, V.C.: Exploiting polyhydroxyalkanoates for tissue engineering. In: Kalia, V. (ed.) Biotechnological Applications of Polyhydroxyalkanoates, pp. 271–282. Springer, Singapore (2019). https://doi.org/10.1007/978-981-13-3759-8_10
14. Rivera-Briso, A.L., Serrano-Aroca, Á.: Poly (3-hydroxybutyrate-co-3-hydroxyvalerate): enhancement strategies for advanced applications. Polymers **10**(7), 732 (2018)
15. Gheibi, A., Khoshnevisan, K., Ketabchi, N., Derakhshan, M.A., Babadi, A.A.: Application of electrospun nanofibrous PHBV scaffold in neural graft and regeneration: a mini-review. Nanomed. Res. J. **1**(2), 107–111 (2016). https://doi.org/10.7508/nmrj.2016.02.007
16. Li, W., Ding, Y., Rai, R., Roether, J.A., Schubert, D.W., Boccaccini, A.R.: Preparation and characterization of PHBV microsphere/45S5 bioactive glass composite scaffolds with vancomycin releasing function. Mater. Sci. Eng. C **41**, 320–328 (2014). https://doi.org/10.1016/j.msec.2014.04.052
17. Monia, T.: β-TCP/DCPD-PHBV (40%/60%): biomaterial made from bioceramic and biopolymer for bone regeneration; investigation of intrinsic properties. J. Appl. Biomater. Funct. Mater. **20**, 22808000221088950 (2022). https://doi.org/10.1177/22808000221088950. PMID: 35410508. J. Nanosci. Nanotechnol. **2**(3–4), 293–312 (2002)
18. Monia, T., Ridha, B.C., Hassib, K.: Composite cement embedded in a biopolymer matrix for bone tissue regeneration. J. Thermoplast. Compos. Mater. **36**(12), 4858–5487 (2023). https://doi.org/10.1177/08927057231153858
19. Trimeche, M., et al.: Elaboration and evaluation of a composite bone substitute based on β-TCP/DCPD and PHBV, preliminary results. Biomed. Eng. Appl. Basis Commun. **28**(05), 1650031 (2016). https://doi.org/10.4015/S1016237216500319
20. Zhou, M., Yu, D.: Cartilage tissue engineering using PHBV and PHBV/Bioglass scaffolds. Mol. Med. Rep. **10**(1), 508–514 (2014)

Smart Materials for Biomedical Application

Monia Trimeche(✉) and Ridha Ben Cheikh

Laboratory of Materials, Optimization and Energy for Sustainability (LAMOED), Department of Industrial Engineering, National School of Engineers of Tunis, Tunis El Manar University, Tunis, Tunisia
monia_trimeche@yahoo.fr

Abstract. In clinical practice, bone defects, often associated with infections or other bone diseases, present substantial challenges. Several implantable biomaterials have demonstrated potential in bone therapy and regeneration, aiding in the repair, replacement, or restoration of both hard and soft tissues. Nevertheless, the presence of residual tumors or severe infections within the original pathological microenvironment of bone defects can impede effective bone regeneration. Consequently, there is a critical demand for the development of versatile biomaterials capable of integrating both bone therapy and regeneration functions. While traditional smart biomaterials respond to stimuli, the evolving field calls for a fresh perspective on biomaterial intelligence. This work delves into the latest breakthroughs and practical applications of intelligent biomaterials in the restoration and regeneration of hard tissues, all while highlighting the challenges associated with developing new engineered biomaterials.

Keywords: Implantable biomaterials · Smart scaffold · Immune-sensitive · Shape-memory · Electromechanical-stimulus

1 Introduction

At present, musculoskeletal disorders leading to tissue degeneration and inflammation are the primary contributors to disabilities and related diseases on a global scale. While bone tissue possesses inherent self-healing and regenerative capabilities, addressing extensive bone defects resulting from trauma, infections, or tumors necessitates external interventions. Consequently, there is a significant demand for biomaterials that facilitate the repair, replacement, or rejuvenation of both hard and soft tissues. This demand continues to rise as the population ages, underscoring the importance of developing and advancing such biomaterials [1].

Various approaches have been employed to develop smart materials with responsive properties for bone therapy and regeneration. These materials can be triggered by various stimuli, including external physical factors, internal disease microenvironments, or a combination of integrated strategies. Common external physical triggers include light irradiation, electric and magnetic fields, ultrasound, and mechanical stimuli. By influencing the internal atomic structures of materials and affecting cell behavior, these

M. Chouchane et al. (Eds.): CMSM 2023, LNME, pp. 195–203, 2024.
https://doi.org/10.1007/978-3-031-65007-9_22

stimuli have the potential to significantly enhance the effectiveness of bone tissue therapy and regeneration.

In addition to externally-triggered strategies, specific pathological microenvironments can also serve as biochemical triggers to activate bone disease therapy and bone regeneration. Notable examples of such microenvironments include the presence of excess reactive oxygen species and mild acidity in tumors, as well as the specific pH reduction and enzymes secreted by bacteria in severe infections. In this chapter, we provide a comprehensive overview and analysis of currently available smart biomaterials with functions related to bone therapy and regeneration.

2 Smart Scaffold Constructs for Bone Tissue Engineering

In the field of bone tissue engineering, scaffolds, cells, and growth factors are the three essential components. Scaffolds not only act as substitutes for the extracellular matrix (ECM) but also function as delivery vehicles for cells and carriers for growth factors. These scaffolds can be composed of natural or synthetic materials, or combinations thereof, providing several advantages [2, 3]. They have the ability to incorporate bone progenitor cells and growth factors, demonstrating osteoconductive and osteoinductive properties that promote the replacement and repair of bone defects. Ongoing advancements in scaffold development have focused on incorporating bioactive molecules and nanoparticles, as well as making specific modifications to their physical and chemical properties [4, 29].

The latest generation of scaffolds, known as smart scaffolds, has been specifically designed to enhance tissue regeneration effectiveness. These scaffolds incorporate smart materials with the remarkable ability to modify their physical and/or chemical properties in response to external stimuli or environmental changes. As a result, they have gained significant recognition in the field of biomedical research. Furthermore, the seamless integration of these materials into the realm of nanomedicine has created unprecedented opportunities for manipulating biological entities at the cellular and even sub-cellular levels [5].

The smart scaffolds presented in this paper fall into two categories: immune-sensitive scaffolds and electromechanical-stimulus scaffolds. Each classification serves a unique purpose and offers distinct advantages in promoting tissue regeneration [6].

2.1 Immune-Sensitive Smart Scaffolds

Upon implantation of biomaterials in the body, they trigger a range of cellular and tissue reactions. These include inflammatory and wound healing responses, foreign body reactions, and the development of fibrous tissue that encloses the implanted materials [30]. Notably, immune cells, particularly neutrophils and macrophages, play a crucial role in responding to tissue damage, foreign substances, and the introduction of both natural and synthetic biomaterials [7].

In the field of tissue engineering, there has been a notable shift away from the conventional approach of designing inert biomaterials to mitigate immune responses [8]. This paradigm change is primarily driven by the realization that numerous chronic diseases involve prolonged inflammatory bone destruction, posing significant challenges in clinical treatment. As a result, the development of smart, immune environment-responsive biomaterials has emerged as a promising avenue for therapeutic interventions in recent years [9] The careful selection of appropriate biomaterials is crucial for the creation of immune-sensitive smart scaffolds. The materials used should exhibit crucial characteristics, including biocompatibility, biodegradability, and the capability to integrate immunomodulatory agents. Ideally, they should possess the potential to steer the host's response, encouraging acceptance of foreign scaffolds, fostering tolerance, and regulating immunological microenvironments, all with the ultimate goal of enhancing cell viability [10].

To achieve precise control over interactions between scaffolds and immune cells, a range of functionalization techniques can be employed. Surface modifications and drug delivery systems are particularly valuable in this regard. Various techniques, including covalent immobilization, alkaline hydrolysis, wet chemical methods such as etching, chemisorption, chemical vapor deposition, and plasma treatments, are extensively employed to effectively alter the surface chemistry of biomaterials. These approaches offer immense value in achieving desired surface properties and enabling efficient drug delivery [11].

Hu et al. introduced a novel example of immune-sensitive smart scaffolds—a bioinspired injectable microsphere designed to modulate macrophages and create a prohealing microenvironment under diabetic conditions. The microsphere formation involved the self-assembly of gelatin nanofibers modified with heparin, along with the inclusion of interleukin 4 (IL4) within the resulting nanofibrous heparin-modified gelatin microsphere (NHG-MS) (refer to Fig. 1). It is worth highlighting that IL4 contains specific binding domains with heparin, which facilitate its stabilization, safeguard it against denaturation and degradation, and enable sustained release for the purpose of regulating macrophage polarization. This innovative injectable microsphere demonstrates great promise as a strategy to enhance bone healing and resolve inflammation in diabetic patients [12].

A novel strategy to enhance the surface chemistry of biomaterials and emulate the extracellular matrix (ECM) of bone involves the application of bone ECM proteins onto synthetic dicalcium phosphate bioceramics. This innovative approach not only demonstrates the ability to modify the material surface, but also exhibits remarkable control over host immune responses and enhances the regenerative capabilities of these biomaterials [13].

By integrating these advancements in biomaterial selection and functionalization techniques, immune-sensitive smart scaffolds can be customized to effectively modulate the immune response and establish favorable microenvironments for tissue regeneration. This approach holds immense potential for addressing the challenges presented by chronic inflammatory bone diseases and significantly improving clinical outcomes in various tissue engineering applications.

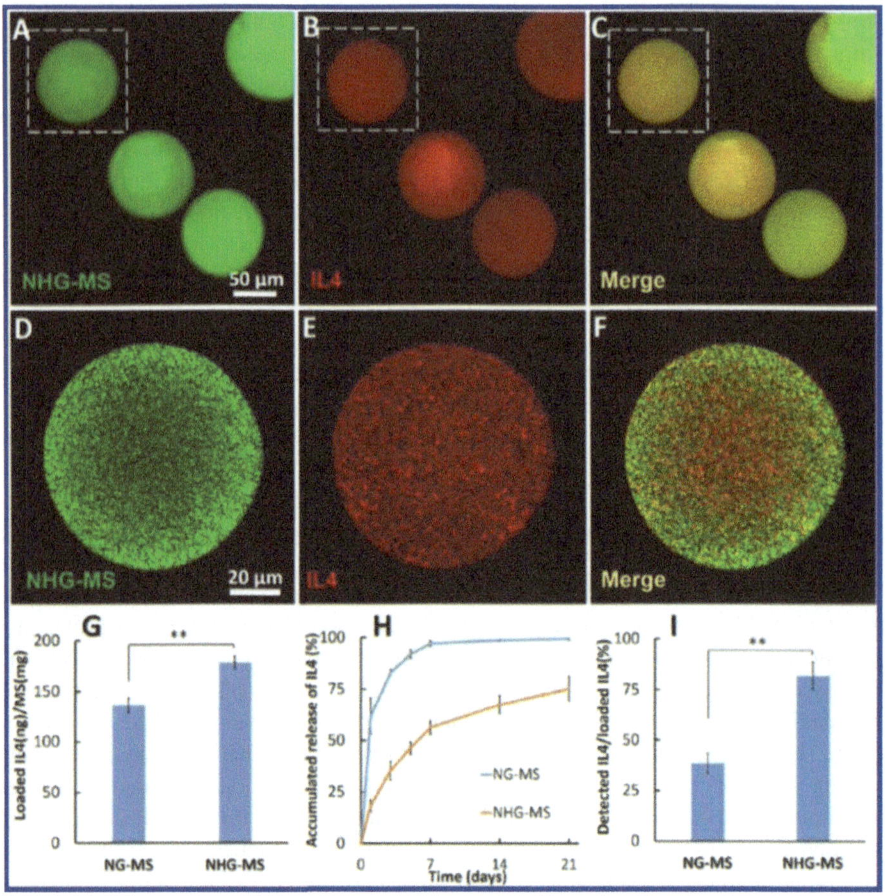

Fig. 1. Characterizations of IL4-loaded NHG-MS and NG-MS and their release profiles. (A–C) Stacked confocal images of IL4-loaded NHG-MS. (D–F) Cross-sectional confocal images at a higher magnification, showing that IL4 (red) was evenly distributed in the NHG-MS (green). (G) Loading efficiency of IL4 in NHG-MS and NG-MS. (H) Release profiles of IL4 from NHG-MS and NG-MS. (I) Total amount of IL4 (released and unreleased) from the NHG-MS and NG-MS detected via an ELISA ($^{**}P < 0.01$) [12].

2.2 Shape-Memory Smart Scaffolds

Extensive exploration has been undertaken to investigate a wide range of materials with diverse physicochemical properties for their potential application in biomedical scaffolds. This comprehensive research encompasses metals, ceramics, polymers, and composites [14]. Notably, shape memory polymers have emerged as a particularly promising area of focus, attracting considerable attention due to their unique attributes that enable easy and minimally invasive surgical delivery. This is achieved through in situ shape activation, leading to reduced trauma and improved patient comfort [15]. Additionally, the inherent shape memory effect of these polymers facilitates seamless integration between the scaffold and the defect, addressing the crucial requirement for enhanced compatibility [15].

Shape-memory polymers (SMPs) are a unique category of intelligent scaffolds that exhibit an extraordinary capacity to regain their initial shape after being deformed, triggered by external stimuli [16]. Such stimuli encompass variations in temperature [17], the application of electric or magnetic fields [18] and exposure to light [19]. The versatility of SMPs has attracted considerable interest in the field of tissue engineering, owing to their extensive applicability and potential [20].

The advantages of the shape-memory behavior in these scaffolds are particularly noteworthy. Firstly, the scaffolds can be pre-designed and conveniently deformed for implantation into bone defects using minimally invasive surgical techniques. Subsequently, they can be expanded to conform to irregularities within the bone defect [21]. This initial implantation is performed with a small-sized scaffold, minimizing damage to host tissues. Following deployment, the scaffold regains its larger shape, effectively filling the bone defect and even precisely matching irregular boundaries [22].

Shida Miao et al. conducted a study to develop biomimetic hierarchical scaffolds using smart polymers that are highly biocompatible, aiming to create dynamic scaffolds for regenerative medicine. The researchers presented an innovative approach, as depicted in Fig. 2. The biomaterial used in this study featured a particulated scaffold structure with a graded microporous architecture, closely resembling the nonuniform porosity found in natural tissues. Additionally, the biopolymers employed showed remarkable biocompatibility, enabling the attachment, proliferation, and differentiation of mesenchymal stem cells (MSCs). This work holds great potential for advancing regenerative medicine [23].

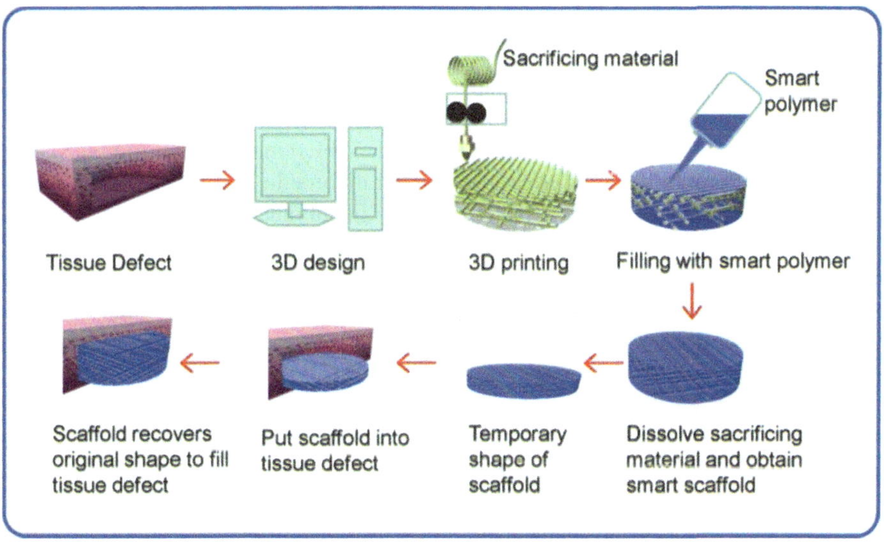

Fig. 2. The illustration of the process for preparing biomedical scaffold according to a specific traumatic defect [23]. Color images available online at www.liebertpub.com/tec

2.3 Electromechanical-Stimulus Smart Scaffolds

The bioelectrical properties inherent in natural bone tissue play a vital role in the process of bone development Specifically, the concept of piezoelectricity describes the conversion of mechanical energy into electrical energy [24]. In normal physiological conditions, bones emit electrical signals when subjected to mechanical stimuli, and these signals actively contribute to bone growth and remodeling [25]. Recognizing the significance of this distinctive characteristic, researchers have harnessed it in the creation of innovative biomaterials known as piezoelectric materials. These materials belong to a category of intelligent mechanoelectrical transducers capable of generating electrical signals when exposed to stress [26]. Numerous studies have demonstrated that piezoelectricity can effectively regulate cell function, stimulate the proliferation of stem cells, and enhance their ability to differentiate into osteogenic lineages. Consequently, piezoelectric biomaterials have garnered considerable attention in the realms of bone healing and tissue engineering [27, 28].

Piezoelectric materials encompass a range of substances, including specific crystals and ceramics, as well as various living tissues and biological macromolecules. Piezoelectric materials encompass various crystals and ceramics, such as barium titanate (BTO) and lead zirconate titanate (PZT). Furthermore, piezoelectric properties can be observed in diverse living tissues like natural bone, tendon, ligaments, cartilage, skin, dentin, and collagen. Remarkably, piezoelectric characteristics can also be found in specific biological macromolecules, including proteins, nucleic acids, and mucopolysaccharides [19].

In the field of BTE (Bone Tissue Engineering) applications, researchers have found that attaching piezoelectric materials to enhance their piezoelectric properties is a highly effective method, as evidenced by previous studies. For instance, Bai et al. [28] applied this approach to address the limited osteoinductivity of existing barrier membranes in classic guided bone regeneration therapy. They developed a novel flexible nanocomposite membrane by uniformly incorporating piezoelectric BaTiO3 nanoparticles into a poly(vinylidene fluoridetrifluoroethylene) (P(VDF-TrFE)) matrix (refer to Fig. 3).

Compared to commercially-available polytetrafluoroethylene (PTFE) membranes, the polarized BTO/P(VDF-TrFE) nanocomposite membranes demonstrated enhanced osteoinductivity, as assessed through immunofluorescence staining of runt-related transcription factor 2 (RUNX-2) expression. Moreover, these membranes exhibited the ability to induce earlier neovascularization and facilitate the complete formation of mature bone structures in critical-sized defects within the mandible of rabbits. This was observed after implantation with Deproteinized bovine bone (DBB) Bio-Oss® granules [28].

In conclusion, electromechanical-stimulus smart scaffolds hold great promise in the field of tissue engineering and regenerative medicine. These scaffolds, incorporating piezoelectric materials, have the ability to convert mechanical energy into electrical signals, providing a dynamic and interactive microenvironment for cells and tissues. The integration of electromechanical stimuli in scaffolds offers exciting opportunities for improving tissue healing, bone development, and other biomedical applications. Further research and development in this area are crucial to unlock the full potential of electromechanical-stimulus smart scaffolds and advance the field of tissue engineering towards more effective and personalized regenerative therapies.

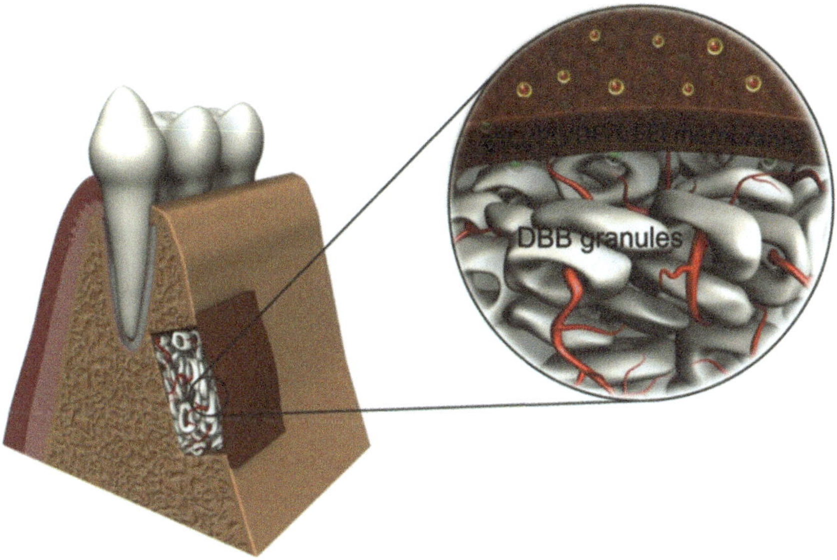

Fig. 3. Synergistic effects achieved by combining piezoelectric BTO/P(VDF-TrFE) nanocomposite membranes with xenogenic DBB grafts in the repair of critical-sized bone defects [28].

Overall, advancing smart stimuli-responsive materials requires in-depth exploration of multifunctional materials, simplified fabrication methodologies, precise control over the effects of biomaterials, suitable animal models for mechanistic studies, and thorough assessments of biodegradation properties. By addressing these areas, we can overcome the existing challenges and concerns, paving the way for successful clinical applications in bone therapy and tissue regeneration.

3 Conclusion

This Paper provides a comprehensive overview of the latest breakthroughs in cutting-edge smart biomaterials designed for the purpose of hard tissue engineering and regeneration. In addition to effectively addressing immediate treatment requirements for severe infections, residual tumor tissue, and other bone diseases, these novel biomaterials play a crucial role in supporting cell adhesion and facilitating rapid cell proliferation and differentiation. Smart stimuli-responsive materials, such as immune-sensitive smart scaffolds, and electromechanical-stimulus smart materials, are key in releasing bioactive components or osteogenic-related elements to drive these processes. As researchers and clinicians further comprehend the potential of immunomodulatory biomaterials, there will be an increasing demand for these intelligent and adaptable biomaterials that can influence the host's immune system and create an environment conducive to healing, regeneration, and repair. Henceforth, it is imperative to direct our endeavors towards the creation of innovative compositions that exhibit amplified and finely tuned intelligent attributes. Moreover, the pivotal goal should be to successfully translate these properties, derived from in vitro experiments, into practical applications for animal models, thereby driving significant advancements in the field of bone tissue repair and regeneration.

References

1. Yi, H., Ur Rehman, F., Zhao, C., et al.: Recent advances in nano scaffolds for bone repair. Bone Res. **4**, 16050 (2016). https://doi.org/10.1038/boneres.2016.50
2. O'Brien, F.J.: Biomaterials & scaffolds for tissue engineering. Mater. Today **14**, 88–95 (2005). https://doi.org/10.1016/S1369-7021(11)70058-X
3. Monia, T., Ridha, B.C.: Polymer-ceramic composites for bone challenging applications: materials and manufacturing processes. J. Thermoplast. Compos. Mater. **37**(4), 1540–1557 (2024). https://doi.org/10.1177/08927057231190066
4. Motamedian, S.R., Hosseinpour, S., Ahsaie, M.G., Khojasteh, A.: Smart scaffolds in bone tissue engineering: a systematic review of literature. World J. Stem Cells **7**, 657–668 (2015). https://doi.org/10.4252/wjsc.v7.i3.657
5. Genchi, G.G., Marino, A., Tapeinos, C., Ciofani, G.: Smart materials meet multifunctional biomedical devices: current and prospective implications for nanomedicine. Front Bioeng. Biotechnol. **5**, 80 (2017). https://doi.org/10.3389/fbioe.2017.00080
6. Zhang, K., et al.: Advanced smart biomaterials and constructs for hard tissue engineering and regeneration. Bone Res. **6**, 31 (2018)
7. Anderson, J.M., Rodriguez, A., Chang, D.T.: Foreign body reaction to biomaterials. Semin. Immunol. **20**(2), 86–100 (2008). https://doi.org/10.1016/j.smim.2007.11.004
8. Xie, Y., et al.: Osteoimmunomodulatory effects of biomaterial modification strategies on macrophage polarization and bone regeneration. Regen. Biomater. **7**, 233–245 (2020)
9. Muire, P.J., Thompson, M.A., Christy, R.J., Natesan, S.: Advances in immunomodulation and immune engineering approaches to improve healing of extremity wounds. Int. J. Mol. Sci. **23**(8), 4074 (2022). https://doi.org/10.3390/ijms23084074
10. Andorko, J.I., Jewell, C.M.: Designing biomaterials with immunomodulatory properties for tissue engineering and regenerative medicine. Bioeng. Transl. Med. **2**(2), 139–155 (2017). https://doi.org/10.1002/btm2.10063
11. Batool, F., et al.: Modulation of immune-inflammatory responses through surface modifications of biomaterials to promote bone healing and regeneration. J. Tissue Eng. **12**, 20417314211041428 (2021). https://doi.org/10.1177/20417314211041428
12. Hu, Z., Ma, C., Rong, X., Zou, S., Liu, X.: Immunomodulatory ECM-like microspheres for accelerated bone regeneration in diabetes mellitus. ACS Appl. Mater. Inter. **10**, 2377–2390 (2018)
13. Mansour, A., Abu-Nada, L., Al-Waeli, H., et al.: Bone extracts immunomodulate and enhance the regenerative performance of dicalcium phosphates bioceramics. Acta Biomater. **89**, 343–358 (2019)
14. Wauthle, R., et al.: Additively manufactured porous tantalum implants. Acta Biomater. **14**, 217 (2015)
15. Huang, W.M., Yang, B., Zhao, Y., Ding, Z.: Thermo-moisture responsive polyurethane shape-memory polymer and composites: a review. J. Mater. Chem. **20**, 3367–3381 (2010). https://doi.org/10.1039/b922943d
16. Zhou, S., Bai, J., Li, T., Gao, X., Xu, R., Shi, Z.: Stress communication between the chain movement and the shape transformation from 2D to 3D. ACS Appl. Mater. Interfaces **14**(1), 2082–2091 (2022). https://doi.org/10.1021/acsami.1c21867
17. Zhang, D., George, O.J., Petersen, K.M., Jimenez-Vergara, A.C., Hahn, M.S., Grunlan, M.A.: A bioactive "self-fitting" shape memory polymer scaffold with potential to treat cranio-maxillo facial bone defects. Acta Biomater. **10**, 4597 (2014)

18. Buckley, P.R., et al.: Inductively heated shape memory polymer for the magnetic actuation of medical devices. IEEE Trans. Biomed. Eng. **53**, 2075–2083 (2006). https://doi.org/10.1109/TBME.2006.877113
19. Lendlein, A., Jiang, H., Junger, O., Langer, R.: Light-induced shape-memory polymers. Nature **434**, 879–882 (2005). https://doi.org/10.1038/nature03496
20. Zulkifli, Z., et al.: Shape memory poly (glycerol sebacate)-based electrospun fiber scaffolds for tissue engineering applications: a review. J. Appl. Polym. Sci. **139**(22), e52272 (2022). https://doi.org/10.1002/app.52272
21. Langer, R., Tirrell, D.A.: Designing materials for biology and medicine. Nature **428**, 487–492 (2004). https://doi.org/10.1038/nature02388
22. Montgomery, M., et al.: Flexible shape-memory scaffold for minimally invasive delivery of functional tissues. Nat. Mater. **16**, 1038–1046 (2017). https://doi.org/10.1038/nmat4956
23. Miao, S., Zhu, W., Castro, N.J., Leng, J., Zhang, L.G.: Four-dimensional printing hierarchy scaffolds with highly biocompatible smart polymers for tissue engineering applications. Tissue Eng. Part C Methods **22**(10), 952–963 (2016). https://doi.org/10.1089/ten.tec.2015.0542
24. Kim, D., et al.: Biomolecular piezoelectric materials: from amino acids to living tissues. Adv. Mater. **32**, 1906989 (2020)
25. Ribeiro, C., et al.: In vivo demonstration of the suitability of piezoelectric stimuli for bone reparation. Mater. Lett. **209**, 118–121 (2017)
26. Przekora, A.: Current trends in fabrication of biomaterials for bone and cartilage regeneration: materials modifications and biophysical stimulations. Inter. J. Mol. Sci. **20**, 435 (2019)
27. Kohata, K., et al.: Enhancement of fracture healing by electrical stimulation in the comminuted intraarticular fracture of distal radius. Bio-med. Mater. Eng. **23**, 485–493 (2013)
28. Bai, Y., et al.: Biomimetic piezoelectric nanocomposite membranes synergistically enhance osteogenesis of deproteinized bovine bone grafts. Int. J. Nanomed. **14**, 3015–3026 (2019)
29. Monia, T.: Sustainable natural biopolymers for biomedical applications. J. Thermoplast. Compos. Mater. **37**(7), 2505–2524 (2024). https://doi.org/10.1177/08927057231214468
30. Trimeche, M.: Biomaterials for bone regeneration: an overview. Biomater. Tissue Technol. **1**, 1–5 (2017). https://doi.org/10.15761/BTT.1000107

Biomimetic Scaffold for Bone Regeneration, Their Manufacturing Techniques and the Applied Materials

Monia Trimeche(✉) and Ridha Ben Cheikh

Laboratory of Materials, Optimization and Energy for Sustainability (LAMOED), Department of Industrial Engineering, National School of Engineers of Tunis, Tunis El Manar University, Tunis, Tunisia
monia_trimeche@yahoo.fr

Abstract. In the realm of clinical practice, bone grafting has evolved to encompass the utilization of various types of grafts, including autogenous, allogeneic, and xenogeneic sources. Additionally, both natural and artificially synthesized materials such as polymers, bioceramics, and composite substances have been integrated into this process. However, these approaches often carry inherent limitations.

To address these shortcomings, the concept of biomimetic scaffolds has emerged. These bone implants can assist in accelerating bone regeneration, similar to that of authentic tissue, which replicates the cellular environment within a living organism.

This chapter aims to provide a comprehensive overview of the diverse array of biomaterials utilized in these scaffolds, coupled with the methodologies employed in the fabrication of biomimetic scaffolds. Moreover, the discussion delves into the intricacies of the limitations posed by intelligent biomaterials.

Keywords: Biomaterial · bone regeneration · Functionalization Biomimetic scaffold · Tissue engineering

1 Introduction

Biomimetic and bionic scaffolds are designed to mimic the natural extracellular matrix (ECM) found in bone tissue [1]. Their development is based on biological inspiration, seeking to replicate the structure, function, and formation of bone. These scaffolds incorporate bioactive molecules and nanomaterials to enhance crucial processes like cell adhesion, proliferation, and differentiation. By creating a favorable microenvironment for bone regeneration, these scaffolds closely resemble the native tissue and promote successful tissue repair and regeneration [2, 3].

In the pursuit of advancing biomaterials for tissue engineering, researchers worldwide are not only focused on developing synthetic materials that mimic the function of native tissue but also emphasize the control of degradation and ensuring minimal toxicity in vivo [4].

M. Chouchane et al. (Eds.): CMSM 2023, LNME, pp. 204–211, 2024.
https://doi.org/10.1007/978-3-031-65007-9_23

This dual approach considers both the functionality and biocompatibility of the materials, seeking to optimize their performance for successful integration and long-term efficacy in the body.

2 Fabrication Techniques for Biomimetic Scaffolds

Various fabrication techniques are employed in the field of materials engineering to create complex structures and functional materials [5]. These techniques include the foam replica method [6], freeze casting, freeze drying [7] phase separation [8], gas foaming [5], rapid prototyping [9], and electrospinning [10]. Each of these methods offers unique advantages and is suitable for different applications. Here is a description of the principles behind each fabrication technique [11].

The foam replica method involves creating a template or mold using a porous material, such as foam, and then filling it with a desired material. Once the material solidifies, the mold is removed, leaving behind a replica of the foam structure. This technique is widely employed in the manufacturing of porous materials, commonly utilized in diverse applications including filtration and tissue engineering [12].

Freeze casting, on the other hand, relies on the controlled freezing of a suspension or slurry. During freezing, ice crystals form and grow, pushing the suspended particles or solutes away. After freeze drying, the ice is removed, leaving a highly porous material with a unique microstructure. This technique is particularly useful for creating scaffolds for tissue engineering and porous ceramics [13].

Freeze drying also known as lyophilization, involves the removal of water from a frozen material through sublimation, bypassing the liquid phase [14]. This process preserves the structure and properties of the material while removing the solvent. It is commonly used in the production of pharmaceuticals, food preservation, and the preparation of delicate biological samples for microscopy [15].

Gas foaming is a process in which a gas is introduced into a polymer matrix, forming bubbles or voids [16]. The polymer is then solidified, resulting in a foam-like structure. Gas foaming is commonly used to create lightweight materials with low density and high porosity, offering applications in insulation, packaging, and aerospace industries [17].

Rapid prototyping also known as additive manufacturing or 3D printing, enables the direct fabrication of three-dimensional objects from computer-aided design (CAD) models. Layer-by-layer deposition or solidification of materials allows for the creation of complex geometries with high precision [18]. This technique has revolutionized the manufacturing industry and has applications in various fields, including product prototyping, medical implants, and custom-designed components [19].

Electrospinning is a technique used to produce ultrafine fibers with diameters in the nanometer to micrometer range. It involves the application of an electric field to a polymer solution or melt causing the elongation and thinning of the material into fibers as it is ejected from a nozzle or syringe. Electrospinning is used in the production of nanofibrous materials for applications such as filtration, tissue engineering, and drug delivery [20].

3 Materials Involved in Biomimetic Scaffolds

Biomimetic materials are fabricated using a range of materials, including polymers, bioceramic phosphocalcique, metals, and composites based on calcium phosphates (CaPs). These materials are specifically chosen for their beneficial characteristics such as osteoconductivity, resorbability, and biocompatibility [21].

In particular, calcium phosphates (CaPs) play a crucial role in biomimetic material fabrication. CaPs exhibit biocompatibility, osteoconductivity, and biodegradability, making them ideal for various biomedical applications. However, their mechanical strength is limited, thereby restricting their usage in load-bearing applications. One prominent type of CaP is β-tricalcium phosphate (β-TCP), which is a high-temperature phase displaying chemical similarities to hydroxyapatite (HAp) but with different resorption capabilities [4]. HAp demonstrates slow resorption in biological environments, while β-TCP exhibits favorable reabsorption properties. Furthermore, the utilization of ionic co-substitutions within calcium phosphates, including the intriguing β-TCP-HA composition, presents an effective means to exert control over the properties of the material, encompassing its structure, microstructure, crystallinity, and dissolution rate [8].

With the latest breakthroughs in the field of biodegradable polymers, glass, and ceramics, we now have the capability to finely tune the rate at which degradation occurs and the in vivo resorption process, all the while enhancing cellular responses at the molecular level [22].

The manufacturing of biocomposite, combining two or more of these biomaterials, allows for the minimization of drawbacks associated with using each material alone. Further investigations conducted by various researchers have explored methods to enhance the biocompatibility of scaffolds. These include the immobilization of specific molecular recognition signals [23], such as peptides [24], as well as the introduction of bone marrow cells, genes [25], and drug delivery [26] into artificial bone scaffolds. Additionally, improvements can be made through modifications in the scaffold's topology, nanostructure, chemistry, and structural aspects [27].

In a separate study conducted by Xiaobo Feng et al., their research focused on exploring the mechanical and biological properties of fully porous PEEK scaffolds with varying pore sizes. The investigation encompassed both in vitro and in vivo experiments. The team utilized fused deposition modeling (FDM) to create PEEK scaffolds with distinct pore sizes of 300, 450, and 600 μm, thereby achieving a porosity of 60%. The primary objective of their study was to determine the optimal pore size that would facilitate improved performance (refer to Fig. 1).

The results of the in vitro tests demonstrated the excellent biocompatibility of the porous PEEK scaffolds produced through FDM. Furthermore, the in vivo analysis revealed that the PEEK-450 scaffolds exhibited remarkable potential for bone in growth and vascular perfusion, surpassing the performance of other scaffold types [28].

Biomimetic scaffolds represent a groundbreaking technological leap forward in the realm of tissue engineering. These scaffolds serve as a sophisticated three-dimensional framework or structure, offering exceptional support for the growth and development of cells and facilitates the regeneration of damaged or missing tissues. With further research and refinement, mimetic scaffold holds promise for revolutionizing regenerative medicine and improving patient outcomes.

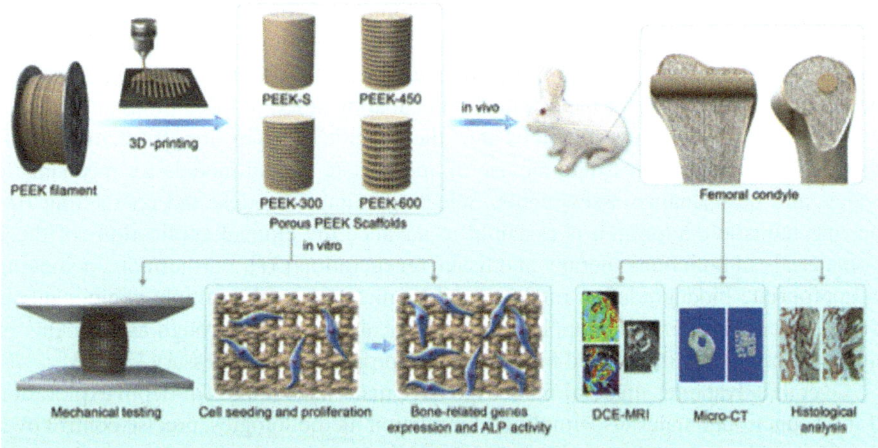

Fig. 1. Schematic illustration of the fabrication of porous PEEK scaffolds and their bioapplication [28].

4 Limit of Smart Biomaterial

While previous research has shown promise, the development of smart stimuli-responsive materials is still at an early stage and requires further investigation to address existing challenges and concerns.

Initially, it is crucial to focus on advancing multifunctional materials that integrate various therapeutic modalities and regenerative components. Presently, smart stimuli-responsive materials are limited to specific modalities such as photothermal ablation, magnetic hyperthermia, SDT, and nanocatalytic therapy. However, future biomaterials could incorporate nanoparticles (NPs) to enhance bone regeneration and facilitate controlled delivery of drugs or genes for precise treatment of bone diseases [29]. Additionally, comprehensive evaluation of the immune responses, metabolic pathways, and biological distribution of newly synthesized biomaterials is necessary to assess their biosafety, given the incorporation of multiple components for therapeutic and regenerative purposes. Furthermore, comprehensive comparisons of strength, toughness, and other physical or chemical properties with state-of-the-art biomaterials are necessary for successful clinical translation, along with addressing potential long-term toxicity concerns.

Furthermore, the synthesis of composite material systems using novel therapeutic nanoplatforms and biomaterial platforms will continue to drive research in this field. To simplify the complex fabrication processes of smart stimuli-responsive materials, it is imperative to explore more facile synthetic methodologies. The integration of multiple components into a single biomaterial platform presents challenges that require the adoption of simplified integrated methodologies [30].

Considering the various strategies employed, such as external stimuli-responsive or internal microenvironment stimuli-responsive approaches, each has advantages and disadvantages in practical biomedical applications. However, achieving precise and rational control over the effects of newly synthesized biomaterials on bone therapy and regeneration remains challenging, particularly in determining optimal parameters for external stimuli and recognizing internal environmental changes. Thorough investigations into these aspects are crucial for the eventual clinical application of these novel strategies.

Secondly, detailed investigations into the specific mechanisms of smart stimuli-responsive materials necessitate the use of appropriate animal models for mechanistic studies and performance assessments. Selecting suitable models and conducting further mechanistic exploration is essential to advance the clinical applications of these biomaterials in both bone therapy and tissue regeneration [31]. Furthermore, achieving an appropriate biodegradation rate of novel biomaterials with multiple components is crucial before their effective application in clinical settings. Thorough assessments of biosafety, biocompatibility, and biodegradation properties are necessary [32, 33].

Overall, advancing smart stimuli-responsive materials requires in-depth exploration of multifunctional materials, simplified fabrication methodologies, precise control over the effects of biomaterials, suitable animal models for mechanistic studies, and thorough assessments of biodegradation properties. By addressing these areas, we can overcome the existing challenges and concerns, paving the way for successful clinical applications in bone therapy and tissue regeneration.

5 Conclusion

In summary, the progress of bone tissue engineering will persistently march towards the path of material innovation and scaffold designs, drawing inspiration from the biomimetic replication of bone tissue's intricate hierarchical structures. Simultaneously, a deeper exploration of the mechanisms at play within the living body will be undertaken, aimed at enhancing bone regeneration. This collective effort ultimately aims to culminate in the successful translation of these advancements into clinical applications.

6 Table Synthesis of Review

See Table 1.

Table 1. Concise synthesis of the review

Section	Summary
Introduction	Biomimetic and bionic scaffolds mimic the normal extracellular matrix (ECM) of bone tissue, enhancing cellular function and its importance in bone regeneration. Biocompatibility and degradation control are key considerations
Fabrication Techniques	A variety of methods are used to create biomimetic scaffolds, including such as foam replica, freeze casting, freeze drying, gas foaming, rapid prototyping, and electrospinning,each offering unique advantages for different applications
Materials for Biomimetic Scaffolds	Biomimetic materials include polymers, bioceramics, metals, and composites, with calcium phosphates (CaPs) being particularly significant due to their biocompatibility and ability to be absorbed by the body
Limit of Smart Biomaterial	Smart stimuli-responsive materials are in early stages and require further investigation for multifunctionality, biosafety assessment, simplified fabrication, precise control, appropriate animal models, and biodegradation properties
Conclusion	Bone tissue engineering is advancing through material innovation based on biomimetic structures. Understanding the mechanisms of the living body aims to improve regeneration and translate it into clinical applications

References

1. Sarrigiannidis, S.O., Moussa, H., Dobre, O., Dalby, M.J., Tamimi, F., Salmeron-Sanchez, M.: Chiral tartaric acid improves fracture toughness of bioactive Brushite-Collagen bone cements. ACS Appl. Bio Mater. **3**(8), 5056–5066 (2020). https://doi.org/10.1021/ACSABM.0C00555
2. Kim, H.D., Amirthalingam, S., Kim, S.L., Lee, S.S., Rangasamy, J., Hwang, N.S.: Biomimetic materials and fabrication approaches for bone tissue engineering. Adv. Healthc. Mater. **6**(23), 1700612 (2017). https://doi.org/10.1002/adhm.201700612
3. Jiang, S., Wang, M., He, J.: A review of biomimetic scaffolds for bone regeneration: toward a cell-free strategy. Bioeng. Transl. Med. **6**(2), e10206 (2020). https://doi.org/10.1002/btm2.10206
4. Monia, T., Ridha, B.C., Hassib, K.: Composite cement embedded in a biopolymer matrix for bone tissue regeneration. J. Thermoplast. Compos. Mater.Thermoplast. Compos. Mater. **36**(12), 4858–4873 (2023). https://doi.org/10.1177/08927057231153858
5. Prakasam, M., Mioara Piticescu, R., Popescu, M.: Chapter fabrication methodologies of biomimetic and bioactive scaffolds for tissue engineering applications. Scaffolds in Tissue Engineering - Materials, Technologies and Clinical Applications. InTech (2017). https://doi.org/10.5772/intechopen.70707

6. Tuna, S., Akkoyun, M.: Chemical foaming of polylactic acid/polypropylene and polylactic acid/polyamide 6: evaluation of changes in their properties. J. Thermoplast. Compos. Mater.Thermoplast. Compos. Mater. **37**(2), 842–863 (2024). https://doi.org/10.1177/08927057231186325
7. Khoo, R.Z., Chow, W.S., Ismail, H.: Tensile, thermal and ultra-violet shielding enhancement of poly(lactic acid) bionanocomposite film using cellulose nanocrystals extracted from sugarcane bagasse. J. Thermoplast. Compos. Mater.Thermoplast. Compos. Mater. **36**(6), 2543–2561 (2023). https://doi.org/10.1177/08927057221098972
8. Trimeche, M., et al.: Elaboration and evaluation of a composite bone substitute based on β-TCP/DCPD and PHBV, preliminary results. Biomed. Eng. Appl. Basis Commun. Indexing **28**(05), 1650031 (2016)
9. Khan, A.S., Ali, A., Hussain, G., Ilyas, M.: An experimental study on interfacial fracture toughness of 3-D printed ABS/CF-PLA composite under mode I, II, and mixed-mode loading. J. Thermoplast. Compos. Mater.Thermoplast. Compos. Mater. **34**(12), 1599–1622 (2021). https://doi.org/10.1177/0892705719874860
10. Abdelhady, S.S., Elbadawi, R.E., Zoalfakar, S.H.: Fabrication of electrospun exfoliated graphite nanosheets/polystyrene composite nanofiber mats. J. Thermoplast. Compos. Mater.Thermoplast. Compos. Mater. **36**(9), 3499–3515 (2023). https://doi.org/10.1177/08927057221129489
11. Sampath, U.G.T.M., Ching, Y.C., Chuah, C.H., Sabariah, J.J., Lin, P.-C.: Fabrication of porous materials from natural/synthetic biopolymers and their composites. Materials **9**, 991 (2016). https://doi.org/10.3390/MA9120991
12. Fiume, E., Ciavattini, S., Verne, E., Baino, F.: Foam replica method in the manufacturing of bioactive glass scaffolds: out-of-date technology or still underexploited potential? Materials (Basel) **14**(11), 2795 (2021). https://doi.org/10.3390/ma14112795
13. Monia, T.: β-TCP/DCPD-PHBV (40%/60%): biomaterial made from bioceramic and biopolymer for bone regeneration; investigation of intrinsic properties. J. Appl. Biomater. Funct. Mater. **20**, 22808000221088950 (2022). https://doi.org/10.1177/22808000221088950
14. Nowak, D., Jakubczyk, E.: The freeze-drying of foods-the characteristic of the process course and the effect of its parameters on the physical properties of food materials. Foods **9**(10), 1488 (2020). https://doi.org/10.3390/foods9101488
15. Hisano, S., Adachi, T., Daikoku, S.: Freeze-drying technique in electron microscopic immunohistochemistry. J. Histochem. Cytochem.Histochem. Cytochem. **33**(5), 485–490 (1985). https://doi.org/10.1177/33.5.3989275
16. Amiri, S., Amiri, S., Mokhtari, S.: Foams for biomedical applications. Polym. Foams Appl. Polym. Foams **2**, 123–143 (2023). https://doi.org/10.1021/bk-2023-1440.ch006
17. Ni, X., Xing, X., Deng, Y., Li, Z.: Applications of stimuli-responsive hydrogels in bone and cartilage regeneration. Pharmaceutics **15**, 982 (2023). https://doi.org/10.3390/pharmaceutics15030982
18. Juda, P., Šmigová, J., Kováčik, L., Bártová, E., Raška, I.: Ultrastructure of cytoplasmic and nuclear inosine-5'-monophosphate dehydrogenase 2 "rods and rings" inclusions. J. Histochem. Cytochem.Histochem. Cytochem. **62**(10), 739–750 (2014). https://doi.org/10.1369/0022155414543853
19. Zhou, S., Bai, J., Li, T., Gao, X., Xu, R., Shi, Z.: Stress communication between the chain movement and the shape transformation from 2D to 3D. ACS Appl. Mater. Interfaces **14**(1), 2082–2091 (2022). https://doi.org/10.1021/acsami.1c2186
20. Jadhav, A., Jadhav, V.S.: A review on 3D printing: an additive manufacturing technology. Mater. Today Proc. **62**, 2094–2099 (2022). https://doi.org/10.1016/j.matpr.2022.02.558
21. Meshram, B.D., Lule, V.K., Vyawahare, S., Rani, R.: Application of edible packaging in dairy and food industry. In: Food Processing and Packaging Technologies-Recent Advances. IntechOpen (2022). https://doi.org/10.5772/intechopen.107850

22. Trimeche, M.: Biomaterials for bone regeneration: an overview. Biomater. Tissue Technol. **1**, 1–5 (2017). https://doi.org/10.15761/BTT.1000107
23. Lu, H.H., El-Amin, S.F., Scott, K.D., Laurencin, C.T.: Three-dimensional, bioactive, biodegradable, polymer-bioactive glass composite scaffolds with improved mechanical properties support collagen synthesis and mineralization of human osteoblast-like cells in vitro. J. Biomed. Mater. Res. A **64**(3), 465–474 (2003). https://doi.org/10.1002/jbm.a.10399
24. Abou-Elkacem, L., Wilson, K.E., et al.: Ultrasound molecular imaging of the breast cancer neovasculature using engineered fibronectin scaffold ligands: a novel class of targeted contrast ultrasound agent. Theranostics **6**(11), 1740–1752 (2016). https://doi.org/10.7150/thno.15169
25. Mittal, A., Negi, P., Garkhal, K., Verma, S., Kumar, N.: Integration of porosity and bio-functionalization to form a 3D scaffold: cell culture studies and in vitro degradation. Biomed. Mater. **5**, 045001 (2010). https://doi.org/10.1088/1748-6041/5/4/045001
26. Power, R.N., Cavanagh, B.L., Dixon, J.E., Curtin, C.M., O'Brien, F.J.: Development of a gene-activated scaffold incorporating multifunctional cell-penetrating peptides for pSDF-1α delivery for enhanced angiogenesis in tissue engineering applications. Int. J. Mol. Sci. **23**(3), 1460 (2022). https://doi.org/10.3390/ijms23031460
27. Hu, Q., Katti, P.S., Gu, Z.: Enzyme-responsive nanomaterials for controlled drug delivery. Nanoscale **6**, 12273–12286 (2014)
28. Giannitelli, S.M., Accoto, D., Trombetta, M., Rainer, A.: Current trends in the design of scaffolds for computer-aided tissue engineering. Acta Biomater. Biomater. **10**(2), 580–594 (2014). https://doi.org/10.1016/j.actbio.2013.10.024
29. Feng, X., et al.: Osteointegration of 3D-printed fully porous polyetheretherketone scaffolds with different pore sizes. ACS Omega **5**(41), 26655–26666 (2020). https://doi.org/10.1021/acsomega.0c03489
30. Wei, H., Cui, J., Lin, K., Xie, J., Wang, X.: Recent advances in smart stimuli-responsive biomaterials for bone therapeutics and regeneration. Bone Res. **10**(1), 17 (2022). https://doi.org/10.1038/s41413-021-00180-y
31. Monia, T.: Sustainable natural biopolymers for biomedical applications. J. Thermoplast. Compos. Mater.Thermoplast. Compos. Mater. **37**(7), 2505–2524 (2023). https://doi.org/10.1177/08927057231214468
32. Wu, R.X., Xu, X.Y., Wang, J., He, X.T., Sun, H.H., Chen, F.M.: Biomaterials for endogenous regenerative medicine: coaxing stem cell homing and beyond. Appl. Mater. Today **11**, 144–165 (2018). https://doi.org/10.1016/j.apmt.2018.02.004
33. Gai, X., et al.: A novel method for evaluating the dynamic biocompatibility of degradable biomaterials based on real-time cell analysis. Regen. Biomater. **7**(3), 321–329 (2020). https://doi.org/10.1093/rb/rbaa017

Improvement of the Numerical Model on the Orthodontic Correction of a Malocclusion Using NiTi Superelastic Alloy

Najla Tebessi, Myriam Laroussi(✉), and Tarak Bouraoui

Laboratory of Mechanical Engineering, National Engineering School of Monastir, University of Monastir, Monastir, Tunisia

meriem.aroussi@enim.u-monastir.tn

Abstract. Optimising the choice of superelastic NiTi orthodontic archwires mainly relies on the force exerted by the orthodontic archwire to correct dental malposition. Therapeutic efficiency and patient comfort largely depend on their applied corrective force. The principal goal of this project is to use numerical simulations to determine the force required to correct these misalignments. A specific configuration of dental malposition is considered. It is an intrusion of a premolar tooth requiring extrusion for correction. The thermomechanical model used for the numerical simulations is based on the Auricchio's approach implemented in Abaqus Finite Elements software. Uniaxial tensile tests were carried out to verify the superelastic behaviour of orthodontic archwires at different temperatures. Optimisation of the correction force is carried out using numerical simulations in the real correction phase, and the results are compared with other numerical simulations and existing experimental results.

Keywords: NiTi orthodontic archwires · Dental misalignment modelling · Thermomechanical simulations

1 Introduction

This project deals with the use of the Finite Element Method (FEM) to analyse the thermomechanical parameters of modern materials, particularly in the fields of bioengineering and dentistry. Although research has been carried out on orthodontic archwires made from NiTi alloys, there are still opportunities for improving these materials, particularly in order to reduce patient pain during the fitting of orthodontic archwires.

The main objective is to provide information on the optimum forces required to minimise pain while maintaining therapeutic efficiency, thereby improving the patient's orthodontic treatment.

A review of the literature is performed, then the model and the numerical methodology as well as the results of those numerical simulations of the thermomechanical behaviour of orthodontic archwires in the correction phase are presented. In addition, this project highlights the importance of FEM in testing several conditions and helping to predict the corrective force of interest in this study.

M. Chouchane et al. (Eds.): CMSM 2023, LNME, pp. 212–221, 2024.
https://doi.org/10.1007/978-3-031-65007-9_24

2 Background

There have been multiple studies in the field of orthodontic archwires modelling. Previous models have primarily focused on the tooth structure, but recent research has highlighted the importance of incorporating alveolar bone into these models to provide a more realistic representation of the oral environment. For instance, Singh *et al.* (2016) [1] conducted a literature review on finite element analysis in orthodontics and emphasized the significance of incorporating alveolar bone into orthodontic models. Similarly, Rossini *et al.* (2021) [2] reviewed the literature on the application of finite element analysis in orthodontics and highlighted the importance of validating FEA models for precise prediction of tooth movement and biomechanics. In addition, Cunha *et al.* (2021) [3] reviewed the recent advances in orthodontic digital setup and 3D modelling, highlighting the potential of incorporating alveolar bone into these models to enhance treatment planning and outcomes.

Gannoun *et al.* (2021) [4] and Gzara *et al.* (2022) [5] utilize a modelling technique where bone behaviour is simulated through the use of springs. This methodology simplifies the bone structure by representing it as interconnected springs. In contrast, our newly developed model goes beyond this approach. Our research builds on those previous studies by considering not just the tooth structure but also key aspects of the alveolar bone, incorporating both its cortical (dense outer layer) and cancellous (spongy inner layer) components leading to a more lifelike depiction of the oral setting.

This advancement allows for more accurate and effective treatment planning, as it can predict the response of oral structures to different orthodontic interventions with greater precision. This model represents a significant step forward in orthodontic research, with the potential to transform clinical practice. By enabling orthodontists to deliver more effective and personalized treatments, our model can improve patient outcomes and satisfaction, ultimately revolutionizing the field of orthodontics.

3 Problem Description

Malocclusion is a state in which at least one tooth has become misaligned. In this case, the use of braces can be a solution. They are a tool to correct this malocclusion by moving the tooth into the correct position in the dental arch.

In this study we are trying to improve the numerical modelling to get as close as possible to the real patient situation. This makes the results more realistic and allows us to take into account the oral conditions of each patient, considering various factors such as the state of the alveolar bone, tooth displacement and oral temperatures.

Note that our results are based on several assumptions about material properties and simplified models. In our case, a fixed orthodontic appliance, called brackets, is bonded to the teeth and an orthodontic wire to ensure tooth alignment. The wire applies a force that causes the movement needed to correct the malocclusion. Tooth movement requires precise force, which should not cause discomfort for the patient.

In order to carry out our study, we start by considering a specific malocclusion configuration, showed in Fig. 1 consisting in a premolar inclusion that requires correction through an extrusion movement.

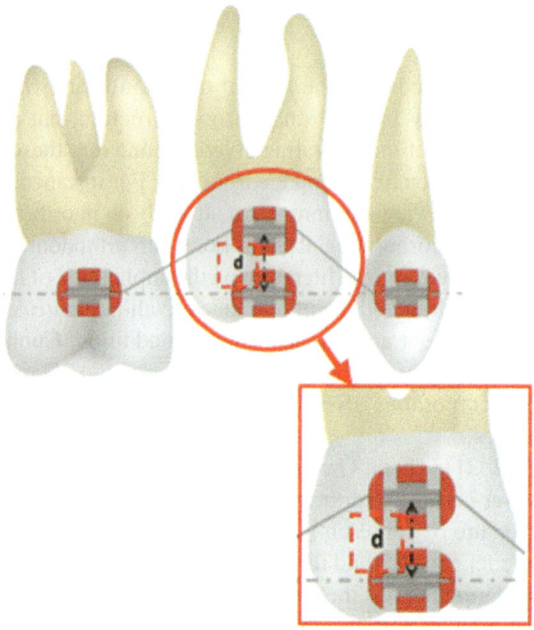

Fig. 1. Premolar inclusion configuration: "d" Dental correction distance.

4 FE Model Creation

4.1 Geometry

In order to have a realistic geometric model of the premolar a dimensional finite element model is constructed using CAD/CAM technology, where the tooth geometry is captured using a 3D cone beam tomography. The resulting FE model of the tooth is shown in Fig. 2(a). The alveolar bone in which the premolar roots are inserted is modmodelled by considering both cortical and cancellous bones surrounding the tooth roots (see Fig. 2(b)). The correction equipment consists in a bracket bonded to the tooth with an orthodontic adhesive (see Fig. 2(c)) and the NiTi wire (see Fig. 2(d)).

4.2 Materials

Except the NiTi alloy, all the other materials are assumed as isotropic, homogenous and linearly elastic. The elastic properties of the different parts are given in Table 1, where the tooth is composed of enamel and dentin.

For the NiTi alloy, three shades are considered: 48.2% Ni [8], 51% Ni [9] and 50.8% Ni [10]. Those shades exhibit superelastic behaviour.Thermomechanical behaviour models are used for the behaviour of superelastic NiTi. Those models take into account the different elastic properties of austenite and martensite. They are able to reproduce the superelastic behaviour of shape memory alloys (SMA), which is characterised by the ability of the material to return to its original shape after deformation, even if the deformation is large. The Auricchio's superelastic model used in this study is a macroscopic

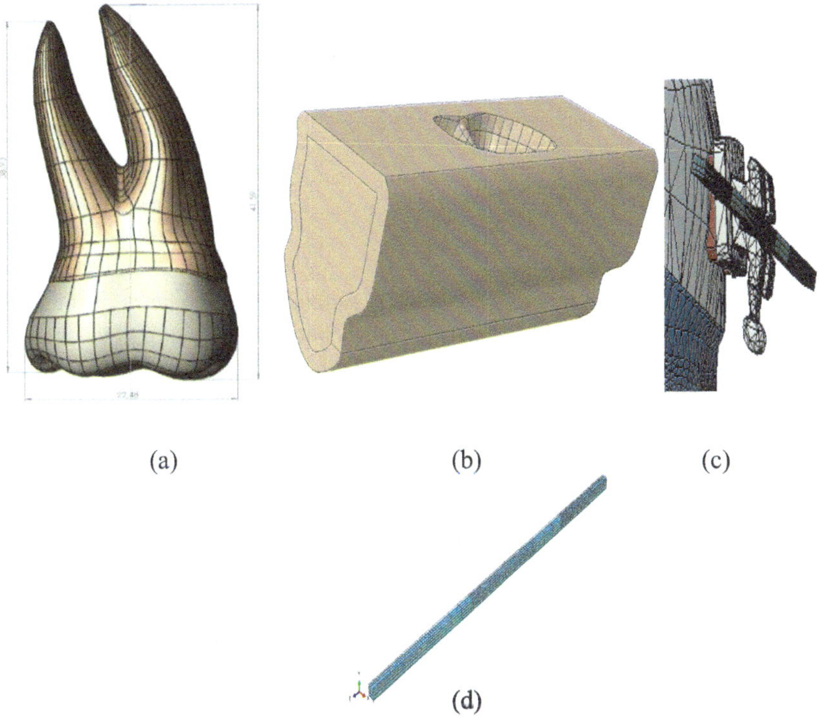

Fig. 2. The different geometric parts: (a) The 3D two-rooted premolar; (b) The alveolar bone; (c) The bracket; (d) The NiTi wire.

Table 1. Properties of the materials used in the FE analysis model [6, 7].

Materials		Young's modulus (GPa)	Poisson's ratio
Dentin		18.3	0.23
Enamel		84.1	0.30
The alveolar bone	Cortical bone	13.7	0.30
	Cancellous bone:	1.4	
Orthodontic adhesive		14.2	0.30
Bracket		197.0	0.27

model.This model takes into account, in addition to the different elastic properties, the transformation temperatures showed in Table 2, where A_s is the Austenite start, A_f is the Austenite finish, M_s is the martensite start and M_f is the martensite finish.

The numerical simulations are performed at temperatures of 35 °C and 50 °C, which are higher than A_f. This indicates superelastic behaviour of the archwire.A comparative study between two numerical models is also carried out: model 1 was created in previous work [4](see Fig. 3(a)) and model 2, as shown in Fig. 3(b), where geometric

Table 2. Transformation temperatures of three shades of NiTi alloys.

Ni (%)	Ti (%)	A_s (°C)	A_f (°C)	M_s (°C)	M_f (°C)
51	49	−2.7	**22.4**	−56.3	−77.6
50.8	49.2	1	**30**	0	−26.1
48.2	50.8	3	**17**	7	−3

improvements are made compared to model 1. Actually, the geometry of the tooth is improved by taking into account not only the shape but also the nature of the dentin and the enamel. In addition, the alveolar bone is modelled taking into account the cortical and the cancellous bones, rather than just adding springs to the roots.

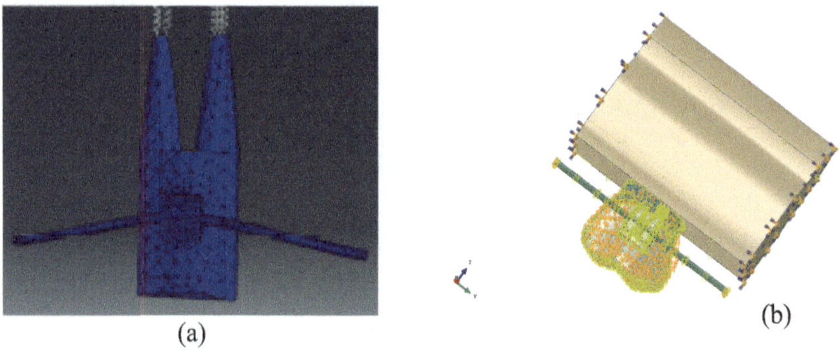

Fig. 3. Finite element models of the correction of a premolar inclusion configuration: (a) model 1 [6]; (b) model 2.

The bone resistance to root displacement in model 1 is modelled by six springs. The equivalent stiffness is computed by applying the mixing law. To determine this stiffness, we computed the modulus of elasticity of the constituent materials as shown in Table 3.

Table 3. Bone density and material properties [11].

Materials		Bone density (HU[a])	Bone density (g cm^{-3})	Young's modulus (GPa)
The alveolar bone	Cortical bone	1765	1.178	14.18
	Cancellous bone	150–850	0.109–0.572	0.259–3.507

[a]Hounsfield units (HU): The HU unit is a measure of radio density commonly used in computed tomography (CT) scans. Measured Bone Density Distribution (UH): This is the most widely used technique in orthodontics.

Hence, the equivalent Young's modulus of the bone is obtained by:

$$E_{total} = \frac{E_{cortical} \times V_{cortical} + E_{cancellous} \times V_{cancellous}}{V_{total}} \quad (1)$$

where V denotes the volume. In our case, considering the high density of the cancellous bone, we have:

$$E_{total} = \frac{14.1810^3 \times 1768.20 + 3.50710^3 \times 2678.27}{4446.47} = 7751.27\,\text{MPa}. \quad (2)$$

Using the Young's modulus already calculated in Eq. (2), Eq. (3) allows to determine the stiffness.

$$\text{K} = \frac{(\text{E} \times \text{A})}{\text{L}} = \frac{(7751.27 \times 227)}{10} = 175954\,\text{N/mm} \quad (3)$$

where:

- K: the stiffness value of the springs;
- A: the root surface area: the average root surface area measurements for the maxillary premolar tooth is between 213 and 227 mm^2 according to [13];
- L: the distance between roots and alveolar bone in the model performed by [10]

As there are six parallel springs, the stiffness of a single spring is hence obtained by:

$$\text{K}_{\text{single}} = \frac{\text{K}}{\text{n}} = \frac{175954}{6} = 29326\,\text{N/mm} \quad (4)$$

4.3 Boundary Conditions and Interactions

Simulation and analysis involve, the study of the bio-model finite elements in real situations in terms of their function and the agents acting on them. In order to achieve this, boundary conditions are applied within two steps:

- The first step corresponds to the placement of the orthodontic wire in the bracket glued to the malpositioned tooth. This is performed by prescribing a displacement "d" of 4 mm along the vertical Z-axis on the middle contour of the wire.
- The second step involves the dental correction phase, during whichthe wire is released in terms of displacement boundary conditions.

In both steps, both ends of the wire as well as both sides of the bone are clamped. This assumption expresses the fact that the orthodontic correction is localized at the malpositioned tooth. Figure 4 shows those boundary conditions. Regarding the temperature, it is assumed to be constant and equal to that of the mouth under normal circumstances, i.e. 35 °C.

Figure 5 shows the different interactions: One between the inner surface of the alveolar bone and the outer surface of tooth roots, another one between the inner surface of the enamel and the outer surface of the dentin adjacent to each other and a last one is the contact zone between the cortical and the cancellous bone.

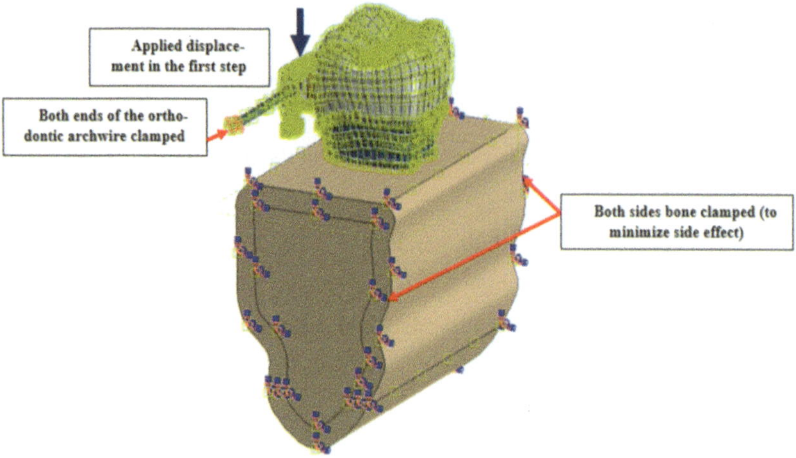

Fig. 4. FEM model showing applied boundary conditions.

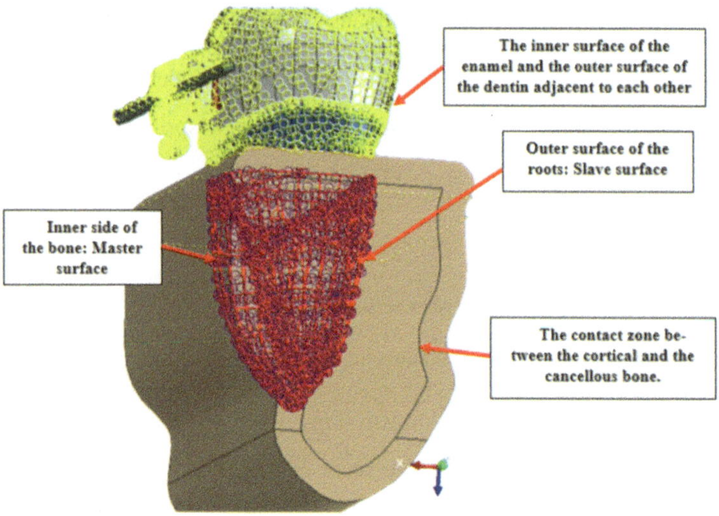

Fig. 5. Model interactions.

4.4 Mesh

All the components of the assembly are considered deformable. 3D continuum finite elementsare chosen. They are tetrahedrons with linear interpolation (C3D4) for bone, tooth and bracket, and with quadratic interpolation and reduced integration (C3D8R) for the orthodontic wire After a convergence study, the maintained numbers of finite elements for each part are summed up in Table 4.

Table 4. Number of elements in each part of the model.

Parts	Number of elements
The alveolar bone	126362
Enamel	4072
Dentin	185267
Orthodontic adhesive	340
Bracket	5657

5 Results

The numerical simulations are performed under the assumption of quasi-static and large strains computations. The response of interest is the reaction force in the clamped ends of the archwire (see Fig. 6).

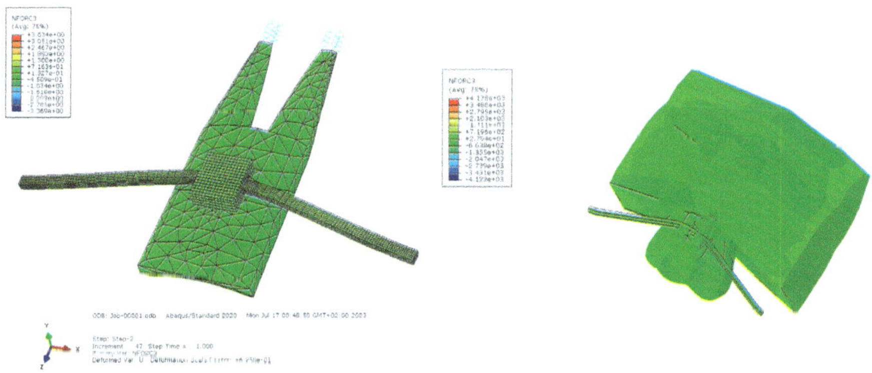

(a) (b)

Fig. 6. Deformed FE models after the first step: (a) model 1; (b) model 2

The NiTi superelastic behavior is observed when plotting this reaction force versus the displacement of the middle of the archwire (see Fig. 7). This figure shows the comparison between the last numerical model performed by [4] and [5] which models the bone by springs, and our new model, which considers the alveolar bone with both parts (cortical and cancellous). Before adjusting the spring stiffness, we carried out a thorough comparison of the results from both models to visually assess the force response and ensure that we had made the right choice to improve this study. In both numerical simulations, the deactivation force differs significantly: 6.4 N in the first model and 1.68 N in the second model, with a target range of [0.05-1N] [8], for effective and painless correction.

In addition, notice that with model 1 the correction is not completed because of the rigidity of the springs whereas with model 2, the 4 mm misalignment is fully recovered.

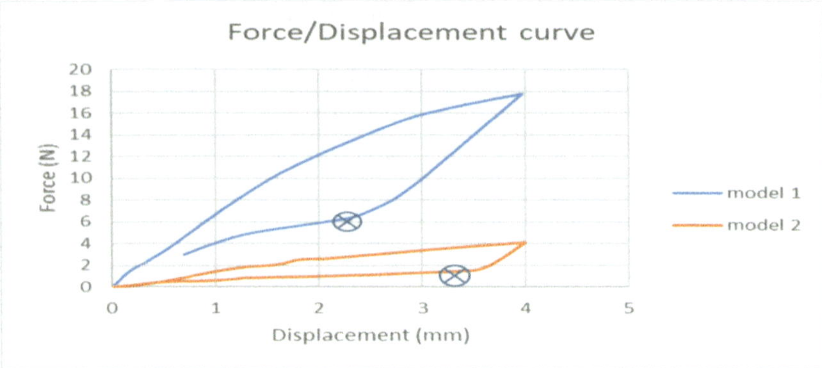

Fig. 7. Force/displacement curves of a 4 mm misalignment for both models

6 Conclusion

This study is in the framework of the numerical simulations aiming the modelling of the use of NiTi alloy orthodontic archwire to correct malocclusions.

It consists in an improvement of a previous model by considering parts of the alveolar bone as well as both parts of the tooth. The comparison of the results of both models showed that due to this more realistic numerical model, we obtain an orthodontic force in the range of the admissible ones.

As for the future of our work, we intend to develop a script that will facilitate the compilation of the selected parameters and the evaluation of the dental correction force. This advance will allow us to further optimize our recommendations for orthodontic treatment and provide a more personalized approach based on real data for each patient.

References

1. Singh, J.R., Kambalyal, P., Jain, M., Khandelwal, P.: Revolution in orthodontics: finite element analysis. J. Int. Soc. Prev. Commun. Dent. **6**(2), 110–114 (2016)
2. Rossini, G., Modica, S., Parrini, S., Deregibus, A., Castroflorio, T.: Incisors extrusion with clear aligners technique: a finite element analysis study. Appl. Sci. **11**(3), 1167 (2021)
3. Cunha, T.D.M.A.D., Barbosa, I.D.S., Palma, K.K.: Orthodontic digital workflow: devices and clinical applications. Dental Press J. Orthod. **26**, 1–34 (2021)
4. Gannoun, M., Laroussi, M., Bouraoui, T.: Détermination de la force générée par un fil orthodontique. In: CFM 2017–23ème Congrès Français de Mécanique. AFM, Maison de la Mécanique (2017)
5. Laroussi, M., Gzara, H., Bouraoui, T.: Design of experiments for numerical simulations of the superelastic behaviour of Ni-Ti orthodontic archwires, Int. J. Mech. Control, In press
6. Seo, J.H., et al.: Comparative analysis of stress in the periodontal ligament and center of rotation in the tooth after orthodontic treatment depending on clear aligner thickness—Finite element analysis study. Materials **14**(2), 324 (2021)
7. Rasmussen, M.J., Togrye, C., Trojan, T.M., Tantbirojn, D., Versluis, A.: Post-gel shrinkage, elastic modulus, and stress generated by orthodontic adhesives. Angle Orthod. **90**(2), 278–284 (2020)

8. Gannoun, M., Laroussi, M., Bouby, C., Ben Zineb, T., Bouraoui, T.: Numerical simulation of the force generated by a superelastic NiTi orthodontic archwire during tooth alignment phase: comparison between different constitutive models. Mater. Res. Express **5**(4), 045405 (2018)
9. Echchorfi, R.: *Dialogue essais-simulation et identification de lois de comportement d'alliage à mémoire de forme en chargement multiaxial* (Doctoral dissertation, Ecole nationale supérieure d'arts et métiers-ENSAM) (2013)
10. Gannoun, M.: *Contribution à une meilleure connaissance du comportement mécanique en service des arcs orthodontiques en Ni Ti superélastiques* (Doctoral dissertation, Ecole nationale des ingénieurs de Monastir-ENIM) (2020)
11. Sugiura, T., Yamamoto, K., Kawakami, M., Horita, S., Murakami, K., Kirita, T.: Influence of bone parameters on peri-implant bone strain distribution in the posterior mandible. Medicina oral, patologia oral y cirugia bucal **20**(1), e66 (2015)
12. Chen, S., Rittel, D., Yona, K.S.: The normal stiffness of the edentulous alveolar process. Bone Rep. **14**, 101066 (2021)
13. Lakhani, K., Vashishth, V., Gugnani, N.: Root surface area measurement of permanent dentition in Indian population–CBCT analysis. Inf. Med. Unlocked **9**, 1–5 (2017)

Finite Element Investigation on Bone Remodeling of FGM Dental Prosthesis

Sameh Elleuch[1(✉)], Hanen Jrad[1,2], Mondher Wali[1,2], and Fakhreddine Dammak[1]

[1] Laboratory of Electrochemistry and Environment (LEE), National Engineering School of Sfax, ENIS, University of Sfax, Sfax, Tunisia
{sameh.elleuch, fakhreddine.dammak}@enis.tn

[2] École Supérieure des Sciences et de la Technologie de Hammam Sousse, University of Sousse, Rue Lamine Abassi, 4 11 Hammam Sousse, Tunisia
hanen.jrad@essths.u-sousse.tn

Abstract. Osseointegrated dental implants have been used in recent years to replace missing teeth. The primary objective of the proposed research is to investigate how functionally graded (FG) implants influence the process of bone remodeling and identify the FG composition that promotes bone adaptation. This study combines a bone remodeling algorithm with finite element analysis to model and assess the bone remodeling process. To achieve this, UMAT subroutines and custom defined materials were integrated into the Abaqus software. The analysis demonstrated the benefits of functionally graded material (FGM) implants when compared to conventional titanium implants. The study revealed that the gradual change in material composition can indeed impact the bone adaptation process. More precisely, the findings suggest that employing functionally graded implants with reduced metal content can improve the bone remodeling process.

Keywords: dental implant · remodeling algorithm · FGM model · 3D EF model

1 Introduction

Dental implant present economic and efficient solution for replacing missing teeth (Joda et al. 2017). In this context, it is necessary to study the relationship between stress distribution and characteristics of the implant near the bone interface. Several finite element (FE) analyses have been conducted to assess the biomechanical performance of dental implants. (Demenko et al. 2014) and (Elleuch et al. 2021) conducted research aimed at assessing the influence of implant type and geometry. Additionally, other studies have explored factors such as implant inclination angle (Wang et al. 2016), bone quality (I Chiang et al. 2014), and loading conditions (Takahashi et al. 2015) on dental implant performance. The stress patterns induced by osseointegrated dental implants have the potential to trigger alterations in bone remodeling, as indicated by studies such as (Jaquiery et al. 2004) and (Elleuch et al. 2023,c). These changes may pose a risk of bone failure. Therefore, it is crucial to investigate the distribution of stress within the bone when dental implants are applied.

M. Chouchane et al. (Eds.): CMSM 2023, LNME, pp. 222–231, 2024.
https://doi.org/10.1007/978-3-031-65007-9_25

Functionally Graded Materials (FGMs) are advanced composites that feature a smooth transition in composition and structure from one material on one surface to a different material on the opposite surface, commonly blending metals and ceramics. This innovative configuration is deliberately engineered to circumvent the discontinuities that typically arise in standard composites. FGMs are characterized by their continuously changing properties along the material's length, which are determined by specific laws and vary in volume fractions. This unique characteristic has garnered considerable interest among researchers and engineers from diverse domains. Notable contributions to the study of FGMs include research by (Van Tung & Duc 2010), (Li et al. 2020), (Zur et al. 2020), (Fu et al. 2021), (Khorasani et al. 2022), (Civalek et al. 2022), (Ghobadi et al. 2022), (Jankowski et al. 2022) and (Elleuch et al. 2023a,b), highlighting the broad and interdisciplinary engagement in this field.

Extensive research efforts have resulted in the introduction of functionally graded materials (FGM) in dental implants, along with the proposal of optimized implant designs aimed at enhancing stress distribution (Hedia et al. 2014). In this context, the primary objective of this study is to propose a three-dimensional finite element model of the dental bone system, incorporating varying compositions of implant materials such as pure titanium (Ti) and Titanium/Hydroxyapatite(Ti/HAP) functionally graded materials (FGM) with different gradations. The aim is to assess how the implant structure influences bone remodeling and its consequent effect on stress distribution within the dental implant system.

2 Materials and Methods

2.1 FGM Model

This study assesses the impact of functionally graded materials (FGM) dental implants on bone remodeling in the context of osseointegrated dental prostheses. The FGM implant configuration used in this study follows the methodology proposed by (Lin et al. 2009). Both the implant and abutment are modeled using a Titanium/Hydroxyapatite composition. The mechanical properties are assumed to exhibit a continuous and gradual variation along the axial direction. The volume fractions of the functionally graded material can be determined using the following expressions, Ellouz et al. (2023):

$$V_c = (\frac{y}{h})^n \tag{1}$$

$$V_c = 1 - V_m \tag{2}$$

where

h: the overall height of the abutment and implant structure
y: the vertical position within the implant region
V_m: the volume fraction of titanium (metal)
V_c: the volume fraction of Hydroxyapatite (ceramic)
n: the power gradient parameter

The Poisson's ratio υ and, the Elastic modulus (E) can be calculated as:

$$E = E_c \left[\frac{E_c + (E_m - E_c)V_m^{2/3}}{E_c + (E_m - E_c)(V_m^{2/3} - V_m)} \right] \tag{3}$$

$$\upsilon = \upsilon_c V_c + \upsilon_m V_m \tag{4}$$

In this study, the Young's modulus and Poisson's ratios of the (FGM) dental implant range from 40 GPa to 110 GPa and 0.27 to 0.35, respectively. A 3D finite element model was constructed, including pure titanium as well as different arrangements of Titanium/Hydroxyapatite Functionally Graded Material (FGM) with varying gradations ($n = 1$, $n = 5$, $n = 10$). The goal is to pinpoint the most suitable dental implant design that encourages favorable bone remodeling during the implantation process.

2.2 Finite Element Analysis

The bony structure and implant geometry were created using the Solidworks software package. A comprehensive depiction of the dental structure model can be seen in Fig. 1. The human mandible consists of two parts: the cancellous bone, enclosed by a cortical layer that is 2 mm thick. Within this bone system, there is a cylindrical screw type implant, measuring 10 mm in length and 3.75 mm in diameter. It is assumed that all materials, except for the functionally graded material (FGM) implants, possess isotropic, linear elastic, and homogeneous properties. The mechanical characteristics of the components employed in this research are detailed in Table 1. The numerical models were discretized employing 4-node linear tetrahedral elements, identified as the C3D4 type. Detailed information on element assignments for each model is provided in Table 2.

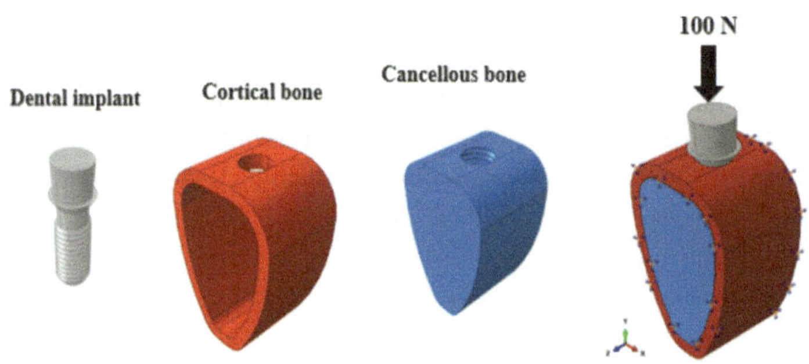

Fig. 1. Three dimensional geometric representations of implant bone system

In this study, it was assumed that the interfaces between the implant and bones and cortical and trabecular bone demonstrate perfect bonding, as elaborated in the research conducted by (Kaman and Celik. 2013). The boundary conditions involved constraining the nodes at both the distal and mesial ends of the bone in all directions to restrict both rotational and translational movements, following the methodology outlined by Lin et al. (2009). A vertical force of 100 N, representing the nominal chewing load as identified in

Table 1. Material properties employed in this research

Material	Young's modulus (GPa)	Poisson's ratio	References
Titanium Ti6Al4V (Ti)	110	0.35	Lin et al. 2010
HAP	40	0.27	Mahmoudi et al. 2018
Cortical bone	14	0.30	Najeeb et al. 2016
Cancellous bone	3	0.30	Kaman and Celik. 2013

the literature (Hedia et al., 2004), was applied to the top of the implant, as illustrated in Fig. 1. To model the bone remodeling process, user-defined subroutines (UMAT) were integrated into the ABAQUS software. The assessment of material properties was based on the local bone density.

Table 2. Structural components and nodal points within the bone implant.

	Dental implant	Cortical bone	Cancellous bone
Elements	8768	13709	37286
Nodes	2029	3519	7412

2.3 Bone Remodeling Algorithm

The bone remodeling process consists of three sequential phases: apposition, equilibrium, and resorption. Mathematical formulations elucidating bone remodeling are detailed in the works of (Lin et al. 2008) and (Elleuch et al. 2023) and presented as follows:

Table 3. Bone remodeling value

	Symbol	Value
Remodeling constant	B	1 (g/cm^3)
Elastic strain energy	U	Calculated
Mechanical stimulus	Ψ	Calculated
Bone density	ρ	Calculated
Remodeling reference	Ξ	0.0036 J/g
Bandwidth of bone remodeling	δ	10%
Mechanical stress	σ	Calculated
Mechanical strain	ε	Calculated

Bone apposition:

$$f(\rho) = \frac{\partial \rho}{\partial t} = B[\Psi - (1 + \delta)\Xi] if \Psi > (1 + \delta)\Xi \tag{5}$$

Bone equilibrium:

$$f(\rho) = \frac{\partial \rho}{\partial t} = 0 if (1 - \delta)\Xi \leq \Psi \leq (1 + \delta)\Xi \tag{6}$$

Bone resorption:

$$f(\rho) = \frac{\partial \rho}{\partial t} = B[\Psi - (1 - \delta)\Xi] if \Psi < (1 - \delta)\Xi \tag{7}$$

The formula relies on constant values (Table 3) sourced from the existing literature (Lin et al. 2008) where,

Bone density values are computed using the remodeling equations, it becomes possible to establish the relationship between bone density and young modulus using the following set of equations, as outlined by (Lin et al. 2010).

$$E = 2.349\rho^{2.15} \text{ for cancellous bone} \tag{8}$$

$$E = -23.93 + 24\rho \text{ for cortical bone} \tag{9}$$

It is assumed that the initial density of cortical bone is 1.74 g/cm^3, accompanied by an elastic modulus of 18 GPa, while the initial density of cancellous bone is assumed to be 0.9 g/cm^3, with a young modulus of 2 GPa, according to (Lin et al. 2009). The bone remodeling equations are solved by linking the increase in bone density to the time increment across different phases, utilizing Euler's forward integration method in conjunction with the Newton-Raphson technique.

The finite element analysis is assessed using Abaqus software, alongside the development of a FORTRAN program crafted to replicate the bone remodeling procedure.

3 Results and Discussion

To validate our findings, we compared the von Mises stress within the implant bone system derived from our present model with those reported in the research conducted by (Djabbar et al. 2010), was composed entirely of pure titanium, as illustrated in Fig. 2. We meticulously recreated the same simulation conditions. As shown in Table 4, the outcomes demonstrate a robust concurrence, with any slight differences potentially attributable to variations in geometry.

Evaluating how stress is distributed plays a pivotal role in evaluating performance of a dental implant system in real clinical situations. In the field of dental implantology, stress distribution can serve as an indicator of whether an implant is likely to succeed or fail over extended periods of use. Therefore, this study primarily focuses on investigating

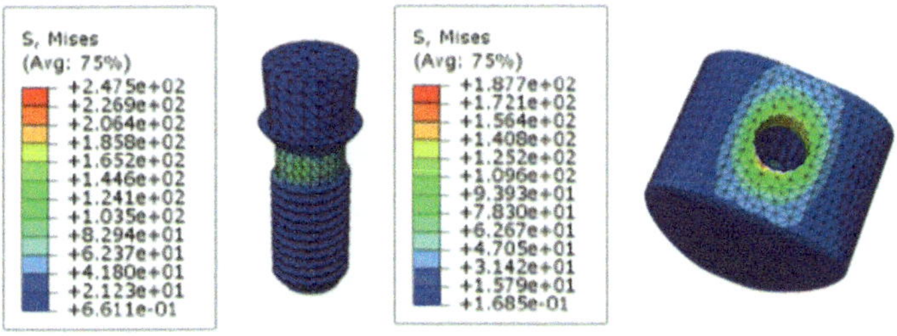

Fig. 2. Distribution of stress within implant- bone system

Table 4. Comparison of the highest von Mises stress within bone-implant

Component	Maximum Von Mises stress (MPa)	
	Present work	Djabbar et al.2010
Dental implant	247.5	245.2
Bone system	187.7	161

how functionally graded materials (FGM) affect an osseointegrated dental implant in a simulated remodeling process. Stress analysis of a pure titanium implant suggests that this particular fixture typically encounters elevated stress levels compared to the FGM (Functionally Graded Material) implant as depicted in Fig. 3. These elevated stress levels could potentially lead to the failure of the dental implant.

In the case of FGM implants, an increase in the metallic component leads to a corresponding rise in the maximum Von Mises stress. The peak stress value is reached when n equals 10.

Throughout the remodeling process, alterations in bone density were consistently monitored to assess the influence of implant materials. Figure 4 showcases the progression of remodeling resulting from FGM dental implants. These images illustrate the evolving bone density over a span of 48 months during the bone remodeling process. The color spectrum ranges from blue, indicating low density regions, to red as density increases. Initially, the bone system displays a uniform distribution of density.

The bone density progression is represented in Fig. 5. It is evident that in case of FGM implants, the bone density is higher than titanium implant.

This suggests that the FGM dental implant transmits more strain energy into the implant bone system owing to its higher stiffness compatibility, consequently mitigating the stress shielding effect typically associated with a stiffer titanium implant. Furthermore, it is evident that a more substantial degree of remodeling takes place when the power gradient parameter n is set to 1, indicating a higher concentration of hydroxyapatite. These results highlight the advantageous remodeling outcomes obtained with FGM implants when contrasted with conventional pure titanium dental implants.

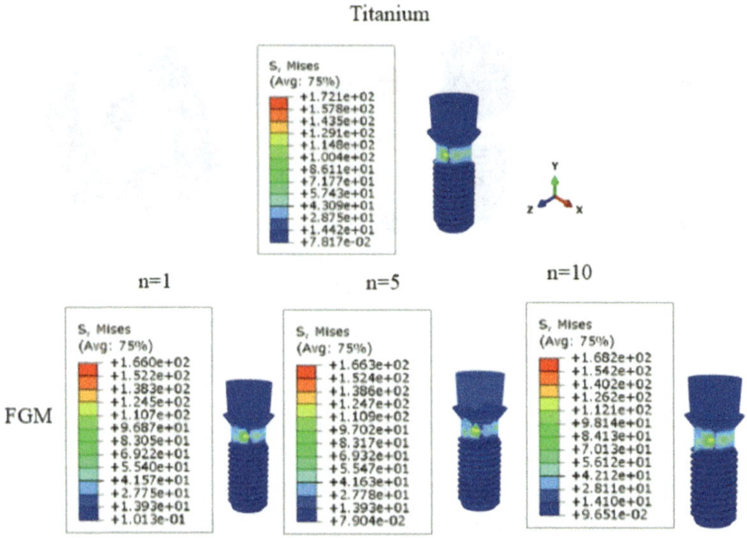

Fig. 3. Stress distribution in Osseo integrated dental implant

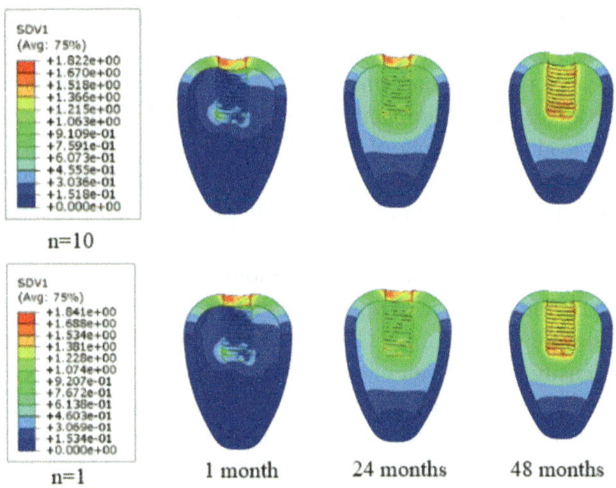

Fig. 4. Bone density contour subjected to FGM implant

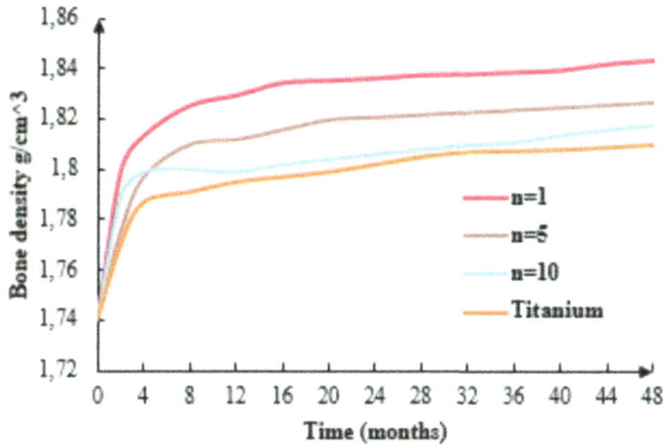

Fig. 5. Change in bone density at bone system during 48 months

4 Conclusions

This study delves into a numerical investigation of bone density and stress distribution to assess the impact of implant materials, specifically focusing on FGM dental implants, on the process of bone remodeling. The material properties of these FGM implants transition from Titanium to Hydroxyapatite along the axial direction. From these simulations of bone remodeling, we can deduce the following conclusions: modifications in ceramic composition, particularly a shift towards lower composition values, significantly impact the bone remodeling process, potentially leading to an ideal implant composition. The results indicate that FGM with reduced metal content ($n = 1$) not only enhances bone density but also reduces stress within the implant. Consequently, this configuration holds promise for a more successful clinical outcome and can be regarded as an optimal model for both remodeling and implant stability.

References

1. Civalek, Ö., Uzun, B., Yaylı, M.Ö.: An effective analytical method for buckling solutions of a restrained FGM nonlocal beam. Comput. Appl. Math. **41**(2), 67 (2022)
2. Demenko, V., Linetskiy, I., Nesvit, K., Hubalkova, H., Nesvit, V., Shevchenko, A.: Importance of diameter-to-length ratio in selecting dental implants: a methodological finite element study. Comput. Methods Biomech. Biomed. Eng. **17**, 443–449 (2014)
3. Djebbar, N., Serier, B., Bouiadjra, B.B., Benbarek, S., Drai, A.: Analysis of the effect of load direction on the stress distribution in dental implant. Mater. Design. **31**, 2097–2101 (2010)
4. Elleuch, S., Jrad, H., Amir, K., Wali, M., Dammak, F.: Design optimization of implant geometrical characteristics enhancing primary stability using FEA of stress distribution around dental prosthesis. Comput. Methods Biomech. Biomed. Engin. **24**(9), 1035–1051 (2021). https://doi.org/10.1080/10255842.2020.1867112
5. Elleuch, S., Jrad, H., Wali, M., Dammak, F.: Finite element analysis of the effect of porosity on biomechanical behaviour of functionally graded dental implant. Proc. Inst. Mech. Eng. Part E: J. Process Mech. Eng. 09544089231197857 (2023)

6. Elleuch, S., Jrad, H., Wali, M., Dammak, F.: Mandibular bone remodeling around osseointegrated functionally graded biomaterial implant using three dimensional finite element model. Int. J. Numer. Methods Biomed. Eng. **39**(9), e3750 (2023)
7. Elleuch, S., Jrad, H., Wali, M., Dammak, F.: Agglomeration effect on biomechanical performance of CNT-reinforced dental implant using micromechanics-based approach. J. Mech. Behav. Biomed. Mater. **145**, 106023 (2023)
8. Ellouz, H., Jrad, H., Wali, M., Dammak, F.: Large deflection analysis of FGM/magneto-electro-elastic smart shells with porosities under multi-physics loading. Mech. Adv. Mater. Struct. 1–25 (2023)
9. Ellouz, H., Jrad, H., Wali, M., Dammak, F.: Numerical modeling of geometrically nonlinear responses of smart magneto-electro-elastic functionally graded double curved shallow shells based on improved FSDT. Comput. Math. Appl. **151**, 271–287 (2023)
10. Fu, Z.J., Yang, L.W., Xi, Q., Liu, C.S.: A boundary collocation method for anomalous heat conduction analysis in functionally graded materials. Comput. Math. Appl. **88**, 91–109 (2021)
11. Ghobadi, A., Beni, Y.T., Żur, K.K.: Porosity distribution effect on stress, electric field and nonlinear vibration of functionally graded nanostructures with direct and inverse flexoelectric phenomenon. Compos. Struct. **259**, 113220 (2021)
12. Hedia, H.S., Mahmoud, N.A.: Design optimization of functionally graded dental implant. Bio-med. Mater. Eng. **14**(2), 133–143 (2004)
13. Hedia, H.S., Aldousari, S.M., Abdellatif, A.K., Fouda, N.: A new design of cemented stem using functionally graded materials (FGM). Bio-med. Mater. Eng. **24**(3), 1575–1588 (2014)
14. I-Chiang, C., Shyh-Yuan, L., Ming-Chang, W., Sun, C.W., Jiang, C.P.: Finite element modelling of implant designs and cortical bone thickness on stress distribution in maxillary type IV bone. Comput. Methods Biomech. Biomed. Eng. **17**, 516–526 (2014)
15. Jankowski, P., Żur, K.K., Farajpour, A.: Analytical and meshless DQM approaches to free vibration analysis of symmetric FGM porous nanobeams with piezoelectric effect. Eng. Anal. Boundary Elem. **136**, 266–289 (2022)
16. Jaquiéry, C., et al.: Reconstruction of maxillary and mandibular defects using prefabricated microvascular fibular grafts and osseointegrated dental implants–a prospective study. Clin. Oral Implant Res. **15**(5), 598–606 (2004)
17. Joda, T., Ferrari, M., Gallucci, G.O., Wittneben, J.G., Brägger, U.: Digital technology in fixed implant prosthodontics. Periodontology 2000, **73**(1), 178–192 (2017)
18. Kaman, M., Celik, N.: Effects of thread dimensions of functionally graded dental implants on stress distribution. World Acad. Sci. Engin. Tech. **78**, 2020–6 (2013)
19. Khorasani, V.S., Żur, K.K., Kim, J., Reddy, J.N.: On the dynamics and stability of size-dependent symmetric FGM plates with electro-elastic coupling using meshless local Petrov-Galerkin method. Compos. Struct. **298**, 115993 (2022)
20. Li, Y.D., Tang, Z.C., Fu, Z.J.: Generalized finite difference method for plate bending analysis of functionally graded materials. Mathematics **8**(11), 1940 (2020)
21. Lin, D., Li, Q., Li, W., Swain, M.: Bone remodeling induced by dental implants of functionally graded materials. J. Biomed. Mater. Res. Part B: Appl. Biomater.: Off. J. Soc. Biomater. Japan. Soc. Biomater. Austr. Soc. Biomater. Kor. Soc. Biomater. **92**(2), 430–438 (2010)
22. Lin, D., Li, Q., Li, W., Swain, M.V.: Functionally graded dental implant and its effect on bone remodeling. In: Advanced Materials Research, vol. 47, pp. 1035–1038. Trans Tech Publications Ltd. (2008)
23. Lin, D., Li, Q., Li, W., Zhou, S., Swain, M.V.: Design optimization of functionally graded dental implant for bone remodeling. Compos. Part B: Eng. **40**(7), 668–675 (2009)
24. Mahmoudi, M., Saidi, A.R., Hashemipour, M.A., Amini, P.: The use of functionally graded dental crowns to improve biocompatibility: a finite element analysis. Comput. Methods Biomech. Biomed. Engin. **21**(2), 161–168 (2018)

25. Najeeb, S., Zafar, M.S., Khurshid, Z., Siddiqui, F.: Applications of polyetheretherketone (PEEK) in oral implantology and prosthodontics. J. Prosthodont. Res. **60**(1), 12–19 (2016)
26. Van Tung, H., Duc, N.D.: Nonlinear analysis of stability for functionally graded plates under mechanical and thermal loads. Compos. Struct. **92**(5), 1184–1191 (2010)
27. Wang, C., Zhang, W., Ajmera, D.H., Zhang, Y., Fan, Y., Ji, P.: Simulated bone remodeling around tilted dental implants in the anterior maxilla. Biomech. Model. Mechanobiol. **15**(3), 701–712 (2016)
28. Żur, K.K., Arefi, M., Kim, J., Reddy, J.N.: Free vibration and buckling analyses of magneto-electro-elastic FGM nanoplates based on nonlocal modified higher-order sinusoidal shear deformation theory. Compos. B Eng. **182**, 107601 (2020)

Mechanical Characterization of Antibacterial Dental Composite PMMA with Nano-particles of the S. Persica Extract and Hydroxyapatite as Fillers

Rihem Chaaben[1(✉)], Ayman Ayedi[2], and Khaled Elleuch[1]

[1] Materials Engineering and Environment Laboratory (LGME), National School of Engineers of Sfax (ENIS), Sfax, Tunisia
rihemchaabene@gmail.com, khaled.elleuch@enis.tn
[2] Laboratory of Advanced Materials (LAMA), National School of Engineers of Sfax (ENIS), Sfax, Tunisia

Abstract. The current study examines the mechanical behavior, specifically focusing on the compressive strength, the bending strength, and the bending modulus of poly (methyl methacrylate) (PMMA) reinforced with the extract of Salvadora persica (S. persica) and the hydroxyapatite (HA). The elaborated composite was designated for use in dental applications (dental restoration and dental prosthesis). Different percentages of S. persica extract nanoparticles and/or HA microparticles were employed to enhance the antibacterial and the mechanical properties, respectively. Compressive strength, bending strength, and bending modulus were investigated using a universal mechanical testing machine. The main results revealed that the optimum percentage of HA reinforcement was 10%. Additionally, the integration of 10% S. persica extract slightly improved these mechanical properties; however, a 20% concentration reduced them. Regarding the hybrid composite, the addition of fillers generally decreased the mechanical properties. As a consequence, the composition (90% PMMA + 10% S. persica extract) exhibited the best antibacterial activity and mechanical properties.

Keywords: PMMA · S. persica extract · Hydroxyapatite · Mechanical properties · Dental application

1 Introduction

Since the 1960s, with the emergence of dental resin composites, significant efforts have been directed towards creating materials with acceptable physical and mechanical characteristics [1]. Moreover, improvements and revolutionary developments have been established in the material compositions. Nevertheless, dental prostheses present intricate challenges, including the impact of mastication forces and the existence of acidic biofilms that can affect their durability. Currently, direct restorations exhibit a yearly failure rate of up to 7.9%, primarily attributed to secondary caries and bulk fracture [2–4]. The oral cavity comprises an intricate environment characterized by high

M. Chouchane et al. (Eds.): CMSM 2023, LNME, pp. 232–239, 2024.
https://doi.org/10.1007/978-3-031-65007-9_26

humidity, adequate temperature, and a plethora of nutrients, which foster the growth of diverse microbial biofilms [5–7].

Hence, it is very important to create a dental composite capable of reducing the occurrence of recurrent caries and preventing the growth of acidic biofilms while maintaining satisfactory mechanical properties. Presently, researchers have been concentrating on enhancing the antimicrobial attributes of biomaterials using various agents, including metal nanoparticles [8], ceramic particles, and particles derived from medicinal plants like Salvadora persica. Previously, S. persica powder was generated and co-polymerized into PMMA dental resin, yielding successful antibacterial results [9]. Nevertheless, there's a need to enhance the antibacterial effectiveness of this composite to limit more and more the formation of diverse bacteria and microbial biofilms. Additionally, there has been considerable discussion about the antibacterial potential of the S. persica extract in the context of pulp tissue regeneration therapy. It has been reported that the S. persica extract possesses various biological properties that are pertinent to combatting the organisms linked to dental plaque and periodontitis [10]. Therefore, extracting S. persica and integrating this extract enhance the antibacterial properties of the composite.

In addition, hydroxyapatite (HA) has been employed as a filler to enhance the mechanical characteristics of dental resin composites. Notably, HA is biocompatible and possesses a chemical composition similar to tooth enamel [11, 12]. The inclusion of HA in restorative dentistry presents a multitude of appealing benefits, such as intrinsic radio-opaque response and better mechanical performance [13].

Consequently, the objective of this work was to elaborate an innovative bioactive composite resin, which has antibacterial activity and suitable mechanical properties that fall within the range recommended by the International Organization for Standardization (ISO), by incorporating the S. persica extract nanoparticles and HA micro-particles into PMMA dental resin.

2 Materials and Methods

A dental composite was elaborated using PMMA as a matrix with varying fillers percentages. The PMMA (Nic Tone, MDC dental, USA) was utilized as the monomer /polymer ratio of 1:2.

To begin with, the elaborated HAP powder was used as a reinforcing material. From nearby slaughterhouses, cortical bovine bones have been gathered. The spongy portions of the bone including the ligaments and tissues attached to the bone have been cut off and the marrow has been removed. In order to eliminate organic matter and collagen, the clean bone specimens have been defatted by boiling them in water for two hours. After that, they have been calcined in an electric furnace (Nabertherm, France) at 800 °C for 6 h using 2 °C/min as a heating rate. Before being crushed with a mortar pestle and sieved, the thermally treated bovine bone samples have been gradually cooled to room temperature [14]. In this investigation, the mean particle size of HA is around 40 μm. Micro-scale HA particles were chosen because nanoscale-HA is a hygroscopic filler, making composites containing it unsuitable for dental restoration due to the significant adverse effect of water. Furthermore, the use of microscopic HA, as opposed to nanoscopic HA, as a reinforcing filler, led to improved mechanical properties [13].

S. persica stems (Miswak, Sewak alBaraka, Pakistan) were air-dried. Then, the stems were processed in a food-processing blender (a coffee-grinder, Moulinex, China). An alcoholic extract was obtained through a cold maceration process, where 20 g of finely powdered S. persica was immersed in ethanol (200 ml) and left undisturbed for 48 h. Subsequently, the filtration of the extract was carried out, and the solvent was totally removed using a rotary evaporator. The mean particle size of the extract is 70 nm. The obtained extract was then stored at 4 °C until further use [15]. The antibacterial activity of the S. persica extract was tested against various oral bacteria at different temperatures, and the results confirmed this activity up to 60 °C. Moreover, the antibacterial activity of the composite containing this extract was also confirmed.

The composites specimens were fabricated by mixing various weight percentages of fillers with PMMA matrix by compression molding. To increase the dispersion of the fillers in the resin, the mixture (PMMA, HA and/or the extract of S. persica) is agitated mechanically at a high shear rate. It is worth noting that many researchers working on nanocomposites at the laboratory scale employ the same manufacturing process (the direct way) based on mechanical stirring, which has been described by the company Ashland [16]. The polymerization of the composites was carried out using a pressure device (Mini Major 2000, Major, Italy) at a temperature of 60 °C for 10 min, applying a pressure of 0.5 MPa. Bending strength, bending modulus and compressive strength of the different samples were measured using a universal mechanical testing machine (TIME WDW-50E). Rectangular bar specimens (64 mm × 10 mm × 3,3 mm) were employed for the three-point bending test (crosshead speed 5 mm/min, span 50 mm) according to the norm ISO (20795-1:2013). To determine the compressive strength, cylindrical specimens (13 mm × 26 mm) were prepared according to ASTM D695 standards for the compressive test, with a loading rate of 1.3 mm/min.

3 Results and Discussion

The long-term clinical application of dental prostheses and dental restoratives requires durable mechanical properties for the resin composite. To evaluate the reinforcing effect of HA, compressive strength (σc), bending strength (σf) and bending modulus (Ef) of the dental resin filled with different weight percentages (5 wt%, 10 wt%, 15 wt%, and 20 wt%) of HA micro-particles and the reference samples (unfilled PMMA) were assessed and the results are exposed in Fig. 1.

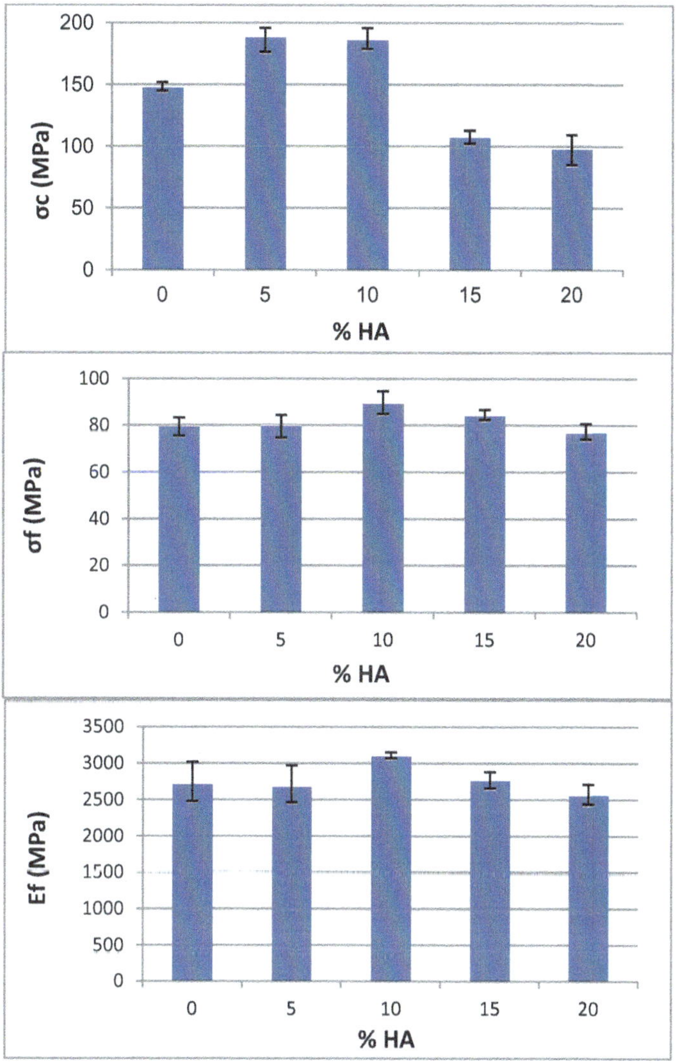

Fig. 1. Compressive strength (σc), bending strength (σf) and bending modulus (Ef) of composites reinforced with different percentages of HA.

The incorporation of HA particles increases the values of σc, σf and Ef compared with the unfilled resin to achieve a maximum at 10% HA. Nevertheless, enhancing the HA concentration beyond 10 wt% did not lead to further improvement in the mechanical properties. Consequently, using HA as a filler successfully improved the mechanical properties of PMMA. This result can be attributed to HA being embedded in PMMA, strengthening the resin and enhancing its strength, while also taking a higher modulus than the matrix, resulting in an increase in the bending modulus of the composite [1]. With higher filler loading, the weakening of the reinforcing effect of HA may have been caused by the undesirable distribution of HA and the formation of agglomerates. Therefore, the stress could be less transferred between the matrix and HA, decreasing the resistance to fracture [17].

As a result, the inclusion of HA micro-particles up to 10% resulted in an enhanced mechanical properties compared to PMMA, then these mechanical properties were decreased. Thus the optimum percentage of HA reinforcement was 10%.

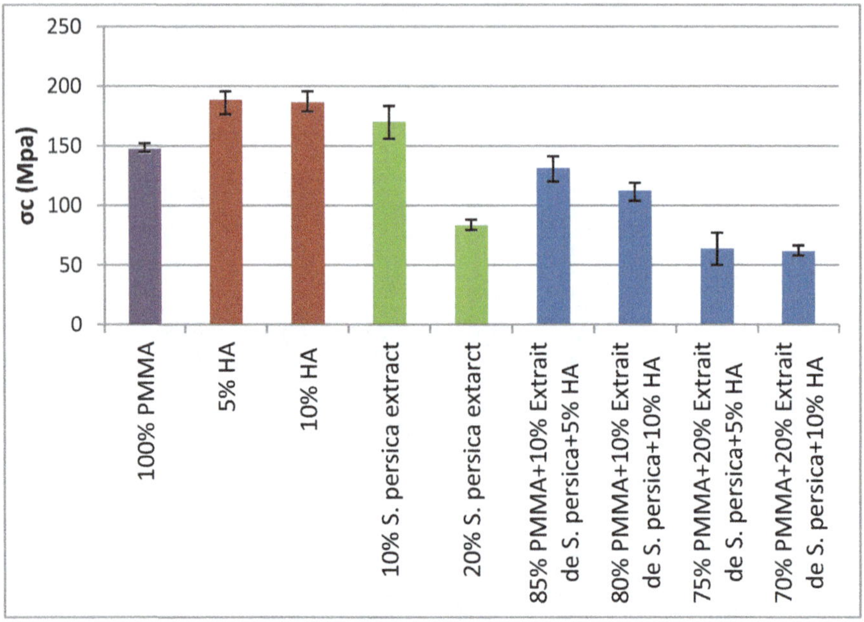

Fig. 2. Compressive strength (σc) of composites reinforced with different percentages of *S. persica* extract and/or HA.

Figures 2 and 3 show the compressive strength (σc), bending strength (σf) and bending modulus (Ef) of composites reinforced with varying percentages of HA and/or S. persica extract. Regarding the incorporation of the S. persica extract, a 10% concentration of the extract increased slightly these mechanical properties, while a 20% concentration decreased them. Generally, the size of filler can make a noticeable change to the composite performance [16, 18]. As the size of the filler decreases and the filler percentage increases, the interface area between the filler particles and the matrix increases, resulting in well transmission of applied load and enhancement in the mechanical performance of the resin-composite [13, 16]. Therefore, reinforcing the matrix with nanoscale filler particles leads to superior performance compared to using micro-scale particles [19, 20]. Furthermore, the degree of interfacial bonding between the filler and the matrix considerably influences the mechanical properties of composite [20]. Indeed, the use of a coupling agent for the filler particles is necessary to enhance the interfacial bonding between the filler and the matrix. This surface treatment is feasible for most kinds of nano-fillers. However, in our case, the antibacterial activity of our extract will be affected,

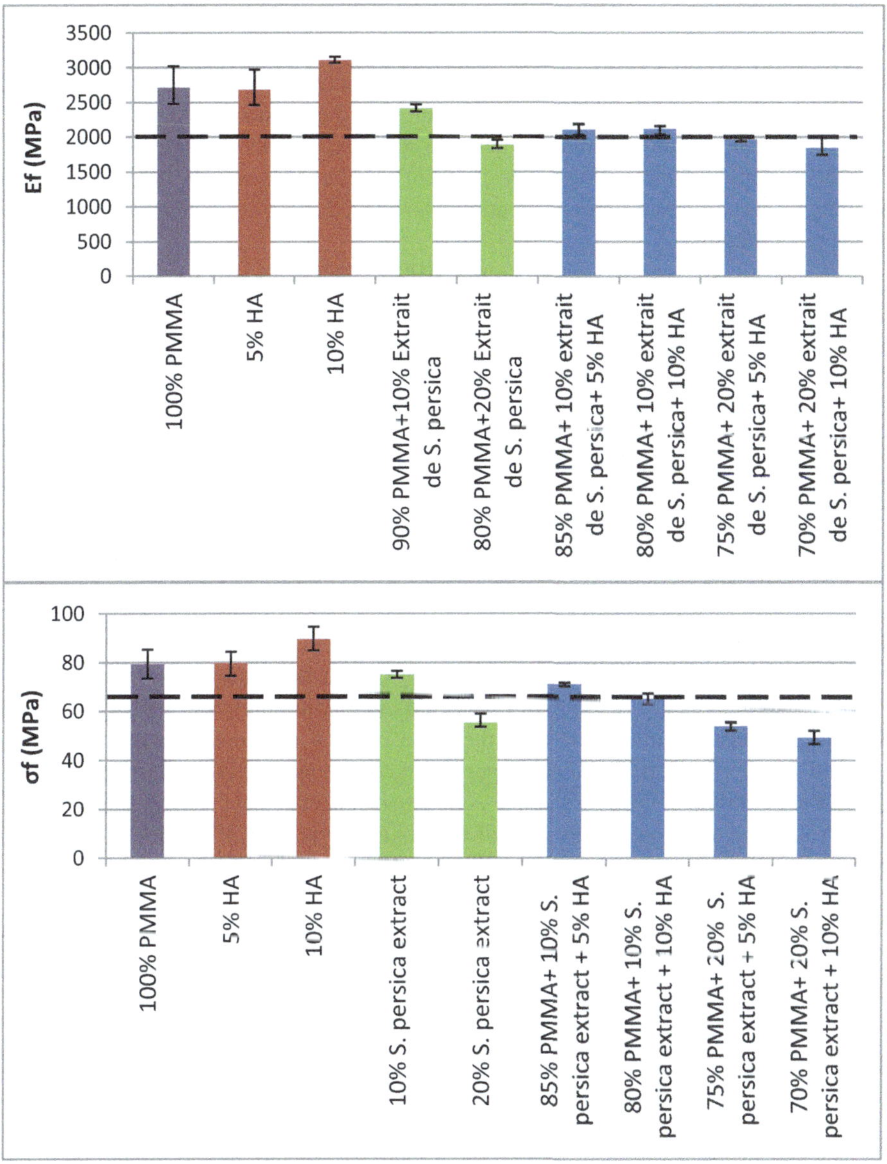

Fig. 3. Bending strength (σf) and bending modulus (Ef) of composites reinforced with different percentages of S. persica extract and/or HA.

even destroyed with this surface treatment. Consequently, the slight improvement in the compressive strength (σc), bending strength (σf) and bending modulus (Ef) of the composite following the incorporation of the nano-particles of the extract can be justified by the lack of coupling agent between fillers and resin. Additionally, the formation of filler particle aggregates may occur within a specific range of filler percentages, reducing the overall mechanical stiffness of the microstructure, which could potentially lead to increased mechanical properties [21]. This phenomenon could also account for the reduction in these mechanical properties when both the extract and HA are combined. Indeed, when both nano and micro-fillers are added simultaneously, they may agglomerate or cluster together, leading to non-uniform distribution within the composite. Filler agglomeration can create stress concentrations and weak points in the material, reducing its overall strength.

As a result, the composite containing 10% S. persica extract achieved the best compromise between mechanical properties and antibacterial activity. Moreover, the compressive strength (σc) of this composite is considered an acceptable choice when compared to commercial dental composites. Research on the mechanical properties of composite materials yields an estimated compressive strength within the range of 70 to 280 MPa, depending on the specific commercial product [22]. In addition, the bending strength (σf) and bending modulus (Ef) were remained above the limits indicated by the standard norm (65 MPa and 2000 MPa, respectively).

4 Conclusion

The development of a nanofilled hybrid composite, achieved by adding nanoparticles from S. persica extract and HA microparticles at various percentages, is carried out with the purpose of enhancing the mechanical performance of the composite and incorporating the antibacterial activity into it. The inclusion of 10% HA or the extract separately resulted in improved mechanical properties in the composites. However, when both fillers were combined, the bending and compressive mechanical properties experienced a decrease. The composite with 10% of S. persica extract struck the best compromise between mechanical properties and antibacterial activity.

Acknowledgements. The authors also gratefully acknowledge the helpful comments and suggestions of the reviewers, which have improved the presentation.

References

1. Fengwei, L., Bin, S., Xiaoze, J., Sultan, S.A., Qinghong, Z., Meifang, Z.: Mechanical properties of dental resin/composite containing urchin-like hydroxyapatite. Dent. Mater. **30**, 1358–1368 (2014)
2. Demarco, F.F., Correa, M.B., Cenci, M.S.: Longevity of posterior composite restorations: not only a matter of materials. Dent. Mater. **28**, 87–101 (2012)
3. Laske, M., Opdam, N.J., Bronkhorst, E.M.: Longevity of direct restorations in Dutch dental practices. descriptive study out of a practice based research network. J. Dent. **46**, 12–17 (2016)
4. Delaviz, Y., Finer, Y., Santerre, J.P.: Biodegradation of resin composites and adhesives by oral bacteria and saliva: a rationale for new material designs that consider the clinical environment and treatment challenges. Dent. Mater. **30**, 16–32 (2014)
5. Dewhirst, F.E., Chen, T., Izard, J.: The human oral microbiome. J. Bacteriol. **192**, 5002–5017 (2010)

6. Souza, J., Mota, R.R., Sordi, M.B., Passoni, B.B.: Biofilm formation on different materials used in oral rehabilitation. Braz. Dent. J. **27**, 141–147 (2016)
7. Chaaben, R.R., Rym, T., Khaled, E., Manel, E., Thomas, K.: Wear behavior of new biomaterial composite for dental application. Polym. Polym. Compos. **28**(8–9), 654–662 (2020)
8. Tripathi, A., Jose, S.M.: Development of nano-antimicrobial biomaterials for biomedical applications. Adv. Biomater. Biomed. Appl. **66**, 479–545 (2017)
9. Chaaben, R., Rym, T., Basma, M., Noamen, G., Khaled, E.: Innovative biocomposite development based on the incorporation of Salvadora persica in acrylic resin for dental material. J. Thermoplast. Compos. Mater. **35**(11), 1–17 (2020)
10. Sofrata, A.H., Claesson, R.L.K., Lingstr¨om, P.K., Gustafsson, A.K.: Strong antibacterial effect of miswak against oral microorganisms associated with periodontitis and caries. J. Periodontol. **79**(8), 1474–1479 (2008)
11. Mitjan, K., Said, J., Lewis, K.I.: Effect of counterface roughness on abrasive wear of hydroxyapatite. Wear **252**, 679–685 (2002)
12. Tihana, T.G., Ionita, M.D., Popescua, R.G., Iordachescu, D.: Effect of hydrophilic–hydrophobic balance on biocompatibility of poly (methyl methacrylate) (PMMA)–hydroxyapatite (HA) composites. Mater. Chem. Phys. **118**, 265–269 (2009)
13. Raul, W.A., et al.: Mechanical properties of visible light-cured resins reinforced with hydroxyapatite for dental restoration. Dent. Mater. **18**, 49–57 (2002)
14. Aymen, A., Jamel, B.: Preparation and microstructural characterization of natural hydroxyapatite extracted from animal bones. JOM **74**(8), 3119–3132 (2022)
15. Nadia, C.C., Aicha, T.T.M., Catherine, M., Abdelkader, A., Boumediene, M.: In vitro and in vivo antimicrobial activity of Algerian Hoggar Salvadora persica L. extracts against microbial strains from children's oral cavity. J. Ethnopharmacol. **144**, 57–66 (2012)
16. Damien, M.M., Éric, G., Carine, C.J.: Properties of Nanofillers in Polymer, Nanocomposites and Polymers with Analytical Methods, Dr. John Cuppoletti (Ed.). ISBN: 978-953-307-352-1, InTech (2011)
17. Deb, S., Wang, M., Tanner, K.E., Bonfield, W.: Hydroxyapatite polyethylene composites: effect of grafting and surface treatment of hydroxyapatite. J. Mater. Sci. - Mater. Med. **7**, 191–193 (1996)
18. Sapiai, N., Aidah, J., Norhashidah, M., Mohd, A.I.U.: Effect of nanofillers dispersion on mechanical properties of Clay/Epoxy and Silica/ Epoxy nanocomposites. Jurnal Teknologi (Sci. Eng.) **76**(9), 107–111 (2015)
19. Aly, A.A., Zeidan, E.S.B., Alshennawy, El-Masry, A.A., Wasel, W.A.: Friction and wear of polymer composites filled by nano-particles: a review. World J. Nano Sci. Eng. **2**, 32–39 (2012)
20. Suman, B.I., Anshuman, D., Dube, N.M., Pratik, R., Sailaja, R.R.N.: Sliding wear and friction characteristics of polymer nanocomposite PAEKPDMS with nano-hydroxyapatite and nanocarbon fibres as fillers. J. Mech. Behav. Biomed. Mater. **86**, 23–32 (2018)
21. Grellmann, W., Seidler, S.: Polymer Testing, 2nd edn. Hanser Publishers, Munich (2013)
22. Shindo, K.: Mechanical analysis and geometric optimization of the restored tooth by indirect method. Doctoral thesis, Université Paris-Saclay (2019)

Surface Finishing and Coating

Optimization of Mass Loss During Abrasive Wear of Nitrided AISI 4140 Steel Parts

Ahmed Daghbouch[1(✉)] and Mohamed Ali Terres[1,2]

[1] Higher Institute of Applied Sciences and Technology of Sousse (ISSATSo), University of Sousse, Ibn Khaldoun, 4003 Taffala City, Sousse, Tunisia
daghbouchdaghbouch@gmail.com

[2] Mechanics, Materials and Processes Laboratory (LMMP), High National School of Engineers of Tunis (ENSIT) - University of Tunis, 5, Taha Hussein Avenue - Montfleury, 1008 Tunis, Tunisia

Abstract. This study aims to reduce the wear of parts made of gas nitrided AISI 4140 steel. This steel has mechanical and physical characteristics superior to most conventional materials, with a very high hardness after gas nitriding. AISI 4140 steels are often subjected to abrasive wear tests, which can result in mass loss. Efforts are therefore necessary to reduce this degradation. An in-depth optimization study is proposed to examine the effects of the hardness of the steel, the normal load, and the speed of wear. Tools such as experimental designs and analysis of variance are used to decipher the impact of each factor on the response variable studied and identify the predominant factors. Among these factors, the normal load is the most significant, with a lower load resulting in less mass loss. Hardness is the second most significant factor, Speed is the least. The hardness also has an interaction effect with the other two factors, which reduces mass loss. The optimal conditions retained are a normal load of 100 N, a Speed of 4.18 m/s, and a Hardness of 1092 HV.

Keywords: Gas Nitriding · Optimization · Wear resistance · Mass loss · AISI 4140 steel

1 Introduction

Nitriding is a thermochemical treatment that involves diffusing nitrogen into the surface of steel to improve its mechanical properties. It plays an important role in various fields by enhancing the mechanical properties of steel parts [1–3]. Gas nitriding is used to minimize the mass loss of materials produced during wear and to increase the hardness of steel parts. This treatment is carried out at relatively low temperatures, thus reducing the risk of deformation of the treated parts. Among the steels that can be treated by gas nitriding, AISI 4140 steel is a popular choice [4, 5]. This construction steel alloyed with chrome and molybdenum is known for its good harden ability and fatigue resistance, making it ideal for applications such as gears and crankshafts. Gas nitriding can further enhance these properties by increasing the hardness of the surface of AISI 4140 steel, making it more resistant to wear [6]. In addition, gas nitriding can be used for a wide range of applications, such as camshafts, injectors, and plastic molding tools.

M. Chouchane et al. (Eds.): CMSM 2023, LNME, pp. 243–253, 2024.
https://doi.org/10.1007/978-3-031-65007-9_27

Optimization is firmly established in various industries, hence the usefulness of these modeling tools that differ from static tools [7–9]. In this work, we present the optimization of experimental conditions to minimize the mass loss produced during abrasive wear tests. We use the MINITAB software to examine the effects of three factors: hardness (HV), normal load (N), and speed (m/s). We use tools such as factorial designs and variance analysis to decipher the impact of each factor on the studied response variable and identify the predominant factors.

2 Material and Treatments

The material studied is low-alloy steel of the 4140 type according to the European standard. The chemical composition of the studied material is reported in Table 1.

Table 1. Chemical composition of 4140 steel.

Element	C	Mn	Si	S	Cr	P	Mo	Ni	Cu	Fe
(%)	0,41	0,77	0,28	0,026	1,02	0,019	0,16	0,16	0,25	Bal

The main parameters of the gas nitriding treatments applied to a series of 4140 steel wear specimens are summarized in Table 2. The depth of nitrogen penetration is assessed by Vickers microhardness tests under a load of 50 gf on cross-sections. The Vickers microhardness filiations under a load of 50 gf (HV0.05), on cross-section, allow to control the effectiveness of each treatment through the determination of the surface hardness (HV max) and the hardened depth. The results corresponding to each nitriding treatment are gathered in Table 2.

Table 2. Characteristics of surface hardening of nitrided layers

State	Treatment Conditions			τ(%)	$e_N(\mu m)$	$e_c \mu m$	HV_{max}
	Type	θn(°C)	t(h)				
NG12	Gaseous	525	12		290	8	890
NG24			24	35	330	11	1092
NG36			36		360	15	930

$\tau = PNH3/(PH)^{1,5}$:Ammonia dissociation rate(%).

3 Methodology

In the present study according to the Taguchi method, a robust design and an L24 are employed for the experimentation. Our study focuses on the influence of wear parameters on mass loss. We are interested in three main parameters, also called factors: Hardness (HV), Normal load (F), and Speed (V). Each of these factors has different levels. These factors, as well as their type and level, are presented in Table 3.

Table 3. Wear Parameter Factors and Levels for Experimentation

Factors	Type	Levels	Values
F (N)	Continuous	4	100-125-150-175
Hardness (HV)	Continuous	3	890-930-1092
V (m/s)	Continuous	2	4,18-8,30

In the context of our study, we have developed a plan comprising 24 tests, each numbered from 1 to 24. Following the selection of the appropriate experimental design based on the input parameters, we have constructed this plan, which will subsequently be completed by the response ΔM/M (%) the mass loss following the wear test. It is calculated using the following formula:

$$\Delta M/M = \frac{(M - M')}{M} \times 100 \tag{1}$$

where **M (g)** is probably the mass of the sample before the wear test. It is measured in grams (g) and **M′ (g)** is probably the mass of the sample after the wear test. It is also measured in grams (g).

Table 4 illustrates the planning matrix as well as the coding of the factors and responses.

Table 4. Order of trials

Order	Hardness(HV)	F(N)	V(m/s)	ΔM/M (%)
1	890	100	8,30	0,1308
2	890	100	4,18	0,0817
3	890	125	8,30	0,1903
4	890	125	4,18	0,1289
5	890	150	8,30	0,2611
6	890	150	4,18	0,2314
7	890	175	8,30	0,2721
8	890	175	4,18	0,3454
9	930	100	8,30	0,0942
10	930	100	4,18	0,0303
11	930	125	8,30	0,1426
12	930	125	4,18	0,0403

(*continued*)

Table 4. (*continued*)

Order	Hardness(HV)	F(N)	V(m/s)	ΔM/M (%)
13	930	150	8,30	0,1855
14	930	150	4,18	0,0403
15	930	175	8,30	0,1790
16	930	175	4,18	0,0661
17	1092	100	8,30	0,0630
18	1092	100	4,18	0,0130
19	1092	125	8,30	0,0870
20	1092	125	4,18	0,0250
21	1092	150	8,30	0,1660
22	1092	150	4,18	0,0381
23	1092	175	8,30	0,3035
24	1092	175	4,18	0,0730

4 Optimization

4.1 Study of significant Factors

Table 5 is a structured way of presenting data that allows for easy comparison and analysis. It's organized into rows and columns, with each row representing a different level and each column representing a different factor. The values in the table are averages, calculated from multiple measurements taken at each level for each factor. The 'Delta' row represents the range of values for each factor. It's calculated by subtracting the smallest average value from the largest average value for each factor. This gives us an idea of how much variation there is in the data for each factor.

By examining these averages, we can deduce that the normal load, hardness, and speed all influence the mass loss during wear tests, but their significance varies. The normal load is the most significant factor, with a lower load resulting in less mass loss. Hardness is the second most significant factor, with increased hardness improving wear resistance. Speed is the least significant factor. Therefore, to minimize mass loss, we should prioritize adjusting the normal load first, then the hardness, and finally the speed (Fig. 1).

Indeed, the literature on the main effects plot for the means shows that the maximum mass loss during abrasive wear is reached for the maximum level of strength, equal to 175 N. This indicates the importance of this factor on mass loss. It is observed that the hardness of 1092 HV, obtained by gas nitriding for 24 h, allows us to minimize mass loss while achieving a satisfactory result. Conversely, the piece nitrided at 930 HV for 36 h yields a less satisfactory result, leading to a loss of time and money. Furthermore, although the piece nitrided at 890 HV for 12 hours is not comparable to those nitrided for 24 and 36 h, it has nevertheless improved wear resistance compared to non-nitrided

Table 5. Answers for averages

Level	Hardness (HV)	F (N)	V (m/s)
1	0.20521	0.06883	0.09279
2	0.09729	0.10235	0.17292
3	0.09607	0.15373	–
4	–	0.20652	–
Delta	0.10914	0.13768	0.08013
Rank	2	1	3

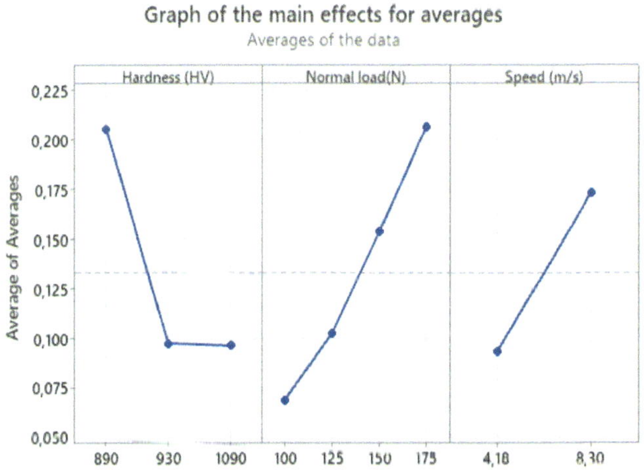

Fig. 1. Main effects graph for averages

steel. This can be beneficial in certain cases to enhance the mechanical properties of steel with minimal loss of time and money.

4.2 Study of Interactions

The interaction graph presented in Figure 2 clearly shows that mass loss decreases significantly when hardness interacts with normal load and speed. This indicates that hardness is a crucial factor in reducing mass loss, whether in combination with normal load or speed. This observation can help us optimize the experimental conditions to minimize mass loss. By adjusting the levels of hardness in combination with the levels of normal load and speed, we can significantly reduce mass loss. The interaction graph provides us with valuable information on the combined effects of the different factors on the studied response and allows us to make informed decisions to improve system performance.

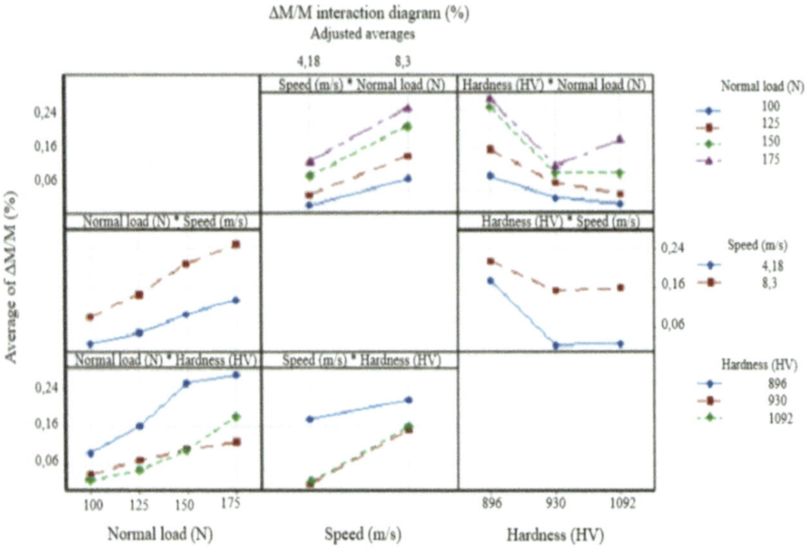

Fig. 2. Interaction diagram for ΔM/M (%)

4.3 Study of Effects by ANOVA

Table 6 gathers the results of a one-way analysis of variance (ANOVA) performed with Minitab to study the effect of different part parameters on mass loss during wear tests. The pooled standard deviation is a measure of the variability of the data for all combined groups. This pooled standard deviation is calculated by taking the square root of the weighted average of the squares of the deviations between each observation and the mean of its group. The weighting is done using the number of observations in each group.

Table 6. Analysis of general model variance

	Value	N	Mean	Standard Error	95%CI
Hardness (HV)	890	8	0.2052	0,0885	(0,1434;0,2670)
	930	8	0.0973	0,0638	(0,0355;0,01591)
	1090	8	0,0961	0,0963	(0,0343;0,1578)
F (N)	100	6	0,0688	0,0431	(-0,0038;0,1415)
	125	6	0,1024	0,0634	(0,0297; 0,1750)
	150	6	0,1537	0,0948	(0,0811; 0,2792)
	175	6	0,2065	0,1194	(0,1338;0,2792)
V (m/s)	4.18	12	0,0928	0,0995	(0.0398;0.1458)
	8.30	12	0,1729	0,0760	(0,1199;0,2259)

It is interesting to note that the average mass loss is higher for a hardness (HV) of 890 than for a Hardness (HV) of 930 or 1092. The increase in normal load leads to a systematic increase in mass loss. This suggests that force has a significant influence on mass loss during abrasive wear tests. The speed data is the most dispersed and the individual observations are the furthest from the mean compared to the other variables.

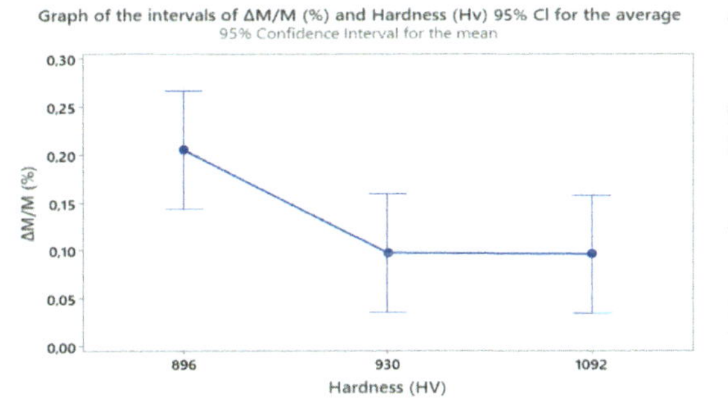

(a)

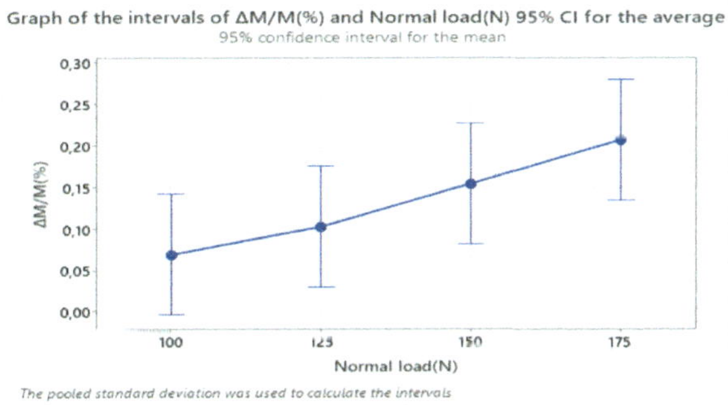

(b)

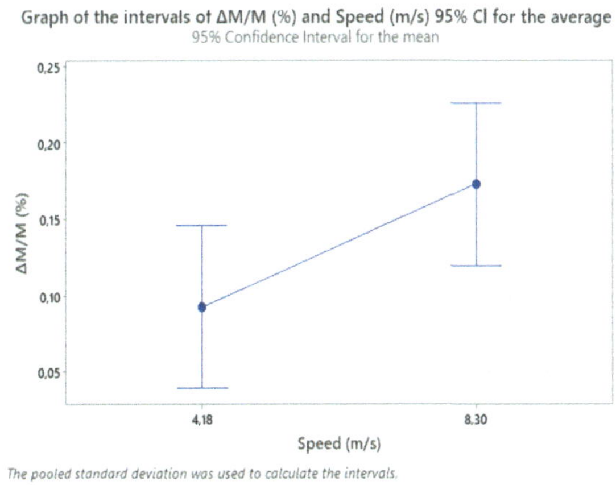

(c)

Fig. 3. Error interval graph: (a) Hardness, (b) Normal load, and (c) Speed.

Where N is the number of observations in each group. We can then add and subtract this error interval from the mean to obtain the confidence interval (Figure 3).

4.4 Analysis of Variance

According to the results presented in Table 7 of the analysis of variance, we can conclude that the overall model is significant with an F-value of 8.86 and a p-value less than 0.001. This means that the independent variables included in the model have a significant effect on the dependent variable.

By examining the individual effects of the independent variables, we can see that the linear effects of hardness (HV), Normal load (N), and speed (m/s) are all significant with a p-values less than 0.001. Normal load (N) has the most important linear effect with an F-value of 24.52. The square effect of hardness (HV) is also significant with a p-value less than 0.001. The interactions between the independent variables do not have a significant effect on the dependent variable.

The "Contribution" column shows the percentage of total variation explained by each source of variation. We can see that the linear effects of the independent variables explain 61.17% of the total variation, the square effects explain 17.79% of the total variation, and the two-factor interactions explain 3.56% of the total variation.

Table 7. Analysis of variance

Source	DF	Sum of Squares	Contrib. (%)	The adjusted sum of squares	Adjusted mean squares	F-value	P-value
Model	8	0,17414	82,53	0,17414	0,02176	8,86	0,00018
Linear	3	0,12908	61,17	0,15178	0,05059	20,58	0,00001
Hardness(HV)	1	0,02584	12,25	0,04764	0,04764	19,38	0,00051
F(N)	1	0,06470	30,67	0,06026	0,06026	24,52	0,00017
V(m/s)	1	0,03852	18,26	0,04387	0,04387	17,85	0,00074
HV^2	1	0,03698	17,53	0,03698	0,03698	15,04	0,00148
Error	15	0,36873	17,47	0,036873	0,00245		
Total	23	0,21101	100,00%				

4.5 Non Parametric Method

The Wilcoxon test is a non-parametric test used to compare paired samples and determine if the differences between them are significant. In our case, the test was used to compare the median values of hardness, normal load and speed.

The median is a measure of central tendency that indicates the value that separates a data set into two equal parts. In other words, half of the data is below the median and the other half is above it. The CI for η (the confidence interval for the median) is an interval that contains the true value of the median with a certain probability, called the confidence level. For example, a 95% confidence interval for the median means that if were peat the experiment several times, 95% of the calculated confidence intervals will contain the true value of the median. The achieved confidence is a measure that indicates the probability that the calculated confidence interval contains the true value of the median. The higher this value, the more reliable the confidence interval.

According to the results in Table 8, we can conclude that the median hardness of nitrided parts is 994 HV, with a 94.97% confidence interval between 913 and 1011. The median normal load is 137.50 N, with a 94.97% confidence interval between 125 and 150. The median speed is 6.24 m/s, with a 94.97% confidence interval. These results indicate that the median values for hardness, normal load, and wear speed are all significantly different from zero. This suggests that these factors have an effect on mass loss during abrasive wear. To minimize mass loss, we can use these results to determine the optimal levels of normal load, speed, and hardness that minimize mass loss.

Table 8. Static description of factors by the Wilcoxon method

Sample	N	Median	Confidence Interval form	Confidence Achieved
Hardness(HV)	24	994,00	(913; 1011)	94,97%
F(N)	24	137,50	(125; 150)	94,97%
V(m/s)	24	6,24	(6,24;6,24)	94,97%

4.6 Regression Equation

The importance of the regression equation in optimization lies in its ability to provide accurate predictions of the response based on the values of the factors. By using this equation, we can identify combinations of factor values that give the best predictions of the response and use this information to optimize experimental conditions. When linear and quadratic regression equations did not yield good results, we decided to use a genetic algorithm to optimize the variables of the adjustment equation provided by Minitab. The genetic algorithm is an optimization method based on biological laws of evolution, particularly natural selection. It allows us to modify the variables of the adjustment equation to obtain the best possible accuracy between experimental and optimal models. The equation of our study is:

$$\Delta M/M\,(\%) = 0{,}0682689 - 0{,}0003175 \times (HV) + 0{,}00184933 \times (F) + 0{,}0197451 \times (V) \quad (2)$$

4.7 Optimal Solution

As part of your optimization, you sought to minimize the ΔM/M (%) response. To do this, you used the response surface design, set a minimization goal, and chose a target of

0.01. You also assigned a weight of 1 and an importance of 1 to this response to reflect its importance in your analysis. The parameters are well identified in Table 9.

Table 9. Optimization parameters

Response	Objective	Target	Upper limit	Weighting	Importance
ΔM/M (%)	Minimum	0,01	0,05	1	1

By using these optimization parameters, you obtained an optimal solution. This solution indicates that the optimal conditions for achieving your goal of minimizing the ΔM/M (%) response are a normal load of 100N, a speed of 4.18 m/s, and a hardness of 1092 HV. These values for the different factors are sufficient to expect to obtain a minimum value for the ΔM/M (%) response. The solution is presented in Table 10.

Table 10. Optimal solution

Solution	Hardness(HV)	F(N)	V(m/s)	Adjusted ΔM/M (%)	Composite desirability
1	1092	100	4,18	0,01	1

5 Conclusion

In conclusion, this study emphasizes the pivotal importance of hardness achieved through gas nitriding in reducing mass loss during abrasive wear. It is clearly demonstrated that normal load is the most influential factor on mass loss, closely followed by hardness, while speed plays a less significant role. The interaction of hardness with normal load or speed has a considerable impact on reducing mass loss.

A gas nitriding reaching a hardness of 1092 HV over a period of 24 h seems to offer the best compromise to minimize mass loss. On the other hand, nitriding at 930 HV over a period of 36 h proves to be less effective, resulting in a loss of time and financial resources. Interestingly, in some cases, nitriding for only 12 h can produce satisfactory results. It improves resistance compared to the non-nitrided state while saving time and money compared to a 24 h nitriding.

The study indicates that the optimal conditions to minimize the ΔM/M (%) response to 0.01 are a normal load of 100 N, a speed of 4.18 m/s, and a hardness of 1092 HV. These results highlight the importance of gas nitriding and its parameters in optimizing the wear resistance of materials.

References

1. Wang, B., Zhao, X., Li, W., Qin, M., Gu, J.: Effect of nitrided-layer microstructure control on wear behavior of AISI H13 hot work die steel. Appl. Surf. Sci. 2–6 (2017)

2. Kuksenova, L.I., Michugina, M.S.: Effect of heating conditions in nitriding on the structure and wear resistance of surface layers of steel 38Kh2MyuA. Metal Sci. Heat Treat. **50**, 68–72 (2008)
3. Terres, M.A., Ammari, L., Chérif, L.: Study of the effect of gas nitriding time on microstructure and wear resistance of 42CrMo4 steel. Mater. Sci. Appl. **8**(06), 493
4. Terres, M.A., Bechouel, R., Ben Mohamed, S.: Low cycle fatigue behavior of nitrided layer of 42CrMo4 steel. Int. J. Mater. Sci. **6**(1), 18–27 (2017)
5. Terres, M.A., Sidhom, H., Larbi, A.B.C., Ouali, S., Lieurade, H.P.: Influence de la résistance à la fissuration de la couche de combinaison sur la tenue en fatigue des composants nitrurés. Matériaux Techn. **89**, 23–36 (9–10)
6. Fernández-Pariente, I., Bagherifard, S., Guagliano, M., Ghelichi, R.: Fatigue behavior of nitrided and shot peened steel with artificial small surface defects. Eng. Fract. Mech. **103**, 2–9 (2013)
7. X. Lu, Z. Jia, H. Wangand S.Y. Liang "The Effect of Cutting Parameters on Microhardness and the Prediction of Vickers Hardness Based on a Response Surface Methodology for Micro-milling Inconel 718". Measurement, (2019), pp.57–62
8. Vikram, S., Pradhan, S.K.: Optimisation des paramètres WEDM à l'aide de la technique Taguchi et de la méthodologie de surface de réponse dans l'usinage délacer AISI D2. Procedia Eng. **97**, 1597–1608 (2014)
9. Pratap Singh, A., Singhv, D.K.: Multi response optimization for micro-EDM machining of AISI D2 die steel using RSM and neural network. Mater. Today: Proc. 1–7 (2020)

Improvement of the Wear Resistance of AISI4140 Steel by Duplex Treatment Combining Gas Nitriding and Physical Vapor Deposition

Abdelkarim Ghanem[1,3](✉), Siwar Toumi[1], and Mohamed Ali Terres[1,2]

[1] Mechanical, Material and Processes Laboratory (LR99ES05) LMMP, University of Tunis (UT), National Higher School of Engineers of Tunis, ENSIT, 5, Avenue Taha Hussein, 1008 Tunis, Tunisia

abdelkarim.ghanem.isetrades@gmail.com

[2] High Institute of Sciences and Applied Technology of Sousse (ISSATSo), University of Sousse, Cité Ettafallah, Avenue Ibn Khaldoun, 4003 Sousse, Tunisia

[3] Higher Institute of Technological Studies of Rades, BP 176 Radès Médina, 2098 Radès, Tunisia

Abstract. Previous studies have noted that gas nitriding and physical vapor deposition coating (PVD) considerably affect the wear resistance, friction coefficient and microstructure of AISI 4140 steel. This work studies the impact of gas nitriding and coating treatment (physical vapor deposition) on the wear resistance, friction coefficient and microhardness of this steel. The microhardness of the nitrided and PVD-coated layer increased with increasing nitrogen concentration, reaching approximately 1750 HV0.05 at a processing temperature of 525 °C and an ammonia content of 35%. Optical microscopy was used to study the structural properties of the nitrided layers. Furthermore, the presence of nitrogen and the TiN coating had a significant influence on the scratch resistance for a single pass, with a critical load (LC1) of approximately (5 N). Multi-pass scratching tests were used to study the wear volume, degradation process and friction coefficient. Furthermore, the multi-pass scratch method was used to obtain valuable data on the mechanical behavior and damage of the layer following duplex treatment, by combining gas nitriding and coating (PVD).

Keywords: AISI 4140 · Single scratch · Gaseous Nitriding · PVD coating · multi-pass scratch

1 Introduction

The main reason for mechanical problems and failure of technical Components is the wear. This requires a multitude of complex surface interactions. These are both chemical and mechanical interactions. The phenomena of wear and friction are interconnected. It is complicated to remove wear. However, it is possible to reduce it by opting for

M. Chouchane et al. (Eds.): CMSM 2023, LNME, pp. 254–263, 2024.
https://doi.org/10.1007/978-3-031-65007-9_28

surface treatments and coating technologies. Its technologies are applied in the fields of aerospace, automobile, oil and gas [1, 2]. Wear and friction are regularly controlled by two basic types of coatings: hard coatings and soft coatings [3]. Nitrides, oxides, borides, and carbon-based coatings (like TIN (CrN), etc.) are hard coatings. Soft coatings are polymers, soft metals, alkaline earth metal halides. [4, 5]. In duplex surface treatment processes, a surface modification method is combined with the coatings in order to obtain optimal tribological properties. The wear resistance and load-bearing capacity of the substrate are improved through duplex treatments, which preserves the surface from cracking and delamination [6]. In recent years, nitriding and coating methods have become increasingly common [7] due to their ability to improve wear and corrosion characteristics, leading to an expansion of their industrial applications. Wear behavior can vary depending on a combination of surface engineering techniques used on the contacting surfaces [8, 9]. The objective of this study is to analyze the possible uses of a solid PVD protective coating on nitrided AISI 4140 steel. Steel pretreated by gas nitriding was coated with a PVD coating in order to improve the tribological characteristics of AISI 4140 steel and to study their impact on the tribological characteristics of duplex treated steel. In this research, AISI 4140 steel samples were treated with gas nitriding treatment and then TIN coating was applied to its already nitrided surface. The different samples were examined using X-ray diffraction (XRD) and optical microscopy to study their structure and composition. Microhardness and nanoindentation tests were also carried out to evaluate the mechanical characteristics. The scratch adhesion strength of the coating on the gas nitrided substrate was discussed Additionally, scratch tests were carried out to evaluate adhesion and wear resistance in progressive and multi-pass loading modes, respectively.

2 Materials and Methods

2.1 Material

In this study, we used commercial low carbon steel for the gas nitriding process of AISI 4140 steel discs (12 mm × 3 mm). The chemical composition is (0.041% C, ≤ 0.05% Si, 0.127% Mn, 0.005% P, 0.006% S). Samples were ground with silicon oxide papers (1000, 1200, 2000 and 2400) and polished with alumina (dimensions). Particle size 0.3 μm) and sprayed onto felt paper. Finally, the samples were degreased in acetone and distilled and deionized (DI) water. The choice of this material is explained by its ability to be nitrided to high surface hardness without compromising strength, and its adaptability to highly stressed machine parts.

2.2 Treatments

The heat treatment process adopted for AISI 4140 steel is as follows: (i) solution treatment at 850 °C for 0.5 h with oil cooling; (ii) tempering at 580 °C for 1 h with air cooling and (iii) nitriding at 550 °C under NH3 = 35% for 24 h to obtain a nitrided layer.

2.3 Coating deposition

The company PV-DÉCOR, located in Raoued-Ariana, is responsible for the development of this coating, being the only Tunisian company to implement PVD coatings, as illustrated in Fig. 1. The materials were placed in a cavity of evaporation under vacuum at a pressure of 2.4 × 10-2 Pa and a temperature of 60 °C. Argon gas was added at a pressure of 7.10-1 Pa. The buffer layer, an intermediate layer of Ti, was applied to the substrates for a period of 3 min. Then, in order to deposit the TiN films (17 min), we gradually increased the flow rate of N2 while maintaining a constant flow rate of Argon gas. TiN coatings with a thickness of 10 μm were obtained.

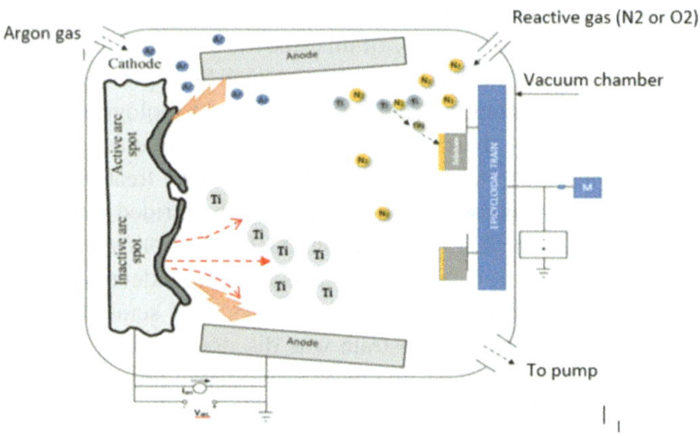

Fig. 1. The technique of deposition of a thin layer by cathodic arc evaporation.

3 Characterization Techniques

3.1 Nano-mechanical Properties

The nanoindentation technique was used to evaluate the nanomechanical properties (hardness (H) and elastic modulus (E) of the coated samples (Fig. 2). A graph representing the applied load and the displacement of the corresponding indenter during the phases loading and unloading is obtained from a nano-indentation test Using the approach of Oliver and Pharr [10, 11], we can evaluate the elastic modulus E and the surface hardness H in. Analyzing the charge-displacement (P-h) curves The discharge rate of the nano-indentations on the TiN coating is shown in Fig. 5.

3.2 Adhesion

The developed layers were tested for adhesion by performing scratch tests with the same equipment. The applied load was progressive from 0.3 N up to 20 N, over a length of 3 mm and with a 10 mm/min sliding speed. Tracing the variation of the tangential force

Fig.2. Nano-indenter with a diamond Berkovitch tip.

and the penetration depth as well as the analysis of the scratch trace by microscopic observations made it possible to determine the critical loads (LC) which represent the interface of the films. In general, three modes of degradation of hard deposits can be induced by the scratch test: cracking at a critical load called LC1, start of spalling at a critical load LC2 and total spalling at an LC3 critical load.

3.3 Tribological Properties

The tribological behavior was studied using continuously loaded single scratches and multi-scratches. The behavior of the film was demonstrated using a Rockwell C diamond indenter with a cone angle of 120° and a tip radius of 5 mm distributed over a stroke of 3 mm, with a sliding speed of 10 mm/min and loads lower than LC1. The wear mark results were analyzed both qualitatively and quantitatively.

4 Results and Discussion

4.1 Microstructural Analysis

Scanning electron microscope (SEM) observed dense and homogeneous morphology, as shown in Fig. 3(a). The decrease in imperfections between these coatings resulted in high adhesion of the coating to the substrate. The structure of the coatings analyzed by EDS can be observed in Fig. 3(b) and Table 1. The composition of the formed layers was similar. These coatings demonstrated satisfactory coating synthesis, which is in agreement with other recent work. The TiN coating provides increased protection against toxic substrate compounds, with an Al% of 0.64% and a Na% of 0%. However, the existence of oxygen in its composition is due to the production method. (absence of ionization, incorrect suction, etc.).

The microstructures of the surface, subsurface and X-ray diffraction pattern of the layer for AISI 4140 steel after gas nitriding are shown respectively in Figs. 3 a-b. Figure 3a shows, the nitride layer consistory of the compound layer and the diffusion layer: a composite layer of irregular thickness ranging from 10 μm to 15 μm on the surface.

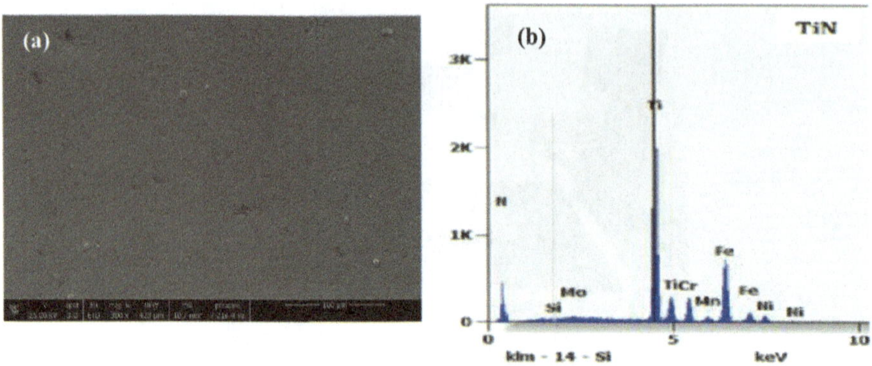

Fig. 3. (a) Microscopic observation of the morphology of thin TiN film by SEM, (b) EDS spectra of thin TiN film.

Table 1. Atom % in composition of TiN Coating.

	C	O	N	Na	Al	Ti	Cr
TiN Coating	11.34	11.93	**24.37**	0	0.64	**47.61**	3.96

A diffusion layer consisting of a solid solution of nitrogen in insertion marked by the carbide precipitation is probably related to carbon gradient reverse diffusion of nitrogen. Figure 3b shows the X-ray diffraction of a layer consisting of a mixture of the ε (Fe 2-3 CN) and γ' (Fe 4 N) phases (Fig. 4).

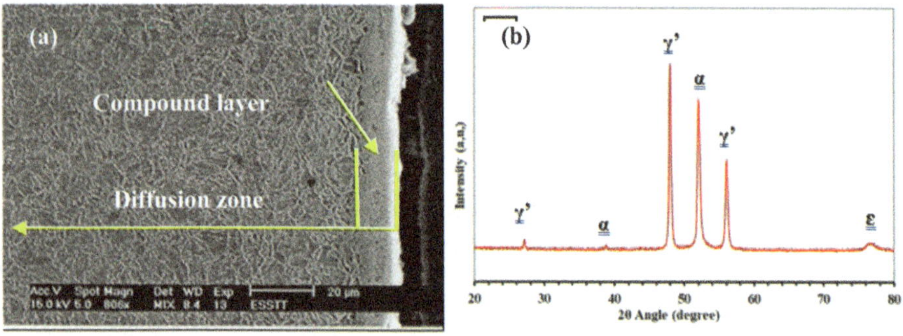

Fig. 4. (a) Microscopic observation by SEM of the cross-section of gas nitrided layer (GN) (b) X-ray diffraction pattern of gas nitrided layer.

4.2 Conventional Nanoindentation

Carrying out the nanoindentation test resulted in a graph that illustrates the relationship between the applied load and the different loading and unloading phases of the indenter movement. During the indentation test, an increase in penetration depth was observed as a function of the applied load. Furthermore, plastic deformation occurred when the maximum load (Pmax) was equal to the maximum penetration depth (hmax). Figure 5 shows the load-displacement curve with the pyramidal nanoindentation trace and Oliver and Pharr's approach. According to the analysis of the load-displacement (P-h) curves, using the Oliver-Pharr model [10], it is possible to evaluate the hardness of the films H and the Young's modulus E. Table 2 presents the results hardness H and modulus E obtained using the Olivier Pharr model. The maximum applied load (Pmax) was 55 mN and the corresponding maximum penetration depth Hmax was 0.8 μm. The maximum indentation thickness was less than 1/10th of the film thickness, which reduced the impact of the substrate on the hardness. The tests were evaluated based on more than 3 variables. Table 2 presents the results which demonstrate that TiN coating deposition improves the mechanical characteristics of the substrate material. The hardness of the thin gas nitrided AISI 4140 TiN layer (TiN/GN) was approximately twice as high as that of the uncoated material The TiN film had a hardness of 23 ± 0.2 GPa and an elastic modulus of 350 ± 14 GPa, respectively. The gas nitrided layer has a lower hardness of 12 ± 0.27 GPa compared to the matrix and an elastic modulus of 220 ± 10.8 GPa. The small particle size of the TiN coating makes it possible to improve the hardness of the TiN deposited on the substrate. The presence of this small grain size was caused by the compressive stress present at the interface between the film and substrate during surface formation. This can also be explained by the lower density and porosity of the TiN deposited on the substrate, as well as by the structural gradient between the TiN film and the gas-nitrided AISI 4140 steel substrate.

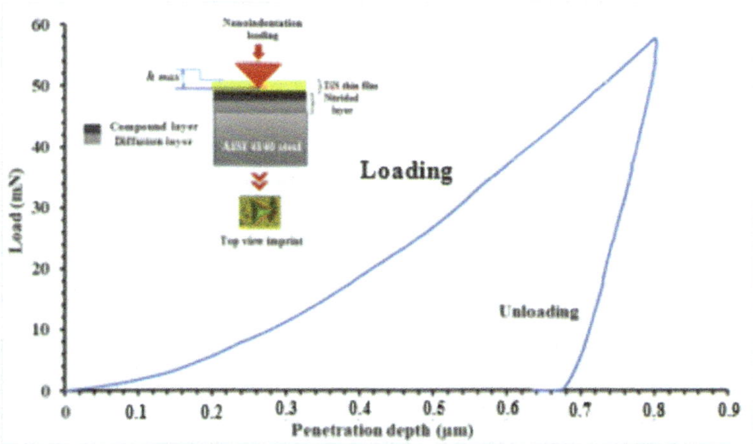

Fig. 5. A typical load-displacement curve of TiN / (GN) AISI 4140

4.3 Failure Mechanisms of Single Scratch

The two states studied (GN) and (TiN/GN) were subjected to a microscratch test in order to determine the resistance of the surface to scratches. Different critical loads were

Table 2. Mechanical properties from the conventional characterization of thin TiN film and Gas nitriding layer.

Specimens investigated	Micro Vickers indentation	Nano-indentation	
	Surface hardness, HV0.05	Hardness (GPa)	Young's modulus (GPa)
TiN/Gas nitriding (GN)	1750	23 ± 0.2	350 ± 14
Gas nitriding (GN)	1090	12 ± 0.25	220 ± 10.8

identified using the scratch test (LC1, LC2 and LC3). The first rupture observed in LC1 is a cohesive rupture.

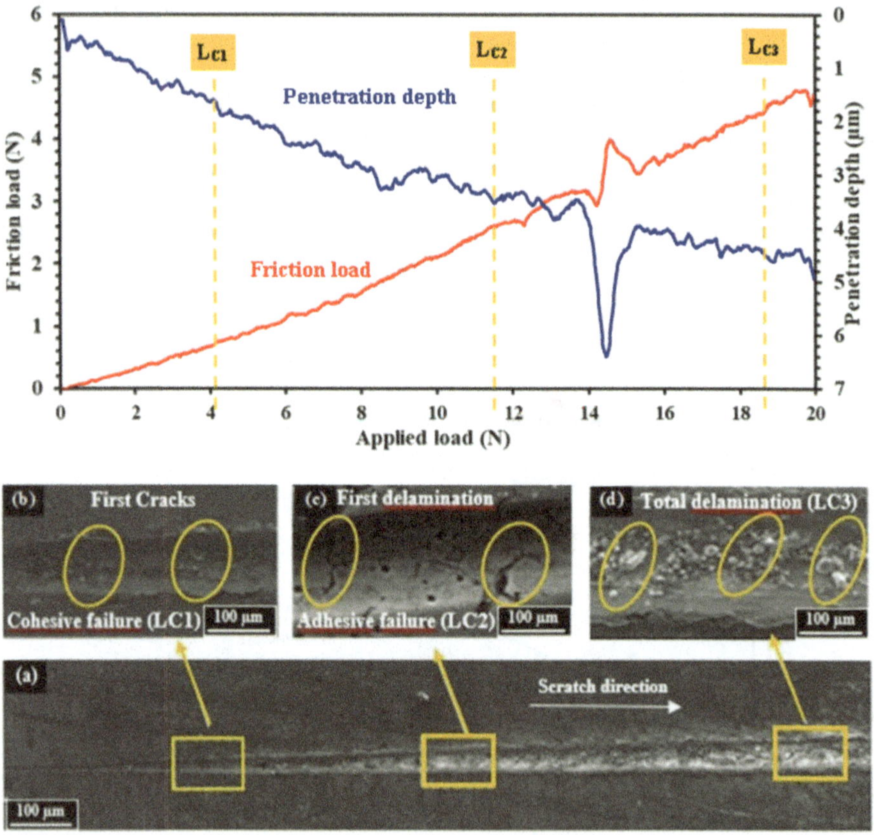

Fig. 6. Optical micrographs and scratch tracks of the TiN/GN layer: (a) scratch track; (b) first cracking; (c) first delamination; (d) total delamination; (e) Evolution of frictional force and penetration depth as a function of normal load of the TiN/GN layer.

LC2 denotes to local flaking of the layer with an adhesion defect. LC3 refers to the highest critical load value. Cohesive scales form the filler LC1. We speak of cohesive tearing because the cracks are located at the level of the film, as illustrated in Fig. 6b. When there is a higher load (LC2), the flaking becomes fixed and the flakes detach from the coating or substrate (Fig. 6c). A complete burst of the layer was observed during a critical load called LC3 (Fig. 6d). The determination of critical loads was carried out by combining optical microscopy data on the scratch marks, measurements of normal and tangential loads, as well as the coefficient of friction during the scratch test. In this study, quantitative and qualitative analyzes were carried out to validate the results of the scratch tests. The TiN/GN and GN layers present critical forces which are summarized in Table 3.

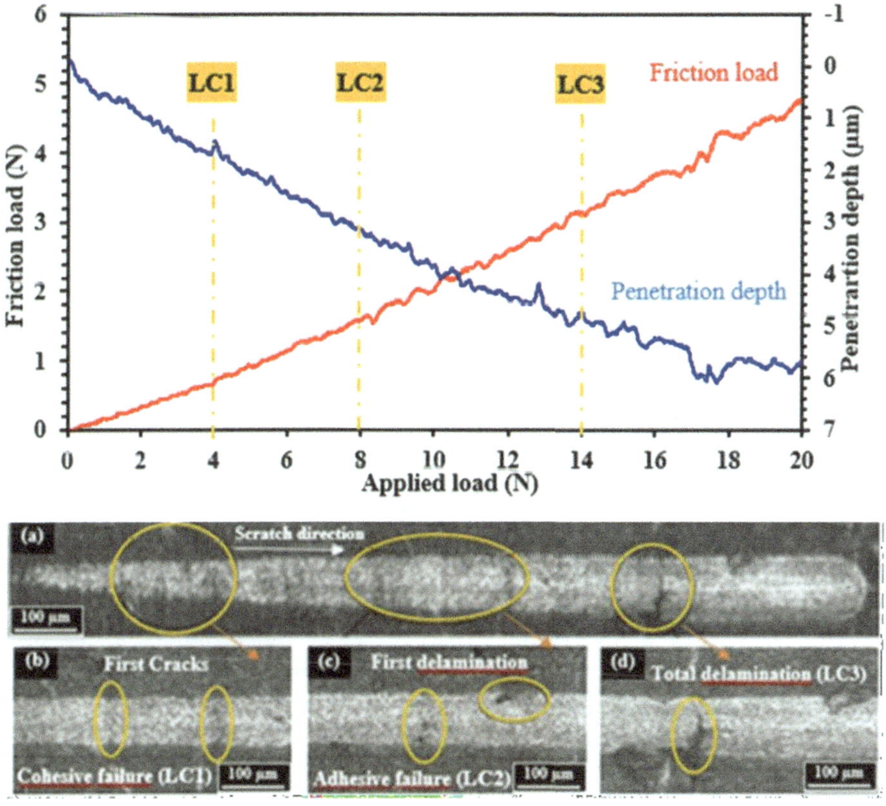

Fig. 7. Optical micrographs and scratch tracks of the GN layer: (a) scratch track; (b) first cracking; (c) first delamination; (d) total delamination; (e) Evolution of frictional force and penetration depth as a function of normal load of the GN layer.

For low loads, the penetration depth increased linearly with the normal load. At a load of LC1 (5 ± 0.17 N for (TiN/GN) and 4 ± 0.20 N for (GN)), the two states suddenly changed in depth, suggesting the presence of a cohesive failure. At an LC2 load (11.7 ± 0.30 N for (TiN/GN) and 8 ± 0.14 N for (GN)), the layer fluctuated substantially revealing

Table 3. Values of critical loads of different layers

Samples	When first cracks appeared, Lc1	Propagation of cracks, Lc2	Delamination, Lc3
TiN/GN	5 ± 0.17	11.7 ± 0.30	18.8 ± 0.25
GN	4 ± 0.20	8 ± 0.14	15 ± 0.19

a partial delamination due to the propagation of mi- crocracks from compression stresses developed by the indenter. At LC3 load, the layer delaminated completely. However, it appears that the (TiN/GN) layer suffered less damage and exhibited greater adhesion than the carbonitrided (GN) layer. According to the results of single layer scratch damage, the failure process of the two layers is shown in Fig. 6–7.

4.4 Multi-pass Scratch

Several scratch tests were carried out on the same track in order to evaluate the resistance of the (GN) and (TiN/GN) layers to friction and wear. For each pass, the friction force, the normal force and the penetration depth were calculated during scratch test in several passes. Figure 8 illustrates how the friction coefficient varies under a normal load of 10 N for the (GN) and (TiN/GN) states as a function of the number of cycles.

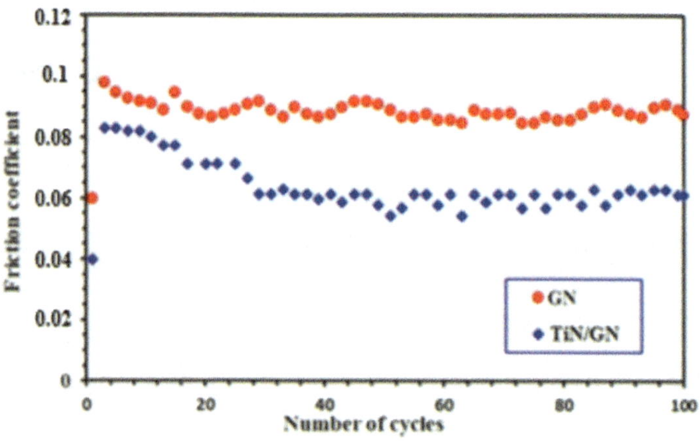

Fig. 8. Variation of friction coefficient at a normal load of 10 N/mn

The frictional behavior also varies from layer to layer. It seems that the layer (TiN/GN) has the lowest friction coefficient value of 0.06.

5 Conclusion

In order to improve their mechanical and tribological performances, a thin film of TiN was applied to AISI 4140 gaseous nitrided alloy substrates by PVD cathodic arc evaporation. Two surface modification techniques, such as cathodic arc PVD (TiN) and gas nitriding

(GN), have been demonstrated to constitute a functional duplex system. The presence of nitrides in the duplex system decreases the contact depth and increases the hardness. During scratch testing, damage was reduced by using a cohesive adhesive. Additionally, during wear testing, the coefficient of friction was reduced by almost half for the duplex system. According to the SEM observations, it was found that the TiN coating has a dense and homogeneous structure. The nano-indentation test demonstrated higher hardness and elastic modulus for the duplex system compared to that of the AISI 4140 gas nitrided alloy. The propagation of microcracks in the scratch pattern was due to the strong compressive stress developed in front of the indenter and the high tensile stress behind the indenter.

References

1. Terres, M.A., Ben Fathallah, B., Ghanem, A.: Effects of surface pre- treatment on the nitrided layers properties. Ame. J. Mech. Industr. Eng. **2**(1), 41–47 (2017). https://doi.org/10.11648/j.ajmie.20170201.17)
2. Terres, M.A., Ammari, L., Cherif. A.K.: Study of the effect of gas nitriding time on microstructure and wear resistance of 42CrMo4 steel. Mater. Sci. Appl. **8**, 493–507 (2017)
3. Gilewicz, R., Jedrzejewski, P., Myslinski, B.: Warcholinski: Structure, morphology, and mechanical properties of AlCrN coatings deposited by cathodic arc evaporation. J. Mater. Eng. Perform. **28**(3), 1522–1531 (2019)
4. Songa, M., et al.: Fe2Ti interlayer for improved adhesion strength and corrosion resistance of TiN coating on stainless steel 316L. Appl. Surf. Sci. **504** (2020)
5. Chen, W., Zheng, J., Meng, X., Kwon, S., Zhang, S.: Investigation on microstructures and mechanical properties of AlCrN coatings deposited on the surface of plasma nitrocarburized cool-work tool steels Vacuum. **121**, 194–201, 10.1016 (2015)
6. Valkov, S., et al.: Comparison of the phase composition and nanohardness of gradient TiN/TiO2 coatings on Ti5Al4V alloy deposited by different PVD methods. Mater. Sci. Eng. **416**, 012–034 (2018)
7. Alkan, S., Gök, M.S.: Influence of plasma nitriding pre-treatment on the corrosion and tri bocorrosion behaviours of PVD CrN, TiN and AlTiN coated AISI 4140 steel in seawater. Lubr. Sci. **34**(2), 67–83 (2022)
8. Quirama, A., Echavarría, A.M., Meza, J.M., et al.: Improvement of the mechanical behavior of the calcium phosphate coatings deposited onto Ti6Al4V alloy using an intermedi TiN/TiO2 bilayer. Vacuum **146**, 22–30 (2017)
9. Mo, L., Zhu, M.H.: Sliding tribological behaviors of PVD CrN and AlCrN coatings against Si3N4 ceramic and pure titanium. Wear **267**, 874–881 (2009)
10. Oliver, W., C., Pharr, G. M.: An improved technique for determining hardness and elastic modulus using load and displacement sensing indentation experiments. J. Mater. Res. **7**, 1564–1583 (1992)
11. Santana, A.E., Karimi, A., Derflinger, V.H., Schütze, A.: Microstructure and mechanical behavior of TiAlCrN multilayer thin films. Surf. Coat. Technol. 177–178 334 (2004)

Tribological Properties of Linear Low Density Polyethylene Reinforced with Almond Shell Particles: Impact of NaOH Treatment

M. Boujelben, M. Abid(✉), Mohamed Kharrat, and M. Dammak

Laboratory of Electromechanical Systems (LASEM), National Engineering School of Sfax, University of Sfax, Sokra Road Km 3.5, PO Box 1173, 3038 Sfax, Tunisia
mohamedabid528@yahoo.fr, {mohamed.kharrat, maher.dammak}@ipeis.rnu.tn

Abstract. In this study the effect of alkaline treatment on the tribological properties of bio-composites (LLDPE/ASP) is examined. Two groups of bio-composites of different mass fractions varying from 0% to 40% and without compatibilizing agent were prepared by single screw extrusion and consolidated by the standard thermo-compression molding process. The first group of bio-composites is developed without treatment of ASP and designated by LLDPE/ASP untreated, while for the second group of bio-composites, the ASP fillers have undergone an alkaline treatment of concentration 1.6 M/L for 24 h, and it is designated by LLDPE/ASP treated. The results found show that the alkaline treatment of ASP particles affects the tribological properties of LLDPE/ASP bio-composites. Likewise, the results of these tests were compared with pure LLDPE. It was also found that the LLDPE/ASP bio-composites presented improve tribological properties compared to pure LLDPE. The results of the friction tests also suggest that a proportion of 20% by weight of ASP in the LLDPE matrix leads to better tribological performances and a clear improvement found in the treated bio-composites compared to the untreated ones.

Keywords: ASP · LLDPE · alkaline treatment · bio-composites · bio-fillers · friction and wear

1 Introduction

In recent years, the use of natural fillers from biomass in the form of granulates or powders of fruit shells (almonds, coconut, apricot, pistachio, ...) as reinforcement in composite materials replacing synthetic reinforcements (glass, carbon, aramid fibers, ...) and mineral (mica, talc or $CaCO_3$) have experienced considerable growth (Dominici et al. 2019) allowing a wide access to several areas of application. For example (Bayer et al. 2017) the automotive industry is a sector which increasingly uses cellulosic fillers, particularly in door panels, seat backs, dashboards and interior parts. This interest comes from the combination of several advantages of natural bio-fillers such as: low cost, low density, non-toxicity, continuous availability and short-term biodegradability (Garcia

M. Chouchane et al. (Eds.): CMSM 2023, LNME, pp. 264–272, 2024.
https://doi.org/10.1007/978-3-031-65007-9_29

et al. 2016). However, these Bio-fillers also have certain disadvantages when using them as reinforcements in composite materials (Hammoui et al. 2015) in particular their hydrophilic which make them incompatible with most organic matrix known by their hydrophobic (Nabinejad et al. 2015). Indeed, many failures can be attributed to poor interactions between the Bio-fillers and the polymer matrix. In order to remedy these drawbacks, several studies have shown that the addition and/or use of coupling agents or compatibilizers can improve interfacial adhesion. Other studies have shown that modifying the filler surface topology can also lead to improved interfacial adhesion. Several methods are usually based on the use of reagents, which contain functional groups that are capable of reacting by forming chemical bonds with the hydroxyl groups of the lignocelluloses material. Generally, when natural bio-fillers are used to reinforce thermoplastics and thermo sets, the most used technique is alkaline treatment (with soda) or mercerizing (Deepak et al. 2021). This treatment reduces the hydrogen bonds between cellulose hemicelluloses and lignin, it makes it possible to eliminate the undesirable constituents of the filler, such as wax, pectin, hemicelluloses and part of the lignin (Mohd et al. 2017). This leads to dimensional, morphological and mechanical changes in the fillers by increasing the filler/matrix contact surface (Ahmed et al. 2021). It also helps promote fibrillation of the bundles. This provides a clear performance with respect to the quality of the filler-matrix interfacial adhesion and therefore promotes the subsequent use of these fillers as reinforcement in polymers. Likewise, the concentration of the solution, the temperature and the treatment time of the fillers are essential parameters of the treatment process. The optimization of these parameters contributes to the improvement of the mechanical properties of the bio-composite (Shehu et al. 2017). Indeed, alkaline treatment can have a detrimental effect on natural fillers. Therefore, there are risks of damage to the fillers once the processing conditions are not optimized (Rajesh et al. 2015). This study constitutes a contribution to the search for new biocomposite materials originating from natural resources (agricultural biomass). It aims to use almond shells after extraction of the fruits (almond kernels) after the shelling operation as reinforcement in a thermoplastic matrix of the Linear Low Density Polyethylene type. It consists of evaluating the effect of alkaline treatment on the morphology of bio-fillers, as well as on the tribological behavior of LLDPE/ASP bio-composites.

2 Experimental Materials and Details

2.1 Materials

The polymer used as a matrix in the preparation of bio-composites is Linear Low Density Polyethylene (LLDPE). It is a semi-crystalline thermoplastic polymer. It was supplied in powder form the company KIMPAS (Turkey) with a density of 0.938 g/cm^3, an average particle size of 300 μm and a melting temperature of 135 °C. The bio-fillers used in the preparation of bio-composites is almond shells collected from the Sfax region (Tunisia) have a lignocellulosic structure. They are mainly composed of 30% cellulose, 25% hemi cellulosic, 30.1% lignin, having a density equal 0.89 g.cm^{-3} (Maaloul et al. 2017).

2.2 Pretreatment of Bio-Fillers

After the crushing operation (stone extraction), the almond shells were washed with water (2% detergent solution) to remove contaminants and adhering dirt, dried in the sun for 72 h, then they are crushed and sieved with a set of sieves of different mesh sizes to obtain homogeneous particles with an average size of 125 μm (Boujelben et al. 2020). After this pretreatment, the ASP powders are referred to as untreated ASP powders (Fig. 1).

2.3 Treatment of Bio-Fillers

The ASP Bio-fillers were kept for 24 h in an aqueous solution of sodium hydroxide with a concentration equal to1.6 mol/L, at room temperature to eliminate a certain quantity of wax and oil covering the external surface of the particles. Then, they were rinsed with water before being immersed in a solution of sulfuric acid of 2% concentration during 10 minutes, washed again with fresh water, then emerged in distilled water for 15 min (El Mechtali et al. 2017), and finally dried in an oven at a temperature of 70 °C for 24 h before being mixed with the LLDPE matrix. After this treatment, the ASPs are designated treated powders (Figure 1). The reaction of sodium hydroxide with the ASP particle is as follows:

$$\text{ASP charge} - \text{OH} + \text{NaOH} \rightarrow \text{ASP charge} - O - \text{Na} + \text{H}_2O + \text{impurities}$$

As shown in this reaction, the important change that occurs is the elimination of the hydrogen bond in the network structure.

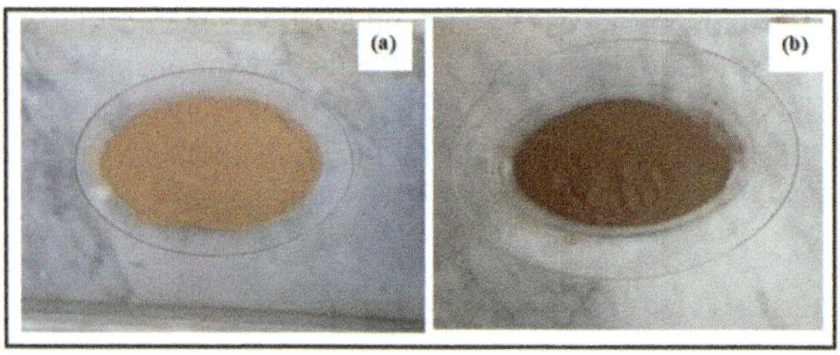

Fig. 1. Treated and untreated ASP powders: (a) Untreated, (b) Treated

2.4 Preparation of Bio-Composites

To examine the effect of alkaline treatment and the rate of bio-fillers on the mechanical, tribological and structural properties of bio-composites, two series of LLDPE bio-composites reinforced each time with 5%, 10%, 20%, 30% and 40% by weight of treated and/or untreated particles were mixed using a mini laboratory single screw extruder during which and each time separately the LLDPE powders/ASP treated or LLDPE/ASP untreated feed the plasticizing screw via the hopper which, by turning, allows them to be brought into the sheath. The heating collars and the mechanical action of the screw (shearing) melt and mix the polymer powders which will become pasty and mix with the ASP powders thus allowing their incorporation. The operating conditions of the mini extruder have been set: screw rotation speed of 8 rpm, the temperature of the screw heating zone is 173 °C ± 4 °C, the flow rate has been adjusted to 100 g/h. The two types of bio-composites LLDPE/ASP treated and LLDPE/ASP untreated thus extruded wire were cut into granules of 2 mm in length (Figure 2).

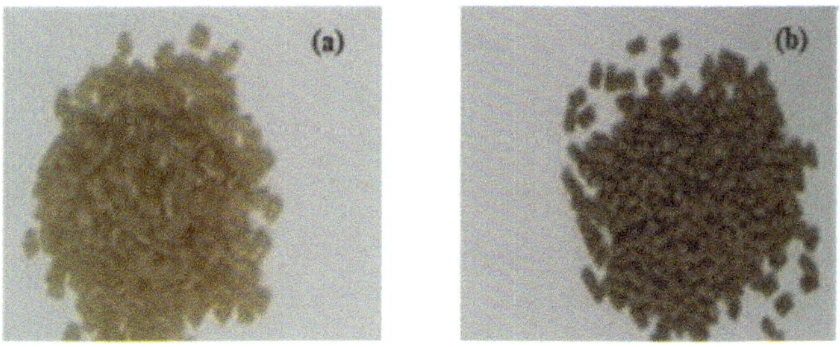

Fig. 2. Bio-composites in the form of granules with 10% ASP : (a) Untreated, (b) Treated

From these granules, test samples were consolidated by the standard thermo compression molding process. The principle consists of heating the two press plates to a temperature of 173 °C during 10 min under a controlled constant pressure of 50 bars (Fig. 3). Each of the plate samples was cooled to room temperature under constant pressure before being removed from the mold than cut into test samples.

(a) (b) (c)

Fig. 3. Bio-composites by the thermo compression molding process (a) LLDPE + 10%ASP in mold, (b) heating press, (c) Bio-composite plate

3 Characterization

3.1 Water Absorption Test

The most important handicap linked to the use of Bio-fillers or natural lignocelluloses fibers as reinforcements in composite materials is their sensitivity to water which considerably reduces their mechanical properties in a humid atmosphere. Water absorption tests of pure LLDPE and the two types of treated and untreated LLDP/ASP Bio-composites were carried out in accordance with ASTM D 570-99. Samples measuring 19 × 19 × 3 mm^3 were cut from the heads of the tensile specimens, dried in an oven at 70°C for 24 h, then cooled in a desiccator and weighed immediately with a micro balance having an accuracy of 10^{-4}g. The samples were immersed in distilled water for 72 h. It should be noted that the choice of this duration mainly aims to clarify the effect of the quantity of water absorbed on the mechanical and tribological behavior of the treated and untreated LLDPE/ASP bio-composites. Five specimens were tested in each sample. Subsequently the samples are removed, the water on the surface of the sample is absorbed with filter paper and the sample is weighed. The water absorption percentage ($W\,\%$) of the samples was calculated using the following equation:

$$W\% = \frac{Wa - Wi}{Wi} \times 100 \tag{1}$$

Wi: mass of the dry sample.
Wa: mass of the sample after absorption.

3.2 Tribological Characterization

Friction and wear tests were carried out on samples of the two types of Biocomposites developed. They were carried out on an alternative tribometer with ball/plane contact. The antagonist used is a high-chromium steel ball (100Cr6) with a radius of 10 mm and an average arithmetic roughness of 0.06 μmm. The sample/antagonist contact is maintained by the application of a normal load of 10 N. An alternating movement of amplitude ± 7.5 mm and frequency 1 Hz is applied to the sample by means of a connecting rod-crank system coupled to a geared motor equipped with an electronic speed variator. During sliding, the tangential force is measured by a force sensor and stored using an acquisition system connected to a computer. Before each friction test, the sample surface was cleaned with ethanol. All tests were carried out dry at a temperature between 20 °C and 25 °C and a relative humidity (RH) between 50 and 60%. The durations of the sliding friction tests were set at 500, 1000, 2500 and 5000 cycles and each test was repeated three times. After each friction test, the depth and section S of the wear trace of each type of treated LLDPE/ASP Bio-composite sample and/or untreated LLDPE/ASP are measured by a profilometer (Surtronic S-100 Series Taylor Hobson England). The volume loss V (mm^3) was calculated using the following formula: V = S × d Where (d) is the sliding stroke equal to 15 mm.

4 Results and Discussion

4.1 Water Absorption Test

Figure 4 shows that water absorption increases with ASP content for both types of treated and untreated LLDPE/ASP Bio-composites. This is due, among other factors, to the highly hydrophilic nature of the ASP Particle compared to the hydrophobic LLDPE matrix. Indeed, the non-cellulose components of the ASP particle have numerous hydroxyl groups (–OH) which interact with the molecules (H_2O) leading to an increase in water absorption as explained by (Bessadok et all. 2007) in their detailed study on the water absorption characteristics of Alfa fibers in which they found that treatments reduced the overall water absorption of Alfa fibers, particularly the styrene treatment. On the other hand, we notice that the alkaline treatment reduces the water absorption for all treated LLDPE/ASP Bio-composites. This leads to the formation of a more rigid interface formed by an improved surface area between the treated ASP particle and the polyethylene matrix. These results are in agreement with the work of (El-Abbassi et al. 2015) who showed that the alkaline treatment of alfa fibers reduced the water absorption rate in polypropylene composites reinforced with alfa fibers, thus improving the interfacial adhesion between the fiber and the matrix and consequently the mechanical properties of these composites.

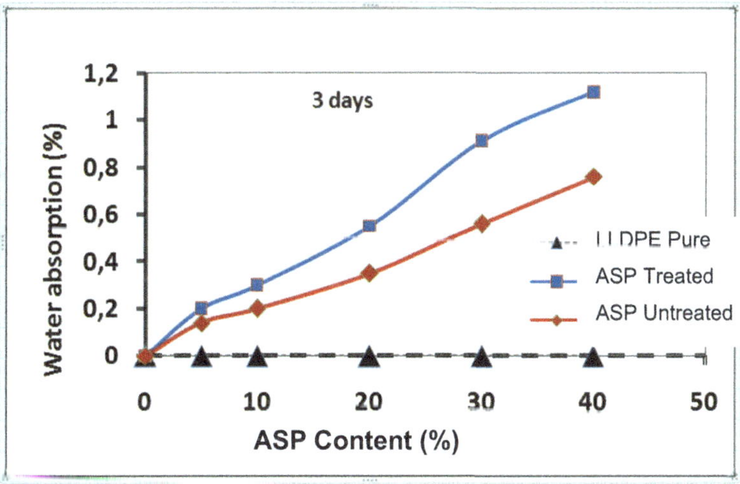

Fig. 4. Effect of ASP content on water absorption of treated and untreated LLDPE/ASP bio-composites

4.2 Tribological Characteristics

Figure 5 shows the evolution of the friction coefficient with the number of sliding cycles for the two types of bio-composites considered treated LLDPE/ASP and untreated LLDPE/ASP. Overall, the coefficient of friction increases continuously at the start of the test, undergoing some fluctuations in the so-called running in phase. Then it gradually reaches an established accommodation regime, stabilizing after a certain number of cycles of approximately 1000 until the end of the test.

Figure 6 shows the evolution of the friction coefficient as a function of the mass percentage of ASP of the two types of bio-composites developed LLDPE/ASP treated and untreated. We note that the friction coefficient of the virgin LLDPE polymer is

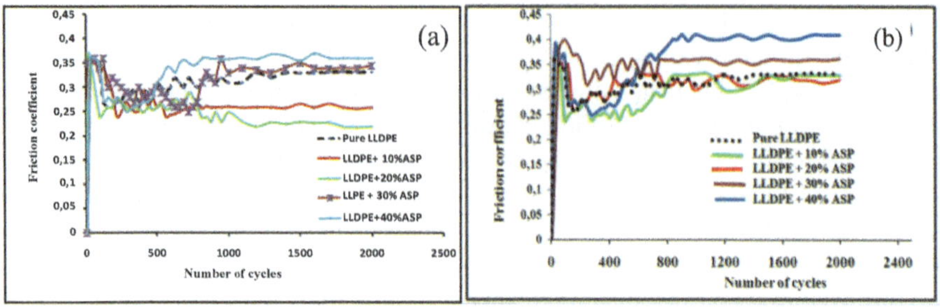

Fig. 5. Typical curves evolutions of friction coefficient with the number of cycles for LLDPE/ASP bio-composites (a) Treated, (b) Untreated

0.33. When the ASP content increases and for the treated LLDPE/ASP bio-composites, the friction coefficient decreases to a value equal to 0.22 for an ASP content of 20%. Whereas for bio-composites produced with untreated ASP bio-fillers, the addition of content less than 20% ASP practically does not affect the friction coefficient. Beyond this percentage, an increase in the coefficient of friction is observed for the two types of bio-composites. This increase is due to the rise in temperature at the interface leading to strong deformations and softening on the support face and the surface topography of the two types of biocomposites leading to an increase in interlocking and interaction between the asperities in contact and consequently the increase in the coefficient of friction.

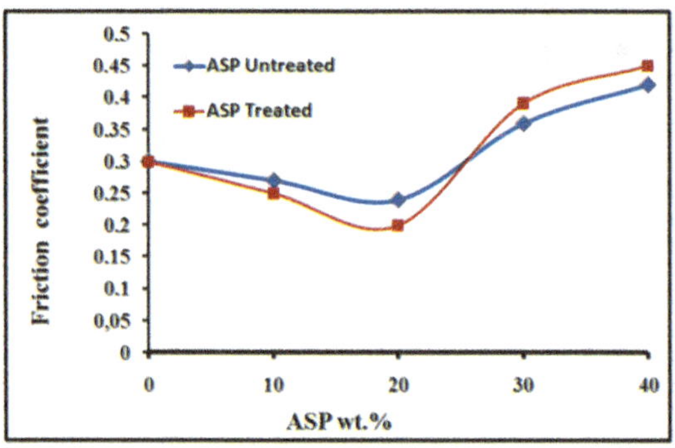

Fig. 6. Evolution of friction coefficient according to the ASP W%

5 Conclusion

In this study the effect of alkaline treatment on the tribological properties of bio-composites (LLDPE/ASP) is examined. Two groups of bio-composites of different mass fractions and their tribological properties were characterized and interpreted. The resulting conclusions can be summarized as follows:

- Alkaline treatment of ASP particles affects the tribological properties of LLDPE/ASP bio-composites.
- Bio-composites LLDPE/ASP presented improve tribological properties compared to pure LLDPE
- A significant and continuous increase in the tribological properties of the treated bio-composites compared to the untreated ones.
- Friction tests also suggest that a proportion of 20% by weight of ASP in the LLDPE matrix leads to better and a clear improvement of tribological performances.
- On the other hand, we can conclude that LLDPE bio-composites reinforced with treated ASP particles appear today as a very interesting alternative to replace synthetic and/or mineral reinforcements in many fields of engineering, such as automotive industry, construction, packaging, furniture,.....

References

1. Dominici, F., et al.: Biopolyethylene-based composites reinforced with alkalis and palmitoyl chloride treated Coffee silver skin. Molecules **24**, 3113 (2019). https://doi.org/10.3390/molecules24173113
2. Bayer, J., Granda, L.A., Méndez, J.A., Pèlach, M.A., Vilaseca, F., Mutijé, P.: Cellulose Polymer Composites (WPC). In: Deans, M., Ed., Advanced High Strength Natural Fiber Composites in Construction, Elsevier, Berkeley, pp. 115-139. https://doi.org/10.1016/B978-0-08-100411-1.00005-4
3. Garcia, D., Carbonell-Verdu, A., Jordá-Vilaplana, A., Balart, R., Garcia-Sanoguera, D.: Development and characterization of green composites from biobased polyethylene and shell of peanut. J. Appl. Polym. Sci. 133, 1-12 (2016). https://doi.org/10.1002/app.43940
4. Hammoui, Y., et al.: Preparation of plasticized wheat gluten/olive pomace powder biocomposite: Effect of powder content and chemical modifications Mater. Des. **87**, 742–749 (2015)
5. Nabincjad, O., Sujan DRahman, M.E., Davies, I.J.: Effect of oil palm shell powder on the mechanical performance and thermal stability of polyester composites. Mater Des. **65**, 823–830 (2015)
6. Deepak, V., Kheng, L.G.: Effect of mercerization/alkaline surface treatment of natural fibres and their utilization in polymer composites: mechanical and morphological studies. J. Compos. Sci. **5**, 175 (2021). https://doi.org/10.3390/jcs5070175
7. Yussni M.H., et al.: The effect of alkaline treatment under various conditions on physical properties of kenaf fiber. In: International Conference on Materials Physics and Mechanics 2017: IOP Conf. Series: Journal of Physics: Conference Series, vol. 914, p. 012030 (2017). https://doi.org/10.1088/1742-6596/914/1/012030
8. Bukar, A.M., Abubakar, M., El-Jummah, A., Alhaji, H.: Development and evaluation of the mechanical properties of coconut fibre reinforced low density polyethylene composite. Open J. Compos. Mater. **12**(3) (2022). https://doi.org/10.4236/ojcm.2022.123007
9. Shehu, U., Isa, M.T., Aderemi, B.O., Bello, T.K.: Effects of NaOH modification on the mechanical properties of baobab pod fiber reinforced LDPE composites. Nigerian J. Technol. (NIJOTECH) **36**(1), 87–95 (2017)
10. Rajesh G., Ratna P., Gupta, A.V.S.S.K.S.: Mechanical and degradation properties of natural fiber-reinforced PLA composites: jute, sisal, and elephant grass. Polym. Compos. (2015). https://doi.org/10.1002/pc.23395

11. Maaloul, N., Arfi, R.B., Rendueles, M., Ghorbal, A., Diaz, M., et al.:Dialysis-free extraction and characterization of cellulose crystals from almond (Prunusdulcis) shells. J. Mater. Environ. Sci. **8**(11), 4171–4181 (2017)
12. Boujelben, M.; Abid, M.; Kharrat, M.; Dammak, M.: Production and mechanical characterization of LLDPE biocomposite filled with almond shell powder. Polym. Polym. Compos. 096739112091086 (2020). https://doi.org/10.1177/0967391120910869
13. El Mechtali, F.Z., et al.: Mechanical and thermal properties of polypropylene reinforced with almond shells particles: impact of chemical treatments. J. Bionic Eng. **12**(3), 483–494 (2015)
14. Bessadok, A., et al.: Effect of chemical treatments of Alfa (Stipatenacissima) fibres on water-sorption properties. Compos. Sci. Technol. **67**, 685–697 (2007)
15. El-Abbassi, F.E., Assarar, M., Ayad, R., Lamdouar, N.: Effect of alkaline treatment on Alfa fibre as reinforcement for polypropylene based ecocomposites: mechanical behaviourandwaterageing. https://doi.org/10.1016/j.compstruct.2015.07.112

Frictional Behavior of the Snake Skin-Inspired Surfaces: Numerical Study

Fatma Elwasli(✉), Slah Mzali, and Salah Mezlini

LGM, Ecole Nationale d'Ingénieurs de Monastir, Université de Monastir, Avenue Ibn Aljazzar, 5019 Monastir, Tunisie
elwaslifatma@gmail.com

Abstract. Bio-inspired textures have recently gained much attention due to their unique properties. Understanding the correlation between bio-inspired textures and friction behavior is critical in optimizing friction response. The study investigates the frictional behavior of surfaces inspired by snake skin, when subjected to sliding. A Finite Element (*FE*) model was implemented using *Abaqus/standard* software, taking the effect of the bio-inspired texture into consideration. During the test, the rigid ball, with a diameter 10 mm, was alternatively sliding on the texture surface. A normal load of 1 N was applied on the ball. Numerical results obtained were consistent with the experimental results. They have shown the significant effect of the sliding direction on the friction behavior. The results depict the anisotropic behavior of thesnake skin-inspired surface. Indeed, friction coefficient recorded when sliding in the positive direction ($\mu+$) was larger than in the negative ($\mu-$) one. The ratio of frictional anisotropy (i.e., $\mu+/\mu-$) is constant (1.77).

Keywords: Friction · Texture · Bio-inspired surface · *Abaqus* · Frictional anisotropy

1 Introduction

For the purposes of reducing wear and energy consumption, improving efficiency and extending lifetime, researchers have made significant efforts in understanding contact-pair behavior[1–3]. Numerous methods, including lubrication, the use of additives, and surface texturing optimization, have been adopted to reduce friction [4].

A great effort has been made to examine the effect of surface texturing on the tribological behavior [5–7]. In recent year, particular focuses were given to the study of bio-inspired surfaces [8–10].

Ballesteros et al. [11] have investigated the effect of texturation inspired by the ventral scales of the P. Regius snake. 3D printer was used to manufacture samples with a bio-inspired surface. Pin-on-disc tests were realized to analyze the tribological behavior. Experimental results have revealed that the pattern geometry of the bio-inspired surface leads to a significant decrease in the friction coefficient. Greiner et al. [12] have investigated the frictional behavior of a biomimetic surface texture inspired by the

M. Chouchane et al. (Eds.): CMSM 2023, LNME, pp. 273–279, 2024.
https://doi.org/10.1007/978-3-031-65007-9_30

snakeskin using a pin on disk test. They have observed a decrease of more than 40% of the friction forces compared to non-textured surface, under dry condition.

Few attentions were paid to the numerical investigation of the effect of bio inspired surfaces on the tribological behavior. Chen et al. [13] have investigated finger-sealing performance using three texture forms diamond, ellipse, and hexagon inspired from snake skins. They have developed a numerical model simulating a textured finger seals under hydrodynamic lubrication. Numerical results have shown that the ellipse textured finger seal presents a greater anti-friction and wear resistance properties compared to other textures.

The present research explores the frictional behaviour of the snake skin-inspired surfaces subjected to friction test under dry condition. A load-controlled model was developed using *Abaqus/standard* software. The established model takes into account the surface texture. The effect of the sliding direction was studied and the friction was examined.

2 Numerical Model

For the purpose of investigatingthe frictional behaviour of the snake skin-inspired surfaces, a numerical model was developed. The model simulates the sliding test using *ABAQUS/standard* software. The 2D finite element (*FE*) model uses plain strain assumption. It was composed of two parts: (i) the ball and (ii) the textured surface sample (Fig. 1). The ball is a rigid body of a half circle shape with a diameter 10 mm. The sample was a deformable body of a rectangular shape with the dimension (15 × 2.4) mm^2. The upper surface of the PLA sample was textured with snake skin-inspired surface [14, 15]. The bottom surface was constrained. First, a normal load was applied on the ball. Then, a displacement with a stroke length of 10mm was imposed on two directions as presented in Fig. 1. Based on a preliminary investigation using a reverse method and Zemzemi's Equations[16], the adhesive friction coefficient for the model was found to be 0.05.

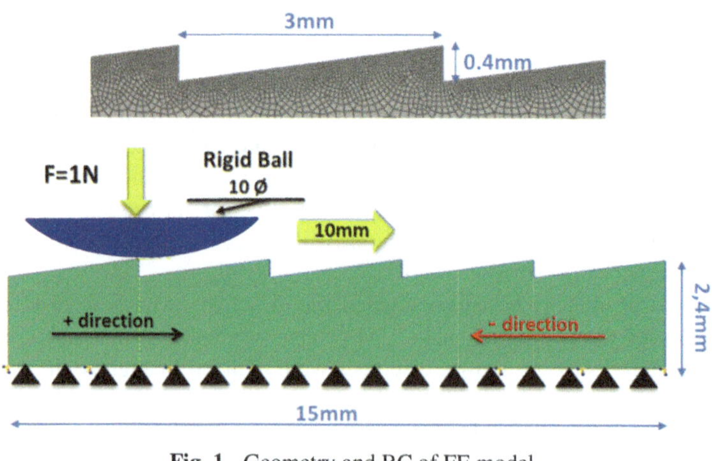

Fig. 1. Geometry and BC of FE model.

The developed model is a load-controlled FE model which leads to different ball depths [17]. This method is crucially significant, spatially when sliding over a rough surface. Tests were realized under a normal load of 1N corresponding to a mean contact pressure of 28.5 MPa.

Once the convergence test was carried out, a mesh with 7900 nodes and 7584 elements was defined. Reduced integration CPE4R elements were used to mesh the PLA sample. The mesh was highly refined along the upper surface of the sample and elements of $30 \times 30\ \mu m^2$ were used. In this study, elastic–plastic-damage model was assumed to describe the PLA sample. The PLA properties adopted in the considered model are outlined in Table 1.

Table 1. PLA material properties [18–20].

Mechanical properties	
Young's Modulus E [MPa]	2800
Poisson ratio ν	0.36
Density ρ [Kg/m^3]	1240
Yield strength σ_y [MPa]	59
Ductile damage parameters	
Fracture strain	1.3
Stress triaxiality	0.33
Strain rate (s^{-1})	0.1

3 Results and Discussion

Friction coefficient was calculated based on the coulomb is low: $\mu = F_t/F_n$, where F_t tangential force and F_n the normal load. Figure 2 illustrates typical results obtained when sliding in the positive direction under a normal load 1 N. Numerical results show that le ball slid over three slopes, this gives rise to three plateaus. The sliding behaviour can be divided in two regions: (i) the first region where a stable friction force was recorded and (ii) the second region where a sharp decrease of the friction force was detected.

The friction coefficient obtained from simulations was confronted to the experimental results obtained from our previous paper published by Mzali et al. [21]. They provided a detailed description of the technical specifications of the experimental approach that was used to carry out the experiments. Simulations and experimental results for the friction coefficient show a good agreement. The numerical curve is similar to that found experimentally. The discrepancy between the numerical and experimental friction coefficient values in the first region and the second region was 0.2% and 29%, respectively.

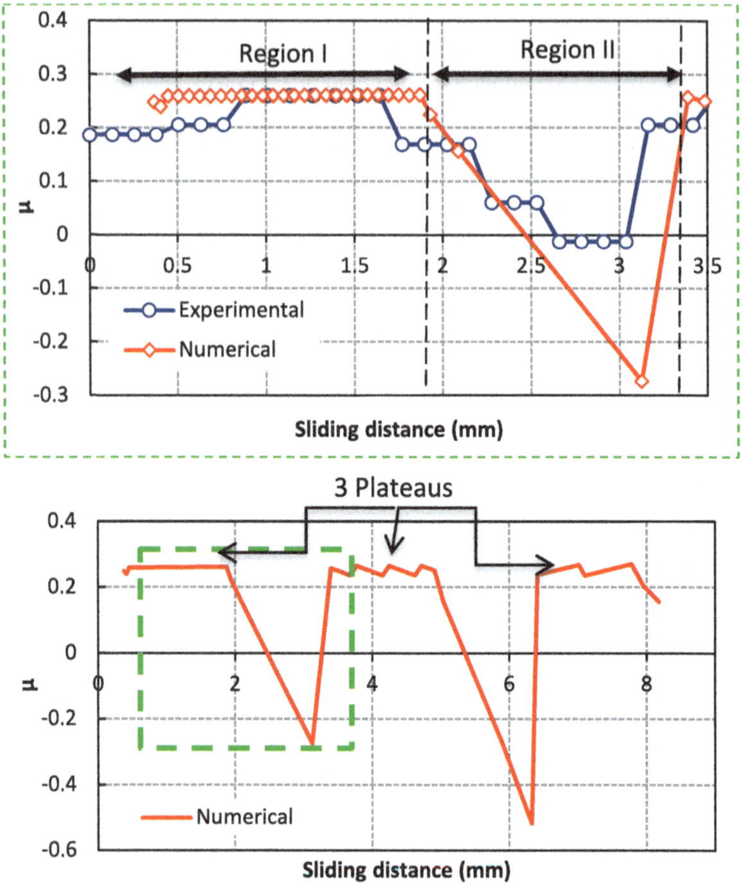

Fig. 2. Numerical and experimental friction coefficient versus the sliding distance.

Figure 3 presents the evolution of the friction coefficient and the depth ball as a function of the sliding direction. Thanks to the load-controlled model, the ball was oscillating between peaks and valleys. When the ball was sliding on slopes (Fig. 3a), a stable friction force with an average value 0.26 was observed. In this first region, the ball was following the slope, thus, the increase of the depth ball. Just after reaching the summit (Fig. 3b), the ball fell to the bottom and the value of the friction force becomes negative. Indeed, as shown in Fig. 3c, the peak pushes ball which generate to the lowest friction value. When the ball reaches the bottom (Fig. 3d), it climbs the slope again and the cycle continues.

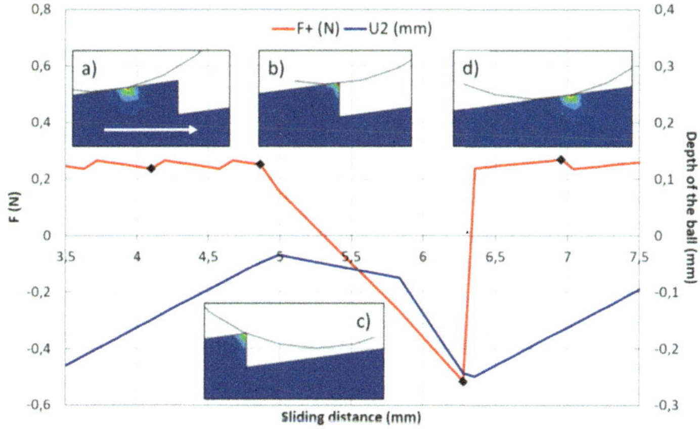

Fig. 3. Friction force and ball depth versus the sliding distance in the positive direction: the ball sliding (a) on the slope, (b) in the summit, (c) after the summit and (d) on the slope.

Figure 4 presents the evolution of the friction coefficient μ versus the sliding distance in the positive (+) and the negative (−) sliding directions. Special focus was put on the comparison of plateaus. The μ recorder in the positive direction is slightly higher than in the negative direction. The studied surface has shown, thus, an anisotropic behaviour. The ratio of frictional anisotropy (i.e., μ+/μ−) is constant (1.77). This may be explained by its geometry. Indeed, during sliding in the positive direction, the ball climbs the slope. This may generate a high tangential force. Contrary to the negative direction, where, the ball was sliding on slopes and no supplementary effort was needed.

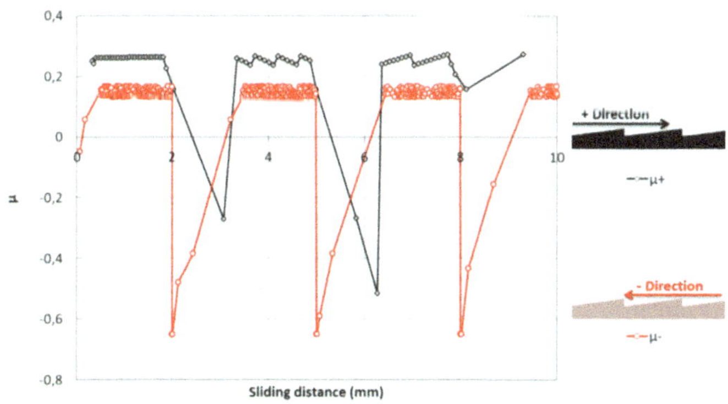

Fig. 4. Friction coefficient μ versus the sliding distance for both directions.

4 Conclusion

This attempt addresses the frictional behavior of a snake skin-inspired surface.

A load-controlled FE model has been developed using ABAQUS/standard. A particular focus was put on the impact of sliding direction on the frictional behavior. Numerical finding leading to the following conclusions:

1. The FE model was found very useful for predicting friction coefficient behavior for a snake skin-inspired surface.
2. The sliding behaviour was divided in two regions:
   (i) 1st region: a stable friction force was recorded during sliding on the slope
   (ii) 2nd region: a sharp decrease of the friction force was detected after reaching the summit.
3. The snakeskin-inspired surfaces have shown an anisotropic behavior. Friction coefficient recorded when sliding in the positive direction was larger than in the negative direction. The ratio of frictional anisotropy (i.e., $\mu +/\mu -$) is constant (1.77).

References

1. Mzali, S., Mkaddem, A., Elwasli, F., Mezlini, S.: Towards an advanced modeling of failure mechanisms' interaction in fiber-reinforced polyester: a mixed-mode loading concept. Compos. Struct. **160**, 70–80 (2017)
2. Elleuch, K., Mezlini, S., Guermazi, N., Kapsa, P.: Abrasive wear of aluminium alloys rubbed against sand. Wear **261**, 1316–1321 (2006)
3. Demirci, I., Mkaddem, A., Khoukhi, D.E.: A multigrains approach to model the micromechanical contact in glass finishing. Wear **321**, 46–52 (2014)
4. Cha, S.C., Erdemir, A.: Coating Technology for Vehicle Applications. Springer International Publishing, Switzerland (2015). https://doi.org/10.1007/978-3-319-14771-0
5. Meine, K., Schneider, T., Spaltmann, D., Santner, E.: The influence of roughness on friction Part I: the influence of a single step. Wear **253**, 725–732 (2002)
6. Demirci, I., Mezghani, S., Yousfi, M., El Mansori, M.: Multiscale analysis of the roughness effect on lubricated rough contact. J. Tribol. **136**(1), 011501 (2014)
7. Spinnler, G.: *Conception Des Machines Principes et Application\r1 Statique. Conception des Machines Principes et Application* vol. 1 (1997)
8. Siddaiah, A., Menezes, P.L.: Advances in bio-inspired tribology for engineering applications. J. Bio Tribocorros. **2**, 1–19 (2016)
9. Richards, C., et al.: Bio-inspired surface texture modification as a viable feature of future aquatic antifouling strategies: a review. Int. J. Mol. Sci. **21**, 1–20 (2020)
10. Sun, F., Xu, H.: A review of biomimetic research for erosion wear resistance. Biodes. Manuf. **3**, 331–347 (2020)
11. Ballesteros, L.M., Zuluaga, E., Cuervo, P., Rudas, J.S., Toro, A.: Tribological behavior of polymeric 3D-printed surfaces with deterministic patterns inspired in snake skin morphology. Surf. Topograph. Metrol. Prop. **9**(1), 014002 (2021)
12. Greiner, C., Schäfer, M.: Bio-inspired scale-like surface textures and their tribological properties. Bioinspir. Biomim. **10**(4), 044001 (2015)
13. Chen, L., Zhang, Y., Cui, Y., Wang, J., Wang, M.: Effects of snake-bioinspired surface texture on the finger-sealing performance under varied working conditions. Machines **10**(7), 569 (2022)
14. O'Hara, K.B., Martinez, A.: Monotonic and cyclic frictional resistance directionality in snakeskin-inspired surfaces and piles. J. Geotech. Geoenviron. Eng. **146**, 04020116 (2020)

15. Martinez, A., Palumbo, S.: Anisotropic shear behavior of soil-structure interfaces: bio-inspiration from snake skin. In: IFCEE 2018 (pp. 94-104) (2018).https://doi.org/10.1061/9780784481592.010
16. Zemzemi, F.: Caracterisation de modeles de frottement aux interfaces piece-outil-copeau en usinage : application au cas de l'usinage des aciers et de l'inconel 718. (l'École Centrale de Lyon -N° d'ordre : 2007–33, 2007). doi:N° d'ordre : 2007–33
17. Pondicherry, K., et al.: Optimization and validation of a load-controlled numerical model for single asperity scratch. Tribol. Lett. **68**, 1–13 (2020)
18. Aldosari, F., Khan, M.A.A., Asad, M., Djavanroodi, F.: Finite element analysis of polylactic acid (PLA) under tensile and compressive loading. In: Journal of Physics: Conference Series (Vol. 2468, No. 1, p. 012094). IOP Publishing (2023)
19. Al-Behadili, A.J.M., Bedaiwi, B.O.: Measurement of absorbed energy of the polylactic acid (PLA) via numerical simulation. J. Mech. Eng. Res. Dev. **43**, 428–435 (2020)
20. Karimi, A., Mole, N., Pepelnjak, T.: Numerical investigation of the cycling loading behavior of 3D-printed poly-lactic acid (PLA) cylindrical lightweight samples during compression testing. Appl. Sci. **12**(16), 8018 (2022)
21. Mzali, S., Elwasli, F., Mezlini, S., Hajlaoui, K., Alrasheedi, N.H.: Tribological behavior of 3D printed biomimetic surfaces. Tribol. Int.. Int. **193**, 109352 (2024)

Laser Pulse Shape Effect on a Laser Shock-Peened Turbine Blade

Manel Ayeb[1,2](✉), Mounir Frija[1,2], and Raouf Fathallah[2,3]

[1] Institute of Applied Sciences and Technology of Sousse (ISSATSo), University of Sousse, Sousse, Tunisia
Ayeb.manel2705@gmail.com

[2] Unit of Mechanical and Materials Production Engineering (UGPMM/UR17ES43), National Engineering School of Sfax (ENIS), University of Sfax, Sfax, Tunisia

[3] National Engineering School of Sousse (ENISo), University of Sousse, Sousse, Tunisia

Abstract. As an innovative surface treatment technique, laser shock processing is used extensively to improve the resistance of metallic workpieces to corrosion, fatigue and wear, particularly in spite of the residual compressive stresses that are generated. The purpose of this paper is to study the geometric shape effect of the laser impact on a turbine blade's thin leading edge. A 3D finite element simulation of Ti-6Al-4V thin leading-edge specimens is performed by means of ABAQUS software. Its main objective consists in studying and comparing the mechanical modifications of the surface generated by circular and square laser spots: the compressive residual stresses, the plastic deformations and the damage.

Keywords: Laser processing · Square spots · Circular spots · Finite element analysis · Very thin leading edge

1 Introduction

Laser processing (LP) is an advanced mechanical surface enhancement process and it is used in different industrial activities like: automotive sector, medical applications, micro-electronic systems (MEMS), military and nuclear applications and precisely the aeronautics sector. LP treatment guarantees a very thin surface process.

This treatment utilizes the mechanical effect of applying intense pulse laser energy of around a decade GW/cm^2 lasting several nanoseconds to a local area of the surface. As a result, work hardening is produced by a plastic strain gradient at depth, and a shock wave is propagated in a surface layer of the structure (Fig. 1). Therefore, these plastic deformations engender deeper compressive residual stresses than in the case of shot peening. These stresses are highly advantageous to enhance resistance to wear and corrosion, and particularly to improve high-cycle fatigue lifetime.

As a main objective of this paper, a laser impact simulation utilizing the finite element analysis (FEA) is performed to develop a representative sample of a thin leading edge of a Ti-6Al-4V turbine superalloy blade by means of the ABAQUS software. This investigation is set up specifically for the purpose of examining the geometric shape effects of laser impact on a thin leading edge for the prediction and comparison of compressive stresses, plastic deformation and damage by Johnson-Cook.

M. Chouchane et al. (Eds.): CMSM 2023, LNME, pp. 280–289, 2024.
https://doi.org/10.1007/978-3-031-65007-9_31

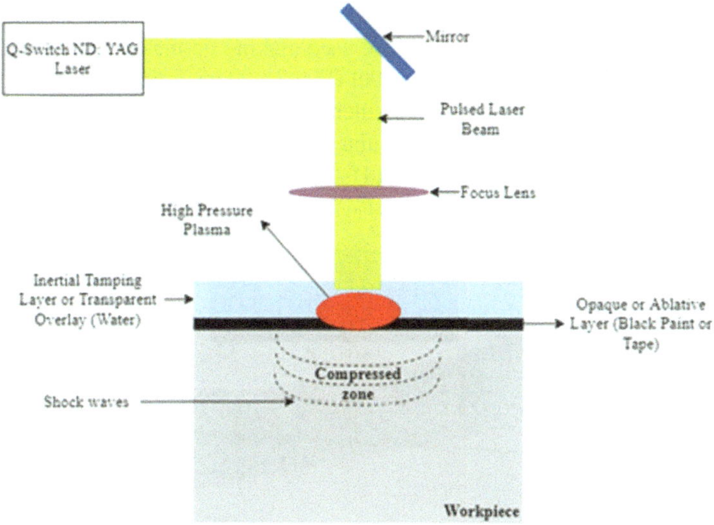

Fig. 1. Laser processing outline.

2 LSP Analysis Using Finite Element Analysis

2.1 Geometrical Pattern

To model a circular laser spot and a square laser spot, we have developed a three-dimensional modeling of a turbine blade sample consisting of a Ti-6Al-4V titanium superalloy with the aim of predicting and comparing the equivalent residual stresses, plastic deformation and damage of Johnson-Cook.

The shape and dimensions of this thin test sample (about 0.75 mm) are indicated in Fig. 2.

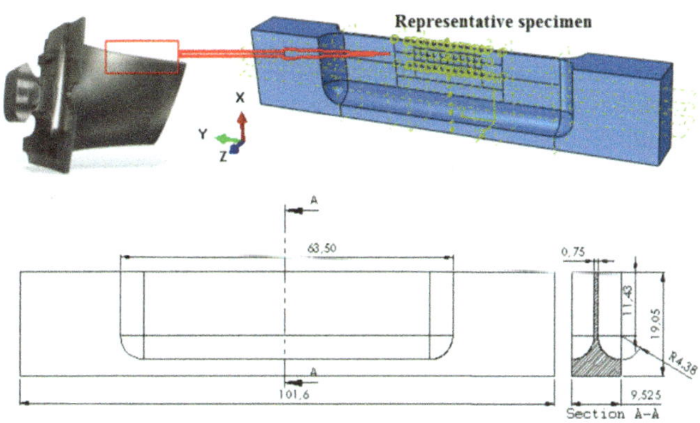

Fig. 2. Representative sample of a turbine blade.

Choosing the most appropriate mesh is a very important step. It is important to ensure that an appropriate number of elements are used making up the area to be treated. Two zones are therefore considered in the model (Fig. 3): Zone 1, to which the LP impacts are applied is meshed with 66 000 Hex elements (C3D8R) for the square laser spots and 66 000 Tet elements (C3D4) for the circular laser spots. As well, the sample's rest is represented by Zone 2 and meshed with 141 418 Tet elements (C3D4).

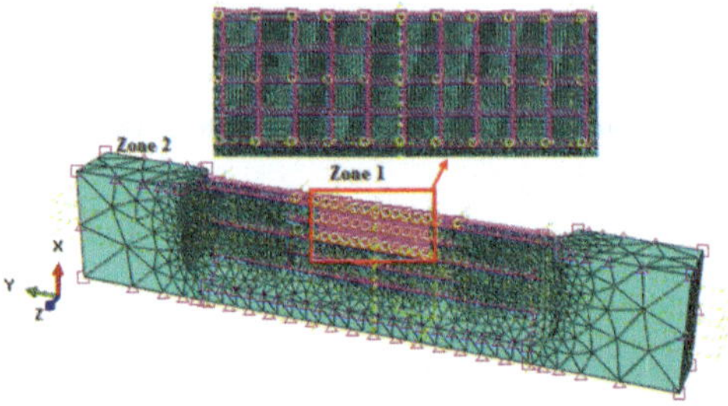

Fig. 3. Detailed mesh of the sample.

2.2 Basic Assumptions

For the purpose of the FEM analysis of the LP process, it is assumed that there are no edge effects, there is a spatially constant pressure across the specimen, the interaction of the material is perpendicular to its surface, there is no increase in temperature within the laser area owing to the application of an ablative thermo-protective coating, a rapid impact is considered, and the Johnson-Cook constitutive law is isotropic.

The shape and dimensions of this thin test sample (about 0.75 mm) are indicated in Fig. 2.

2.3 Typical Material Properties

Using the empirical Johnson-Cook behavior law [1], the Ti-6Al-4V material's thermo-viscoplastic properties are formulated. That law is commonly utilized in the simulation of specific dynamic processes and is applied to the resulting Von Mises equation:

$$\sigma_y(\bar{\varepsilon}_p) = \left[A + B\left(\bar{\varepsilon}^p\right)^n\right]\left[1 + C\ln\left(\frac{\dot{\bar{\varepsilon}}^p}{\dot{\varepsilon}_0}\right)\right]\left[1 - \left(\frac{T - T_0}{T_f - T_0}\right)^m\right] \tag{1}$$

As mentioned in Eq. (1), the immediate deformation rate ranges from $1s^{-1}$ to $_{106s}{}^{-1}$ (LP process can cause deformation rates in excess of $10^6 s^{-1}$ within sample [2]).

As the influence of temperature is ignored, the latter doesn't rise as propagation of the shock wave proceeds. Therefore, it follows that Johnson-Cook's principle has the following formula:

$$\sigma_y(\bar{\varepsilon}_p) = \left[A + B(\bar{\varepsilon}^p)^n\right]\left[1 + C\ln\left(\frac{\dot{\bar{\varepsilon}}^p}{\dot{\varepsilon}_0}\right)\right] \tag{2}$$

where *A*, *B*, *C*, *m* and *n* illustrate the plastic parameters of Johnson-Cook and are equal to 1098 MPa, 1092 MPa, 0.014, 1.1 and 0.93, respectively [3].

Certainly, the damage that is based on the Johnson-Cook law [2], is assumed according to the subsequent formula:

$$\bar{\varepsilon}^p_{rupture} = \left[d_1 + d_2\exp\left(d_3\frac{\sigma_H}{\sigma_{eqVM}}\right)\right]\left[1 + d_4\ln\left(\frac{\dot{\bar{\varepsilon}}^p}{\dot{\varepsilon}_0}\right)\right]\left[1 + d_5\frac{T - T_0}{T_f - T_0}\right] \tag{3}$$

$$D = \sum \frac{\Delta\bar{\varepsilon}^p}{\bar{\varepsilon}_f} = \sum \frac{\Delta\varepsilon^p_{eq}}{\bar{\varepsilon}_f} \tag{4}$$

where d_1, d_2, d_3, d_4 and d_5 are damage values for Johnson-Cook and are equal to −0.09, 0.27, 0.48, 0.014 and 3.87, respectively [3].

According to Eq. (3), the corresponding plastic deformation at fracture $\bar{\varepsilon}_{rupture}$ depends on the reference strain $\dot{\varepsilon}_0$, the equivalent Von Mises stress σ_{eq} and the hydrostatic stress σ_H. As well, the last one is equal to $1s^{-1}$.

In Eq. (4), the damage accumulation *D* depends on the corresponding plastic deformation up to failure $\bar{\varepsilon}_f$ and the plastic incremental deformation and $\Delta\varepsilon^p_{eq}$. By *D* reaching *1*, there is no stress and the fissure appears.

2.4 Load and Boundary Assumptions Applied to the LP Process

In Fig. 4, the boundary assumptions and the load engendered by LP treatment.

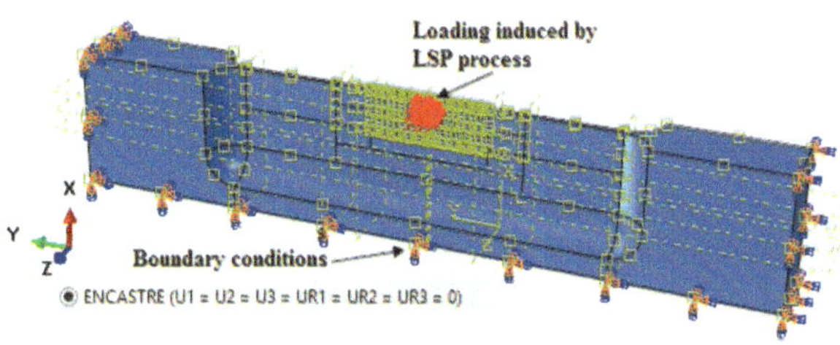

Fig. 4. LP process loading and boundary assumptions.

The relevant boundary assumptions include the following requirements:

- The two edges and the base are blocked ($U_r = U_\theta = U_z = UR_r = UR_\theta = UR_z = 0$).
- The middle vertical line is marked as a line of symmetry.
- The temperature of the boundary zone is 298 K.

Within this study, the dynamic response will be treated and it is based on the application of a time profile of a pressure that is spatially constant over the surface of the specimen.

As indicated in Fabbro's model [4], maximum pressure values (expressed in gigapascals (GPa)) are determined for various power densities, where the pulse width and wavelength are 20 ns and 1064 nm respectively, and expressed as follows:

$$P_{max} = 0.01\sqrt{\frac{\alpha}{2\alpha + 3}}\sqrt{Z}\sqrt{AI_0} \tag{5}$$

According to Eq. (5), the maximum pressure P_{max} depends on the laser power density I_0 expressed in gigawatt per square centimeter (GW/cm^2), the equivalent impedance Z, as expressed in grams per square centimeter per second ($g/cm^2/s$), the efficiency coefficient α, and the absorbing factor A.

In this paper, the full width at half maximum (FWHM), the peak pressure (P_{max}), the square laser spot dimension and the circular laser spot diameter are respectively about: *20* ns, *5* GPa, *3*3* mm ^{2} and *3* mm. The normalized profile of the peak pressure is shown in Fig. 5 for FWHM equal to *20 ns*.

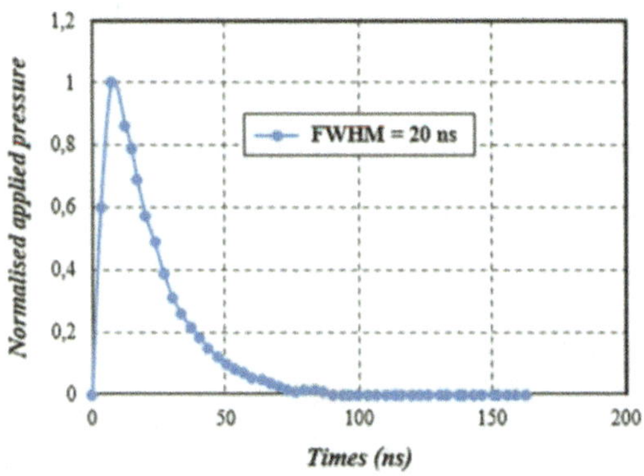

Fig. 5. Normalized pressure versus time for FWHM equal to 20 ns.

2.5 Main Computation Stages

In the recent study, the calculation steps [5–9] is divided over two stages: the loading step 'CHARG' that is operated over 1000 ns, and the relaxation step 'RELAX' that takes 0.5 ms.

3 Findings and Discussions

In recent work, the ABAQUS/EXPLICIT finite element code is utilized to achieve these results. Indeed, the explicit method guarantees a high degree of robustness. Similarly, it provides excellent resolution of fast shocks. For this reason, it is used to perform a simulation of the laser impact with the objective of analyzing its geometric shape and its influence on compressive stresses, plastic strains and damage by Johnson-Cook. Thus, two different geometric shapes of laser impact are employed: the first is a single circular impact (Fig. 6 (a)) with a diameter of 3 mm and the second is a single square impact (Fig. 6 (b)) of dimension 3*3 mm2 which are each applied to a single face of a thin leading edge representing a turbine blade.

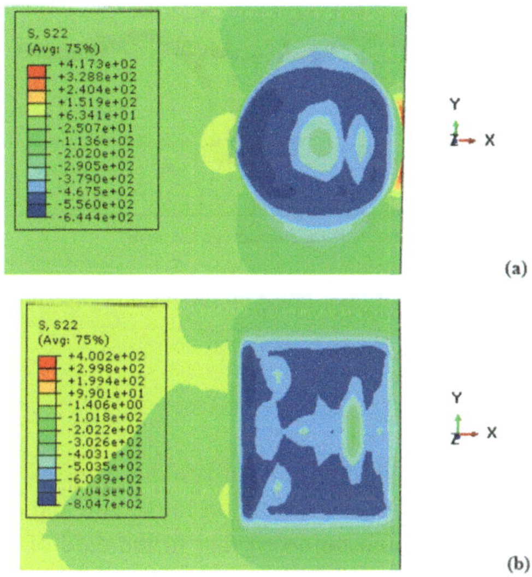

Fig. 6. Residual stress distribution for (a) circular laser impact and (b) square laser impact.

3.1 Residual Stresses

According to Fig. 7(a), the distribution of compressive residual stresses has a similar profile when both circular and square laser impacts are used. Therefore, it can be seen that for a circular laser impact, the residual compressive stresses are not localized at the center of the impact (the stress hole) (Fig. 6(a)). However, those generated by the impact of a square laser are located at the center by the treated area and spread over an area of approximately 1.5 mm from the impact center in the y-axis. (Fig. 6(b)). As well, the square laser impact induces the most demanding impact to generate a more uniform surface (in particular by the use of multiple layers), nevertheless, the circular laser impact can generate a uniform surface only with a sufficiently high coverage rate in order to restrict the untreated surface areas.

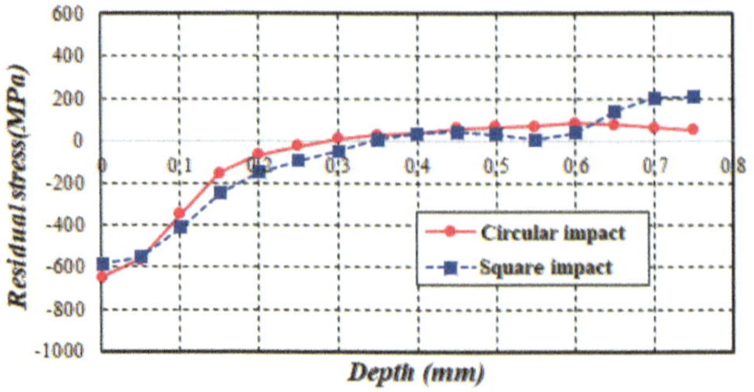

(a)

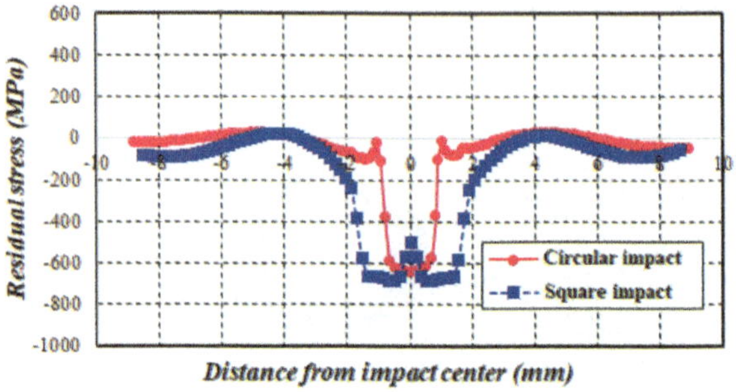

(b)

Fig. 7. Geometric shape influence of laser pulse on the residual stresses (a) in depth and (b) at the surface.

3.2 Plastic Strains

Plastic deformation occurs as a direct result of residual stresses. Thus, the profiles of plastic deformation display a similar distribution for square and circular laser impacts (Fig. 8(a)). Nevertheless, it should be noted that in the case of circular laser pulses, the difference in the depth of plastic behavior between square and circular laser pulses can be attributed to dissimilarities in the spatial distribution of the energy deposition. In addition, the peak plastic deformation is localized at the center of the laser pulse (Fig. 8(b)). Whereas that generated by the square laser impact is spread over the entire treated surface. Indeed, for a circular laser impact, the geometrical shape for the laser pulse influences its concentration around the central zone treated. In the case of a square laser pulse, the pressure waves are orthogonal and cover the whole treated zone.

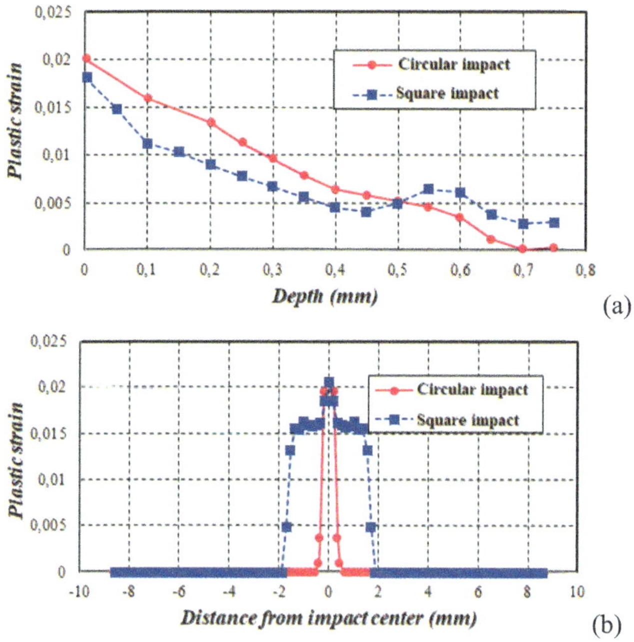

Fig. 8. Geometric shape influence of laser pulse on the plastic strain (a) in depth and (b) at the surface.

3.3 Johnson-Cook Damage

With circular laser pulses, the highest levels of damage by Johnson-Cook (Fig. 9) can be found centered in the laser's impact center. However, for square laser spot, they are spread over the entire impacted surface area.

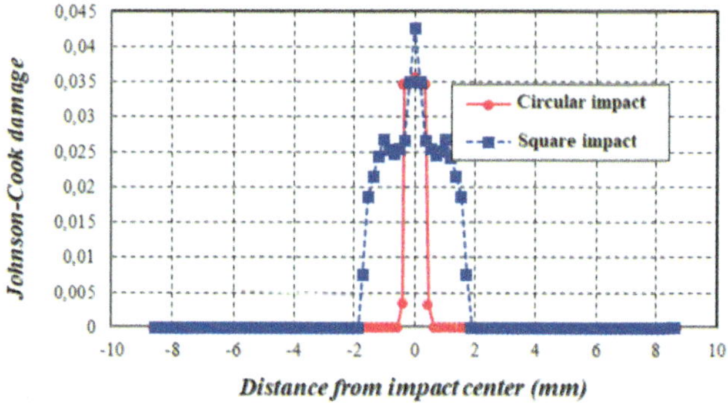

Fig. 9. Geometric shape influence of laser pulse on the damage by Johnson-Cook (a) in depth and (b) at the surface.

4 Conclusion

In this paper, the effects of the geometric shape of laser impact on thin leading-edge surface made with Ti-6Al-4V are investigated. It has been observed there is no compressive residual stress remaining in the middle zone of the processed part when a circular laser impact is applied. This is due to the waves simultaneously being focused around edges of the impact zone, leading to a non-uniform superficial stress. However, the use of a square laser pulse can avoid this problem. In fact, this kind of laser spot enables a large surface area to be treated effectively with a coverage ratio that is below 5% in comparison with circular laser spots. In addition, the morphology of the surface formed by square pulses is smoother than that created by circular pulses [10].

The importance of this article will be to show precisely the geometric shape effect of the laser impact on the residual stresses at depth, the corresponding plastic deformation and the damage of Johnson-Cook generated as a result of processing by LP.

In addition to its immediate implications for surface treatment processes, this study sets the stage for further investigations into optimizing laser impact geometry to enhance mechanical characteristics and capabilities of aerospace components and other high-stress applications.

Acknowledgements. This research is being performed through the funding and support granted to the Unit of Mechanical and Materials Production Engineering (UGPMM/UR17ES43) by the Tunisian Ministry of Higher Education and Scientific Research.

References

1. Johnson GR, Cook WH: A constitutive model and data for metals subjected to large strains, high strain rates and high temperatures. In: Proceedings of the 7th international symposium on ballistics 54, 541–547 (1983)
2. Smith, P.R., Shepard, M.J., Prevéy, P.S., III., Clauer, A.H.: Effect of power density and pulse repetition on laser shock peening of Ti-6AI-4 V. J. Mater. Eng. Perform. **9**, 33–37 (2000)
3. Kay, G.: Failure modeling of Titanium 6Al-4V and Aluminum 2024–T3 with the Johnson-Cook material model. Office of Aviation Research Washington, DC (2003)
4. Peyre, P., et al.: FEM simulation of residual stresses induced by laser peening. Eur. Phys. J. Appl. Phys. **23**, 83–88 (2003)
5. Ayeb, M., Frija, M., Fathallah, R.: Influence of multiple laser impacts on thin leading edges of turbine blade. Proc. Inst. Mech. Eng., Part L: J. Mater.: Des. Appl. **234**(1), 130–143 (2020)
6. Frija, M., Ayeb, M., Seddik, R., Fathallah, R., Sidhom, H.: Optimization of peened-surface laser shock conditions by method of finite element and technique of design of experiments. Int. J. Adv. Manuf. Technol. **97**(1–4), 51–69 (2018)
7. Ayeb M : Développement d'un modèle de prévision de l'état induit du traitement, efocsa, Editions universitaires européennes, ISBN-13:978–613–8–39716–8, 92 (2018)
8. Ayeb, M., Frija, M., Fathallah, R.: Prediction of residual stress profile and optimization of surface conditions induced by laser shock peening process using artificial neural networks. Int. J. Adv. Manuf. Technol. **100**(9–12), 2455–2471 (2019)
9. Ayeb, M., Frija, M., & Fathallah, R.: Effect of Multiple Impacts on Thin Leading Edges of Turbine Blade Treated by Laser Shock Peening Process. In: Lecture Notes in Mechanical Engineering. pp. 498–506 (2020)

10. Wen Cao, Z., Gang Che, Z., Kun Zou, S., Xing Fei, Q.: Numerical simulation of residual stress field induced by laser shock processing with square spot. J. Shanghai Univ. (English Edition) **15**(6), 553–556 (2011)

Predicting and Optimizing the Mechanical Effects Generated by a Laser-Treated Turbine Blade Using Artificial Neural Networks and ANFIS Techniques

Manel Ayeb[1,2](✉), Mourad Turki[3,4], Mounir Frija[1,2], and Raouf Fathallah[2,5]

[1] Institute of Applied Sciences and Technology of Sousse (ISSATSo), University of Sousse, Sousse, Tunisia
Ayeb.manel27705@gmail.com

[2] Unit of Mechanical and Materials Production Engineering (UGPMM/UR17ES43), National Engineering School of Sfax (ENIS), University of Sfax, Sfax, Tunisia

[3] Higher Institute of Technological Studies of Mahdia (ISETMA), Hiboun, Tunisia

[4] Research Laboratory "Automatic, Electrical Systems and Environment" LAS2E, National Engineering School of Monastir (ENIM), University of Monastir, Monastir, Tunisia

[5] National Engineering School of Sousse (ENISo), University of Sousse, Sousse, Tunisia

Abstract. This article discusses the application of artificial neural networks (ANNs) and adaptive neuro-fuzzy inference systems (ANFIS) to the prediction and optimization of mechanical effects resulting from laser shock processing (LSP) applied to turbine blades made of Inconel 718. LSP treatment is used in the aerospace industry to improve the durability and performance of turbine blades. The use of ANNs and ANFIS offers a data-driven approach to modeling and optimizing the mechanical properties of these treated blades. The development of predictive models that take into account various parameters such as: maximum applied pressure, full-width at half maximum and dimension is explored. In addition, this paper discusses optimization of LP process that harness the power of ANNs and ANFIS to maximize the desired mechanical properties, thereby contributing to the advancement of turbine blade technology in aerospace applications.

Keywords: Laser shock peening · Finite element method · Residual Stress · Plastic Strain · Johnson-Cook Damage · Inconel 718 · Artificial Neural Networks (ANNs) · Adaptive Neuro-Fuzzy Inference (ANFIS)

1 Introduction

Laser peening (LP) as an advanced surface treatment has proven its value in the field of materials engineering. This treatment consists in applying intense laser pulses to the material surface to induce compressive residual stresses within the treated zone, by means of high-pressure shock waves. The LP is characterized by its dynamic plasticity, whereby the laser-induced shock waves interact with the material to impart plastic strain

M. Chouchane et al. (Eds.): CMSM 2023, LNME, pp. 290–304, 2024.
https://doi.org/10.1007/978-3-031-65007-9_32

to the surface and compressive stresses to the subsurface, providing effective tensile stress mitigation and fatigue performance [1–14]. One of the LP's key characteristics is its contactless nature, eliminating the danger of damage or contamination, and enabling accurate control of the depth and intensity of the treatment (Fig. 1).

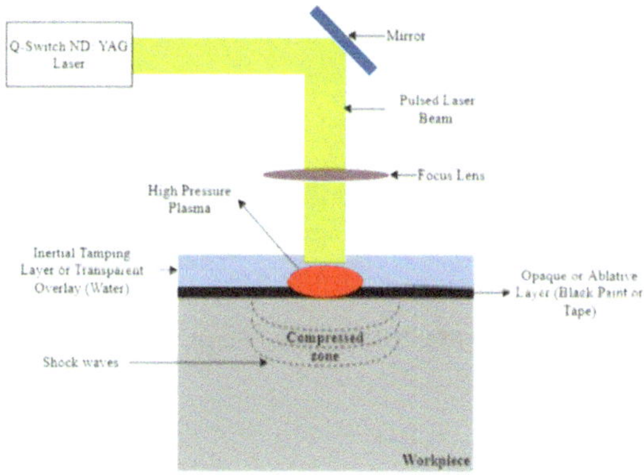

Fig. 1. Principle of LSP process [14].

This paper uses the ABAQUS software to perform numerical simulation for predicting the residual stress, plastic stain and damage profiles induced when the LP process is applied on a thin representative specimen of the leading edge of an Inconel 718 turbine blade. After that, we focus on using artificial neural networks (ANNs) and adaptive neuro-fuzzy inference systems (ANFIS) approaches to predict and optimize the various effects induced by the LP treatment.

2 Numerical Simulation

2.1 Basic Hypotheses

For the finite element method (FEM) analysis of the LP process, we assume that the edge effects are not considered, the pressure is applied in a spatially uniform manner across the processed zone, the movement of the material is orthogonal to the surface, the temperature effect is ignored as an ablative thermo-protective layer is employed, the shock is fast, and the behavioral law of Johnson-Cook is isotropic.

2.2 Geometric Model

A 3D model of a thin representative sample (approx. 0.75 mm) of an Inconel 718 turbine blade is developed to simulate the square laser impacts applied to the superior surface of the material.

Two zones are considered in the model: the first, subjected to LP treatment, must be meshed finer, and the second represents the remainder part of the sample (Fig. 2). We use the following mesh to represent each zone:

- Zone 1: 66 000 hexahedral elements with reduced integration, C3D8R type.
- Zone 2: 141 418 tetrahedral elements of C3D4 type.

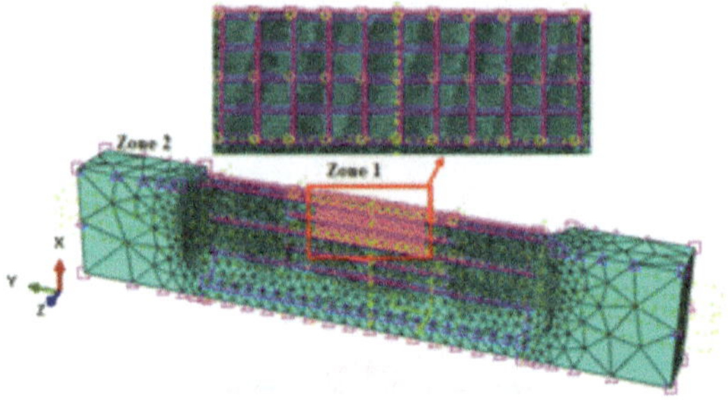

Fig. 2. 3D model mesh.

2.3 Behavior Law Properties

To model the thermo-elasto-visco-plastic behavior of the Inconel 718, we employ the Johnson-Cook empirical behavior law [15] that has been proven in several physical investigations. The Johnson-Cook law is defined as follows:

$$\sigma_y(\bar{\varepsilon}^p) = \left[A + B(\bar{\varepsilon}^p)^n\right]\left[1 + Cln\left(\frac{\dot{\varepsilon}}{\dot{\varepsilon}_0}\right)\right]\left[1 - \left(\frac{T - T_0}{T_f - T_0}\right)^m\right] \quad (1)$$

Based on Eq. (1), the instantaneous strain rate ranges between $1s^{-1}$ and $10^6 s^{-1}$ (With LP, deformation rates can exceed $10^6 s^{-1}$ into the desired material [16]).

The temperature is assumed to be negligible in this study, since there is no increase in temperature during the propagation of the shock wave. As a result, Johnson-Cook's law becomes the following:

$$\sigma_y(\bar{\varepsilon}^p) = \left[A + B(\bar{\varepsilon}^p)^n\right]\left[1 + Cln\left(\frac{\dot{\varepsilon}}{\dot{\varepsilon}_0}\right)\right] \quad (2)$$

Table 1 describes the five material constants of Johnson-Cook's law: *A*, *B*, *C*, *m* and *n*.

Table 1. The parameters of Johnson-Cook's law [17].

Young's modulus E [MPa]	Poisson's ratio ν	Density ρ [t/mm3]	A [MPa]	B [MPa]	C	m	n
185 000	0.33	8.19*10–9	1 200	1 284	0.006	1.2	0.54

Johnson and Cook developed a rupture criterion involving cumulative plastic deformation [15]. By this, the damage is then taken into consideration using the Johnson-Cook law (Eqs. 3 & 4):

$$\overline{\varepsilon}^{p}_{rupture} = \left[d_1 + d_2 exp\left(d_3 \frac{\sigma_H}{\sigma_{eqVM}}\right)\right]\left[1 + d_4 ln\left(\frac{\dot{\overline{\varepsilon}}^{p}}{\dot{\varepsilon}_0}\right)\right]\left[1 + d_5 \frac{T - T_0}{T_f - T_0}\right] \tag{3}$$

where:

$$D = \sum \frac{\Delta \overline{\varepsilon}^{p}}{\overline{\varepsilon}_f} = \sum \frac{\Delta \varepsilon^{p}_{eq}}{\overline{\varepsilon}_f} \tag{4}$$

d_1, d_2, d_3, d_4 and d_5 characterize the material constants (Table 2). In Eq. 3, σ_H denotes the hydrostatic stress and σ_{eqVM} represents the Von Mises equivalent stress.

Table 2. The Johnson-Cook damage parameters [17].

d1	d2	d3	d4	d5
0.11	0.75	– 1.45	0.04	0.89

In Eq. 4, D characterizes the accumulated damage and is described as a function of the equivalent plastic strain at fracture $\overline{\varepsilon}_f$ and the incremental plastic strain $\Delta\varepsilon^{p}_{eq}$. The stresses are set to zero and the rupture begins, when D is equal to 1.

2.4 Loading and Boundary Conditions

In our investigation, the boundary conditions are considered as follows:

- The base and two extreme faces are fixed ($U_1 = U_2 = U_3 = UR_1 = UR_2 = UR_3 = 0$).
- The vertical central plane is a symmetry plane.
- The temperature of the specimen's boundaries is about 298 K.

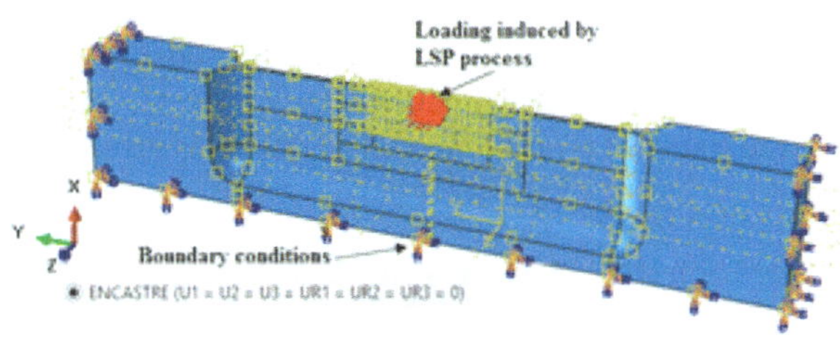

Fig. 3. Loading and boundary conditions.

In terms of loading, we consider only the mechanical effect of the LP treatment, expressed in the form of a quasi-instantaneous, time-varying pressure *P(t)*(Fig. 3).

Figure 4 depicts the load distribution induced by the LP process.

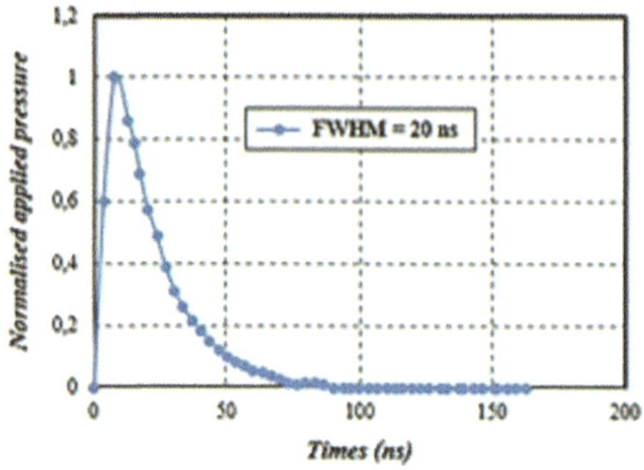

Fig. 4. Normalized load time profile (full-width at half-maximum (FWHM) = 20 ns).

2.5 Loading and Boundary Conditions

The calculation is structured in two steps:

- The first step is the "CHARG" loading step that lasts 1,000 ns.
- The second step represents the "RELAX" relaxation stage and takes 0.5 ms.

3 Computational Analysis

Taking the lead in artificial intelligence, ANNs and ANFIS technologies are getting considerable interest for their remarkable ability to solve complicated problems and manipulate vast datasets.

3.1 ANNs Technique

The training algorithm is developed to achieve a minimum overall error, including: standard prediction error "root mean square" (RMS), "absolute fraction of variance" (R2) and "mean absolute percentage error" (MAPE). The hidden layer error is estimated on the back-propagation of the estimated output layer error.

While learning, the standard error of prediction (RMS) is given by:

$$RMS = \left(\left(\frac{1}{p}\right)\sum_{j}\left|t_j - o_j\right|^2\right)^{1/2} \tag{5}$$

In addition, the absolute fraction of variance (R2) and the mean absolute percentage error (MAPE) are determined and expressed, respectively, by the following formula:

$$R^2 = 1 - \left(\frac{\sum_j \left(t_j - o_j\right)^2}{\sum_j \left(o_j\right)^2}\right) \tag{6}$$

$$MAPE = \frac{\sum_j o_j - \sum_j t_j}{\sum_j t_j} * 100 \quad (7)$$

The learning algorithm is defined as a procedure for adjusting the weights of a network in order to minimize the error between the outputs predicted using the ANNs technique and the numerical outputs obtained using the FEM analysis.

The ANNs structure adapted in our study is illustrated in Fig. 5.

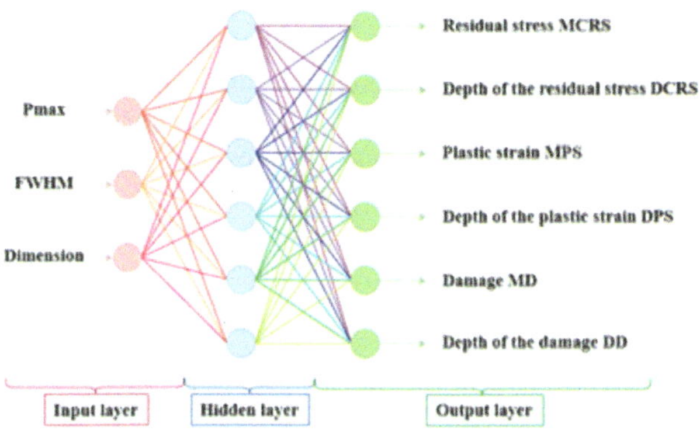

Fig. 5. The ANNs structure.

The various parameters used in our research are presented in Table 3.

Table 3. The different parameters.

The parameters	The choice
The used algorithm	Backpropagation
The network inputs	The maximum pressure (P_{max}), the full width at half-maximum (FWHM) and the edge dimension of the square impact (Dimension)
The network outputs	The stress MCRS, the plastic strain MPS, the damage MD and the depths DCRS, DPS and DD
The learning function	Levenberg-Marquardt (LM)
The number of neurons in the hidden layer	6
The 6-layer transfer function	linear

Three different input parameters are chosen in the ANNs learning step, so as to predict the desired LSP-induced state. The input parameters must have three different levels (low, medium and/or high). The selection of these parameters is listed in Table 4.

Table 4. Selection of the adopted parameters.

Input parameters	Units	Level 1 Low	Level 2 Medium	Level 3 High
P_{max}: The maximum pressure	GPa	3	5	7
FWHM: The full width at half-maximum	ns	10	20	30
Dimension: The edge dimension of a square laser impact	mm	3	4.5	6

To perform the training phase, 27 simulations using the values given to the input parameters are performed, as shown in Table 4. The focus in this work is on the prediction of residual stress, plastic strain and damage. The residual stress profile is depicted in Fig. 6.

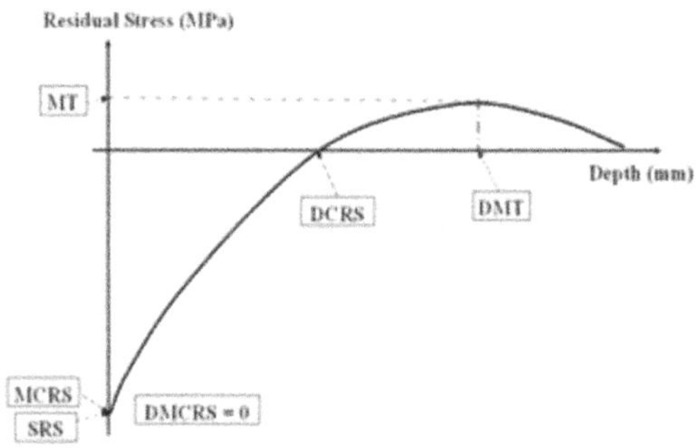

Fig. 6. Residual stress profile.

The plastic strain profile is shown in Fig. 7.

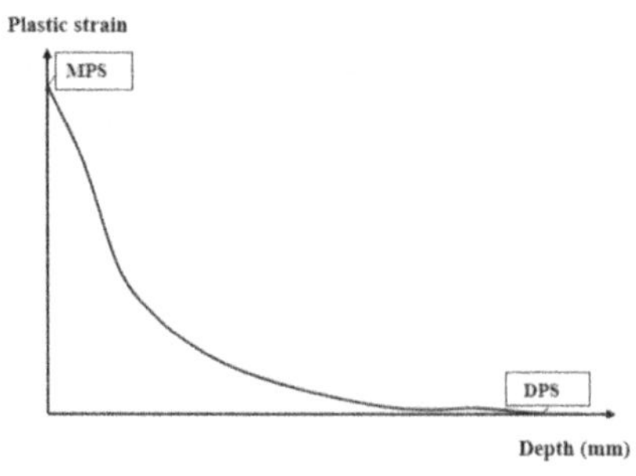

Fig. 7. Plastic strain profile.

The Johnson-Cook damage profile is showed in Fig. 8.

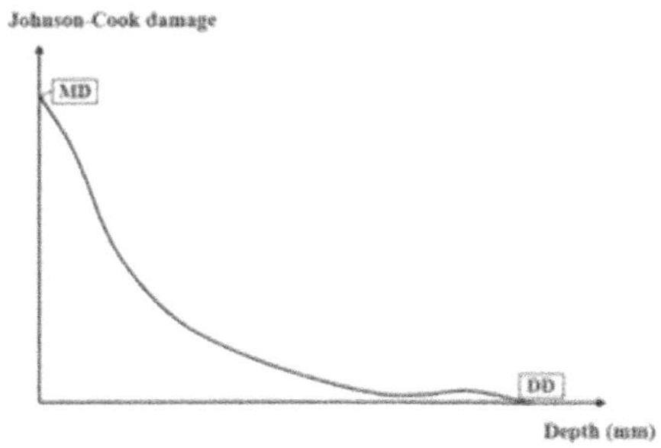

Fig. 8. Plastic strain profile.

Next, the obtained profiles are studied and the key values are noted (Table 5) and we limit our study to the prediction of MCRS, DCRS, MPS, DPS, MD and DD values to visualize these different profiles.

Table 5. Output simulation results.

Factors	Output results	Units
MCRS	Maximum compressive residual stress (negative value)	MPa
DCRS	The depth of compressive residual stress	mm
MPS	Maximum plastic strain	-
DPS	The depth of plastic strain	mm
MD	The maximum Johnson-Cook damage	-
DD	The depth of Johnson Cook damage	mm

3.2 ANFIS Approach

Figure 9 illustrates the ANFIS structure, which consists of five layers.

The outputs of layer 1 are presented as follows:

$$L_i^1 = \mu_{A_i}(x), i = 1, 2 \tag{8}$$

$$L_i^1 = \mu_{B_i}(x), i = 3, 4 \tag{9}$$

where: x represents the input to node i, and Ai depicts the linguistic function label. μAi(x) and μBi(y) represent Gaussian functions given by:

$$\mu_{A_i}(x) = exp\left[-\left(\frac{x - c_i}{a_i}\right)^2\right] \tag{10}$$

where: ai, ci depict the deviation (or width) and center (or mean) parameters of the membership function, respectively.

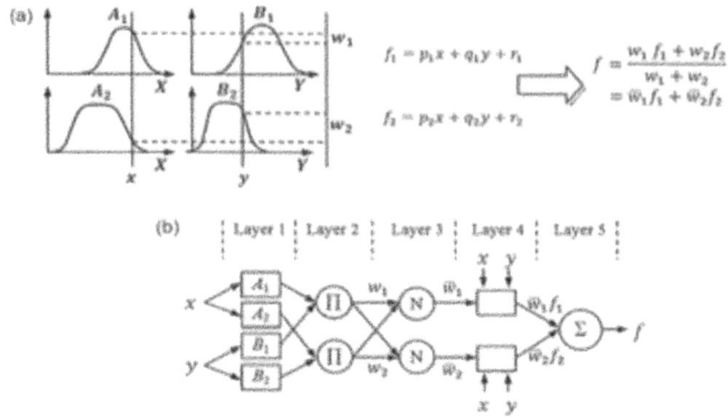

Fig. 9. The ANFIS structure.

The outputs of layer 2 are given as follows:

$$L_i^2 = w_i = \mu_{A_i}(x)\mu_{B_i}(y), i = 1, 2 \tag{11}$$

where: w_i represents the strengths of rules.

The outputs of layer 3 are given as:

$$L_i^3 = \overline{w_i} = \frac{w_i}{w_1 + w_2}, i = 1, 2 \tag{12}$$

where: $\overline{w}_i$ are called the normalized weights.

The outputs of layer 4 are given by the following equation:

$$L_i^4 = \overline{w_i}(p_i x + q_i y + r_i), i = 1, 2 \tag{13}$$

where: pi, qi and ri represent consequent parameters.

In the layer 5, the single node computes the inputs sum which depict the model's output and is given as:

$$L_i^5 = \sum_{i=1}^{2} \overline{w_i}(p_i x + q_i y + r_i) = \frac{\sum_{i=1}^{2} w_i(p_i x + q_i y + r_i)}{w_1 + w_2}, i = 1, 2 \tag{14}$$

The above description illustrates that the ANFIS structure is composed of two types of parameters: premises {ai, ci} and consequences {pi, qi, ri}.

4 Results and Discussion

For the ANNs approach, the learning algorithm is executed in MATLAB. During the training step, the number of neurons in the hidden layer is progressively incremented to achieve the optimum results. In this way, the best approach with minimum error is reached with the Levenberg-Marquardt (LM) algorithm based on six neurons in the hidden layer.

With the ANFIS approach, the number of MF members for each input is 4. Here, we increase the number of epochs till achieving the ideal results. The best number of epochs is 200. Also, an ANFIS model is provided for each output with joint third-party inputs.

Table 6 illustrates the statistical error values for the two approaches.

Table 6. The statistical error values for ANNs and ANFIS techniques.

Response	Errors					
	RMS		R^2		MAPE	
	ANNs	ANFIS	ANNs	ANFIS	ANNs	ANFIS
MCRS	2.86	2.53	0.93	0.99	37.68	10.69
DCRS	0.09	0.06	0.58	0.99	– 39.28	– 7.63
MPS	0.02	0.01	0.92	1	– 22.36	1.6
DPS	0.06	0.01	0.98	1	14.72	0.17
MD	0.04	0.01	0.95	1	– 18.94	– 0.53
DD	0.05	0.01	0.98	1	14.72	0.17

Figures 10, 11 and 12 illustrate the residual stress, plastic strain and damage profiles, respectively, predicted with the ANNs and ANFIS approaches and those predicted numerically with FEM analysis with the selected input data.

We can clearly observe that the differences are very small for the different selected input parameters, and are almost negligible.

On the basis of this improved correlation, it is obvious that these models can be used to predict numerical values for residual stress, plastic strain and damage. This highlights the importance of ANNs and ANFIS for estimating the induced effects values, as well as their profiles.

This optimization step focuses on characterizing the values of the optimum range of Pmax, FWHM and Dimension, that can be formulated by the following system:

$$\begin{cases} Max_{(P_{max},FWHM\,,Dimension)} MCRS \\ Max_{(P_{max},FWHM\,,Dimension)} DCRS \end{cases}$$

The model equations are used to predict optimum response values using the following input parameters: Pmax is 7 GPa, FWHM is 25 ns and the dimension is equal to 6 mm. Table 7 lists the values predicted using the ANNs and ANFIS methods (with MATLAB software) and the numerical results obtained using FEM analysis (with ABAQUS software) for MCRS, DCRS, MPS, DPS, MD and DD under these optimum predicted conditions. The numerical values generated by the FEM analysis and those calculated by the ANNs method are relatively close, with the maximum calculated error equal to 8.45% at the DPS and DD depths. In addition, the numerical values obtained by FEM analysis and those predicted by the ANFIS method are similar, and the maximum error obtained is equal to 11.13% at the DPS and DD depths.

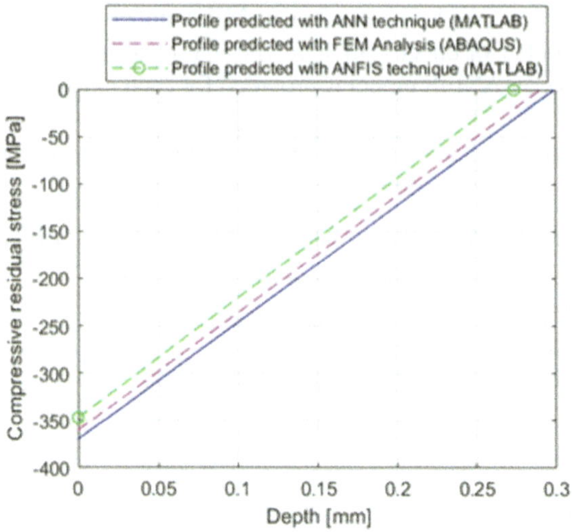

Fig. 10. Residual stress profile for the following parameters: Pmax = 3G Pa, FWHM = 25 ns, Dimension = 4.5 mm.

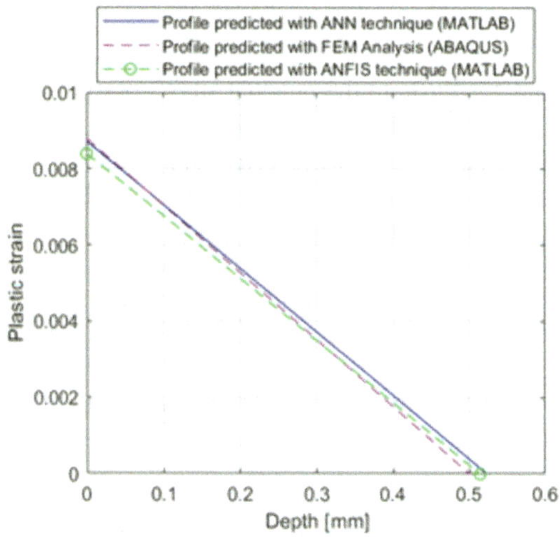

Fig. 11. Plastic strain profile for the following parameters: Pmax = 3 GPa, FWHM = 25 ns, Dimension = 4.5 mm.

These new patterns are therefore highly effective for predicting and optimizing all responses.

Figures 13, 14 and 15 illustrate, respectively, a comparison of the residual stress, plastic strain and damage profiles obtained using ANNs and the ANFIS method, and the numerical method using FEM.

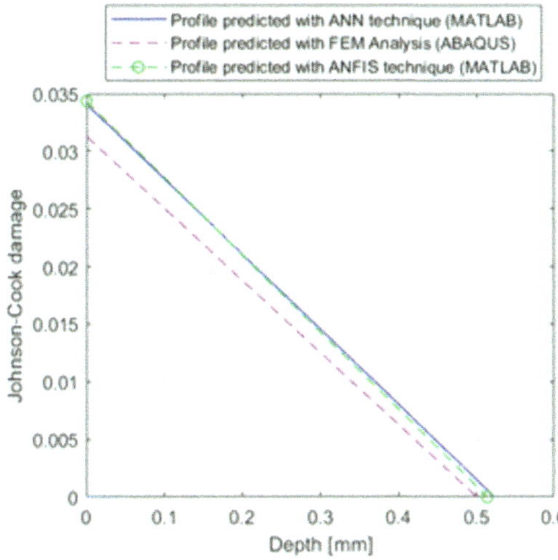

Fig. 12. Johnson-Cook damage profile for the following parameters: Pmax = 3 GPa, FWHM = 25 ns, Dimension = 4.5 mm.

Table 7. The statistical error values for ANNs and ANFIS techniques.

Response	ANNs predicted value	ANFIS predicted value	Numerical value (ABAQUS)	Difference for ANNs (%)	Difference for ANFIS (%)
MCRS (MPa)	– 1077.8	– 1119.2	– 1040.22	3.63	7.61
DCRS (mm)	0.4954	0.52	0.54	8.26	3.70
MPS	0.0446	0.0462	0.0470684	5.24	1.84
DPS (mm)	0.8134	0.8335	0.75	8.45	11.13
MD	0.2077	0.2222	0.219533	5.39	1.21
DD (mm)	0.8134	0.8335	0.75	8.45	11.13

Indeed, these figures provide excellent agreement between the three methods used to predict residual stress, plastic stain and damage profiles. Moreover, the ANNs and ANFIS approaches perform better than the numerical technique.

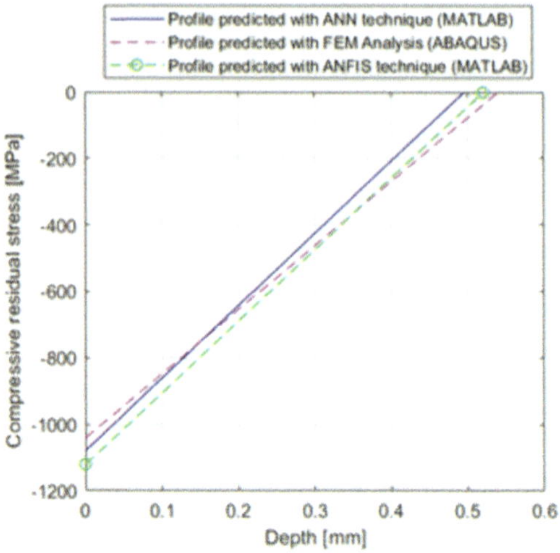

Fig. 13. Residual stress profiles predicted by ANNs, ANFIS and FEM for optimum conditions: Pmax = 7 GPa, FWHM = 25 ns, Dimension = 6 mm.

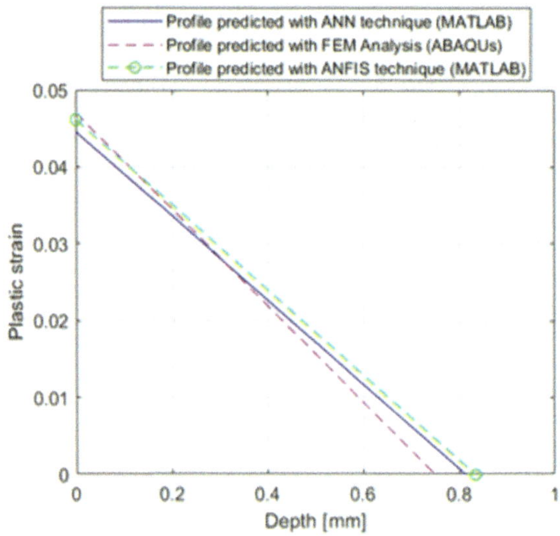

Fig. 14. Plastic strain profiles predicted by ANNs, ANFIS and FEM for optimum conditions: Pmax = 7 GPa, FWHM = 25 ns, Dimension = 6 mm.

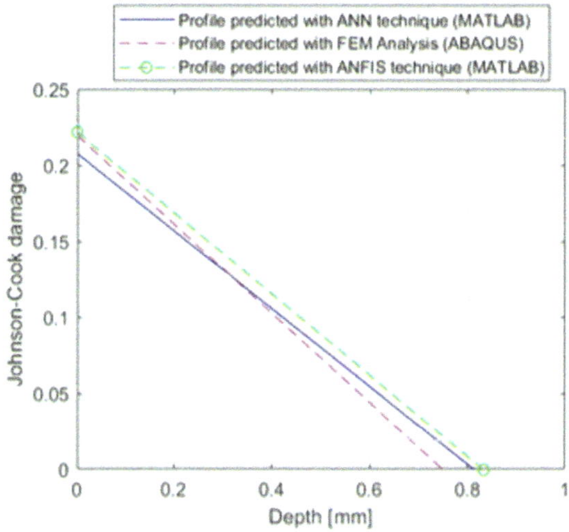

Fig. 15. Plastic strain profiles predicted by ANNs, ANFIS and FEM for optimum conditions: Pmax = 7 GPa, FWHM = 25 ns, Dimension = 6 mm.

5 Conclusion

In this paper, the main objective of predicting and optimizing the residual stress, plastic strain and damage profiles induced by the LP process has been achieved using innovative approaches: ANNs and ANFIS.

Using FEM analysis, a thin representative sample of an Inconel 718 turbine blade is treated by the LP process

The second stage of our study demonstrated the remarkable predictive capabilities of ANN and ANFIS. By training these intelligent models with data obtained from FEM simulations, we obtained highly accurate predictions of residual stresses, plastic strains and damage profiles induced by LP. Thus, ANNs and ANFIS techniques are capable of learning from examples and predicting different results in a short time, by comparison with conventional methods. Hence, we use the learning algorithm to predict induced responses without needing to define several parameters, like material parameters, behavior law and mesh size, that are necessarily selected in the FEM technique.

Acknowledgements. This work is carried out thanks to the support and funding allocated to the Unit of Mechanical and Materials Production Engineering (UGPMM/UR17ES43) by the Tunisian Ministry of Higher Education and Scientific Research.

References

1. Frija, M., Ayeb, M., Seddik, R., et al.: Optimization of peened-surface laser shock conditions by method of finite element and technique of design of experiments. Int. J. Adv. Manuf. Technol. **97**, 51–69 (2018)

2. Manel Ayeb : Développement d'un modèle de prévision de l'état induit du traitement, efocsa, Editions universitaires européennes, ISBN-13 :978–613–8–39716–8, 92 pages (2018)
3. Ayeb, M., Frija, M., Fathallah, R. : Simulation de la surface des bords d'attaque d'une aube de turbine traitée par le procédé de grenaillage par choc laser. 4ème Congrès Tunisien de Mécanique (CoTuMe), Hammamet, Tunisia (2018)
4. Ayeb, M., Frija, M., Fathallah, R.: Prediction of residual stress profile and optimization of surface conditions induced by laser shock peening process using artificial neural networks. Int. J. Adv. Manuf. Technol. **100**, 2455–2471 (2019)
5. Frija, M., Ayeb, M., Seddik, R., Fathallah, R.: Prediction of high-cycle fatigue behavior of parts treated by laser shock peening. In: International Conference on Innovative Materials, Manufacturing and Advanced Technologies (IMMAT'2019), Monastir, Tunisia (2019)
6. Ayeb, M., Frija, M., Fathallah, R.: Influence of overlapping rate on thin leading edges of turbine blade treated by laser peening. In: International Conference on Innovative Materials, Manufacturing and Advanced Technologies (IMMAT'2019), Monastir, Tunisia (2019)
7. Ayeb, M., Frija, M., Fathallah, R.: Effect of Multiple Impacts on Thin Leading Edges of Turbine Blade Treated by Laser Shock Peening Process. In: Aifaoui, N., et al. Design and Modeling of Mechanical Systems - IV. CMSM 2019. Lecture Notes in Mechanical Engineering. Springer, Cham. https://doi.org/10.1007/978-3-030-27146-6_54 (2020)
8. Ayeb, M.: Simulation et optimisation du procédé du grenaillage par choc la-ser: Simulation active et optimisation des conditions opératoires du procédé du grenaillage par choc laser, efocsa, Editions universitaires européennes, ISBN-13: 978–620–2–54959–2, 156 pages (2021)
9. Frija, M., Ayeb, M., Seddik, R., Fathallah, R.: Reliability prediction of high-cycle fatigue behavior of parts treated by laser shock peening. Fatigue Fract. Eng. Mater. Struct. **2**, 1–16 (2023). https://doi.org/10.1111/ffe.14091
10. Ayeb, M., Frija, M., Fathallah, R.: Influence of multiple laser impacts on thin leading edges of turbine blade. Proc. Inst. Mech. Eng., Part L: J. Mater.: Des. Appl. **234**(1), 130–143 (2019). https://doi.org/10.1177/1464420719873936
11. Ayeb, M., Frija, M., Fathallah, R.: Prediction of Residual Stress Profile Induced by Laser Shock Peening Using Artificial Neural Networks. In: 6th International Conference on Laser Peening and Related Phenomena, South Africa (2016)
12. Ayeb, M., Frija, M., Fathallah, R.: Prediction of residual stress profile and optimization of surface conditions induced by laser shock peening process using artificial neural networks. Int. J. Adv. Manuf. Technol. **100**, 2455–2471 (2019). https://doi.org/10.1007/s00170-018-2883-z
13. Ayeb, M., Frija, M., Fathallah, R.: Laser peening: a review of the factors, effects, applications, comparison with shot peening and state-of-the-art. Met. Mater. Int. (2023). https://doi.org/10.1007/s12540-023-01517-4
14. Frija, M., Ayeb, M., Seddik, R., Fathallah, R.: Reliability prediction of high-cycle fatigue behavior of parts treated by laser shock peening. Fatigue Fract. Eng. Mater. Struct. **46**(9), 3514–3529 (2023). https://doi.org/10.1111/ffe.14091
15. Johnson, G., Cook, W.: A constitutive model and data for metals subjected to large strains, high strain rates and high temperatures. In: Proceedings of the 7th International Symposium on Ballistics, 54: 541–547 (1983)
16. Amarchinta, H., Grandhi, R., Langer, K., Strangel, D.: Material model validation for laser shock peening process simulation. Modell. Simul. Mater. Sci. Eng. **17**(1), 15 (2009)
17. BORJA ERICE ECHAVARRI: Flow and Fracture Behaviour of High-Performance Alloys, 144 (2012)

Experimental Study of the Effects of Laser Treatment Settings on the Mechanical Properties of AISI 4340 Steel

Meriem Aloui[1], Salma Slama[2(✉)], Asma Belhadj[2], and Noureddine Barka[1]

[1] Mathematics, Software and Engineering Department, University of Quebec at Rimouski, Rimouski, Canada
{Meriem.aloui,Noureddine_Barka}@uqar.ca

[2] Applied Mechanics and Engineering Laboratory (LMAI-LR-11ES19), National Engineering School of Tunis (ENIT), University of Tunis El Manar (UTM), Tunis, Tunisia
{selma.slama,asma.belhadj}@enit.utm.tn

Abstract. This study presents an experimental parametric analysis that used the design of experiments methodology to evaluate the effect of laser heat treatment parameters on the surface mechanical properties of AISI 4340 steel parts. The research employed a Taguchi design consisting of three factors at three levels: laser power (P), scanning speed (V), and focal distance (DF). This study presents an experimental analysis of the impact of various parameters on the lines generated by the laser during the treatment process. The analysis is based on hardness tests and metallographic observations.

Keywords: Parametric analysis · laser heat treatment · AISI 4340 steel · hardness tests · metallographic observations

1 Introduction

Laser heat treatment is an advanced engineering technique that provides significant advantages in modifying material properties. It allows targeting specific areas of the part without altering the overall characteristics of the material. Compared to other heat treatment processes, this method offers precise control over the depth of the affected area, enabling tailoring of surface properties to the particular needs of the application. Laser heat treatment induces a transformation of the surface microstructure of steel parts, resulting in improved hardness, wear resistance, corrosion resistance, thermal conductivity, and electrical conductivity (Laurens et al., 1996, Gouton, 2015).

Numerous scientific studies have been conducted on laser heat treatment. For instance,(Seghier, 2021) study predicted the hardness profile of gears made of 4340 steel after laser surface heat treatment. Furthermore, (Tarchoun et al., 2020) research aimed to develop a predictive optimization strategy for the performance of 4340 steel after heat treatment based on experimental results. The research conducted by (Fakir,

M. Chouchane et al. (Eds.): CMSM 2023, LNME, pp. 305–314, 2024.
https://doi.org/10.1007/978-3-031-65007-9_33

2018) aims to develop a model to predict temperature distribution in a cylindrical sample of AISI-4340 steel that has undergone laser thermal treatment. This analysis combines numerical and experimental approaches to accurately determine heat transfer and absorption parameters, enabling the prediction of the depth of the thermal treatment.

Most of the research in this area has focused on laser quenching treatment, which does not involve reaching the liquid state. However, this research primarily focuses on laser melting treatment, a technique used to improve the resistance of metal surfaces to aqueous corrosion (Peyre et al., 2012). The objective of this study is to evaluate how laser surface treatment parameters affect the mechanical properties of AISI 4340 steel components. To accomplish this, we conducted a parametric experimental investigation using a Taguchi plan. The parameters studied were laser power (P), scanning speed (V), and focal length (DF). We selected the specific values for the different parameter levels based on prior research and experimentation carried out at the UQAR laboratory.

2 Principle of Laser Surface Treatment

Laser surface treatment is a process that enhances the mechanical characteristics of materials. A laser beam is directed onto the target surface, as shown in Fig. 1, resulting in various surface changes, such as melting, hardening, and modifications of the metallurgical structure of the treated area.

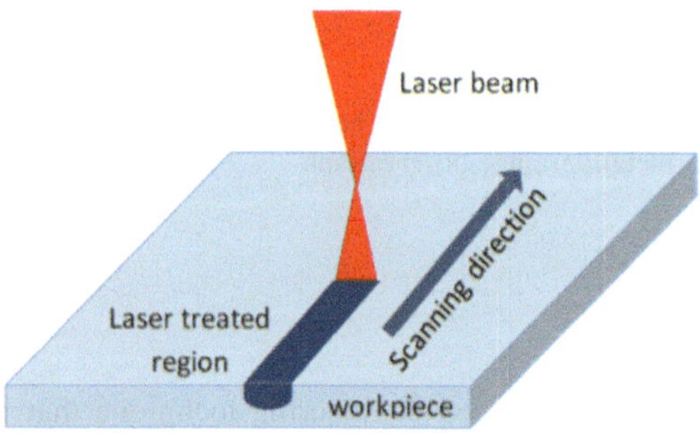

Fig. 1. Laser surface treatment.

3 Experimental Study

In this parametric study, we chose to use a design of experiments to establish the relationships between the main parameters of the laser treatment and its effect on the mechanical properties of the treated components, specifically the dimensions of the area affected by the heat generated by the laser beam. To achieve this, we varied the laser power (P), the scanning speed (V) and the laser focal distance (FD). We then carried out hardness measurements and observations to define the area affected by the treatment, i.e. its depth and width. The steps carried out in this study are illustrated in Fig. 2.

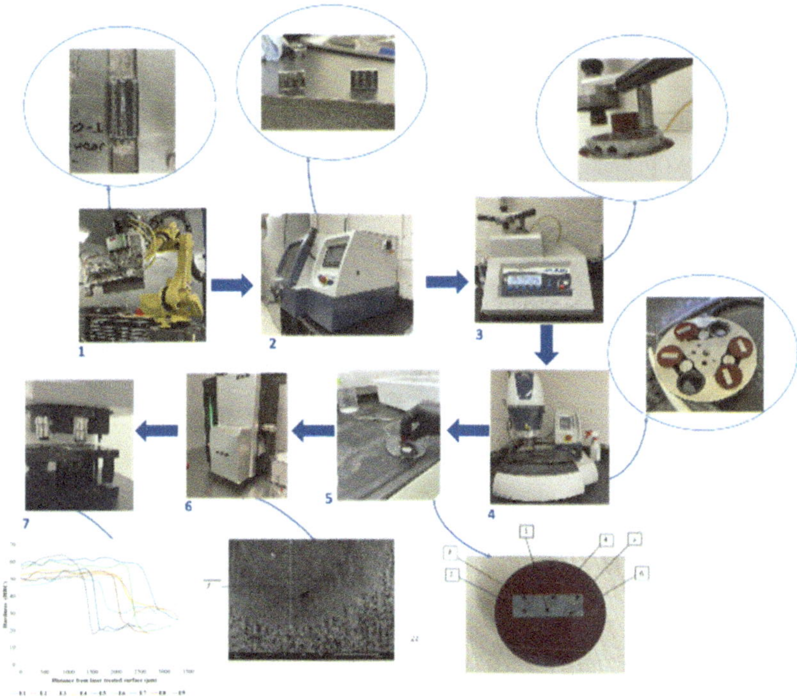

Fig. 2. Experimental steps.

3.1 Steel Properties

AISI 4340 steel is a low alloy steel with high strength and hardness, widely used in the aerospace, oil and gas industries and for the manufacture of highly stressed mechanical parts. Table 1 shows the chemical composition of AISI 4340 steel.

Table 1. Chemical composition of AISI 4340 steel (SAE International, 2007)

Element	*C*	*Cr*	*Ni*	*Mo*	*Mn*	*Si*	*S*
(%)	0.38–0.43	0.7–0.9	1.65–2.0	0.2–0.3	0.6–0.8	0.15–0.3	0.04

3.2 Design of Experiments

This parametric investigation is based on a Taguchi experimental design with three factors, each with three levels. The selection of these levels, as shown in Table 2, is based on previous research carried out in the UQAR laboratory. Table 3 shows the 9 parameter combinations to be investigated.

Table 2. Factors levels

Factor	*Level*		
Power P(W)	500	600	700
Scanning speed V(mm/s)	2	3	4
Focal distance DF (mm)	8	12	16

Table 3. Design of experiments

Test	*Focal Distance DF (mm)*	*scanning speed V (mm/s)*	*Power P(W)*
E1	8	2	500
E2	8	3	600
E3	8	4	700
E4	12	2	600
E5	12	3	700
E6	12	4	500
E7	16	2	700
E8	16	3	500
E9	16	4	600

3.3 Heat Treatment Parts

The parts are processed using a UQAR laser machine (Fig. 3.a), which has a class 4 Nd:YAG laser cell. This laser operates at a maximum voltage of 600 V and emits light at a wavelength of 1070 nm. AISI 4340 steel samples measuring 40*20*6 mm were used to process the nine laser lines (Fig. 3.b).

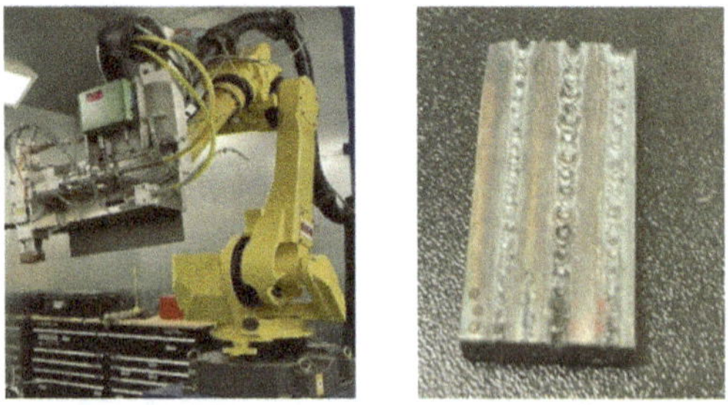

(a) (b)

Fig. 3. (a) UQAR laser machine (b) Treated sample.

4 Experimental Investigations

The treated parts were cut in order to carry out observations and hardness measurements on the section of the treated lines. Following this, the components were coated, polished, and subjected to Nital etching before being examined under a scanning electron microscope.

4.1 Micrographic Analysis

Figure 4 shows a macrograph of the cross-section of a laser line. There are two zones, Z1 and Z2, for all laser lines. The first zone, Z1, has been affected by the heat generated by the laser beam and is further subdivided into two sub-zones: the melted zone (MZ) and the heat-affected zone (HAZ). The second zone, Z2, corresponds to the unaffected zone, i.e. the base metal (BM).

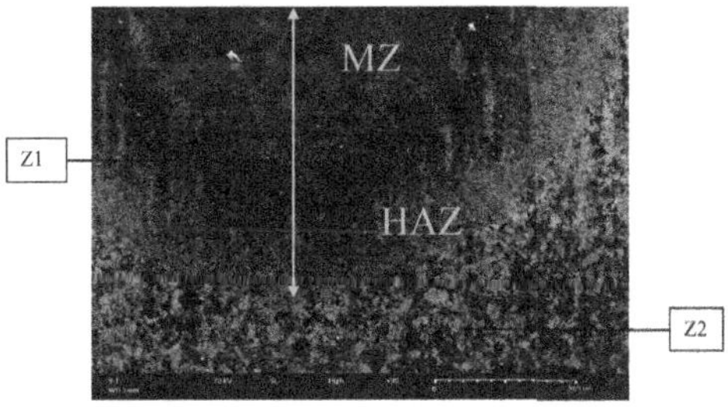

Fig. 4. Treated sample macrograph.

Micrographs of all laser lines in Zone 1 (Fig. 5) reveal a martensitic structure, which is typical of rapid cooling. In Zone 2 (Fig. 6), the microstructure consists of ferrite and pearlite, similar to that of the base metal. This variation in microstructure suggests rapid cooling in Zone 1, leading to the formation of martensite due to the intense thermal energy generated by the laser. Zone 2 has not undergone any microstructural transformation.

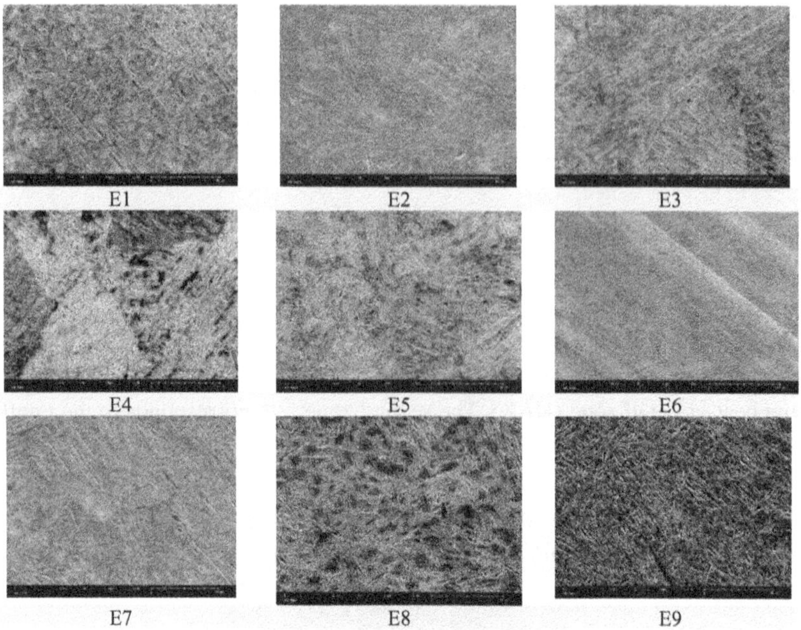

Fig. 5. Micrographs of Z1.

4.2 Hardness Test

The hardness of various tests was measured along the Y-axis, as shown in Fig. 7. The hardness profiles for tests E1 to E9, depicted in Fig. 8.a, exhibit a similar curve shape. Each test features three distinct zones: Z1, Z', and Z2 (Fig. 8.b). Zone 1 is defined by a relatively constant but significantly high hardness value ranging from 48 HRC to 61 HRC, confirming the presence of the martensitic structure. Zone 2 represents a minimum hardness ranging from 20 HRC to 31 HRC, corresponding to the ferrito-perlitic structure associated with the base metal (MB). Zone Z' represents an intermediate zone where the hardness undergoes a significant decrease.

4.3 Laser Line Dimensions

The metallurgical observations and hardness measurements enabled us to define the zones and determine the dimensions of the laser lines. Table 4 shows the widths and depths of each laser line.

Laser line depth

The regression equation modeling the depth as a function of the parameters DF, P and V is as follows:

$$Depth = -7.908 + 1.382DF - 0.3154V + 0.007475P - 0.04962DF^2 - 0.2253V^2 + 0.000009P^2 + 0.1302DF * V - 0.001086DF * P$$

with $R^2 = 100\%$

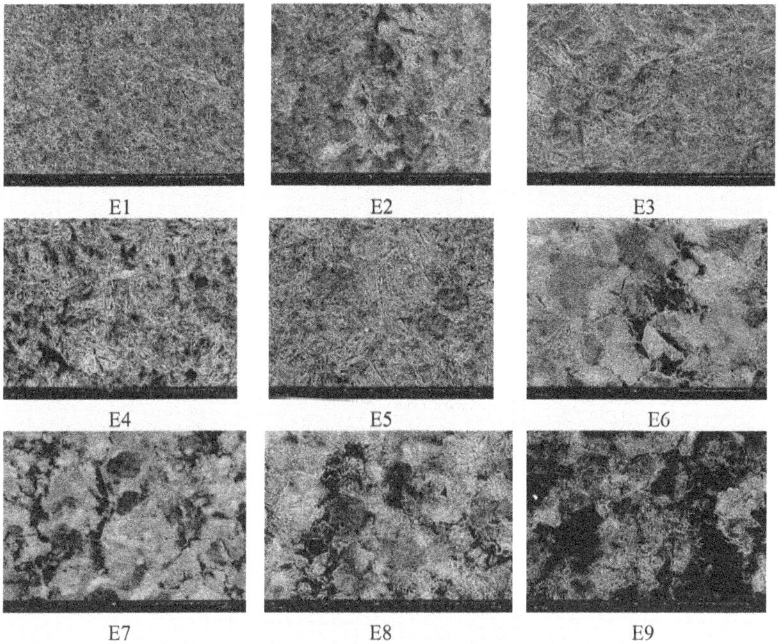

Fig. 6. Micrographs of Z2.

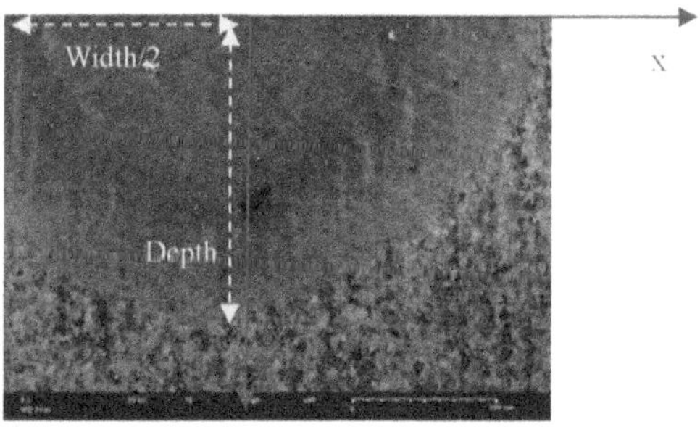

Fig. 7. Laser line dimensions.

Laser line width

The regression equation modeling the width as a function of the parameters DF, P and V is as follows:

$$Width = 8.679 + 1.302DF - 3.762V - 0.02847P - 0.04998DF^2 + 0.1525V^2 + 0.000042P^2 + 0.1844DF * V - 0.001184DF * P$$

with $R^2 = 100\%$

Figures 9 and 10 demonstrate that the focal distance is identified as the most influential factor on the depth and width of the laser lines. Specifically, varying the focal length from 8 to 12 mm leads to an increase in both the width and depth of the laser lines, while changing it from 12 to 16 mm results in a significant reduction in the dimensions of the

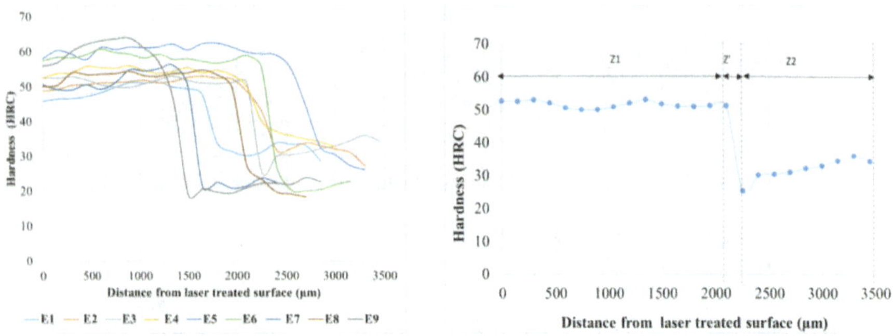

Fig. 8. (a) Hardness profiles (b) Hardness zones.

Table 4. Laser line dimensions

Test	Focal Distance DF (mm)	scanning speed V (mm/s)	Power P(W)	Depth (mm)	Width (mm)
E1	8	2	500	2.1639	3.3934
E2	8	3	600	2.6311	2.6639
E3	8	4	700	2.8276	3.0737
E4	12	2	600	3.0245	4.0327
E5	12	3	700	3.7577	4.4016
E6	12	4	500	2.3814	2.4432
E7	16	2	700	1.6082	2.96
E8	16	3	500	2.0721	2.3814
E9	16	4	600	2.2611	2.485

laser lines. Furthermore, the figures reveal that an increase in laser power corresponds to an increase in both the width and depth of the laser line. Conversely, any increase in scanning speed leads to a reduction in line width.

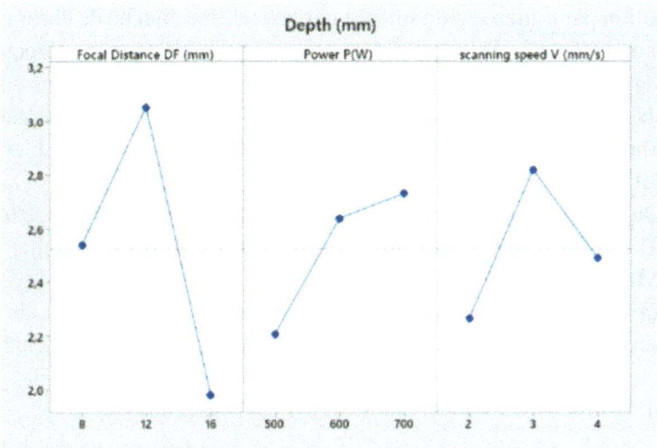

Fig. 9. Main effects graphics for Depth.

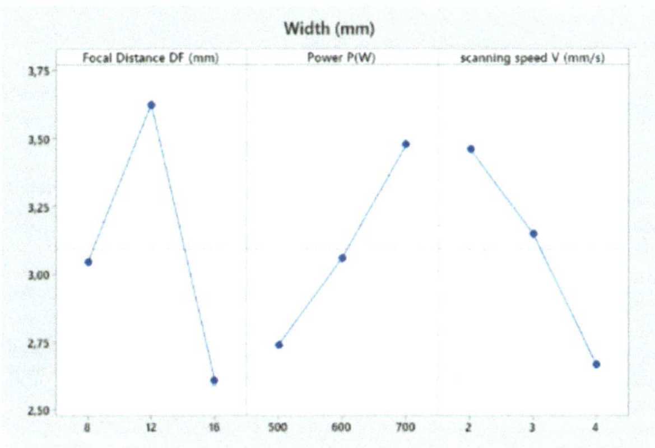

Fig. 10. Main effects graphics for Width.

5 Conclusion

This study conducts a parametric analysis to investigate the influence of laser power, scanning speed, and focal distance on the mechanical properties of AISI 4340. The findings of the study reveal that, in all configurations, the dimensions of the laser lines exhibited significant variations based on these parameters. Indeed, laser line dimensions are primarily dependent on focal distance.

References

Laurens, P., Dubouchet, C: Application des lasers aux traitements de surface. Ed. Techniques Ingénieur (1996)

Peyre, P., Vignal, V.: Traitements de surface par laser et tenue à la corrosion aqueuse. Techniques de l'Ingenieur (2012)

Gouton, L.: Formation par traitement de surface par laser d'une couche de chromine sur un alliage base nickel pour limiter le relâchement des cations en circuit primaire (Doctoral dissertation, Paris, ENSAM) (2015)

Fakir, R., Barka, N., Brousseau, J., Caron-Guillemette, G.: Numerical investigation by the finite difference method of the laser hardening process applied to AISI-4340. J. Appl. Math. Phys. **6**(10), 2087–2106 (2018)

Tarchoun, B., El Ouafi, A., Chebak, A.: Numerical investigation of laser surface hardening of AISI 4340 Using 3D FEM model for thermal analysis of different laser scanning patterns. Model. Num. Simul. Mater. Sci. **10**(03), 31 (2020)

Seghier, A: Prédiction du profil de dureté dans le cas de traitement thermique de surface par laser pour des engrenages en acier 4340 (Doctoral dissertation, Université du Québec à Rimouski) (2021)

SAE International. Norme AMS6414K. Norme, AMS6414K (2007)

Optimization of Operating Parameters for Laser Shock Peening Surface Treatment on Components Made from Nickel-Based Super-Alloy INCONEL 718

Mounir Frija[1,4](✉), Taha Dali[2], Manel Ayeb[1,4], and Raouf Fathallah[3,4]

[1] Institute of Applied Sciences and Technology of Sousse (ISSATSo), University of Sousse, Sousse, Tunisia
mounir.frija@issatso.u-sousse.tn
[2] Borj El Amri Aviation School, Ministry of National Defence Air Force, Borj El Amri, Tunisia
[3] National Engineering School of Sousse (ENISo), University of Sousse, Sousse, Tunisia
[4] Unit of Mechanical Production Engineering and Materials (UGPMM), UR17ES43, B.P. 1173, 3038 Sfax, Tunisia

Abstract. This study is focused on the optimization of laser shock surface treatment for components made from the Nickel-based super alloy INCONEL 718. The objective is to enhance the efficiency and effectiveness of this treatment process by leveraging the Design of Experiments (DoE) methodology.

In our investigation, we delve into understanding the intricate relationship between the various treatment parameters and the distinct effects they induce during the process. By isolating and examining each parameter's influence on the outcomes, we aim to identify the critical factors that significantly impact the results. This systematic analysis allows us to pinpoint the optimal settings for laser shock treatment, thus minimizing waste and maximizing performance.

Furthermore, a key aspect of this study involves the development of empirical mathematical models to approximate the results of numerical simulations. These mathematical formulations serve as valuable tools for aeronautical engineers tasked with modeling laser shock treatment. By translating complex simulation data into practical and actionable mathematical expressions, engineers can more accurately predict the outcomes of laser shock peening surface treatment under varying conditions.

The integration of meticulous experimental design, statistical analysis, and empirical modeling offers a holistic approach to optimizing laser shock treatment. The findings from this research hold promise for refining the accuracy and dependability of this vital process in manufacturing components with INCONEL 718, thereby providing valuable advancements for the aerospace industry and its associated domains.

Keywords: Laser Sock Peening · INCONEL 718 · Optimization · Design of Experiments (DoE) · Numerical simulations · Aeronautical engineering

M. Chouchane et al. (Eds.): CMSM 2023, LNME, pp. 315–325, 2024.
https://doi.org/10.1007/978-3-031-65007-9_34

1 Introduction

We are focused on optimizing the operating conditions for laser shock surface treatment on components manufactured from Nickel-based super alloy INCONEL 718 [1, 2]. The application of Design of Experiments (DoE) has greatly facilitated this approach. Mathematical models, derived from finite element simulation results, will be established.

To explore the impacts and interplay of various process control factors, one can utilize Design of Experiments (DOE) techniques. These involve manipulating input parameters across a full factorial design, allowing researchers to objectively assess the effects of changing specific factors, irrespective of other settings. While this approach is generally effective, it often demands a considerable number of runs in numerical simulations [3, 4]. Nevertheless, it proves invaluable in uncovering how significant factors shape the response, paving the way for the creation of polynomial models, both linear and quadratic, that capture these influences and their statistical significance. This methodology advantages statistical tools like response surface plots and analysis of variance (ANOVA), typically facilitated by software such as MINITAB® 14 [5].

The methodology adopted comprised several sequential steps: (i) identifying the significant factors influencing the process, (ii) determining the dispersion range for each factor through their coefficient of variation, (iii) constructing a matrix for numerical modeling, (iv) executing numerical computations based on the planned modeling matrix, (v) formulating mathematical models, (vi) calculating factor coefficients and their interactions. To simplify data recording and processing, the highest, middle, and lowest levels of factors were coded as + 1, 0, and -−1 respectively, with intermediate levels being derivable. Subsequently, (vii) the significance of factors was tested using ANOVA, and (viii) the adequacy of mathematical models was verified, if necessary, by computing the correlation coefficient.

In our investigation, the Design of Experiments centered on a complete factorial design incorporating three process factors: Pmax, FWHM, and D, each at three levels (Low, Medium, High) as outlined in Table 1. A total of 27 simulations were conducted to design the laser shock process.

Table 1. Design of Experiments Input Parameters.

Input parameters	Unity	Level 1 Low	Level 2 Medium	Level 3 High
Pmax: Maximum Pressure	GPa	**3**	**5**	**7**
FWHM: Full width at Half Maximum	ns	**10**	**20**	**30**
D:Laser Impact Diameter	mm	**3**	**5**	**7**

Building upon the validated 2D finite element simulation model established in the preceding study, every numerical simulation underwent processing under distinct conditions to ascertain the profiles of residual stresses, equivalent plastic strain, and damage resulting from the process. Figure 1 depicts the residual stress profile model. It's crucial to note that this model holds validity solely within the selected feasibility domain outlined in this study.

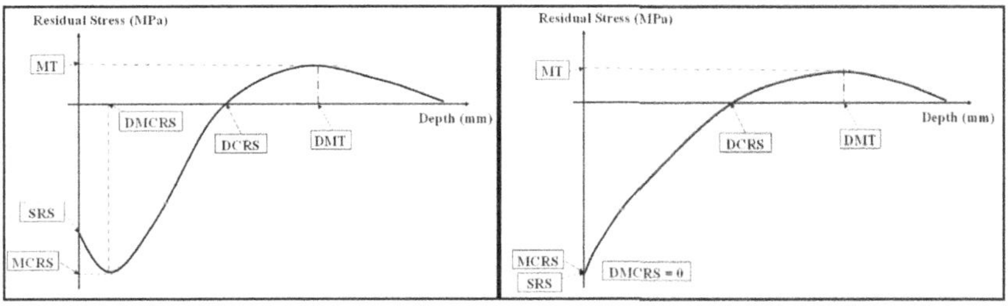

Fig. 1. Illustration of the shape of the residual stress profile induced by laser shock Surface treatment.

Next, the acquired profiles underwent analysis, and pivotal values were documented. The subsequent table presents the six key values extracted from the residual stress profile:

Table 2. Key values of a residual stress profile.

Factors	Output Results	Unity
CRSmax	Maximum compressive residual stress (Negative value)	MPa
DMCRS	Depth of the maximum compressive residual stress	mm
DCRS	Depth of compressive residual stresses	mm
SRS	Surface residual stress after laser shock (Negative value)	MPa
MT	Maximum tensile stress (Positive value)	MPa
DMT	Depth of the maximum tensile stress	mm

In our research, the primary output results derived from the design of experiments (DoE) simulations are limited to the following parameters: CRSmax, DCRS, SRS, MT, DMCRS, DMT, and the maximum equivalent plastic strain value, denoted as PEEQ (Refer to Table 2). These parameters will serve as responses, elucidated later in the context of the three process parameters.

Moreover, we are exploring the impact of treatment parameters on each resultant outcome. To approximate simulation outcomes through empirical mathematical laws for optimization, we will utilize statistical tools, integrating response surface methodology (RSM).

Response surface modeling techniques have been devised to analyze experimental outcomes and construct an empirical model of the observed response. Generally, this approach entails establishing a transfer function linking input treatment parameters with output values, resulting in a descriptive geometric representation of the desired output's behavior.

2 Evolution of the Residual Stress Profile According to the Treatment Parameters

2.1 Evolution of the Maximum Compressive Residual Stress (CRSmax)

From Fig. 2, it is evident that CRSmax is contingent upon both the maximum pressure Pmax and the impact diameter D, whereas the Full Width at Half Maximum (FWHM) exhibits negligible influence on the outcome of the process. Consequently, we can disregard FWHM to streamline the variables in the CRSmax evolution function.

The stabilization of the CRSmax response is notably evident at a pressure value of 5 GPa, as illustrated in Fig. 3.

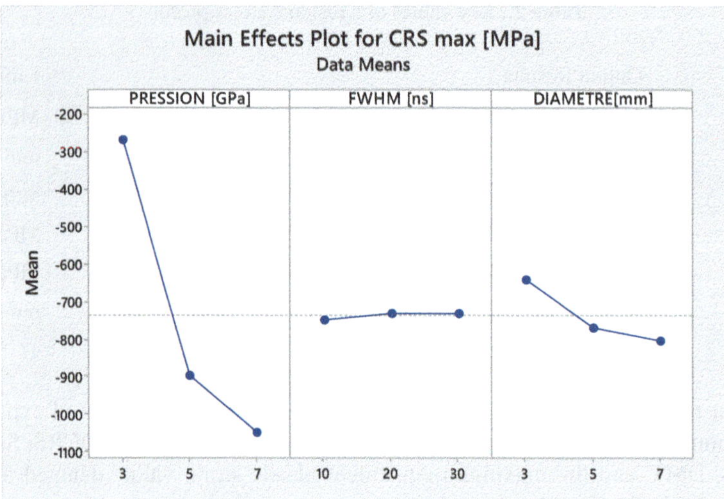

Fig. 2. Factors influencing maximum residual stress.

Utilizing the Design of Experiments (DoE) methodology enables us to ascertain the absolute value of the response function CRSmax, as defined in Eq. 1, subsequent to eliminating non-significant effect parameters.

$$\begin{aligned}\textbf{CRS max [MPa]} = {} & 365 - 189\,\text{Pmax(GPa)} - 1.8\,\text{FWHM(ns)} \\ & + 29\,\text{D(mm)} + 0.83\,\text{Pmax(GPa)} * \text{FWHM(ns)} \\ & - 12.6\,\text{Pmax(GPa)} * \text{D (mm)} - 2.36\,\text{FWHM(ns)} * \text{D(mm)} \\ & + 0.41\,\text{Pmax(GPa)} * \text{FWHM(ns)} * \text{D(mm)} \\ & \textbf{Adjustment rate} = \textbf{82.99\%}\end{aligned} \tag{1}$$

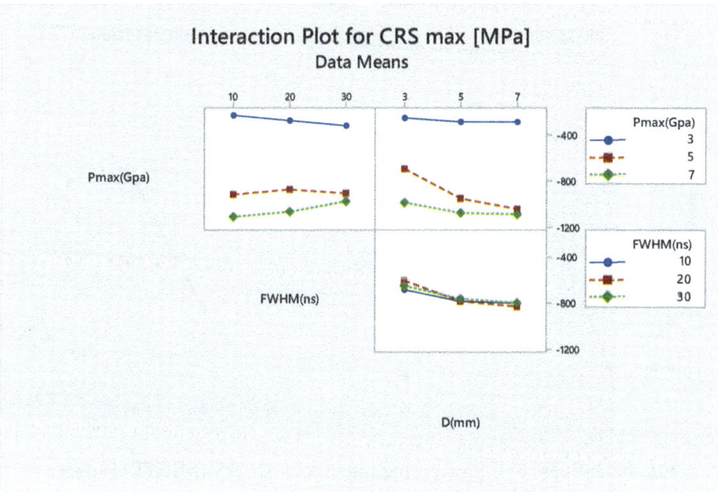

Fig. 3. Interaction curve for CRSmax (MINITAB data).

2.2 Influence of Process Parameters on DCRS Output Parameter

From Fig. 4, we notice that the depth of compressive residual stresses depends on the Full Width at Half Maximum in the first place and on the maximum pressure and the diameter but the interactions between these parameters have no effect on DCRS.

$$\begin{aligned}\textbf{DCRS[mm]} = & -0.094 + 0.0457\ \text{Pmax(GPa)} + 0.0093\ \text{FWHM(ns)} \\ & - 0.0040\ \text{D(mm)} - 0.00035\ \text{Pmax(GPa)} * \text{FWHM(ns)} \\ & + 0.0004\ \text{Pmax(GPa)} * \text{D(mm)} + 0.00027\ \text{FWHM(ns)} * \text{D(mm)} \\ & + 0.000312\ \text{Pmax(GPa)} * \text{FWHM(ns)} * \text{D(mm)} \\ & \textbf{Adjustment rate} = \textbf{85.31\%}\end{aligned} \tag{2}$$

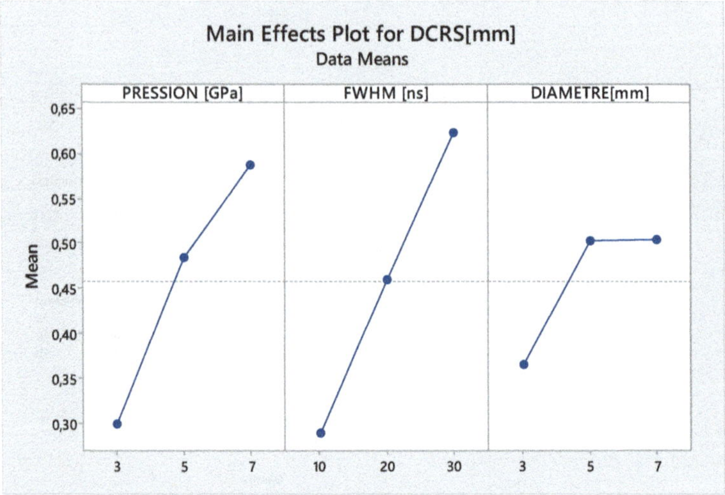

Fig. 4. Influence of process parameters on DCRS (MINITAB data).

2.3 Influence of Process Parameters on Surface Residual Stress (SRS)

Figure 5 shows that the surface residual stress value depends on the maximum pressure and impact diameter and FWHM does not have a significant effect on SRS.

$$\begin{aligned}\textbf{SRS[MPa]} = & 298 - 179\ \text{Pmax(GPa)} - 4.1\ \text{FWHM(ns)} + 36\ \text{D(mm)} \\ & + 2.04\ \text{Pmax(GPa)} * \text{FWHM(ns)} - 13.4\ \text{Pmax(GPa)} * \text{D(mm)} \\ & - 2.09\ \text{FWHM(ns)} * \text{D(mm)} + 0.26\ \text{Pmax(GPa)} * \text{FWHM(ns)} * \text{D(mm)} \\ & \textbf{Adjustment rate} = \textbf{86.21\%}\end{aligned} \tag{3}$$

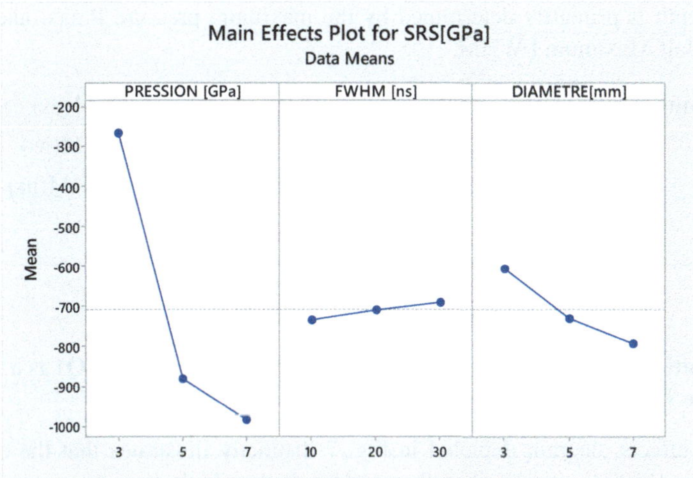

Fig. 5. Influence of process parameters on SRS(MPa).

2.4 Evolution of DMCRS Output Parameter

Out of 27 simulations, we have 16 cases where the maximum residual stress is located on the surface of the impacted target (59% of simulations). For the rest of the cases, it is in an underlayer and its depth is less than 150 μm (see Fig. 6). This represents an advantage for laser shock treatment, since the surface condition considerably influences the polycyclic fatigue behaviour and stress corrosion phenomena.

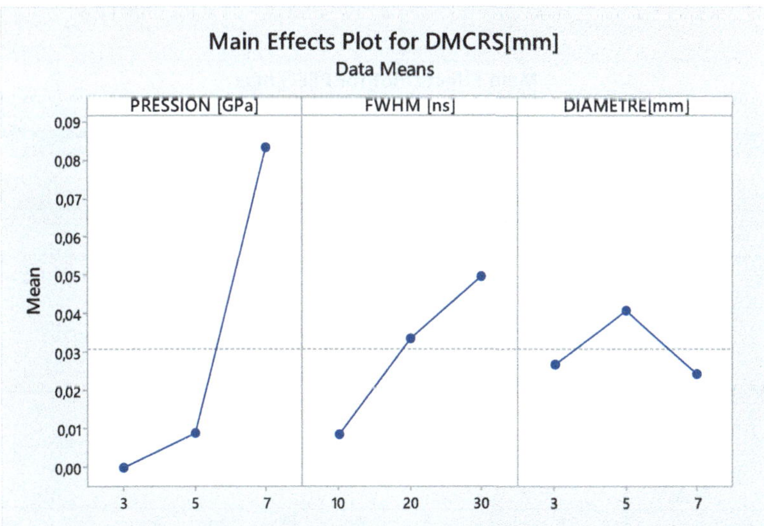

Fig. 6. Variation of the DMCRS value depending on the treatment parameters.

The depth is primarily determined by the maximum pressure Pmax, and the Full Width at Half Maximum FWHM.

$$\begin{aligned}\mathbf{DMCRS[mm]} = & -0.033 + 0.0055\,\text{Pmax(GPa)} - 0.00175\,\text{FWHM(ns)} \\ & + 0.0080\,\text{D(mm)} + 0.00074\,\text{Pmax(GPa)} * \text{FWHM(ns)} \\ & - 0.00179\,\text{Pmax(GPa)} * \text{D(mm)} - 0.00045\,\text{FWHM(ns)} * \text{D(mm)} \\ & + 0.000094\,\text{Pmax(GPa)} * \text{FWHM(ns)} * \text{D(mm)} \\ & \mathbf{Adjustment\ rate} = \mathbf{89.99\%} \end{aligned} \tag{4}$$

2.5 Evolution of the Value of the Equivalent Plastic Strain (PEEQ) as a Function of the Treatment Parameters

The main effects diagram depicted in Fig. 7 distinctly illustrates that the equivalent plastic strain (PEEQ) is primarily influenced by changes in the maximum pressur Pmax.

According to Fig. 7 and Fig. 8 we notice that the diameter has no effect on the equivalent plastic deformation and the interaction between the factors is not really significant except between FWHM and Pmax when the maximum pressure is greater than 5 GPa.

$$\begin{aligned}\mathbf{PEEQmax} = & -0.0142 + 0.00434\,\text{Pmax(GPa)} + 0.000032\,\text{FWHM(ns)} \\ & + 0.00048\,\text{D(mm)} + 0.000023\,\text{Pmax(GPa)} * \text{FWHM(ns)} \\ & - 0.000076\,\text{Pmax(GPa)} * \text{D(mm)} - 0.000032\,\text{FWHM(ns)} * \text{D(mm)} \\ & + 0.000008\,\text{Pmax(GPa)} * \text{FWHM(ns)} * \text{D(mm)} \\ & \mathbf{Adjustment\ rate} = \mathbf{83.74\%} \end{aligned} \tag{5}$$

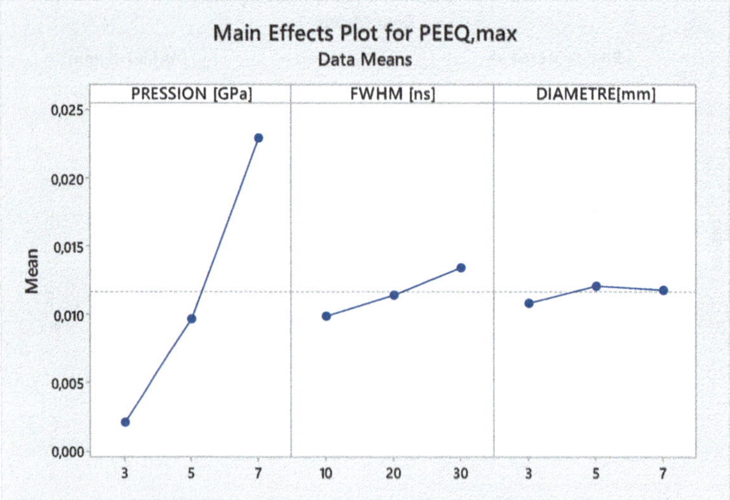

Fig. 7. The impact of process parameters on the equivalent plastic strain, as indicated by the MINITAB data.

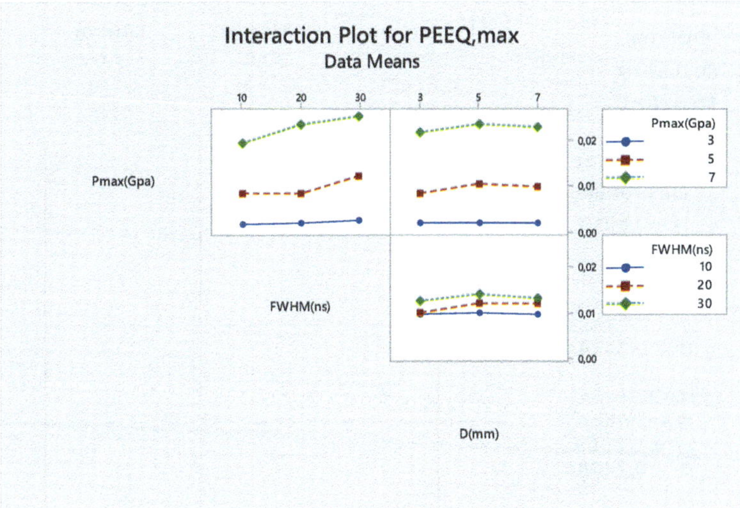

Fig. 8. Interaction curves for PEEQmax (MINITAB data).

3 Optimization of Laser Shock Processing Parameters

The objective of our project is to maximize the residual compressive stresses (CRSmax and SRS in absolute value) and to minimize the residual tension stress (MT) and the equivalent plastic deformation so we introduced these conditions into our MINITAB calculation program so that it gives us the optimal parameters for this treatment.

Figure 9 shows us that the optimal value of maximum pressure Pmax is 4.6 GPa, the optimal value of FWHM is 30ns and the optimal diameter is 7 mm.

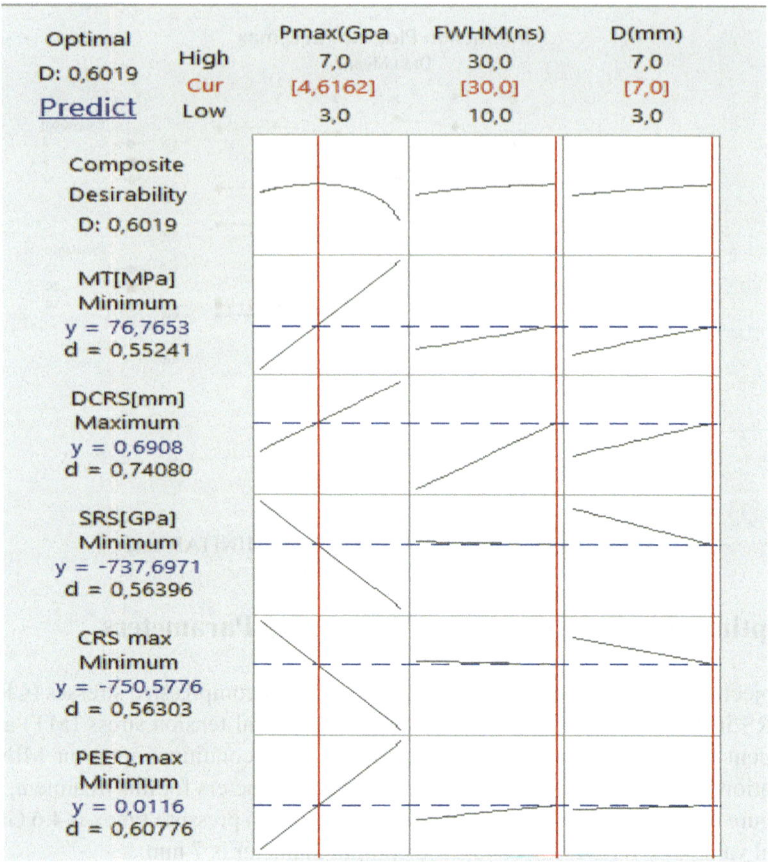

Fig. 9. Optimization of laser shock treatment parameters (MINITAB).

4 Conclusion

In this paper, we investigated the impact of laser shock treatment parameters on key values within the residual stress profile (CRSmax, DCRS, SRS, MT, DMCRS, DMT, and the maximum equivalent plastic strain value: PEEQ). This investigation was conducted by constructing a designed experiment using the MINITAB computational software.

The optimization process led to the determination of the optimal values for the maximum pressure Pmax, which was found to be 4.6 GPa. Additionally, the optimal values for FWHM were determined to be 30 ns, and the optimal diameter was calculated to be 7 mm.

Acknowledgements. The authors extend their gratitude for the valuable comments and suggestions provided by the reviewers, which have enhanced the quality of the paper.

References

1. Gill, A.S., Telang, A., Vasudevan, V.K.: Characteristics of surface layers formed on inconel 718 by laser shock, peening with and without a protective coating. J. Mater. Process. Technol. **225**, 463–472 (2015)
2. Fairand,B.P. and Clauer, A.H.: Laser generated stress waves: their characteristics and their effects to materials. In: Conference Proceeding No.50: Laser-Solid Interactions and Laser Pro-cessing, pp. 27–42 (1978)
3. Frija,M.: Traitement superficiel des composants aéronautiques par choc laser, Book ISBN978-620-2-28826-2, éditions universitaires européennes (2018)
4. Jacques, A.: Pratique industrielle de la méthode Taguchi: les plans d'expériences, AF-NOR, 1995, ISBN 2-12-465008-4
5. Minitab User Manual (Release 15), Making data analysis easier, USA (PA), MINITAB Inc. State College, (2001)

Ball Burnishing Parameters Affecting the Residual Stress of the $Zr_{65}Cu_{15}Al_{10}Ni_{10}$ Metallic Glass

Sami Bouzayeni[1(✉)], Fathi Gharbi[1,2], Tarek Benameur[1], and Khaled J. Al-Fadhalah[3]

[1] Mechanical Engineering Laboratory, LR99ES32, ENIM, University of Monastir, Monastir, Tunisia
Bouzayeni.sami@gmail.com, t.benameur@enim.rnu.tn

[2] Institut Supérieur des Sciences Appliqués et de Technologie de Kairouan, ISSATKR, University of Kairouan, Kairouan, Tunisia

[3] Mechanical Engineering Department, College of Engineering and Petroleum, Kuwait University, 5969 Safat, Kuwait
khaled.alfadhalah@ku.edu.kw

Abstract. Ball burnishing (BB) is a superfinish operation performed to improve surface, subsurface quality and mechanical proprieties. In the present study, a bi-dimensional (2D) finite element model (FEM) was developed to confirm that the BB of $Zr_{65}Cu_{15}Al_{10}Ni_{10}$ bulk metallic glass (BMG) affects its microstructures, roughness, and residual stress. As a part of the process, the influence of different parameters on the residual stress surface in the axial and transverse directions was investigated. The topography of the burnished samples is analyzed by an atomic force microscope (AFM) and scanning electron microscope (SEM). The ball burnishing process was performed, and the improvements in the surface and subsurface integrity were found to be significant. Residual stress after BB was found to be affected by burnishing force to a statistically significant extent. The Zr-based BMG displayed significant improvements in surface roughness and residual stress following the burnishing process. The topographical and morphological results of the treated samples show a variation of irregularity of the burnished faces and the formation of the shear bands (SBs) below the hardened layer. The FE simulation results indicate that the residual stress values and depth are increased by the increase of burnishing force. The ball burnishing process enhanced the surface topography and produce the residual stress on the Zr-BMG samples.

Keywords: Bulk Metallic Glass · Ball Burnishing · Residual stress · Surface topography · Finite Element Modeling

1 Introduction

The life of many components used in engineering and structural functions can be extended through surface treatment and modification. The use of thermal, mechanical, chemical, and coatings (both hard and soft) is all part of the surface engineering

M. Chouchane et al. (Eds.): CMSM 2023, LNME, pp. 326–337, 2024.
https://doi.org/10.1007/978-3-031-65007-9_35

process (Barshilia, 2021). In order to polish off metal after machining, the ball burnishing (BB) technique is usually employed (Revankar et al., 2014). This method is unique in that it produces surface hardening and beneficial residual stresses with large compressive strengths, which together minimize surface roughness (Okada et al., 2017). It is referred to by Hassan as the last step in the fabrication of a mechanical component, which involves pressing down while rolling a hard ball on the metal surface (Denkena et al., 2017). Low temperature plastic deformation is used in this non chip method to produce a flat, hardened surface (Konefal et al., 2013). BB offers a number of benefits, including the ability to be done in standard machine shops and the fact that it is a reasonably inexpensive process that requires less experienced personnel than other surface processing options (Bruijn et al., 2021). The BB parameters that affect residual stress are burnishing pressure, burnishing feed, burnishing speed, and number of passes (Bourebia et al. 2021). The pressure and feed rate during BB significantly affect the residual stress of AISI 8620 steel (Cappilla et al., 2021).

Many studies have looked at how BB affects mechanical properties. For instance, a recent study by Jawahir et al. found that burnishing surfaces reduces surface roughness and creates high-compressed residual stress, which affects corrosion and fatigue resistance (Siward et al. 2021). After conducting a slide diamond burnishing experiment on 42CrMo4 alloy steel shafts, Korzynski et al. observed that the surface layer's compressive stress increased by as much as 400 MPa, improving surface roughness by 0.05–0.18 μm (Korzynski et al., 2011). BB is a finishing technique that involves plastic deformation on the workpiece surface. It has been found to improve the surface quality and wear resistance of various materials, including AISI 1010 (Gharbi et al., 2023), AISI 316L stainless steel (Attabi et al. 2021), and Mg-4Zn-1Si alloy (Ramesh et al. 2021). The process has been shown to reduce the average roughness of the surface, increase ductility, modify the yield stress and work hardening properties, and reduce planar anisotropy (Siward et al., 2020). Residual stress is an important factor in simulations of mechanical components as it affects fatigue life. Rodriguez et al. developed a 2D finite element model to simulate the axial and tangential residual stress distribution, which showed qualitative agreement with experimental data (Schoop, 2021). Yen et al. used both 2D and 3D finite element models to analyze the surface geometry after BB on hardened AISI 52100 steels. They found that the 3D model provided a more realistic representation, although the 2D model had better predictions of residual stress (Yang et al., 2020).

The presence of residual stress plays a crucial role in the mechanical performance of metal alloys, as various studies have shown (Kikuchi et al., 2018). Although bulk metallic glasses (BMGs) have attracted attention as a potential new class of structural materials with high impact strength, good wear and corrosion resistance, making them suitable for various industrial products such as microelectromechanical systems (Ryu et al. 2021). The main objective is to predict the residual stress states produced by the ball burnishing (BB) process on $Zr_{65}Cu_{15}Al_{10}Ni_{10}$ BMG as a result of the effect of selected burnishing parameters. The data obtained from the experiments are then used in a two-dimensional (2D) finite element model (FEM) with a constitutive micromechanical law to predict a relationship between burnishing parameters and residual stress.

2 Experimental Procedures

2.1 Sample Preparation and Post Experiment Characterization

$Zr_{65}Cu_{15}Ni_{10}Al_{10}$ amorphous alloy of composition (in at. %) was made by the copper mold casting method by applying high heat and pressure to the mixed particles. The size of the sample is $8 \times 4 \times 2$ mm^3, no mechanical polishing was performed on the surfaces. X-ray diffraction (XRD) measurements of the alloys were obtained to confirm the amorphous structure of the sample. In addition, we examine the surfaces of burnished samples using the atomic force microscope (AFM) and the scanning electronic microscope (SEM).

2.2 Ball Burnishing Setup

A schematic illustration of the BB procedure is shown in (Fig. 1). We created a calibration curve by compressing the tool with an electromechanical universal testing machine in order to determine a link between the burnishing ball's movement and the applied force. Double-sided Scotch tape was used to firmly keep the specimens in place so that they could be positioned with precision within the CNC milling machine. The lubricant used in the burnishing process is called ESSOLUBE HD 15W-40, and it has a kinematic viscosity of 113 mm^2/s at 40 °C and 15.4 mm^2/s at 100 °C. A single pass and a BB feed of 0.5 mm were used in each experiment.

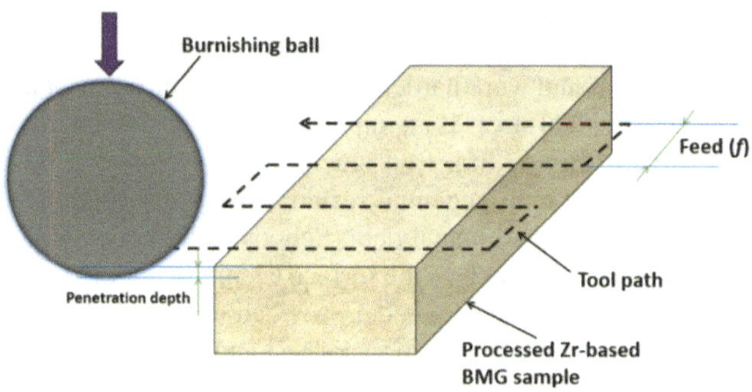

Fig. 1. The BB process applied in $Zr_{65}Cu_{15}Ni_{10}Al_{10}$ bulk metallic glass sample.

Table 1 shows the burnishing parameters and their levels.

Table 1. Burnishing conditions and their levels.

Parameters		Levels	
Burnishing force (N)	300	400	500
Penetration depth (mm)	0.10	0.15	0.20
Burnishing feed rate (mm/min)	40	60	80

2.3 Analytics Residual Stress Calculations

Monotonic nanoindentation experiments with a diamond Berkovich tip were carried out at ambient temperature in order to determine the axial residual stress. According to the (Lee et al. 2004)'s model, the non-equibiaxial residual stresses from the indentation tests were computed and classified into four groups: uniaxial stress ($\sigma_x^r \neq 0, \sigma_y^r = 0$, 3 and 5), equibiaxial stress ($\sigma_x^r = \sigma_y^r = 0$, 1 and 6), biaxial stress ($\sigma_x^r \neq \sigma_y^r \neq 0$, 2), and pure shear stress ($\sigma_x^r = -\sigma_y^r \neq 0$, 4), respectively. A stress ratio, or k, can be used to describe the ratio between biaxial stress, i.e., k = σ_y^r/σ_x^r, where k is a number between −1.0 and 1.0. Equibiaxial stress and pure shear stress are two subtypes of biaxial stress. The measurement of biaxial stress may be reduced to an equibiaxial issue since pure shear stress has no impact on the indentation load. As seen below, one may compute the biaxial stress.

$$\sigma_x^r = \frac{3(P_0 - P_1)}{(1 + K)A_C} \tag{1}$$

where P_0 and P_1 are the maximum loads for the materials with and without residual stress, respectively. The region of contact between the material and the strained indenter is called A_c. The equibiaxial stress condition, which occurs when the stress ratio $k = 1$.

3 FE Simulation

3.1 Constitutive Micro-mechanical Law

Based on the availability of potential sites permitting atomic leaps, Spaepen presented a microscopic deformation process that takes into consideration the free volume (FV) as an ordering parameter. The area where the free volume exceeds the effective volume v* of a hard sphere representing an atom is the potential site. According to Spaepen's theory, a flow is induced by a single atomic leap under the influence of heat or mechanical stress. There must be enough unoccupied space close to the atom for this leap to occur. The presence of a thermal activation energy Gm is necessary for this leap. Atomic jumps happen in the direction of the applied force when shear stress is applied. Therefore, following the pattern of the force applied, the dominating flow leaps over the potential barrier (Spaepen, 1976).

The following equation can be used to indicate the plastic strain rate:

$$\dot{\gamma}_p = \exp\left(-\frac{\gamma v*}{vf}\right) 2\upsilon_G \sinh\frac{\tau\Omega}{2kBT}\exp\left(-\frac{\Delta G^m}{kBT}\right) \tag{2}$$

This equation can be simplified as follows:

$$\dot{\gamma}p = \dot{\gamma}_0 \sinh\left(\frac{\tau\Omega}{2kBT}\right) \tag{3}$$

where:

$$\dot{\gamma}_0 = Cf\,2\upsilon_G \exp\left(-\frac{\Delta G^m}{kBT}\right) \tag{4}$$

and,

$$C_f = \exp\left(-\frac{\gamma v^*}{vf}\right) \tag{5}$$

Using the user-defined material (UMAT) subroutine, the aforementioned equations are implemented in the FE software ABAQUS/Standard calculation code. Table 2 shows the mechanical and physical characteristics of $Zr_{65}Cu_{15}Ni_{10}Al_{10}$ MG utilized in these simulations:

Table 2. $Zr_{65}Cu_{15}Ni_{10}Al_{10}$ material properties used in the FE simulation (Kawamura et al., 1999).

Parameter	value
Young Modulus, E	95 GPa
Poisson ration, v	0.371
Density, ρ (10^{-9} Ton/mm^3)	6.271
Free volume activation, σ_0 (GPa)	0.420
Number of atomic jumps, n_D	3
Geometrical coefficient, α	0.15
Parameter, β	1

3.2 2-D Model Meshing and Boundary Conditions

The generally accessible ABAQUS nonlinear FE program was used to model the ball polishing procedure. Simulations for 2D FE were performed using ABAQUS/standard. The following assumptions form the basis of the FE models:

1. The ball tool is modeled as an analytically rigid component, since it is considered rigid.
2. The shape of the surface roughness is neglected.
3. The isothermal conditions (ambient temperature) and the absence of friction were predicted, as the pressurized fluid acts as both a cooler and a lubricant during the process.
4. The influence of ball polishing speed is ignored.

The survey was limited to the cross-sections cut along the longitudinal direction of the parallelepiped in order to represent the 2D burnishing process. In fact, when simulating the longitudinal object, more fundamental factors, such as feed speed and surface roughness profile, are taken into account only when simulating the cross-section. The treated part is shaped by a rectangle fixed on its lower edge and without horizontal movement on its side edges, since the length of the part is significantly larger than the diameter of the ball tool. The part is built in ABAQUS CPE8 using eight-node biquadratic 2D solid plane deformation components. Three repetitive actions make up this load (Fig. 2):

- Vertical displacement (loading) associated with the penetration phase of the material
- During unloading, the ball returns to its starting position.
- The ball is transported horizontally over a distance after being unloaded.

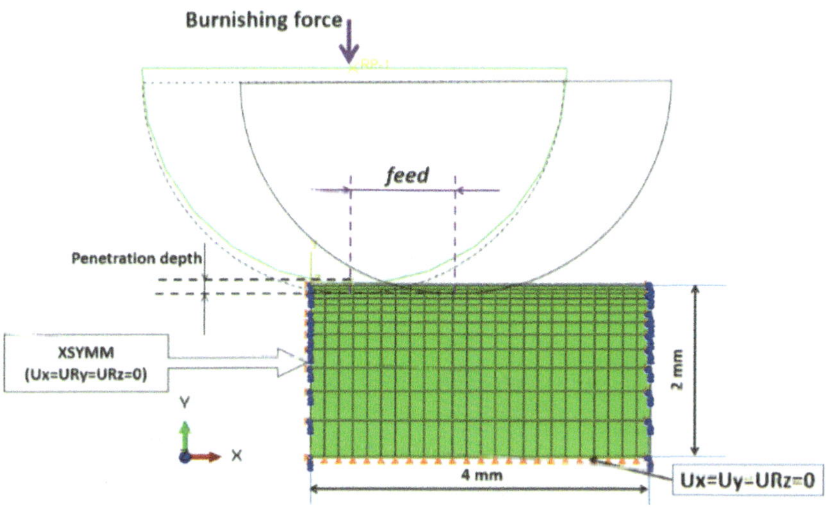

Fig. 2. 2D geometrical model and finite elements mesh used for ball burnishing simulations

4 Results and Discussions

4.1 Surface Topography Analysis

Figure 3 displays the 3D AFM pictures of the examined material under various burnishing parameters. The surface of the burnished samples can be seen in these photos to have a maximum roughness value of approximately 56 nm, showing a high level of surface treatment from the ball burnishing process. With a frequency of around 10 µm, the surface morphology displays a characteristic pattern of parallel lines that is consistent with the browning direction. The surface topography exhibits a regular pattern of shallow scratches and grooves, which is compatible with the BB procedure.

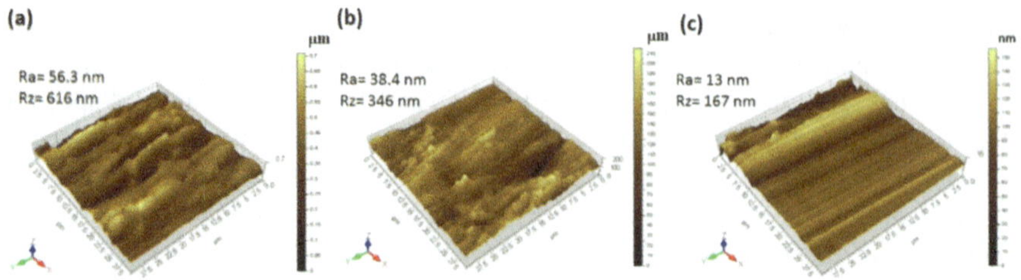

Fig. 3. Comparison of AFM images of burnished surfaces topography within an area of 30 × 30 μm for different ball burnishing conditions: (a) $F = 500$ N, $p = 0.10$ mm, $f = 80$ mm/min, (b) $F = 500$ N, $p = 0.15$ mm, $f = 40$ mm/min, (c) $F = 500$ N, $p = 0.20$ mm, $f = 60$ mm/min.

Figure 4 displays SEM pictures of the burnished sample's cross-section and burnished surface with the following parameters: F = 300 N, p = 0.15 mm, and f = 40 mm/min. These photos reveal the development of sliding bands inside the metallic glass sample, demonstrating how the burnishing process has caused an amorphous alloy to deform plastically. As observed in Fig. 4(c), there are major bands (depicted with yellow arrows) that are dominating, with some discernible secondary bands on the lateral surfaces (depicted with blue arrows). We can observe that, the single SB is deflected with a semi-circular path through the sample in the area underneath the passing bead. Some spread to the inner region during the shear banding process, undergoing further deformation to take into consideration the total plasticity of all samples. In fact, when the SBs move through the burnished BMGs, they can be seen deflecting and branching on the lateral surfaces (see picture corresponding to the dashed box).

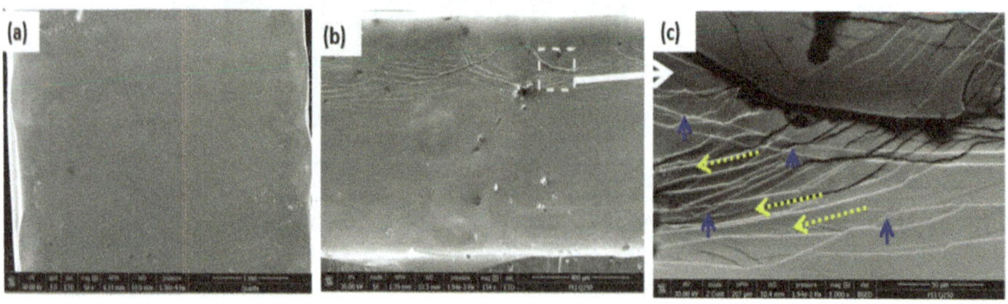

Fig. 4. SEM images of the studied surfaces on the (a); burnished surface, (b); cross section and (c); magnified image of dashed box, with burnishing conditions (F = 300 N, p = 0.15 mm, f = 40 mm/min).

4.2 Residual Stress Analysis

The results obtained from the FE simulation of equivalent strain under burnishing conditions (F = 300 N, p = 0.15 mm, f = 40 mm/min) are shown in Fig. 5(a). The deformed shape shows a peak in the samples' longitudinal limit zone and an increase in the equivalent deformation field and free volume distribution (SDV1) values in the area where SBs develop, as indicated in Fig. 5(b).

SEM micrographic images were captured in the cross-section of the burnished sample, indicating the existence of a spherical cap below the ball penetration, in which there is a proliferation pattern of SBs in the form of semi-arcs, as illustrate in Fig. 5(c). The agreement between the simulations and micrographs is good. The contours of the numerical simulation satisfactorily represent the physical phenomena captured by experimental observation.

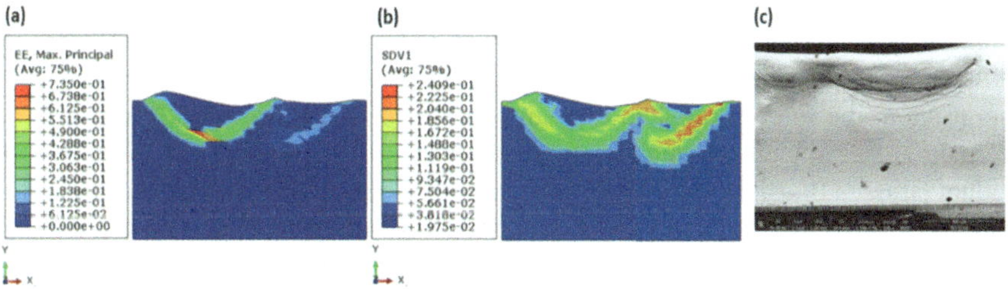

Fig. 5. (a); Distribution of equivalent strain, (b); free volume concentration SDV1, (c); SEM images of cross section for burnishing condition ($F = 300$ N, $p = 0.15$ mm, $f = 40$ mm/min).

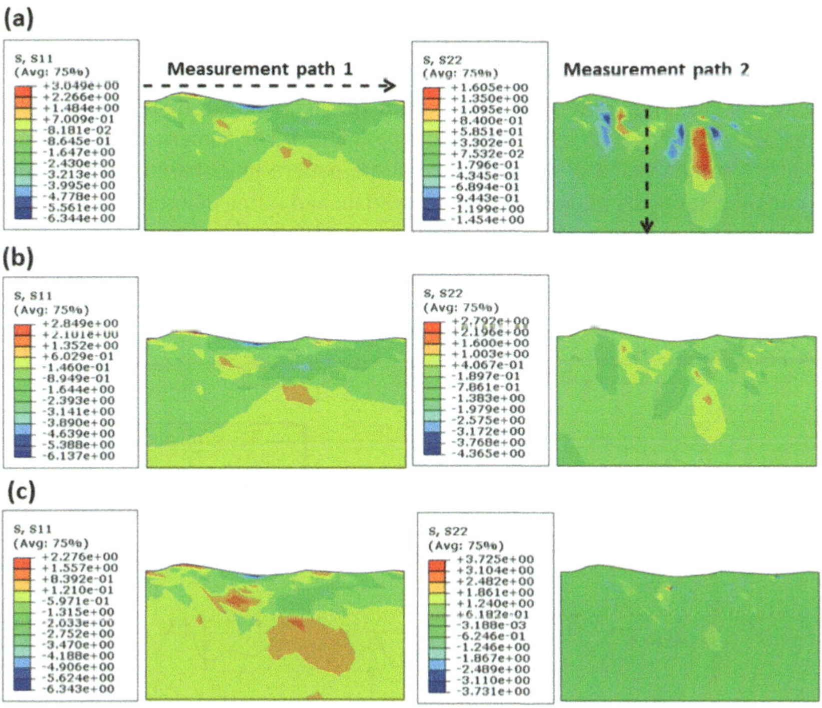

Fig. 6. Distributions contours of axial residual stress S11 and transversal residual stress S22 for burnishing conditions ($p = 0.15$ mm, $f = 40$ mm/min) with different burnishing force applied (a); $F = 300$, (b); 400 and (c); 500 N.

Figure 6 depicts the simulated model of residual stress fields in the tool's depth and feeding directions, S11 (X-direction) and S22 (Y-direction), using two measurement trajectories. After the full discharge of mechanical loads in the current work. Figure 7 illustrates the extraction of the simulated axial and transverse residual stresses from measurement path 1 and 2, respectively. For BB parameters (F = 300 N, p = 0.15 mm, and f = 40 mm/min), a comparison between the projected simulation results and the data derived analytically was done.

The surface layer of the workpiece exhibits the largest deformation values throughout all simulations, as can be shown. Every simulation has different predicted deformation values depending on the burnishing force value. The simulations turned out to be the most responsive to the force value, and they grew in line with that. Since BB involves little force, the material's elastic qualities take center stage. The surface's minimal positive residual stresses (tensile) seem to be at odds with the specifications of the test and the method. Negative residual stresses (in compression) were recorded as depth from the surface increased, although this is still not a realistic result.

Figure 7 shows the residual stresses in the S11 axial direction, as determined by experimental measurements and numerical simulations using the parameters (F = 300 N, p = 0.15 mm, and f = 40 mm/min), respectively. It should be noted that the distribution of residual stresses along the ball contact zone are of negative sign, which explains the existence of compressive state stresses, as are the values obtained experimentally. By comparing the two profiles, we found that the difference between the two curves is around 40%. On the other hand, numerical simulations, often performed using finite element analysis, offer a powerful means of predicting residual stress distributions following BB treatment. The integration of experimental and numerical approaches can provide a comprehension of axial residual stress pattern in Zr-based metallic glasses, exploiting the strengths of each method to validate and refine theoretical models based on nanoindentation testing while providing experimental validation and insight into real behavior.

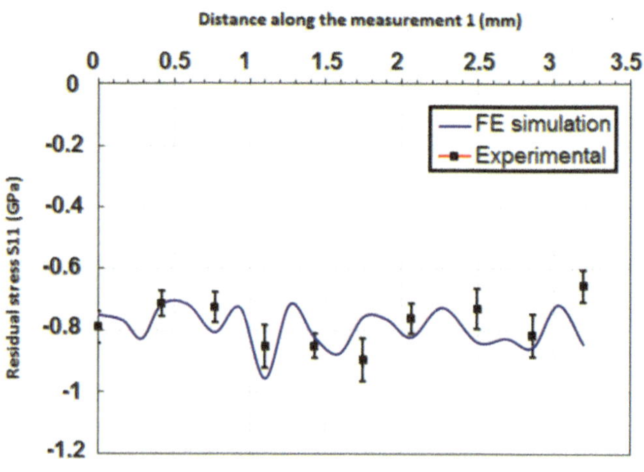

Fig. 7. Comparison between experimental and simulated profiles of axial residual stress S11 for burnishing conditions (F = 300 N, p = 0.15 mm, f = 40 mm/min).

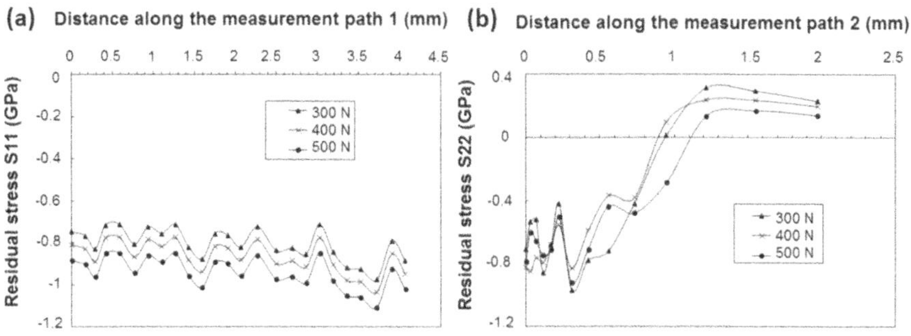

Fig. 8. Comparison between simulated profiles of: (a) Axial residual stress S11 and (b) transversal residuals stress S22 for burnishing conditions (p = 0.15 mm, f = 40 mm/min) with different burnishing force applied (F = 300, 400 and 500 N).

The residual stress profiles are affected by the burnishing force, as seen in Figs. 8(a) and (b). As can be observed, the suggested model depicts the advancements brought about by force increases. It has been shown that when BB force rises, the compression residual stress layer also grows. For 500 N, this layer, which is regarded as one of the most crucial factors in surface treatment, may reach a depth of up to 1.15 mm. It should be mentioned that the ideal thickness selection must always take into account the state of the material and the loading conditions. For instance, a deep layer is far more important in components that are thrust tensile loaded than it is in the top layer that is thrust compression loaded. Furthermore, it is demonstrated how the burnishing force affects the maximum axial and transverse residual stresses. However, because of the material's constraints on how much it can deform, the maximum tangential residual stress only gradually changes with high pressure, leading to an expansion of the compression layer. By examination of these profiles shows that the increase in the burnishing force causes an increase in the transverse residual stress S22 of about 20%.

5 Conclusions

In this paper, topographic analysis by scanning electron microscope (SEM) and atomic force microscope (AFM) shows an improvement in surface roughness and the production of shear bands (SB) in the samples following the plastic deformation created by the ball burnishing (BB) process on amorphous metal $Zr_{65}Cu_{15}Al_{10}Ni_{10}$. In addition, a 2D FE model of the burnishing process was presented. The current numerical simulation provides only a qualitative description of axial and transverse residual stress variations. The maximum axial residual stress obtained by the numerical calculations differs from the experimental measurement by around 40%. The effects of BB force on residual stresses, and plastic strain distribution were investigated using numerical prediction. These results can be summarized as follows:

- A distribution of compressive residual stresses in the sub-surface and tensile stresses within the burnished samples.
- Increasing the burnishing force increases the magnitude of residual compressive stress.

An increase in transverse residual stresses of about 20% with the increase of burnishing force (300–500 N).

Continued research could explore the use of advanced characterization techniques, such as atomic force microscopy (AFM) and scanning electron microscopy (SEM), to further analyze the topography and microstructural changes induced by ball burnishing in metallic glasses.

References

Barshilia, H.C.: Surface modification technologies for aerospace and engineering applications: current trends, challenges and future prospects. Trans. Indian Natl. Acad. Eng. **6**(2), 173–188 (2021)

Revankar, G.D., Shetty, R., Rao, S.S., Gaitonde, V.N.: Analysis of surface roughness and hardness in ball burnishing of titanium alloy. Measurement **58**, 256–268 (2014)

Okada, M., et al.: Development and characterization of diamond tip burnishing with a rotary tool. J. Mater. Process. Technol. **244**, 106–115 (2017)

Denkena, B., Grove, T., Maiss, O.: Surface texturing of rolling elements by hard ball-end milling and burnishing. Int. J. Adv. Manuf. Technol. **2017**(93), 3713–3721 (2017)

Konefal, K., Korzynski, M., Byczkowska, Z., Korzynska, K.: Improved corrosion resistance of stainless steel X6CrNiMoTi17-12-2 by slide diamond burnishing. J. Mater. Process. Technol. **213**(11), 1997–2004 (2013)

Bruijn, A.C., Gómez-Gras, G., Pérez, M.A.: On the effect upon the surface finish and mechanical performance of ball burnishing process on fused filament fabricated parts. Addit. Manuf. **46**, 102133 (2021)

Bourebia, M., et al.: Effect of ball burnishing process on mechanical properties and impact behavior of S355JR steel. Int. J. Adv. Manuf. Technol. **116**, 1373–1384 (2021)

Capilla-González, G., Martínez-Ramírez, I., Díaz-Infante, D., Hernández-Rodríguez, E., Alcantar-Camarena, V., Saldaña-Robles, A.: Effect of the ball burnishing on the surface quality and mechanical properties of a TRIP steel sheet. Int. J. Adv. Manuf. Technol. **116**, 3953–3964 (2021)

Swirad, S., Pawlus, P.: The effect of ball burnishing on dry fretting. Materials **14**(22), 7073 (2021)

Korzynski, M., Lubas, J., Siward, S., Dudek, K.: Surface layer characteristics due to slide diamond burnishing with a cylindrical-ended tool. J. Mater. Process. Technol. **211**(1), 84–94 (2011)

Gharbi, F., Al-Fadhalah, K.J., Tharwan, M.: Improvement of surface quality of AISI 1010 steel plates by ball burnishing. J. Mater. Eng. Perform. **86**, 1–13 (2023)

Attabi, S., Himour, A., Laouar, L., Motallebzadeh, A.: Effect of ball burnishing on surface roughness and wear of AISI 316L SS. J. Bio Tribo-Corrosion **7**, 1–11 (2021)

Ramesh, S., Anne, G., Kumar, G., Jagadeesh, C., Nayaka, H.S.: Influence of ball burnishing process on equal channel angular pressed Mg-Zn-Si alloy on the evolution of microstructure and corrosion properties. SILICON **13**, 1549–1560 (2021)

Swirad, S., Pawlus, P.: The effect of ball burnishing on tribological performance of 42CrMo4 steel under dry sliding conditions. Materials **13**(9), 2127 (2020)

Schoop, J.: In-situ calibrated modeling of residual stresses induced in machining under various cooling and lubricating environments. Lubricants **9**(3), 28 (2021)

Yang, L., Zhang, J., Chen, T., Lei, W., & Liu, X.: Predicting residual stress by finding peak shape using artificial neural networks. In: 2020 IEEE International Conference on Parallel & Distributed Processing with Applications, Big Data & Cloud Computing, Sustainable Computing & Communications, Social Computing & Networking (ISPA/BDCloud/SocialCom/SustainCom, pp. 633–640 (2020), IEEE

Kikuchi, S., et al.: Statistical fatigue properties and small fatigue crack propagation in bimodal harmonic structured Ti-6Al-4V alloy under four-point bending. Mater. Sci. Eng. A **711**, 29–36 (2018)

Ryu, W.H., Kim, K.J., Yoo, G.H., Park, E.S.: Alloy design strategy to improve fluidity of Zr-based bulk metallic glass for near-net-shape manufacturing. J. Alloy. Compd. **896**, 162680 (2021)

Lee, Y.H., Kwon, D.: Estimation of biaxial surface stress by instrumented indentation with sharp indenters. Acta Mater. **52**, 1555 (2004)

Spaepen, F.: A microscopic mechanism for steady state inhomogeneous flow in metallic glasses. Acta Metall. **25**, 407–415 (1976)

Kawamura, Y., Shibata, T., Inoue, A., Musomoto, T.: Stress overshoot in stress-strain curves of Zr65Ni10Cu15Al10. Mater. Trans. JIM **40**, 335–342 (1999)

A Method to Extract the Elastoplastic Properties of the Constituent Layers of Multilayer Coatings

Yamen Ben Ammar[1(✉)], Khalil Aouadi[2,3], Faker Bouchoucha[1], Corinne Nouveau[4], Aurélien Besnard[4], and Alex Montagne[5]

[1] Laboratoire de Recherche LR18ES45 sis à l'institut Préparatoire aux études d'Ingénieur de Nabeul, Université de Carthage, Campus Universitaire, 8000 Mrezga, Tunisia
Yamen_benammar@yahoo.fr

[2] Applied Mechanics and Systems Research Laboratory (LR03ES06), Tunisia Polytechnic School, University of Carthage, Rue El Khawarizmi, BP 743, 2078 La Marsa, Tunisia

[3] Moroccan Foundation for Advanced Science, Innovation and Research (MAScIR), Rue Mohammed El Jazouli, Rabat Design Center, 10100 Rabat, Morocco

[4] Arts et Metiers Institute of Technology, LABOMAP, HESAM Université, UBFC, 71250 Cluny, France
{corinne.nouveau,aurelien.besnard}@ensam.eu

[5] Arts et Metiers Institute of Technology, MSMP, 8, Rue Boulevard Louis XIV, 59046 Lille Cedex, France
alex.montagne@ensam.eu

Abstract. This paper presents an approach to characterize the elastoplastic properties of the distinct layers, constituting multilayer coating. The proposed procedure utilizes nanoindentation load-displacement (P-h) curves and a non linear least squares fitting analysis to extract the elastoplastic properties of each layer in the multilayer coating. The accuracy of the optimization results is achieved by choosing initial guess parameters closer to the target values using the modified Jönsson and Hogmark model. The methodology is validated on a CrN/CrAlN multilayer coating systems with varying layer thicknesses from 1 to 0.5 μm, from which the optimal elastoplastic properties: Young's modulus (E), yield stress (σ_Y), and strain hardening exponent (n) of each individual layer were determined. The results show good agreement between the simulated and the experimental (P-h) curves. Furthermore, the results revealed a reduction in the material parameters (E, H and σ_Y) of the constituent layers when the layer thickness decreases. These findings suggest that decreasing the coating layer thickness lead to an increase in the plastic deformation within the coatings, which reduces the stress concentration in this area and improves the adhesion properties of CrN/CrAlN multilayer coatings.

Keywords: Instrumented-indentation testing · Elastoplastic properties · finite element modeling

M. Chouchane et al. (Eds.): CMSM 2023, LNME, pp. 338–349, 2024.
https://doi.org/10.1007/978-3-031-65007-9_36

1 Introduction

In the last decade, the concepts of multilayer coatings have been proposed as a strategy to improve properties such as hardness, resistance to crack propagation, and toughness of traditional monolayer coatings [1, 2]. In general, the better performance of multilayer coatings against wear damage, scratch and abrasion emanates from incorporation of alternating multilayer interface systems [3].

Due to their remarkable capabilities, multilayer coatings have become an important research objective in the last decade. Among these studies, a particular interest is given to the optimization of the structure and the type of multilayer coatings in order to achieve coatings with the desired mechanical properties. Based on the work of (Kataria et al. 2012) [4], the performance of multilayer coatings is substantially dependent on the elastoplastic properties of the monolayers that constitute them. Furthermore, the distribution of residual stress and deformation at the coating/substrate interface, which control the tribological and adhesion properties of the multilayer coatings, is mainly related to the elastoplastic properties of the monolayers of multilayer [4, 5]. Therefore, the determination of mechanical properties such as hardness, Young's modulus, yiled stress, strain hardening coefficient and fracture toughness of individual monolayer is necessary to achieve the design purposes [6].

Previous works [6, 7], have shown that the assessment of the mechanical properties of multilayer coatings must be performed through nanoindentation under high indentation loads. This approach arises from the fact that the indenter must penetrate through the multiple layers of the coating to separate the contribution of each individual layer from the others ones and also from the substrate. In this context, numerous analytical models have been developed to predict the hardness and young's modulus of monolayer coatings systems from experimental nanoindentation load-displacement curves and composite hardness and composite modulus profiles [7]. However, for the case of multilayer coatings, there has limited development of comparable models that enable the determination of (H) of each individual layers. Notable among these models are the modified Johnsson and hogmark model [6], the modified Puchi-Cabrera model and the modified Korsunsky model [7]. Among these three models, a special attention has been directed award the modified Johnsson and hogmark (JH) model. This model stands out for its minimal number of fitting parameters (just 2), and its high predictive performance compared to the others models [7].

Nevertheless, these analytical models are unable to provide information about the elastoplastic properties of the constituent layers of multilayer coatings. In this case, finite element simulation, especially when integrated with the experimental nanoindentation data, have been successfully used to characterize the true elastoplastic properties of coatings on substrate, such as (E), (σ_y) and (n) [8–12].

The aim of the present work is to evaluate the elastoplastic material parameters of the layers of multilayer coatings. To achieve this goal, an optimization procedure based on the experimental nanoindentation and finite element analysis has been devised for deducing the elastoplastic properties of each individual layer of the multilayer coating (E, σ_y and n). The proposed methodology has been successfully applied to a CrN/CrAlN multilayer coating system with various layers thickness. The results obtained from this work serve as valuable reference for the structural optimization of multilayer coatings.

2 Experimental Details

2.1 CrAlN/CrN Multilayer Coating Deposition

In this study, CrN/CrAlN multilayer coatings with various layer thicknesses were deposited on X50CrMoV8-1 steel samples (20 × 20 × 5 mm^3) using DC reactive magnetron sputtering. During deposition the substrate temperature and the substrate bias voltage were fixed at 300 °C and – 500V, respectively, under 0.5 Pa of working pressure. Nitrogen and Argon gas with flow rate of 33.3 and 68.8 sccm, respectively were used. Then a high-purity Cr and Al target (99.99%) was used to deposit CrN/CrAlN multilayer coatings for a deposition time of 2 h. Figure 1 shows the different multilayer coatings deposited in this study. Two multilayer coatings, denoted as C1 and C2 were developed with a thickness gradient, choosing the same layer thicknesses used in the other coatings (1 and 0.5 μm) and keeping the total thickness fixed at 2 μm.

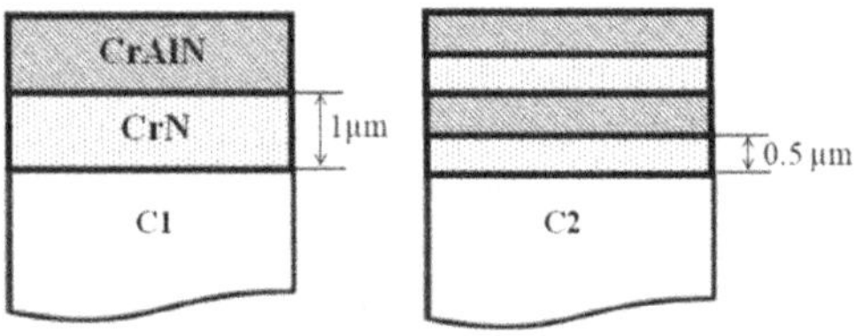

Fig. 1. Scheme of the CrN/CrAlN multilayer systems

2.2 Nanoindentation Tests

The evaluation of the mechanical properties (H) and (E) of CrN/CrAlN multilayer coatings was performed with a nanoindentation system (MTS, USA) equipped with a Berkovich indenter. To explore the impact of the substrate on the elastoplastic properties of the coating layers, large indentation depths, greater than the film thickness were conducted using the CSM mode with a constant strain rate of 0.05 s^{-1}. Each sample underwent nine indents with a maximum load of 750 mN. The estimation of (H) and (E) of the layers and the substrate was performed using the analytical modified (JH) model. The principle of the modified (JH) model will be detailed in the next section.

2.3 Estimation of the Mechanical Properties of Multilayer Coatings by the Modified (JH) Model

The modified (JH) model [7] was used to estimate the (H) and (E) of each monolayer composing the multilayer coatings. In this model, the hardness of the composite structure (H_c) of a multilayer consisting of N layers is evaluated by a mixture law that relates the hardness of i-layer ($H_f^{(i)}$) and the substrate (H_s):

$$H_c = \sum_{i=1}^{N} a_f^{(i)} H_f^{(i)} + a_f^{(s)} H_s \tag{1}$$

where, $a_f^{(i)}$, measure the contribution of the i-layer of the multilayer coating to the composite hardness (H_c). This variable is equal to the ratio of the flow pressure area of the i-layer to the total flow pressure area. For the first layer (i = 1), this ratio can be expressed as a function of indentation depth, h, and the thickness of the first layer, $t_f^{(1)}$, by:

$$\begin{cases} a_f^{(1)} = 1 & \text{if } h < C^{(1)}\, t_f^{(1)} \\ a_f^{(1)} = 1 - \left(1 - \dfrac{C^{(1)}\, t_f^{(1)}}{h}\right)^2 & \text{Otherwise} \end{cases} \tag{2}$$

where $C^{(1)}$, represents a constant depending on the indenter geometry and the indentation behavior of the layer under consideration [7]. In this model, the volume fraction (a_f) of the coating material should be between 0 and 1. Consequently, if h is lower than $C^{(1)}$ $t_f^{(1)}$, the substrate does not influence the composite hardness, and $a_f^{(1)}$ is equal to 1. Whereas, for the jth layer of the coating, the volume fraction (a_f) is expressed as follows:

$$\begin{cases} a_f^{(j)} = 1 - \sum_{i=1}^{j-1} a_f^{(i)} & \text{if } h < \sum_{i=1}^{j} C^{(i)}\, t_f^{(i)} \\ a_f^{(j)} = \left\{1 - \left\{1 - \dfrac{\sum_{i=1}^{j} C^{(i)}\, t_f^{(i)}}{h}\right\}^2\right\} - \left\{1 - \left\{1 - \dfrac{\sum_{i=1}^{j-1} C^{(i)}\, t_f^{(i)}}{h}\right\}^2\right\} & \text{Otherwise} \end{cases} \tag{3}$$

Therefore, once the volume fraction of each layer has been evaluated, the volume fraction corresponding to the substrate material can be obtained:

$$a_f^{(s)} = 1 - \sum_{i=1}^{N} a_f^{(i)} \tag{4}$$

In order to consider the indentation size effect (ISE) in the calculation, the model was then enhanced by incorporating this factor. Hence, the experimental composite hardness data can be described, and the macrohardness can be determined independently of the load for the coating the i-layer ($H_{f0}^{(i)}$) and substrate (H_{S0}), respectively, assuming [7]

$$H_f = H_{f0}^{(i)} + \frac{B_f^{(i)}}{h} \quad and \quad H_s = H_{s0} + \frac{B_s}{h} \tag{5}$$

where, B_f and B_s represents the parameters which characterize the (ISE).

Furthermore, the yield stress (σ_Y) of the constituent layers and the substrate can be evaluated based on their hardness values using the Tabor relationship [13], H/c, where c is a constant equal to 3 for ideal plastic materials.

Additionally, the same model was used to determine the Young's modulus of the i-layer ($E_f^{(i)}$) and the substrate (E_s), with C considered as a variable [14]. Thus, for a multilayer coating, the composite modulus, (E_c), can be expressed by:

$$E_c = \sum_{i=1}^{N} a_f^{(i)} E_f^{(i)} + a_f^{(s)} E_s \tag{6}$$

Therefore, the modified (JH) model will be used to describe the change in the hardness and Young's modulus of the composite structure with indentation depth. Also, it allows the calculation of (H) and (E) of monolayers composing the multilayer, as well as that of the substrate by means of least square fitting analysis.

2.4 Finite Element Modeling

In this study, the commercial finite element software Abaqus Standart 2017 was used to simulate the indentation process. The composite structure was composed of 8806 4-node 2D axisymmetric elements (CAX4R). The dimension of the composite system was set to $25 \times 70\ \mu m^2$. This model used X50CrMoV8-1 as the substrate and featured CrAlN/CrN multilayer coatings with various layers number (2 and 4 layers) and thicknesses (1 and 0.5 μm), as illustrated in Fig. 2. The total thickness of the multilayer coatings remained unchanged at 2 μm. The mesh of the contact area was refined to improve the simulation accuracy (0.1 μm × 0.1 μm). The Berkovich indenter was modelled using 4-node 2D axisymmetric deformable conical body (CAX4R) with a face angle of 70.3°, giving the same nominal indental volume as a Berkovich tip. All the interfaces between materials in the composite structure were assumed to be perfectly bonded. The contact between the indenter tip and the top surface layer are assumed to be frictionless [10]. All nodes at the bottom of the model are constraints. Indentation was applied by displacement control with a linear ramp in time and a final tip displacement equal to the experimental depth measurement at maximum load.

2.5 Optimizations Analysis

2.5.1 Optimization Algorithm

In this work, an optimization algorithm is implemented using the MATLAB software. The connection between ABAQUS and MATLAB is provided using Abaqus2Matlab software tool [15], enabling ABAQUS to be run from MATLAB and the results to be post-processed. The proposed optimization algorithm seeks to identify the optimal elastoplastic properties of the constituent layers in a multilayer coating by reducing the gap between the nanoindentation (P-h) curves and the simulated curves using the MATLAB function LSQNONLIN [12]. This function, based on the trust-region algorithm [16] is efficient in solving bound-constrained nonlinear minimization problems, and to ensure the accuracy of the results by using bound constraints for each materials variables. Hence, the prediction error in indentation depth is evaluated iteratively, summed and minimized to obtain the better coincidence between the experimental (P-h) curve and the corresponding simulated ones. To achieve this goal, a cost function F(X), which

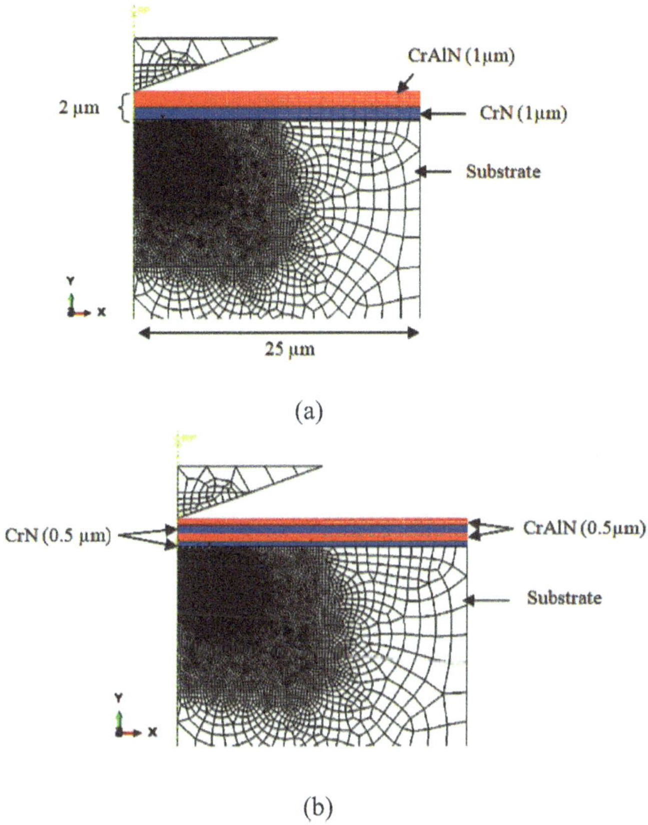

Fig. 2. FEM geometric model: (a) 2 layers model (C1) (b) 4 layers model (C2).

describes the error criteria, minimized until a minimum convergence value of 10^{-2} is reached. The cost function is expressed as a function of the predicted force $P(X)^{pre}$ and the experimental one $P(X)^{exp}$ by:

$$F(x) = \frac{1}{2}\sum_{i=1}^{N}\left[P(x)_i^{pre} - P_i^{exp}\right]^2 \rightarrow \min \qquad LB \leq x \leq UB \tag{7}$$

where X are the elastoplastic properties $X = \left[material\,properties\,(E, \sigma_y, n, ...)\right]^T$, which are starting from initial guess parameters $X_0 = \left[material\,properties\,(E_0, \sigma_{y0}, n_0, ...)\right]^T$. N is the total number of experimental points, and LB and UB are the lower and upper boundaries constraints of each material parameter X.

2.5.2 Optimization Procedure

In this study, the elastoplastic properties (E, σ_Y and n) of layers, forming the multilayer coating, were parameterized in the ABAQUS input files and adjusted at each iteration of the optimization to generate a new input files. In the simulation, a perfectly-plastic material model was used for the Berkovich tip, with Young's modulus, E = 1024 GPa, a Poisson ratio, $\upsilon = 0.07$, and a yield stress, $\sigma_y = 35.7$ GPa [10]. The deformation of the monolayers and the substrate is modelled using a power law strain hardening model. The (σ-ε) relationship is divided into elastic and plastic parts as follows:

$$\sigma = \begin{cases} E\varepsilon \ \ for \ \ \varepsilon \leq \dfrac{\sigma_y}{E} \\ K\varepsilon^n \ for \ \ \varepsilon > \dfrac{\sigma_y}{E} \end{cases} \tag{8}$$

where E, σ_y, n and K are the Young's modulus, the yield stress, the strain hardening coefficient, and the strength coefficient, respectively.

By decomposing the total strain into elastic and plastic ones, the stress can be expressed as a function of plastic strains as follow:

$$\sigma = K\left(\frac{\sigma_y}{E} + \varepsilon_{pl}\right)^n \tag{9}$$

where K can be expressed by:

$$K = E^n \sigma_y^{1-n} \tag{10}$$

Thus, to define the plastic behaviour of the constituent layers in the ABAQUS input file, the optimization algorithm calculates firstly the coefficient K through Eq. (10) and updates through Eq. (9) the stress data value. The principle of the optimization procedure used in this work is shown in Fig. 3

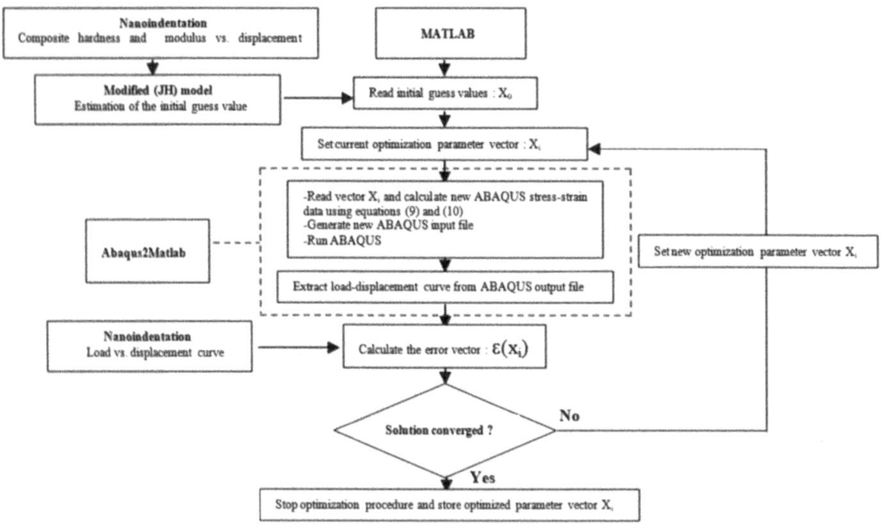

Fig. 3. Principle of the optimization procedure

In this approach, the intial guess parameters $X_0 = \left[elastoplastic\ properties\ \left(E_0, \sigma_{y_0}, n_0\right)\right]$ of CrN and CrAlN monolayers were substituted into the finite element model calculation, and the gap between the simulated and the experimental (P-h) curves was calculated through the cost function F(X), and minimized iteratively by modifying the elastoplastic properties $X = \left[elastoplastic\,properties\,\left(E, \sigma_y, n\right)\right]$ of each monolayer. Thus, when the convergence condition reached the specified tolerance (10^{-2}), the optimization process would be completed, and the optimal elastoplastic material parameters of each layer would be determined. To guaranteed the accuracy of the results and the uniqueness of the solution, the initial guess parameters $X_0 = \left[elastoplastic\,properties\,\left(E_0, \sigma_{y_0}, n_0\right)\right]$ of the two monolayers are taken close to the target values [12, 17] by using the estimated mechanical properties provided by the modified (JH) model (E and σ_Y). Only (n) is chosen arbitrary. Thus, two optimization processes will be performed to determine the optimal elastoplastic properties (E, σ_y and n) of CrN and CrAlN monolayers for each multilayer coating systems: C1 and C2. In the simulation, the Poisson ratio υ of CrN and CrAlN is assumed to be equal and constant of magnitude of 0.22 [9]. To reduce the number of unknown parameters in the model, the elastoplastic properties of the substrate are fixed in the simulation, assuming that its (E) and (σ_y) are equal to the estimated values measured by the modified (JH) model and the Tabor assumption, respectively. For the strain hardening coefficient (n) and the Poisson ratio (υ) of the substrate are assumed to be equal to 0.1 and 0.3 [17], respectively.

3 Results and Discussion

3.1 Estimation of the Initial Guess Parameters of the Optimization Procedure Using the Modified (JH) Model

Figure 4 and 5 illustrate the changes in the experimental (Hc) and (Ec) of the two CrN/CrAlN multilayer coating systems: C1 and C2, with the indentation depth. The modified (JH) model is used to describe the experimental composite hardness, assuming that the 2 layers undergo plastic deformation by taking C equal to 0.1746 [7]. Thus, the hardness of the constituent layers of multilayer and the substrate is evaluated by means of least squares fitting analysis of the experimental data (see Fig. 4(a) and Fig. 5 (a)) through the MATLAB function fminsearch. The same approach is used to evaluate their (E) (see Fig. 4(b) and Fig. 5 (b)). Furthermore, the yield stress (σ_y) of each monolayer and substrate is estimated using the estimated value of hardness through the Tabor assumption. Table 1 summarizes the values of the various mechanical properties obtained by the modified (JH) model. Results show that the modified (JH) model provides a good description of the variation in (Hc) and (Ec) of CrN/CrAlN multilayer coatings with indentation depth.

Table 1. Modified (JH) model results

	C1 condition			C2 condition		
	CrAlN	CrN	substrate	CrAlN	CrN	substrate
H (GPa)	9.68	5.92	7.72	7.63	4.86	8.32
E (GPa)	221.41	121.12	218.51	127.2	92.1	233.21
σy (GPa)	3.22	1.98	2.57	2.54	1.65	2.77

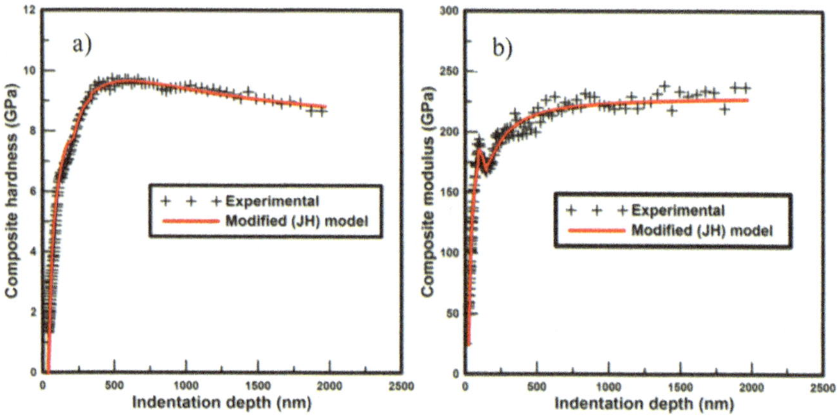

Fig. 4. Variation of (a) (Hc) and, (b) (Ec) with the indentation depth for C1 condition

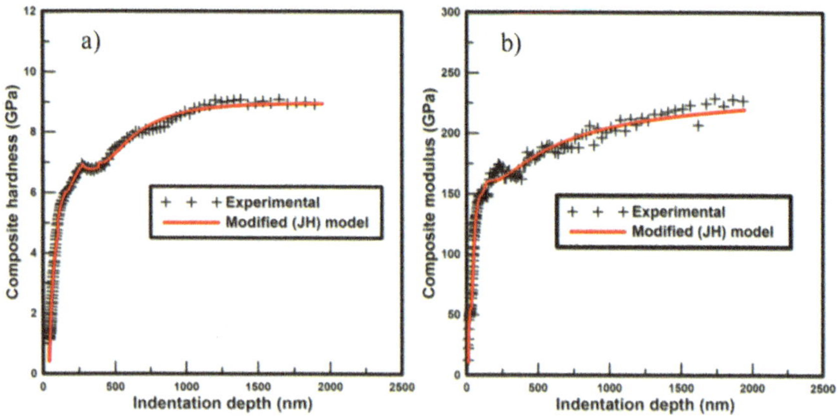

Fig. 5. Variation of (a) (Hc) and, (b) (Ec) with the indentation depth for C2 condition

3.1.1 Optimization Results

In this study, the proposed optimization procedure is validated on CrN/CrAlN multilayer coating systems: C1 and C2, and as a result the optimal elastoplastic properties (E, σ_y

and n) of CrN and CrAlN layers is extracted. The optimization results are summarized in Tables 2 and 3. The initial value parameters of the optimization analysis are defined based on the results of the modified (JH) model (Table 1). Hence, the estimated value of (E) and (σ_y) of the two layers (CrN and CrAlN) (Table 1) are taken as initial values in the optimization analysis (Tables 2 and 3). For (n), the correspondent initial value was arbitrary taken. Furthermore, the bound constrains of each variable are defined based on previous works [9]. Figure 6 represents the comparison between the experimental (P-h) curves and the corresponding simulated ones, using the optimal solution of elastoplastic material parameters of CrN and CrAlN layers, respectively. The comparison results demonstrate a good agreement between the experimental and the simulated results, indicating the efficiency of the proposed methodology. It's clear that the gap between the optimized values of (E) and (σ_y) and their initial values (see Tables 2 and 3), was generally low (20%), indicating the accuracy of the proposed approach.s

Table 2. Optimization result for C1 condition

	Parameters	Initial values	Boundaries	Optimized values
CrAlN layer	E	221.41 [GPa]	50 < E [GPa] < 240	192.17 [GPa]
	σy	3.22 [GPa]	1 < σy [GPa] < 14	3.15 [GPa]
	n	0.2	0 < n < 0.5	0.33
CrN layer	E	121.12 [GPa]	50 < E [GPa] < 240	128.32 [GPa]
	σy	1.98 [GPa]	1 < σy [GPa] < 14	2.42 [GPa]
	n	0.2	0 < n < 0.5	0.29

Table 3. Optimization result for C2 condition

	Parameters	Initial values	Boundaries	Optimized values
CrAlN layer	E	127.2 [GPa]	50 < E [GPa] < 240	131.1 [GPa]
	σy	2.54 [GPa]	1 < σy [GPa] < 14	2.97 [GPa]
	n	0.2	0 < n < 0.5	0.23
CrN layer	E	92.1 [GPa]	50 < E [GPa] < 240	98.4 [GPa]
	σy	1.7 [GPa]	1 < σy [GPa] < 14	2.21 [GPa]
	n	0.2	0 < n < 0.5	0.21

The results show a decrease in the material properties (H, σ_Y and E) of CrN and CrAlN layers when the layer thickness decreases. These findings suggest that decreasing the coating layer thickness lead to an increase in the plastic deformation within the coatings, which reduces the stress concentration in this area and improves the adhesion properties of CrN/CrAlN multilayer coatings. Indeed, the CrN/CrAlN multilayer coating system with layer thicknesses of 0.5 μm (C2 system) generated the lowest stress and the highest plastic deformation, improving the adhesion properties of the multilayer structure.

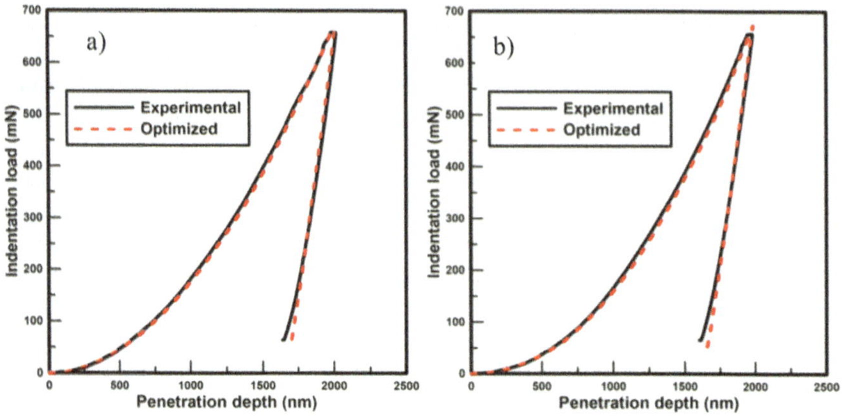

Fig. 6. Comparison of predicted and experimental (P-h) curves of CrN and CrAlN layers: (a) C1 condition; (b) C2 condition

4 Conclusion

In this study, CrN/CrAlN multilayer coatings with various layer thicknesses were prepared using PVD technology. The assessment of the elastoplastic properties of the constituent layers of multilayer coating was performed through nanoindentation using a trust-region algorithm combined with numerical indentation models. The optimization results show that the material properties (H, σ_Y and E) of the two layers decrease when decreasing the layer thickness. The CrN/CrAlN multilayer coating system with monolayer thicknesses of 0.5 μm produced the lowest stress and the highest plastic deformation during the indentation process, leading to the best adhesion properties compared with the other systems.

References

1. Yuan, Z., Han, Y., Zang, S., Chen, J., He, G., Chai, Y.: Analysis of the mechanical properties of TiN/Ti multilayer coatings using indentation under a broad load range. Ceram. Int. **47**, 10796–10808 (2021)
2. Kovalev, A., et al.: Damage accumulation phenomena in multilayer (TiAlCrSiY)N/(TiAlCr)N, monolayer (TiAlCrSiY)N coatings and silicon upon deformation by cyclic nanoindentation. Nanomaterials **12**, 1312 (2022)
3. Bauyrzhan, R., et al.: Effect of bilayer thickness and bias potential on the structure and properties of (TiZr/Nb)N multilayer coatings as a result of acr-pvd deposition. Materials **15**, 7696 (2022)
4. Kataria, S., Goyal, S., Dash, S., Sandhya, R., Mathew, M.D., Tyagi, A.K.: Evaluation of nano-mechanical properties of hard coatings on a soft substrate. Thin Solid Films **522**, 297–303 (2012)
5. Bull, S.J.: Size effects in the mechanical response of nanoscale multilayer coatings on glass. Thin Solid Films **571**, 290–295 (2014)
6. Puchi-Cabrera, E.S., Staia, M.H., Iost, A.: Modeling the composite hardness of multilayer coated systems. Thin Solid Films **578**, 53–62 (2015)

7. Rahmoun, K., Iost, A., Keryvin, V.: A multilayer model for describing hardness variations of aged porous silicon low-dielectric-constant. Thin Solid Films **518**, 213–221 (2009)
8. Li, Z., et al.: Research on determining elastic-plastic constitutive parameters of materials from load depth curves based on nanoindentation technology. Micromachines **14**, 1051 (2023)
9. Ammar, Y.B., Aouadi, K., Nouveau, C., Besnard, A., Montagne, A.: Identification of the elastic-plastic properties of CrN coating on elastic-plastic substrate by nanoindentation using finite element method-reverse algorithm. Thin Solid Films **756**, 139356 (2022)
10. Xing, X., Wang, Y., Xiao, G.: Identifying the elastoplastic properties of ductile film on hard substrate. Vacuum **189**, 110252 (2021)
11. Karimpour, M., Balint, D.S., Rzepiejewska-Malysca, K.A., Szerling, A., Michler, J., Lin, J.: An inverse method for extracting the mechanical properties of the constituent materials of a multilayer from nanoindentation data. Comput. Mater. Sci. **68**, 384–390 (2013)
12. Kang, J., Becker, A., Sun, W.: Determining elastic plastic properties from indentation data obtained from finite element simulations and experimental results. Int. J. Mech. Sci. **62**, 34–46 (2012)
13. Tabor D: The hardness of metals. Oxford classic texts in the physical sciences, Oxford: Oxford University Press (1921)
14. Abib, H., Iost, A., Montagne, A., Rahmoun, K.: Investigations on the mechanical properties of the elementary thin films composing a CuIn1-xGaxSe2 solar cell using the nanoindentation technique. Thin Solid films **756**, 139356 (2009)
15. Papazafeiropoulos, G., Muñiz-Calvente, M., Martínez-Pañeda, E.: Abaqus2Matlab: a suitable tool for finite element post-processing. Adv. Eng. Softw. **105**, 9–16 (2017)
16. Coleman T, Branch MA, Grace A: Optimization toolbox for use with MATLAB: user's guide, Version 2. Math Works, Incorporated (1999)
17. Iracheta, O., Bennett, C., Sun, W.: Characterization of material property variation across an inertia friction welded CrMoV steel component using the inverse analysis of nanoindentation data. Int. J. Mech. Sci. **107**, 253–263 (2016)

On the Wear Response of Ni-B-TiO_2 Composite Coating Against Sliding of Barite Counterparts

Mouna Kallel(✉) and Khaled Elleuch

Laboratory of Materials Engineering and Environment, Material Engineering Department, Sfax National School of Engineering, Sfax, Tunisia
mouna.kallel@enis.tn

Abstract. Wear poses a significant challenge in various industries, particularly in mineral industries that rely largely on crusher machines. Within this sector, crusher hammers are among the components most susceptible to severe wear due to repetitive impact and abrasion during crushing operations. In this study, a Ni-B-TiO_2 composite coating is electrodeposited onto heat-treated P20 mold steel to enhance the wear resistance of these hammers. To better replicate the elements of the tribosystem at the laboratory scale, the investigated coating undergoes repetitive sliding with both hemisphere- and cone-shaped counterparts, made from barite rocks, using a reciprocating sliding wear test. The findings reveal that, despite being much softer than the coating surface, barite counterparts effectively induce wear on the Ni-B-TiO_2 coating and experience material loss during the wear test. The influence of counterpart geometry on the wear behavior of the coating surface is notable, especially for longer sliding durations. Compared with the hemisphere-shaped counterpart, the cone-shaped counterpart results in greater wear on the coating surface, which progressively increases with sliding time. In contrast, the hemisphere-shaped counterpart reduces by half the wear of coating while allowing for nearly steady-state wear damage as sliding time increases.

Keywords: Hard surface · soft counterparts · wear behaviors

1 Introduction

Numerous wear issues in various machine parts that move relative to each other have been discovered in different industrial machinery [1–5]. For instance, hammer crushers utilized within the mineral industry have encountered significant wear damage in their hammer components due to the continuous abrasion and impact of barite rocks during the crushing process. Barite rocks consist primarily of barite sulfate ($BaSO_4$) with impurities such as Al, Ca, Sr, and Si. Barite is predominantly utilized in the oil and gas sectors. Furthermore, apart from the silica content within the composition of barite rock, a notable characteristic of these rocks is their sharp points, resulting in earlier wear damage to the hammer [6]. Several attempts have been made to extend the hammer's service life. First, heat-treated AISI P20 (P20 HT) was used as the hammer material instead of High Chromium White Cast Iron (HCWCI) since it is more resistant to wear [7]. As

M. Chouchane et al. (Eds.): CMSM 2023, LNME, pp. 350–359, 2024.
https://doi.org/10.1007/978-3-031-65007-9_37

coatings and films are designed to improve the wear of bulk materials in some cases [8], Kallel et co-workers [9, 10] have effectively deposited the Ni-B-TiO_2 composite coating on P20 HT by both conventional and novel methods to further enhance its durability against wear. It is interesting to note that Ni-B-TiO_2 composite coating has recently been developed by electrodeposition and electroless processes [11–13]. The incorporation of TiO_2 nanoparticles into the Ni-B matrix was aimed at improving its corrosion resistance and wear protection.

On the other hand, the tribological behavior of the Ni-B-TiO_2 coating was thoroughly assessed under abrasion and/or impact loadings [9, 10]. The results underlined a fascinating wear resistance of this coating. However, these previous works dealing with the wear resistance of Ni-B-TiO_2 coating have used either alumina balls or Rockwell diamond indenters as antagonist materials for all the performed tests. Therefore, in this study, counterparts made of barite rocks were exclusively employed to assess the wear behavior of Ni-B-TiO_2 coating, aiming to better simulate the wear issues of hammers that occurred during barite rock crushing. Given that barite rocks characterized by sharp and round edges, cone and hemisphere barite counterparts are designed. Contributions from this research are accomplished through a direct comparison of the ability of Ni-B-TiO_2 composite coating to withstand repetitive sliding of both hemisphere- and cone-shaped counterparts over time. In this regard, the impact of sharpness on the wear behavior of industrial components is discussed.

2 Materials and Methods

The Ni-B-TiO_2 composite coating was electrodeposited on the heat-treated AISI P20 (Böhler M201) substrate [7]. The Ni-B bath solution is a modified Watts bath, into which TiO_2 nanoparticles were incorporated at a concentration of 5 g/L. Table 1 presents the chemical composition of the electroplating bath.

Table 1. Chemical composition of the electroplating bath.

Bath composition	Concentration (g/L)
Nickel sulfate, $NiSO_4 .6H_2O$ (98%)	280
Nickel chloride, $NiCl_2.6H_2O$	40
Boric acid, H_3BO_3 (99,8%)	38
Trimethylamine borane, (TMAB)	3
Anionic Surfactant	0.2
TiO_2 nanoparticles (ø 21 nm)	5

The coatings were electrodeposited for 1 h at a current density of 1 A/dm^3 and a temperature of 37 °C. The pH of the plating solution was set at 3.5. The used substrate was cut into square specimens with dimensions of 40 mm × 40 mm × 4 mm. Thereafter, the substrates were tempered after being quenched in oil. Prior to the electrodeposition process, the specimens were ground with SiC papers until they achieved a mirror-like shine, after which they were cleaned with acetone and pickled in a sulfuric acid solution. The electrodeposition process was comprehensively described in a previous publication [10]. The thickness of the coating measures approximately 16 ± 2 μm [10].

The morphology of the coating was analyzed using scanning electron microscopy (SEM) coupled with an Energy Dispersive X-ray Spectroscopy (EDS) device. Vickers micro-indentation was carried out to estimate the microhardness and elastic modulus of both the Ni-B-TiO_2 composite coating and the barite counterpart under a load of 0.5 N. To ensure the reliability of the measurements, the test was repeated five times.

The tribological behavior of the studied coating was assessed using a reciprocating sliding test. Despite being a two-body abrasive test, this method allows for a simpler determination of the effect of counterpart geometry on the wear resistance of the coating. Hemisphere- and cone-shaped counterparts, made from barite rock, were specifically designed to slide against the Ni-B-TiO_2 coating. The counterparts were mounted to the machine pin holder in the manner that the hemisphere end and the cone tip were placed in contact with the coating surface. Considering the real operating conditions of hammers, the test parameters were selected, particularly focusing on the contact pressure, while also considering potential adjustments to the wear testing to ensure safe functioning. The test parameters are detailed in Table 2.

Table 2. Operating parameters of tribological sliding wear tests.

Parameters	Values
Load (N)	3
Contact pressure of HSC (MPa)	**0.022**
Contact pressure of CSC (MPa)	**0.038**
Sliding speed (mm/s)	50
Track length (mm)	10
sliding duration (min)	5; 10; 30; 60 and 120

HSC: Hemisphere-shaped counterpart.
CSC: Cone-shaped counterpart.

The reciprocating sliding test was paused at regular intervals to allow for the acquisition of images of the coated surface using an optical microscope and the measurement of wear depth in various areas of the track using a 2D profilometer.

An electronic laboratory scale balance with a precision of 0.001 mg was used to measure the weight of both counterparts before and after the sliding wear test to estimate weight loss.

3 Results and Discussions

The surface morphology of the Ni-B-TiO_2 composite coating appears discontinuous and displays numerous pores (Fig. 1). The porosity of the coating is attributed to the incorporation of TiO_2 nanoparticles into the Ni-B matrix, which acts as a barrier, preventing the formation of further Ni grains [9]. Table 3 presents the EDS analysis results for the elaborated coating, which confirms the incorporation of TiO_2 nanoparticles into the Ni-B matrix.

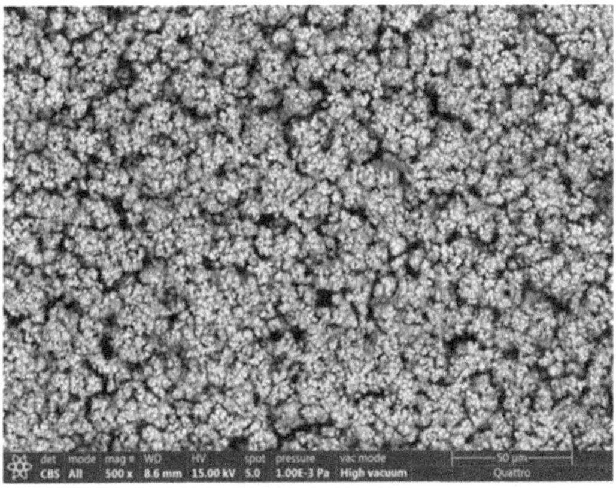

Fig. 1. SEM morphology of Ni-B-TiO_2 composite coating.

Table 3. EDS analysis of Ni-B-TiO_2 composite coating.

Chemical element	B	O	Ti	Ni
Wt%	2.64	14.18	3.58	79.60

Table 4 reveals various mechanical properties and Ni grain size of the Ni-B-TiO_2 coating.

Table 4. Mechanical properties of Ni-B-TiO_2 composite coating.

	Microhardness (Hv)	Elastic modulus (GPa)	Roughness Ra (μm)	Grain size (nm)
Ni-B-TiO_2 coating	748 ± 59	203 ± 12	0.2	22
Barite counterpart	161 ± 3	63 ± 1	–	–

Figure 2 presents the hemisphere- and cone-shaped counterparts before and after the reciprocating sliding test. These counterparts are made from barite rock. Noting that the chemical composition of barite figures out the presence of impurities, including silica, known for its abrasive nature [6]. Both counterparts were meticulously designed in the laboratory using a polishing machine. They share a common circular cylinder shape, with one featuring a hemispherical end to form the hemisphere-shaped counterpart, and the other presenting a cone tip to produce cone-shaped counterpart. The radius of the cylindrical part measures 10 mm for both counterparts, while the cone possesses an opening angle of 60 degrees. Upon close examination of the counterparts before and after sliding, it is noticeable that the hemisphere one seems to be comparable in both cases, except its end flattens a bit as a result of the sliding effect (Fig. 2a). Nevertheless, the vertex of the cone-shaped counterpart is shattered, resulting in a less acute tip (Fig. 2b).

Furthermore, the weight measurements of both counterparts depict a weight loss of 16.4 mg and 24 mg, which correspond to the hemisphere counterpart and cone one, respectively. This result confirms that the cone-shaped counterpart experiences greater wear compared to the hemisphere-shaped one. This outcome is predictable and logical, given the distinctive tip characteristics of each specimen.

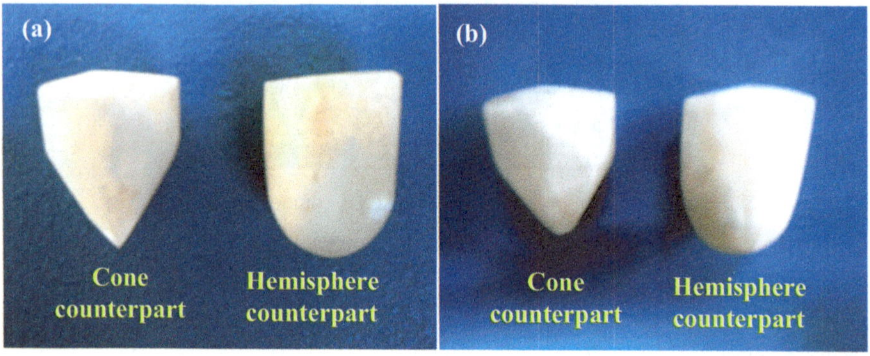

Fig. 2. Hemisphere- and cone-shaped counterpart (a) before and (b) after reciprocating sliding test.

Figure 3 displays the wear evolution on the coating surface by the use of hemisphere and cone counterparts over time test. Despite being less hard than Ni-B-TiO_2 coating, both barite counterparts contribute to wear damage on the coating surface. The surface damaged by hemisphere counterparts shows significant wear up to a sliding time of 60 min. Thereafter, the surface appears to undergo a polishing effect, with only minor damage occurring. However, the cone-shaped counterpart produces a more pronounced level of damage compared to its hemisphere-shaped counterpart. The wear track formed by the cone counterpart progressively widens and deepens over the duration of the test. Notably, the surface exhibits extensive damage, characterized by deeper grooves. This can be attributed to the sharpness of the counterpart as well as the presence of residual barite particles acting as wear debris, thereby accelerating the rate of wear. These findings align with the observations of both counterparts after the test, as discussed in Fig. 2.

In order to achieve a comprehensive understanding of the evolution of coating wear surfaces subjected to motion from hemisphere- and cone-shaped counterparts, the wear depth of the tracks was assessed across the width at various positions along the track length. The wear depth in each wear track for every sliding time was averaged and subsequently presented in a graph (Fig. 4). These wear track depth measurements align with the optical images of worn surfaces over sliding time. The wear depth resulting from the hemisphere-shaped counterpart was initially higher than that from the cone-shaped

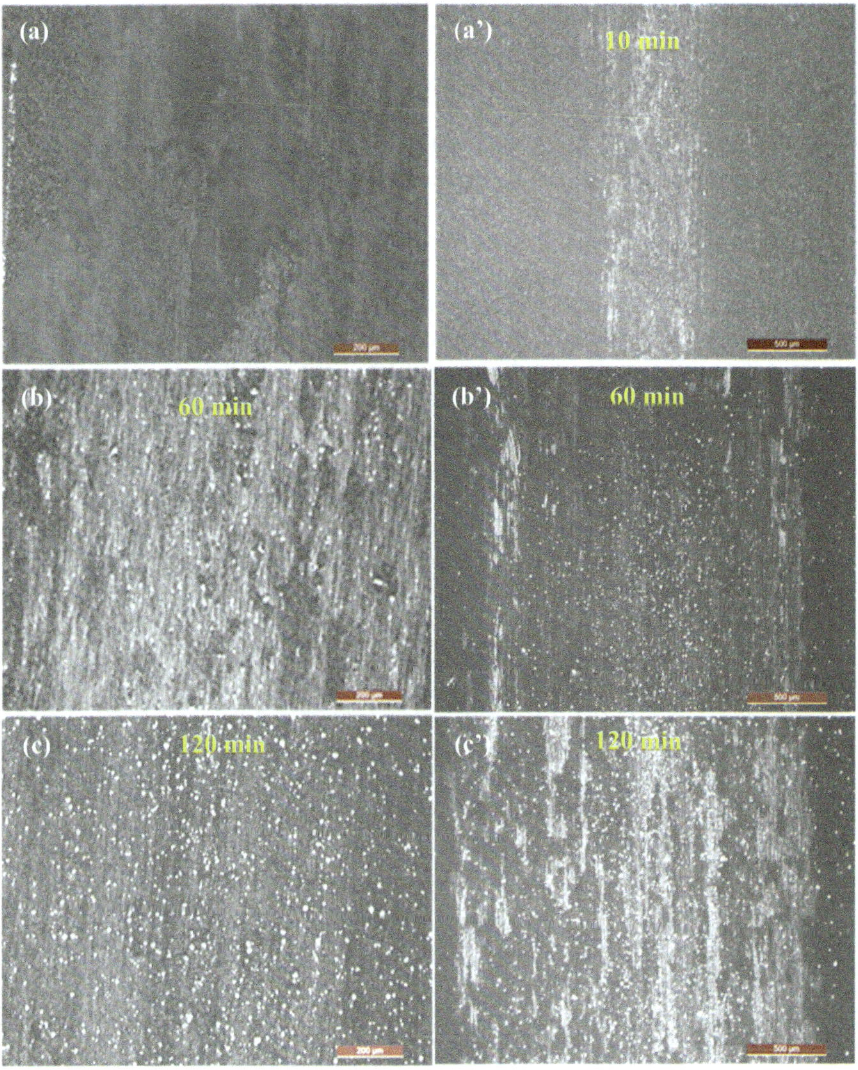

Fig. 3. Worn surface coating by (a, b and c) hemisphere- and (a', b' and c') cone-shaped counterparts after reciprocating sliding test in different sliding duration.

one. For instance, after a sliding duration of 5 min, the maximum wear depth induced by the hemisphere counterpart measured around 5.7 μm, while that caused by the cone-shaped counterpart was 4.7 μm. After a sliding time of 10 min, the wear behavior of counterparts is entirely reversed. The coating surface demonstrated greater resistance to wear from the sliding of the hemisphere counterpart compared to the cone counterpart. Furthermore, the continuous sliding of the cone-shaped counterpart led to a progressive increase in wear with the duration of sliding.

By the end of the test, the wear depth and maximum track width reached approximately 16.6 μm and 2218 μm, respectively, for the cone-shaped counterpart, whereas the wear depth resulting from the hemisphere counterpart was approximately 8.8 μm. Over time, the wear depth generated by the hemisphere counterpart gradually increased after repetitive sliding from 5 to 30 min. Subsequently, the wear depth appeared to stabilize, remaining relatively constant for the remainder of the test duration, with an average value of around 8.5 μm.

Initially, it may seem unexpected that the wear depth when using a cone-shaped counterpart is less significant than when using a hemisphere-shaped one. This phenomenon could be attributed to the heterogeneous structure of barite rock materials, including the designed counterparts. Additionally, considering that barite rocks contain silica [6], it seems that the contact surface of the hemisphere-shaped counterpart is more enriched with silica content than the other counterpart. This enrichment likely leads to aggressive scratching of the coating surface, consequently resulting in greater wear. This is because silica is an abrasive material, and therefore, it primarily controls the wear behavior at the start of surface contact [14]. The evolution of wear damage is predominantly governed by stress concentration. In this context, the sharpness of the cone tip becomes pivotal since it significantly raises stress concentration. Consequently, this increase in stress concentration has the potential to show the contribution of counterpart geometry in coating wear behavior. Thus, it is clear that the shape of the counterpart is a significant factor that influences its abrasive efficiency, as highlighted in numerous previous research studies [15–17].

Furthermore, in this study, it was discovered that the sharpness of counterparts is a critical factor, that limits the lifespan of the studied coating. Over 2 h, the cone-shaped indent exceeds the depth of the coating, unlike the hemisphere-shaped one, which exhibits a wear depth ranging about half the coating thickness.

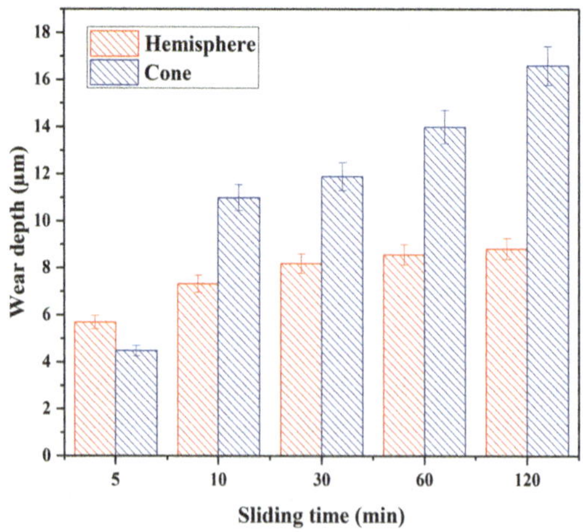

Fig. 4. Wear depth evolution of Ni-B-TiO_2 composite coating by hemisphere- and cone-shaped counterparts with sliding duration.

Furthermore, as the counterpart was broken into small particles, especially in the case of cone-shaped one, these barite particles remain entrapped between the counterpart and the coating surface, acting as third-body abrasives and thereby further wear damage occurs [18]. This phenomenon contributes to additional wear, notably in the case of cone-shaped counterpart.

4 Conclusion

Since the wear phenomenon still preoccupies a lot of industries, it is imperative that scientists deal with some wear industrial problems. In this study, the wear behavior of Ni-B-TiO_2 coating under repetitive sliding of hemisphere- and cone-shaped barite counterparts was investigated in order to simulate an industrial issue being encountered by a Tunisian company. The intriguing conclusions drawn from this investigation are outlined below.

- Despite the low hardness of the barite counterparts and the elevated contact pressure, the reciprocating sliding test was successfully conducted.
- After the wear test, it was observed that the barite counterparts experienced weight loss, notably the cone-shaped counterpart. The broken particles from the counterparts could act as third-body abrasives, leading to additional wear damage on the coating surface.
- Irrespective of the counterpart geometry, the Ni-B-TiO_2 coating failed to withstand the repetitive sliding of barite, despite its hardness being five times greater than that of barite.
- The wear behavior of the coating at the beginning of the test was largely controlled by the silica content in the barite rocks.
- As the sliding time increased, the geometry of the counterparts emerged as the primary influencer on the tribological behavior of the Ni-B-TiO_2 coating.
- The repetitive sliding of the cone-shaped counterpart aggressively scratched the surface, resulting in the greatest wear depth. This wear depth progressively increased with test duration, eventually slightly exceeding the coating thickness after a sliding duration of 2 h. Meanwhile, the hemisphere-shaped counterpart caused less damage to the coating surface, resulting in a lower wear depth that tended to stabilize after 30 min of testing.

References

1. Bahri, A., Guermazi, N., Elleuch, K., Ürgen, M.: On the erosive wear of 304 L stainless steel caused by olive seed particles impact: modeling and experiments. Tribol. Int. **102**, 608–619 (2016). https://doi.org/10.1016/j.triboint.2016.06.020
2. Ben Saada, F., Elleuch, K.: Damage of stainless steel components by olive paste. Tribol. Trans. **59**, 856–864 (2016). https://doi.org/10.1080/10402004.2015.1115569
3. Khalladi, A., Elleuch, K.: Tribological behavior of wheel-rail contact under different contaminants using pin-on-disk methodology. J. Tribol. **139**, 011102 (2017). https://doi.org/10.1115/1.4033051

4. Elleuch, K., Mezlini, S., Fouvry, S., Kapsa, P.: Friction damage of aluminium alloys. Ind. Lubr. Tribol. **55**, 279–286 (2003). https://doi.org/10.1108/00368790310496428
5. Zeidi, A., Ben Saada, F., Elleuch, K., Atapek, H.: On the failure of punching process. Eng. Fail. Anal. **120**, 105035 (2021). https://doi.org/10.1016/j.engfailanal.2020.105035
6. Kallel, M., Zouch, F., Antar, Z., Bahri, A., Elleuch, K.: Hammer premature wear in mineral crushing process. Tribol. Int. **115**, 493–505 (2017). https://doi.org/10.1016/j.triboint.2017.06.025
7. Kallel, M., Elleuch, K.: Wear behavior of high chromium cast iron and heat treated AISI P20 steel under round and sharp counterbodies. In: Walha, L., et al. (eds.) Design and Modeling of Mechanical Systems – V, pp. 704–714. Springer International Publishing, Cham (2023). https://doi.org/10.1007/978-3-031-14615-2_79
8. Findik, F.: Latest progress on tribological properties of industrial materials. Mater. Des. **57**, 218–244 (2014). https://doi.org/10.1016/j.matdes.2013.12.028
9. Kallel, M., Antar, Z., Masseoud, M., Vesco, S., Barletta, M., Elleuch, K.: Comparative investigation of scratch resistance and tribological performance of Ni–B–TiO2 composite coatings prepared by conventional and novel processing methods. Ceram. Int. **47**, 14438–14454 (2021). https://doi.org/10.1016/j.ceramint.2021.02.023
10. Kallel, M., Bahri, A., Elleuch, K.: Investigation on the wear resistance of Ni-B-TiO2 composite coatings for dry crushing application. In: Pakseresht, A., Sharifahmadian, O. (eds.) Advances in Chemical and Materials Engineering, pp. 218–244. IGI Global (2022). https://doi.org/10.4018/978-1-7998-9683-8.ch010
11. Niksefat, V., Ghorbani, M.: Mechanical and electrochemical properties of ultrasonic-assisted electroless deposition of Ni–B–TiO2 composite coatings. J. Alloy. Compd. **633**, 127–136 (2015). https://doi.org/10.1016/j.jallcom.2015.01.250
12. Wang, Y., Wang, S.-J., Shu, X., Gao, W., Lu, W., Yan, B.: Preparation and property of sol-enhanced Ni–B–TiO 2 nano-composite coatings. J. Alloy. Compd. **617**, 472–478 (2014). https://doi.org/10.1016/j.jallcom.2014.08.060
13. Wang, S.-J., et al.: Preparation and property of duplex Ni – B – TiO_2 / Ni nano-composite coatings. Int. J. Mod. Phys. B **29**, 1540022 (2015). https://doi.org/10.1142/S0217979215400226
14. Ara, M.H., Ferreira, R.O., Pereira Agudelo, J.I., Pintaúde, G., Azevedo, C.R.D.F.: Effect of silica size on the relative wear of microstructures containing solidification carbides and tempered martensite using a dry sand rubber wheel test. Tribol. Trans. **67**, 70–86 (2024). https://doi.org/10.1080/10402004.2023.2291109
15. Jourani, A., Hagège, B., Bouvier, S., Bigerelle, M., Zahouani, H.: Influence of abrasive grain geometry on friction coefficient and wear rate in belt finishing. Tribol. Int. **59**, 30–37 (2013). https://doi.org/10.1016/j.triboint.2012.07.001
16. Szala, M., Szafran, M., Macek, W., Marchenko, S., Hejwowski, T.: Abrasion resistance of S235, S355, C45, AISI 304 and Hardox 500 steels with usage of garnet, corundum and carborundum abrasives. Adv. Sci. Technol. Res. J. **13**, 151–161 (2019). https://doi.org/10.12913/22998624/113244

17. Mezlini, S., Kapsa, P., Abry, J.C., Henon, C., Guillemenet, J.: Effect of indenter geometry and relationship between abrasive wear and hardness in early stage of repetitive sliding. Wear **260**, 412–421 (2006). https://doi.org/10.1016/j.wear.2005.02.106
18. Zhang, G., Zhang, Z., Xuan, J., Chen, B., Jiang, D., Song, X.: The elevated temperature oxidation and wear behavior of Fe20Co20Ni20Cr8Mo12B10Si10 high-entropy alloy coating by laser cladding. J. Market. Res. **29**, 4216–4231 (2024). https://doi.org/10.1016/j.jmrt.2024.02.135

Preparation and Characterizations of NiCrFeSiAlBC Coatings via a HVOF Process

Fida Harabi[1], Basma Ben Difallah[1](✉), Mohamed Kharrat[1], Clizia Aversa[2], Massimiliano Barletta[2], Yasin Mohamed El Sayed[3], Marco Fortunato[3], Stefano Lionetti[3], and Roberta Valle[3]

[1] Laboratory of Electromechanical Systems, National School of Engineers of Sfax, University of Sfax, Sfax, Tunisia
fida.harabi@enis.tn, basma.bendhaifallah@ipeis.usf.tn, mohamed.kharrat@ipeis.rnu.tn

[2] Mechanical Department, University of Roma Tre, Roma, Italy
{clizia.aversa,massimiliano.barletta}@uniroma3.it

[3] RINA Consulting - CSM S.p.A, RINA Company, Roma, Italy
{yasin.mohamedelsayed,marco.fortunato,stefano.lionetti, roberta.valle}@rina.org

Abstract. Thermal spray composite coatings have been widely used in a diversity of industrial applications, particularly, the high-velocity oxy-fuel (HVOF) thermal spraying nickel alloy coatings have been used to enhance the wear resistance of various engineering components especially for the machining tools. In the present work, NiCrFeSiAlBC coating was deposited on C45 substrate with the HVOF thermal spray process and the microstructure, mechanical and tribological properties of the deposit was investigated. The microstructure of the HVOF-sprayed NiCrFeSiAlBC coating was characterized by scanning electron microscope (SEM). The micro hardness was determined in the substrate and in the coating by using Vicker's micro-indenter, also, tribological tests were conducted via a reciprocating ball on flat tribometer at 10 N normal load, the ball being from Si_3N_4 material. All the results were discussed in relation to the hardness and wear resistance. It was found that the NiCrFeSiAlBC coating exhibited a very good tribological performance when compared to the uncoated substrate (C45).

Keywords: C45 substrate · NiCrFeSiAlBC coating · HVOF process · mechanical and tribological properties · microstructure

1 Introduction

High-velocity oxygen fuel (HVOF) is one of the thermal spray processes in which the initial powders constituting the coating material is heated at high temperatures through high-velocity spraying. High-velocity oxygen fuel process is widely used in the industrial applications, power plants (Sadeghi et al. 2019), aerospace and machine tools (Singh et al. 2023) because of the functional advantages offered by the deposits such as high

M. Chouchane et al. (Eds.): CMSM 2023, LNME, pp. 360–367, 2024.
https://doi.org/10.1007/978-3-031-65007-9_38

hardness, good adhesion to the substrate and excellent wear performances (Yu et al. 2022). The nickel-alloy thermal spraying coating has shown to be one of the best coatings for depositing high structural, mechanical and tribological properties, because of their enhanced friction and wear properties and resistance to corrosion (Saaedi et al. 2010). In our NiCrFeSiAlBC coating, the Chromium and Boron compositions improve high strength and hardness. Moreover, Silicon has a propriety to increase self-fluxing properties. The Carbon enhances the hardness level of the coatings as well as their wear resistance (Miguel et al. 2003), (Gómez-del Río et al. 2008), (Eskandari et al. 2023). The aim of this work is to prepare NiCrFeSiAlBC coating using a HVOF technique and to study their characteristics comparing to C45 substrate. The microstructure, mechanical and tribological properties of NiCrFeSiAlBC coatings were investigated. Since hardness is an important parameter that affects the tribological performances of sliding pairs in contact, this property was studied and the results were correlated to the tribological behavior of uncoated and coated C45 samples.

2 Experimental

2.1 Coating and Substrate Materials

In this study, a NiCrFeSiAlBC metalic powder is used as feedstock. The powder is agglomerated and sintered spheroids. It is a self-fluxing alloy consisting of nick-el and cobalt based products with a proven track record for tribological applica-tions. NiCrFeSiAlBC has a fusing temperature more than 1450 °C. The coating was deposited on the medium carbon steel C45 (AISI 1045 steel) with 25 × 25 × 20 mm^3 dimensions using an industrial thermal spray process "PRAXAIR TAFA HVOF-JP5000" with specified parameters presented in Table1. The surface of 25 × 25 mm^2 was sandblasted to be coated in proper conditions. The average thickness of the coating is equal to 155 μm. Figure 1 shows the steel substrate. The substrate was used without any heat treatment.

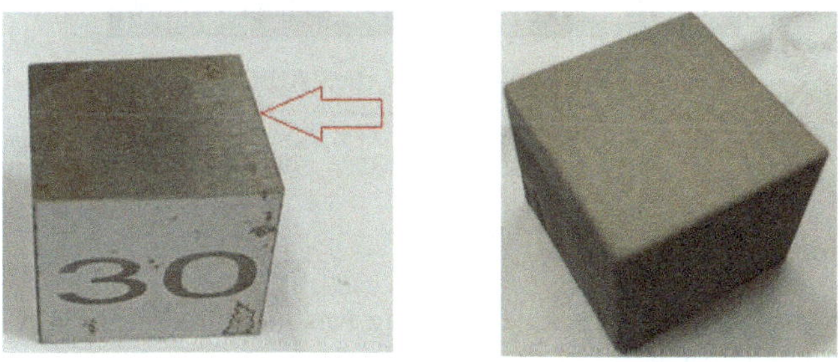

(a) (b)

Fig. 1. C45 substrate (a) before coating and (b) after coating deposition.

Table 1. HVOF thermal spray parametrs.

Parameters	Values
Flow rate O_2 (scfh)	1950
Flow rate K (gph)	5.75
Distance from substrate (mm)	380
Combustion pressure (psi)	96

2.2 Microstructure Properties

The powder and coating microstructures were characterized using a scanning elec-tron microscope 'SEM' "COXEM-CX200 Plus". Figure 2 shows the morphology for the used NiCrFeSiAlBC powder revealing spherical shape particles with a huge difference in size distribution ranging from 3 µm to 45 µm. The coated samples were cut using a Laser method and then embedded in a resin. Thereafter, the cross-sectional surface was polished and sputtered by a small layer of gold to investigate the coating cross-section.

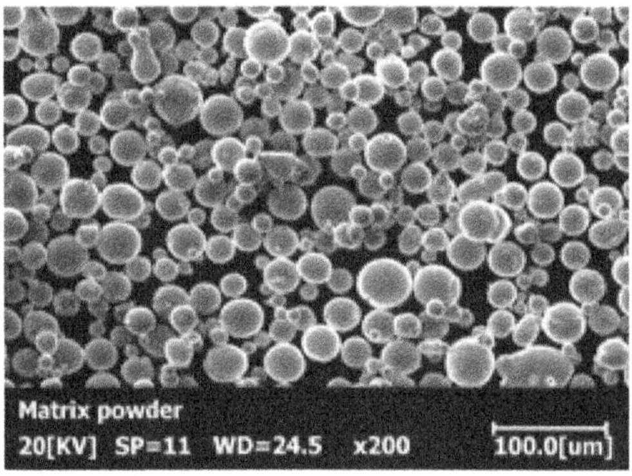

Fig. 2. SEM image of NiCrFeSiAlBC powder.

2.3 Mechanical and Tribological Tests

Vickers micro-hardness of the sprayed coating was measured using a Leitz RZD-DO apparatus equipped by a diamond micro-indenter under a normal load equal to 3 N with a dwell time of 10 s. In this case 9 measurements per sample on the substrate and the NiCrFeSiAlBC coating were performed. A gap between two consecutive imprints, in both lateral and transverse directions, is higher than twice the diagonal of the indenter footprint.

The tribological characteristics were studied by using a ball on flat linear reciprocating tribometer "CSM Plus instruments" as shown in Fig. 3. The tribological tests were performed at room temperature using Si_3N_4 ball as antagonist with 8 mm diameter. The tests were conducted at the same conditions; a normal load equal to 10 N, a sliding stroke equal to 5 mm, a sliding speed equal to 50 mm/s and a test duration equal to 5000 cycles. Before weighing the mass losses, samples were carefully removed from the machine, the surface on which friction occurs was washed with alcohol, cleaned under air pressure and then dried.

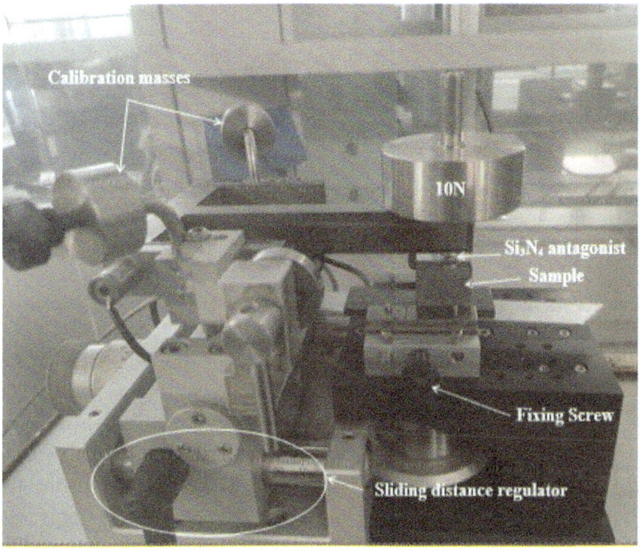

Fig. 3. Linear reciprocating tribometer "CSM Plus instruments".

3 Results and Discussion

3.1 Microstructural Characteristics

The microstructure of HVOF sprayed coating deposited on C45 steel substrate obtained by SEM is shown in Fig. 4. A few number of unmelted particles is seen on the top surface of the NiCrFeSiAlBC coating (Fig. 4a). The coating exhibits a dense and continuous structure. Regarding the microstructure in cross-section no cracking was observed with the presence of a few number of pores (Fig. 4b), some contamination are also seen in the interface between the coating and the substrate.

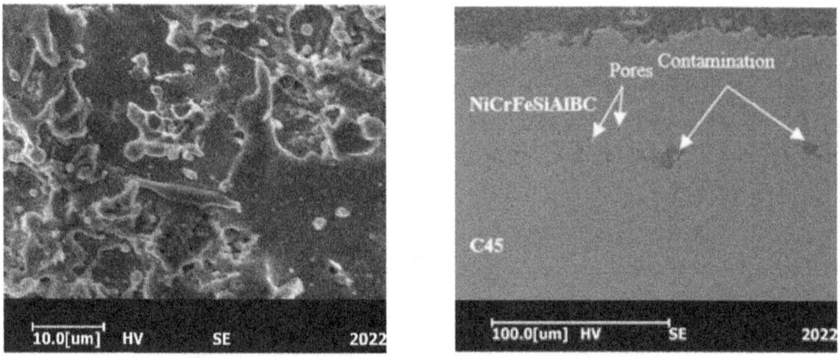

(a) (b)

Fig. 4. SEM images of (a) top surface of NiCrFeSiAlBC coating and (b) cross-section of NiCrFeSiAlBC coating.

3.2 Hardness Results

The microhardness measurements were performed on the surface of C45 substrate as well as on the cross-section surface of NiCrFeSiAlBC coating as shown in Fig. 5(a) and (b). The mean values of the microhardness values are given in Fig. 5(c). Features like the amount of un-melted or semi-molten particles, porosities and their morphologies, fast cooled particles and the morphology of grain and their sizes affect the microhardness of thermal sprayed coatings (Afsous et al., 2020). The average microhardness of the substrate is about 178 $HV_{0.3}$. The hardness of NiCrFeSiAlBC is 6 times higher than those reported on the substrate. The maximum hardness was obtained in the coating is equal to 1017 $HV_{0.3}$. In this case and according to the literature review higher hardness means that the coating has higher friction and wear resistance.

3.3 Friction Tests Results

Dry tribological tests were repeated 3 times for both materials during 5000 cycles. Mass loss measurements were carried out by using a micro-balance type RADWAG AS 220.R2 PLUS with an error value equal to 1 mg. As we can see in Fig. 6, the evolutions of the friction coefficient in the two cases are almost similar during the first 500 cycles of the friction test, which it is defined as a running-in phase. After this phase the friction coefficient have already started to get a significant difference. The average friction coefficient of the substrate is equal to 0.73 ± 0.08. While The NiCrFeSiAlBC Coating had the lowest mean value of the friction coefficient, it is equal to 0.44 ± 0.05. As can be seen from Fig. 7, the mass loss of the substrate is higher than the value obtained in the coating. It is equal to 11.2 mg in the substrate while, it become equal to 4.7 mg for the coating. These significant differences are mainly due to the high hardness of the coating (Tyagi et al.2022).

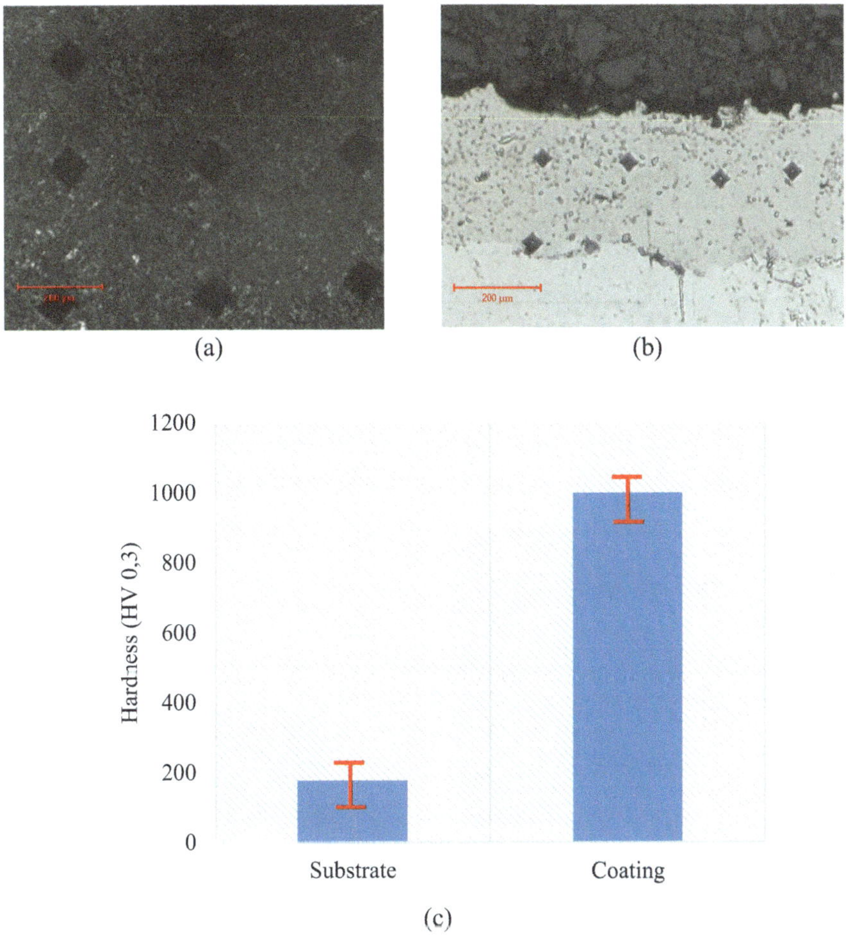

Fig. 5. SEM images of (a) the substrate and (b) NiCrFeSiAlBC coating after indentation test (c) Vickers microhardness of C45 substrate and NiCrFeSiAlBC coating.

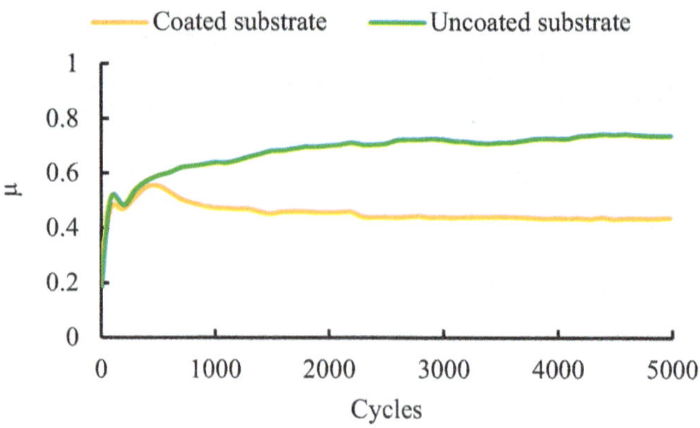

Fig. 6. Typical evolutions of the friction coefficient as a function of the number of sliding cycles.

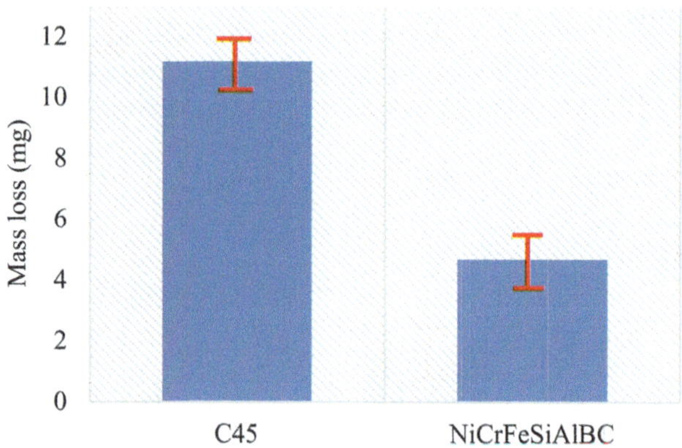

Fig. 7. Mass loss average values after tribological tests.

4 Conclusion

In this paper, the microstructural, mechanical and tribological properties of advanced NiCrFeSiAlBC coatings deposited by HVOF thermal spray process were evaluated and compared to those of the C45 substrate. From the results, the following conclusions can be drawn:

- The microstructure of the NiCrFeSiAlBC coating is continuous and dense with no obvious microstructural defects.
- Based on micro-hardness analysis, the NiCrFeSiAlBC coating is a very hard material when compared to the substrate.
- Tribological tests were carried out on both substrate and coating. A very low friction coefficient was obtained with the NiCrFeSiAlBC coating.
- According to the mass loss analysis, the NiCrFeSiAlBC coating offered an excellent wear resistance performance.

References

Singh, S., Berndt, C., Singh Raman, R.K., Singh, H.: Applications and developments of thermal spray coatings for the iron and steel industry. Materials **16**, 516–518 (2023)

Sadeghi, E., Markocsan, N., Joshi, S.: Advances in corrosion-resistant thermal spray coatings for renewable energy power plants. Part I: effect of composition and microstructure. J. Therm. Spray Technol. **28**, 1749–1788 (2019)

Yu, J., Liu, X., Yu, Y., Li, H., Liu, P., Huang, K., Sun, R.: Re-search and application of high-velocity oxygen fuel coatings. Coatings **12**, 828 (2022)

Saaedi, J., Coyle, T.W., Arabi, H., Mirdamadi, S., Mostaghimi, J.: Effects of HVOF process parameters on the properties of Ni-Cr coatings. J. Therm. Spray Technol. **19**, 521–530 (2010)

Miguel, J.M., Guilemany, J.M., Vizcaino, S.: Tribological study of NiCrBSi coating obtained by different process. Tribol. Int. **36**, 181–187 (2003)

Gómez-del Río, T., Garrido, M.A., Fernández, J.E., Cadenas, M., Rodríguez, J.: Influence of the deposition techniques on the mechanical properties and microstructure of NiCrBSi coatings. J. Mater. Process. Technol. **204**, 304–312 (2008)

Eskandari, A., Shafyei, A., Sayyedan, F., Afsous, M., Adamzadeh, M.: Inves-tigating the microstructural, mechanical, and tribological properties of NiCrBSi high-velocity oxygen fuel and high-pressure high-velocity oxygen fuel coatings. J. Mater. Eng. Perform. **32**, 1671–1682 (2023)

Afsous, M., Shafyei, A., Soltani, M., Eskandari, A.: Characterization and evaluation of tribological properties of NiCrBSi-Gr composite coatings de-posited on stainless steel 420 by HVOF. J. Therm. Spray Tech. **29**, 773–788 (2020)

Tyagi, A., Chourasia, S., Vats, K., Murtaza, Q., Walia, R.S., Dhawan, V.: Tribological behavior of HVOF based composite coatings for wear resistance applications. Mater. Today **50**, 2376–2380 (2022)

Manufacturing Engineering and Additive Manufacturing

Investigation of the Mechanism of Cutting by the Ball End Mill Tip and Its Influences on the Surface Topography and the Cutting Force

Amine Regaieg[1,2](✉), Rami Belguith[3], and Lotfi Sai[1,4]

[1] Unité de Génie de Production Mécanique et Matériaux, ENIS, Université de Sfax, Sfax, Tunisie

[2] École Polytechnique de Tunisie, EPT, Université de Carthage, Carthage, Tunisia
amine.regaieg@ept.u-carthage.tn

[3] LGCgE, FSA de Béthune, Université d'ARTOIS, Arras, France
rami.belguith@univ-artois.fr

[4] Institut Préparatoire aux Etudes d'Ingénieurs de Nabeul, IPEIN, Université de Carthage, Carthage, Tunisia

Abstract. The modeling of the cutting process is one of the important roads to optimize cutting parameters through computer-aided manufacturing simulations. However, many models, whether analytical, numerical, or geometrical, are plagued by inaccuracies, consistently exhibiting a percentage of errors when compared to experimental tests. This paper investigates and analyzes certain interpretations and observations to illustrate inaccuracies in the existing models for roughness, topography, and cutting forces, specifically in the context of 3-axis ball end milling. An approach is proposed to enhance the accuracy of predicting these models.

Keywords: 3-axis ball end milling · tool tip · topography · cutting forces · rubbing

1 Introduction

The ball end milling process is used in manufacturing industries to produce complex parts such as dies, molds, and cavities. Usually, these parts are made up of rounded corners and free-form surfaces that require ball-nose end mill cutters, particularly for finishing operations. Adding to these properties, the ball end mill produces good-quality surfaces for which finishing operations are the final machining step. The researchers developed models of different criteria as a function of the cutting parameters, with the goal of finding the optimal combination of each cutting process. The most developed criteria are the roughness of the surface topography and the formation of the cusps and scallops on the machine surface profiles, the cutting forces, the limits of the cutter workpiece engagement region, the instantaneous chip thickness, and the material removal rate. The principal remarks observed in all these studies are the non-good correlations between the theoretical simulations of models and experimental validations. The percentage of errors is quantified, discussed, by some researchers.

M. Chouchane et al. (Eds.): CMSM 2023, LNME, pp. 371–381, 2024.
https://doi.org/10.1007/978-3-031-65007-9_39

For all types of models (analytical, geometrical, numerical, etc.) and their differences in accuracy, the difference between the predictions made by the simulations and the results of the experiment tests always exists. The researchers attributed these differences to the multiple phenomena that can't be modeled and affect the experiment tests, such as the vibrations, the reliability of the machine, the accuracy of the measuring instruments, the thermomechanical effect, and the behavior of the machined materials and the machine-tool. The models of the cusps and the scallop forms produced by the ball-end milling process are cited.

Mizugaki et al. (2003) developed a model to calculate the scallop height for a different tool axis inclination. The difference between the computer simulation and the experiment results fluctuated between 2.5 μm in the tree axis and reached 5 μm for an inclination angle equal to 5°. Chen et al. (2005) developed a regression model to predict the feed interval scallop for two tool radius of 3 mm and 5 mm as a function of the federate. The difference between the simulation with the regression model and the experiment reached is 20 μm for a pick interval of 0.75 mm and a feed equal to 0.5 mm. When the tool radius is equal to 5 mm, the differences reach 10 μm. Batista et al. (2017) predicted mathematically the formation of the cusps and scallops in ball end milling. The experimental tests show that the part material is crushed or plowed near the tool tip path, where the cutting speed is very low compared to the feed rate. The end mill marks form cusps, which cannot be modeled because they depend on the quantities of material removed by the end of the cutter. This phenomenon causes surface damage and dimensional errors on the milled surface. In the work of Regaieg et al. (2024), it was proved that there is a region near the tool tip where the cutter scraped the material without removing it. The main cause is that the cutting speed in this region less than the feed rate where the tooth scraped the materials and rubbed without cutting.

For the surface qualities, Arizmendi et al. (2008) developed a geometrical model to predict surface topography considering a tool parallel axis offset for two milling modes and two tests in the zig-zag. The second mode is in one direction of cut for the same feeds and picks. The results of the topography of the milled surfaces showed a difference between the measured pitch, which contained a cusp in the feed direction that was not presented on the simulated one.

In addition, there are some observations that essentially affect the cutting geometries, such as the instantaneous chip thickness, the cutting length, and the limits of the cutter workpiece engagement region. Regarding the work of Sai et al. (2018), that demonstrated that the cutter workpiece engagement region is larger in the tool tip. This largest engagement induces the engagement of the next flute before the first is disengaged, which makes a supplementary cutting force not modeled by the previous works. An analytical model was developed as a function of the cutting parameters. The precision of the model shows an inaccuracy in the previous cutting force models precisely at the end of the mill region. Regaieg et al. (2023) demonstrated the existence of the rubbing zone in this region where the flank face of the tooth rubs with the uncut materials. This phenomenon causes a thermomechanical load, residual stress, flank wear, disturbs the topography, heats tool and workpiece contacts, and increases the cutting forces. The rubbing was not considered in the previous works, which developed cutting forces by a thermomechanical or mechanistic model, but it can be the source of the inaccuracy between the simulations and the experimental results. This difference was noticed by Chiang et al. (1995) in their developed cutting force model, the product of the cutting area and the specific cutting coefficient obtained by third-order curve-fitting of data measured in experiments. This difference between the simulations and the results of the experiments is 15%. The authors explain the mean error as a neglected deflection of the cutter and the inaccuracy in the specific cutting force coefficient measurements. In ball-end milling, Lee and Altintas (1996) shown that there is an inevitable error in the cutting force prediction when compared to experiment data near the tool tip. The plowing action between the cutting edge and the material dominates the cutting force in the axial direction near the tool tip pattern. Additionally, Batista et al. (2017) showed that when cutting takes place below the critical effective diameters, cutting forces are relatively low; the particular pressure created in this region increases the workpiece surface's plowing and crushing.

In this paper, the study is focused on the region near the ball-end mill tip, which is discussed at infinitesimal points. The true trajectories of the points in the cutting region are analyzed and show that some points around the tip zone, when the cutting speed becomes less than the feed rate, cannot remove the materials, but they rub and crush them at the end of each cycloid, forming a cusp.

2 Un-modeled Cusps Formation Process

The cutting edge is discretized on successive points M, their trajectories are formulated by the Eq. (1):

$$\begin{cases} x_M(\theta(z)) = V_f t + e\sin(\omega t) + R(z)\sin\left(\omega t + \rho - \varphi(z) + \frac{(j)2\pi}{N_f}\right) \\ y_M(\theta(z)) = e\cos(\omega t) + R(z)\cos\left(\omega t + \rho - \varphi(z) + \frac{(j)2\pi}{N_f}\right) \end{cases} \quad (1)$$

(X_M, Y_M, Z_M) are the coordinates system of the point M. V_f is the feedrate according to X direction. The path of each point M in the coordinate system related to the workpiece is represented in Fig. 2. Based on the tooth trajectories aspect, the cutter workpiece engagement region can be divided into three zones as shown in Fig. 1. These zones are presented as follows: The zone-I: the path of point M since the cutting edge intersect previous path. The zone-II: the point M rotates without intersecting previous path. The zone-III: the trajectory of point M presents a cycloid. The limit of the third zone is characterized by the fact that the trajectory of the point M is no longer a trochoid, but a cycloid presented in red in the Fig. 1. If we calculate the speed vector of the point M:

$$\begin{cases} V_{XM} = \frac{dx_M(t,z)}{dt} = \frac{N_f f_z \omega}{2\pi} + e\omega\cos(\omega t) + R(z)\omega\cos\left(\omega t + \rho - \varphi(z) + \frac{(j)2\pi}{N_f}\right) \\ V_{YM} = \frac{dy_M(t,z)}{dt} = -e\omega\sin(\omega t) - R(z)\omega\sin\left(\omega t + \rho - \varphi(z) + \frac{(j)2\pi}{N_f}\right) \end{cases} \quad (2)$$

At the bottom of the cycloid when the time is t_c, the speed is canceled, so the X-component and Y-component of the speed must be null: $V_{YM}(t_i) = 0$, $V_{XM}(t_i) = 0$:

$$\begin{cases} \frac{N_f f_z}{2\pi} + e\cos(\omega t_c) + R(z)\cos\left(\omega t_c + \rho - \varphi(z) + \frac{(j)2\pi}{N_f}\right) = 0 \\ e\sin(\omega t_c) + R(z)\sin\left(\omega t_c + \rho - \varphi(z) + \frac{(j)2\pi}{N_f}\right) = 0 \end{cases} \quad (3)$$

On another side, and according to Eq. 1-b and Eq. 3-a, the Y-composant of the point M is always constant at the bottom of the cycloid regardless the runout values e andρ, and according to Eq. 1-a and Eq. 3-b, the X-composant can be expressed with the understanding that $\theta_c = \omega t_c + \rho - \varphi(z) + \frac{(j)2\pi}{N_f} = \pi$:

$$\begin{cases} x_M(t_c) = \frac{N_f f_z}{2\pi}\omega t_c = \frac{N_f f_z}{2}(\pi - \rho + \varphi(z) - \frac{(j)2\pi}{N_f}) \\ y_M(t_c) = -\frac{N_f f_z}{2\pi} \end{cases} \quad (4)$$

On the other hand, the Eq. 3-a can resolve the limit radius R_{II-III} of the beginning of the cycloid i.e. the limit between the zone-II and the zone-III, with neglecting the term $\varphi(z)$:

$$R_{II-III} \approx \frac{N_f f_z}{2\pi} + e\cos\left(\rho + \frac{(j)2\pi}{N_f}\right) \quad (5)$$

For given values of e and ρ, the feed rate must be greater than a minimum value so that the zone III and the cycloid can be generated: $f_z > \frac{2\pi e}{N_f}\cos\left(\rho + \frac{(j)2\pi}{N_f}\right)$, with the understanding that the radius must be always positive.

Treating the zone-II, and to avoid having an intersection between the paths, we have to follow this condition:

$$x_M(pi/2) = \frac{N_f f_z}{2\pi}\left(\frac{\pi}{2} - \rho - \frac{(j)2\pi}{N_f}\right) + e\sin\left(\frac{\pi}{2} - \rho - \frac{(j)2\pi}{N_f}\right) + R(z)\sin\left(\frac{\pi}{2} + \frac{(j)2\pi}{N_f}\right) = N_f f_z \quad (6)$$

Finally, the limit radius R_{I-II} between the zone-II and the zone-I:

$$R_{I-II} \approx \left(\frac{3}{4} + \frac{\rho}{2\pi} - \frac{(j)}{N_f}\right)N_f f_z - e\cos\left(\rho + \frac{(j)2\pi}{N_f}\right) \quad (7)$$

The three zones according to the feed rate and runout parameters e and ρ, are calculated and presented in Table 1:

Table 1. Three zones modeling

Parameters	In term of $R(z)$	In term of z
Zone-I	$R_{I-II} < R(z)$	$R - \sqrt{R^2 - R_{I-II}{}^2} < z$
Zone-II	$R_{II-III} < R(z) < R_{I-II}$	$R - \sqrt{R^2 - R_{II-III}{}^2} < z < R - \sqrt{R^2 - R_{I-II}{}^2}$
Zone-III	$0 < R(z) < R_{II-III}$	$0 < z < R - \sqrt{R^2 - R_{II-III}{}^2}$

To explain the path of the cutting part at the end of the tooth, we considered a point M of the tooth belonging to zone-III. The radial position of point M is presented at instantaneous positions by an arrow beginning from the discretized disc center and

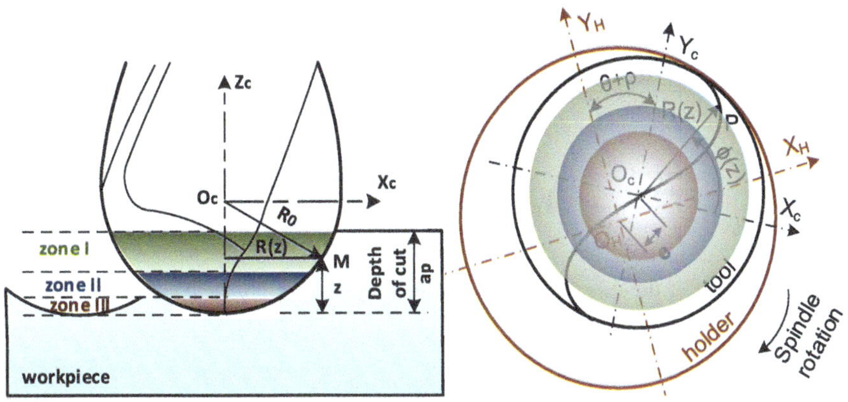

Fig. 1. Cuter workpiece engagement zones

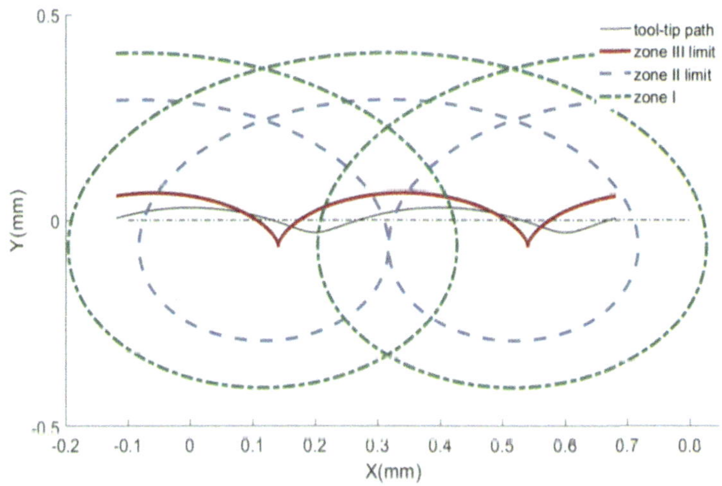

Fig. 2. Trajectory of point M in three zones, $f_z = 0.2$ mm/tooth, $e = 30$ µm, $\rho = 80°$, $R_{I-II} = 53.5$ µm, $R_{II-III} = 0.286$ mm

pointed at point M. The tooth removes the material at point M in the position of the red arrowhead, as shown in Fig. 3. When the tooth is at the end of the instantaneous revolution, it is rolling on the workpiece and is unable to form the chip trajectory, which presents a cycloid. The material scrapped all along is abandoned and stays stoked. This phenomenon causes many problems because there is no cutting and chip formation near the tool tip. That has a direct consequence on the surface topography and the depth of the cut to be removed.

Another problem that could be noticed is that when the tool turns, its flank face interferences with the uncut material, which causes the heating and the wear of the tool as well as a non-uniformly hardened surface.

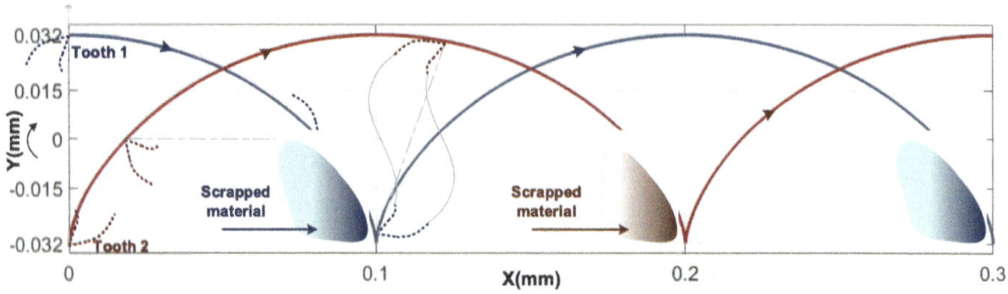

Fig. 3. Simulation of the cusps formation mechanism for an arbitrary point M in the zone-III (Regaieg et al. 2024)

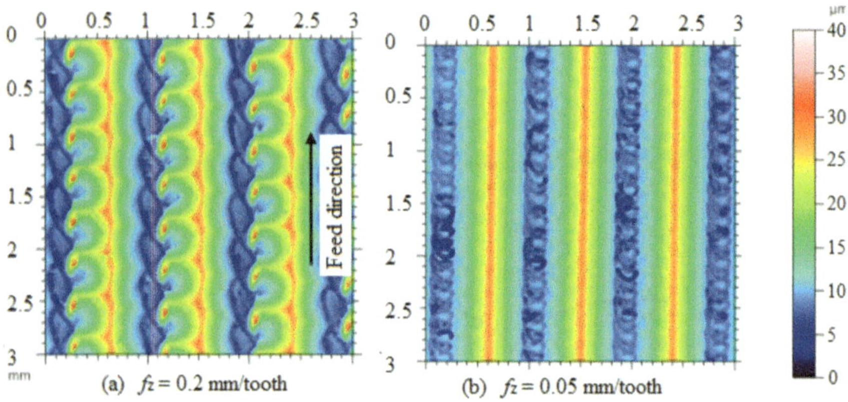

Fig. 4. Influence of the feed per tooth on surface topography, $f_z = 0.2$ mm/tooth, (Guillemot 2010)

Based on the experiment work of Guillemot (2010), the Fig. 4a, shows the real cycloid trajectories of the teeth and their effect on the surface qualities. The cusps on the end of each cycloid cannot be modeled or simulated. From the observation, the cutting speed has not a noticeable effect on the size of cusps, but only the feed per tooth can influence them when it decreases. The second pitch shows that for a feed equal to $f_z = 0.05$ mm/tooth, the cusps are less regular in their forms and very numbered on the same pitch size.

The experimental findings and the simulated milled surface in up-milling are contrasted in Fig. 5. In the simulation, the ploughed material is not visible in Fig. 5c, according to the studies of Batista et al. (2017), but in the measured surface topography, the minimum chip thickness is not reached near the tool tip (around 2 µm), which favors the apparition of crushed and non-machined materials. This observation proves the existence of accumulated materials at the end of each cycloid. A simulated trochoidal trajectory of the teeth is plotted and superposed to the Fig. 5b in order to show the position of the studied cusps. When the path became a cycloid in the third zone (from a depth of $z = 0.38$ mm), the tool scrapped the material from point A and accumulated it in point B to form the unmodeled cusps.

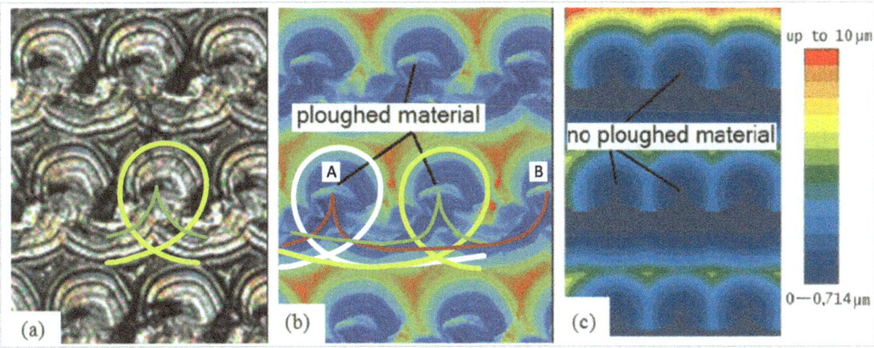

Fig. 5. (a) Optical microscopy, (b) Measured surface topography (trajectory plotted by author) (c) simulated surface topography ($a_e = 0.346$ mm, v = 7200 mm/min, $f_z = 0.15$ mm/tooth and up-milling) (Batista et al. 2017)

3 Rubbing of the Tool Flank Face on the Machined Surface

A second phenomenon observed in the end milling tip mechanism is the rubbing, which is modeled by Regaieg et al. (2023). The points M on the flank face of a tooth become in contact with the unmachined surfaces when the cutting speed of point M is less than the feed rate. This phenomenon causes rubs and induces resultant cutting forces that pushed the cutter in the inverse direction of the cut. The rubbing-limit radius R_r for each tooth (j) is expressed by the Eq. 8, where α is the clearance angle (Fig. 6):

$$R_r = \left(\frac{N_f . f_z}{2\pi} - e.sin\left(\alpha - \rho - \frac{(j)2\pi}{N_f} \right) \right) / sin\alpha \quad (8)$$

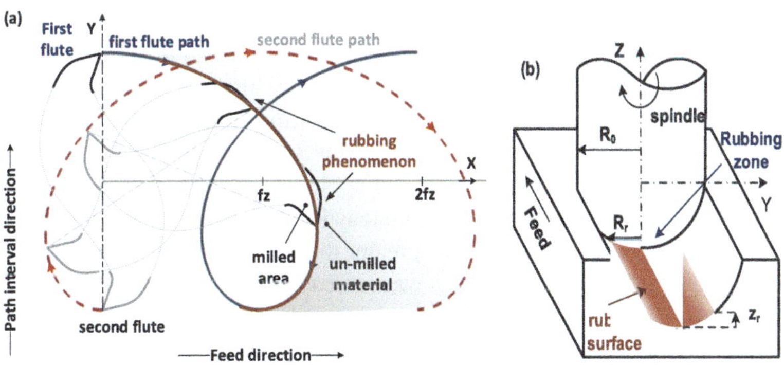

Fig. 6. Rubbing parameters (a) the trochoidal cutting paths, (b) rubbing areas. (Regaieg et al. 2023)

Rubbing shortens the tool's life and impacts surface roughness by increasing cutting pressures. The rubbing phenomenon has been studied by numerous researchers who are interested in chatter stability and dumping. The primary cause of process damping has been identified in the literature as the tool's flank face penetrating into the workpiece (Ahmadi and Ismail 2011). Budak and Tunc (2009) presented a tool interference coefficient that links the degree of damping to the indentation volume penetrated on the tool flank face. This coefficient is determined by energy analysis. They discovered that the specific process of dampening decreased as the clearance angle increased. Furthermore, the contact mostly happens at the tool's edge rather than the flank face at greater clearance angles. The limit of the rubbing zone's effective radius is shown in Fig. 7b for varying values of the clearance angle α, based on the feed per tooth. It is clear that using a same tool with a larger clearance angle and a lower feed value is required to reduce the rubbing zone.

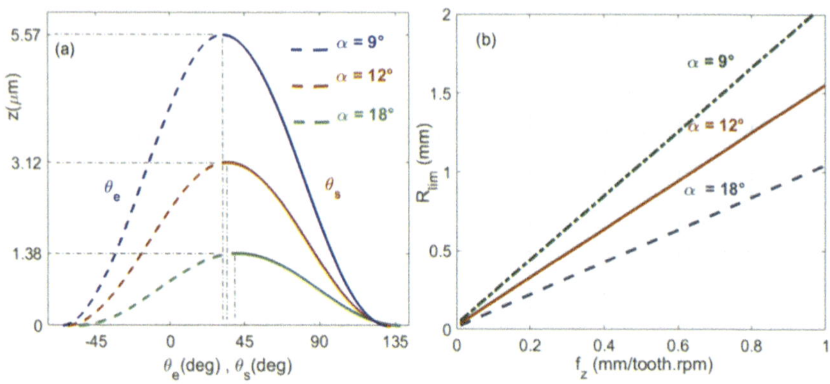

Fig. 7. (a) Start and finish angles of rubbing (b) limit radius of interference for different clearance angles , $R_0 = 5$ mm, $e = 6$ μm, $\rho = 67°$ (Regaieg et al. 2023)

This region exhibits a significant area of rubbing related to the tool-workpiece normal force. It produces a tangential force that opposes the cutter's relative motion. This rubbing produces a lot of heat and significantly increases the wear and breakage of tools. The surface quality is impacted when the flank face collides with the non-machined materials.

Chen et al. (2016) provided the curve of the limit of rubbing depth (Fig. 8) for clearance angles α between 3° and18° in order to investigate the impact of this parameter. The fluctuation of the peak stress for the same clearance angle is correlated with the decreasing form of the curve. The maximum tensile stress for the cutting force changed somewhat when the clearance angle was greater than 9° and decreased as the clearance angle increased from 0° to 9°. This effect was more significant for clearance angles close to 0°. When the clearance angle was greater than 9°, the maximum rubbing depth barely altered from its minimum value at that relief angle. Therefore, in order to reduce the rubbing effect, clearance angles greater than 9° were appropriate.

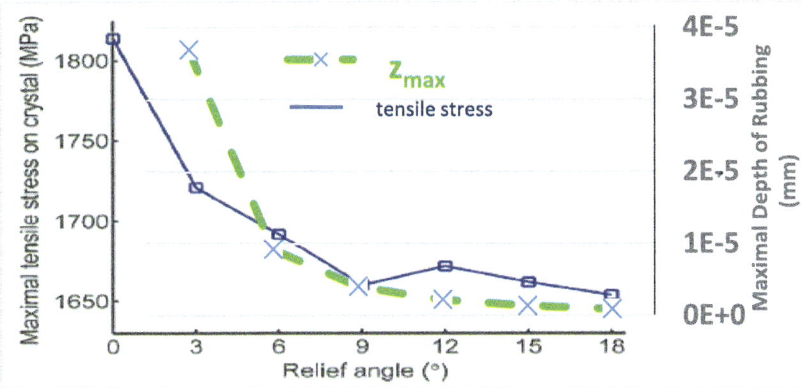

Fig. 8. Maximum tensile stress and maximum depth of rubbing for different relief angles (Regaieg et al. 2023)

4 Modeling Validation

To validate the developed model, a comparison of the predicted cutting force was conducted with two tests from (Ghorbani and Movahhedy 2019) using two different depths of cut in slotting milling. The results indicate a minor error of 4.83% on the maximal value in the case of an axial cutting depth of $a_p = 3$ mm. However, for a lower axial depth of cut $a_p = 1$ mm, this error increases noticeably to 20.4%. This is mainly due to the parasitic forces induced by the rubbing zone, where the tool generates cusps on the machined surface. These additional cutting forces have a minimal impact for high axial cutting depths, but for lower values, their effect becomes considerably more noticeable (Fig. 9).

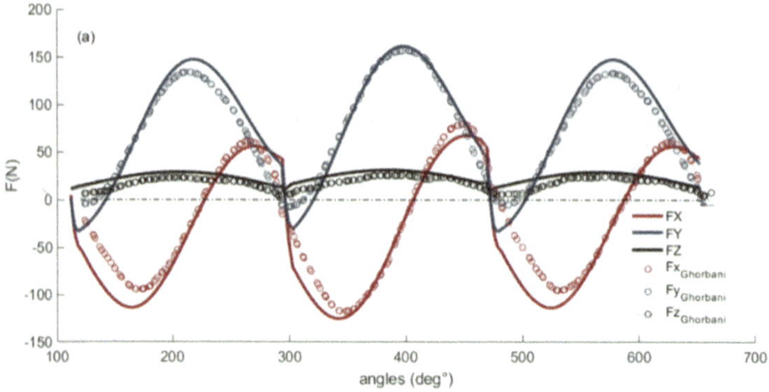

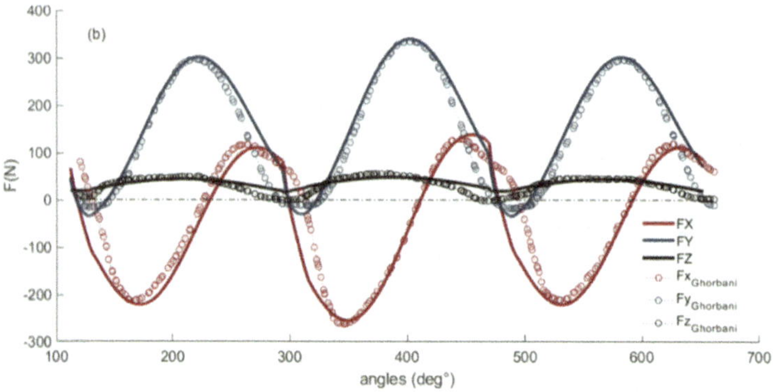

Fig. 9. Cutting forces of predicted and experimental tests for slot milling (f_z = 0.02 mm/tooth, $R_0 = 5$ mm, $e = 1.6$ µm, $\rho = 68°$) (a) $a_p = 1$ mm (b) $a_p = 3$ mm

5 Conclusion

In conclusion, this paper highlighted the limitations of current models in predicting roughness, topography, and cutting forces during ball end milling. By analyzing discrepancies between experimental results and model predictions, the authors identified shortcomings related to unmodeled cusp formation and the influence of tool flank rubbing on the machined surface. These phenomena contribute significantly to the irregularities observed in predicted topographies and cutting forces. To bridge this gap, the paper proposes a novel approach that incorporates these factors. By accounting for un-modeled cusp formation and tool flank rubbing, future models can achieve more accurate simulations of cutting forces and machined surface topography, ultimately leading to improved optimization of cutting parameters through computer-aided manufacturing.

Acknowledgements. The work is carried out thanks to the support and funding allocated to the Unit of Mechanical and Materials Production Engineering (UGPM2/UR17ES43) by the Tunisian Ministry of Higher Education and Scientific Research.

References

Ahmadi, K., Ismail, F.: Analytical stability lobes including nonlinear process damping effect on machining chatter. Int. J. Mach. Tools Manuf **51**, 296–308 (2011). https://doi.org/10.1016/j.ijmachtools.2010.12.008

Arizmendi, M., Fernández, J., de Lacalle, L.N.L., et al.: Model development for the prediction of surface topography generated by ball-end mills taking into account the tool parallel axis offset. Experimental validation. CIRP Ann. **57**, 101–104 (2008). https://doi.org/10.1016/j.cirp.2008.03.045

Batista, M.F., Rodrigues, A.R., Coelho, R.T.: Modelling and characterisation of roughness of moulds produced by high-speed machining with ball-nose end mill. Proc. Inst. Mech. Eng. Part B J. Eng. Manuf. **231**, 933–944 (2017). https://doi.org/10.1177/0954405415584898

Budak, E., Tunc, L.T.: A new method for identification and modeling of process damping in machining. J. Manuf. Sci. Eng. **131**, 051019 (2009). https://doi.org/10.1115/1.4000170

Chen, J.-S., Huang, Y.-K., Chen, M.-S.: Feedrate optimization and tool profile modification for the high-efficiency ball-end milling process. Int. J. Mach. Tools Manuf **45**, 1070–1076 (2005). https://doi.org/10.1016/j.ijmachtools.2004.11.020

Chen, N., Chen, M., Wu, C., et al.: The design and optimization of micro polycrystalline diamond ball end mill for repairing micro-defects on the surface of KDP crystal. Precis. Eng. **43**, 345–355 (2016). https://doi.org/10.1016/j.precisioneng.2015.08.015

Chiang, S.-T., Tsai, C.-M., Lee, A.-C.: Analysis of cutting forces in ball-end milling. J. Mater. Process. Technol. **47**, 231–249 (1995). https://doi.org/10.1016/0924-0136(95)85001-5

Ghorbani, M., Movahhedy, M.R.: Extraction of surface curvatures from tool path data and prediction of cutting forces in the finish milling of sculptured surfaces. J. Manuf. Process. **45**, 273–289 (2019). https://doi.org/10.1016/j.jmapro.2019.07.008

Guillemot, N.: Prise en compte de l'intégrité de surface pour la prévision de la tenue en fatigue de pièces usinées en fraisage. (2010)

Lee, P., Altintas, Y.: Prediction of ball-end milling forces from orthogonal cutting data. Int. J. Mach. Tools Manuf **36**, 1059–1072 (1996). https://doi.org/10.1016/0890-6955(95)00081-X

Mizugaki, Y., Kikkawa, K., Terai, H., et al.: Theoretical estimation of machined surface profile based on cutting edge movement and tool orientation in ball-nosed end milling. CIRP Ann. **52**, 49–52 (2003). https://doi.org/10.1016/S0007-8506(07)60528-9

Regaieg, A., Belguith, R., Sai, L., Bouzid Saï, W.: A new approach to investigate the formation mechanism of the not modelable cusps in 3-axis ball end milling. In: Proceedings of the 2nd International Conference on Innovative Materials, Manufacturing, and Advanced Technologies. Springer Nature Switzerland, Cham, pp. 241–249 (2024)

Regaieg, A., Belguith, R., Sai, L., Bouzid, W.: Geometric modeling of the rubbing zone limits in 3-axis ball-end milling. Int. J. Adv. Manuf. Technol. (2023). https://doi.org/10.1007/s00170-023-11037-5

Sai, L., Belguith, R., Baili, M., et al.: An approach to modeling the chip thickness and cutter workpiece engagement region in 3 and 5 axis ball end milling. J. Manuf. Process. **34**, 7–17 (2018). https://doi.org/10.1016/j.jmapro.2018.05.018

Effect of the Programmed Smoothing Parameters on the Kinematic Behavior of the NC Machine in HSM

Rania Sbaa[1,2(✉)], Rafik Tarchoun[2], Mondher Ben Ahmed[2], Mohamed Essid[1,3], and Wassila Saï Bouzid[1]

[1] Unit of Mechanical Production Engineering and Materials (UGPMM), National Engineering School of Sfax, University of Sfax, UR 17ES43, B.P. 1173, 3038 Sfax, Tunisia
kesrania@gmail.com

[2] Higher Institute of Technological Studies of Mahdia, Av. Mourouj, 5100 Mahdia, Tunisia

[3] Higher Institute of Technological Studies of Sfax, Route de Mahdia Km 2.5, B.P. 88 A, 3099 El Bustan, Sfax, Tunisia

Abstract. The accuracy of the complex machined part is one of the key aspects of High-Speed Milling (HSM) performance. CAM software tends to meet this requirement by approximating the real tool path by short, linear blocks discontinuous at tangents. The smoothing of these discontinuities would be necessary to minimize the NC machine vibration as it passes through these discontinuities. This paper aims to explore the effect of programmed smoothing parameters (G641, G643) on the kinematic behaviour of the NC machine, by a set of experimental tests. The experimental test shows that G641 provides a smoother curvature variation compared to G643. This difference is reflected in the feed rate variation, where G641 shows high fluctuations, while G643 maintains a nearly constant feed rate.

Keywords: High-Speed Milling · Local smoothing · Programmed smoothing parameters · Kinematic behavior

1 Introduction

While the CAM software has been improved to meet the high-speed machining performance requirements and complex part production, the CNC design remains limited.

Indeed, in HSM, the tool path still relies on G01 linear segments with tangential and curvature discontinuities [1, 2]. These discontinuities lead to variations in kinematic machine behavior (speed, acceleration and jerk) and resulting in frequent occurrences of acceleration and deceleration phenomena.

These aspects contribute to a decrease in machining efficiency and the generation of intense vibrations, ultimately affecting both machining quality and accuracy [3, 4].

To overcome these problems, a smoothing block should be integrated in each transition to ensure the continuity of the tool path [5].

M. Chouchane et al. (Eds.): CMSM 2023, LNME, pp. 382–390, 2024.
https://doi.org/10.1007/978-3-031-65007-9_40

In light of this, two smoothing modes have been implemented in the CNC unit, which are respectively the global smoothing mode and the local smoothing mode [6, 7].

The global smoothing mode approximates the discrete linear tool paths by high-order continuous parametric curves such as the polynomial spline [8], B-spline [9] and NURBS [10].

The local smoothing mode inserts a parametric spline at each control point of the programmed toolpath [11–13].

The parametric spline is generally defined in the NC program with one of two geometric parameters (see Fig. 1): the Transition length Tl_p or the Contour tolerance ρ [14]. Tl_p is programmed with the G641 smoothing function, while ρ is programmed with the G643 smoothing function.

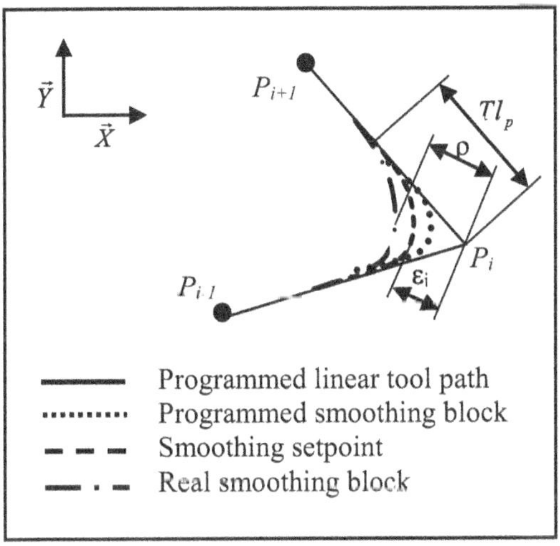

Fig. 1. Geometric parameters of a transition element in local smoothing

Investigating the effect of programmed local smoothing parameters on the kinematic behavior of the NC machine in HSM (actual path, axis feed rate) is the concern of this work.

2 Identification of the Programmed Local Smoothing Parameters

The model presented here is based on a toolpath running over two linear segments $([P_{i-1}P_i], [P_iP_{i+1}])$ and joined by a smoothing element $S_i(u)$ passing through points $\left(P_{IN_i}, Q^*_{pi}, P_{OUT_i}\right)$ and considering the maximum contour tolerance of the tool path setpoint ε_i represented in Fig. 2 by the distance between Q^*_{pi} and P_i.

U_{p_i} and V_{p_i} are the unit vectors tangent to smoothing block and defined by Eq. (1):

$$\begin{cases} U_{p_i} = \frac{-T_{p_i}}{\|T_{p_i}\|} = \left[U_{pX_i}, U_{pY_i}\right]^T \\ V_{p_i} = \frac{-T_{p_{i+1}}}{\|T_{p_{i+1}}\|} = \left[V_{pX_i}, V_{pY_i}\right]^T \end{cases} \quad (1)$$

With:

$(T_{p_i}, T_{p_{i+1}})$: The vectors tangent to smoothing element at point P_{IN_i} and P_{OUT_i} respectively, with:

$$\|\overrightarrow{T_{p_{i+1}}}\| = \|\overrightarrow{T_{p_i}}\| = Tl_p \tag{2}$$

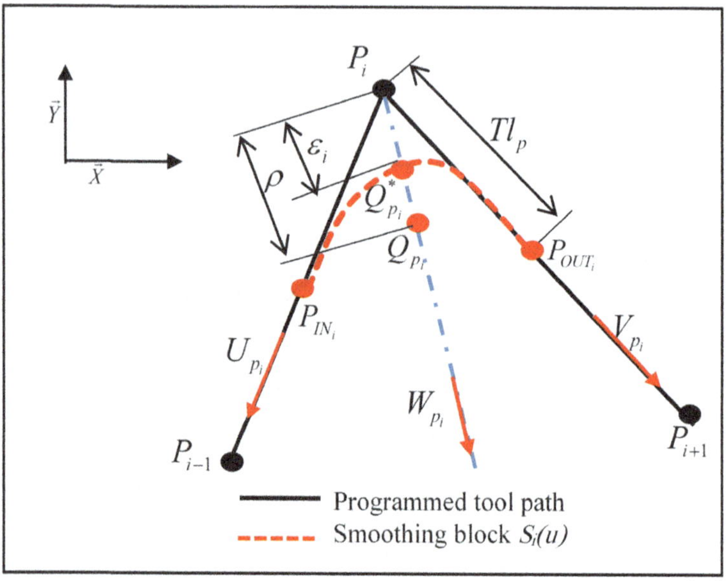

Fig. 2. Geometric modeling of the smoothing block in 3-axis HSM

Referring to [15, 16], the smoothing block is described by a Hermite spline of degree 3, and represented by Eq. (3):

$$S_i(u) = (u^3, u^2, u, 1)\begin{bmatrix} 2 & -2 & 1 & 1 \\ -3 & 3 & -2 & -1 \\ 0 & 0 & 1 & 0 \\ 1 & 0 & 0 & 0 \end{bmatrix}\begin{bmatrix} P_{IN_i} \\ P_{OUT_i} \\ \mu T_{p_i} \\ \mu T_{p_{i+1}} \end{bmatrix}, \quad 0 \le u \le 1 \tag{3}$$

With:

- μ: The weight of two tangent vectors.

By taking the two Eqs. (1) and (3), the Transition length Tl_p is expressed as a function of μ:

$$Tl_p = \begin{cases} \dfrac{8\tau\rho W_{pX_i}}{4(U_{pX_i} + V_{pX_i}) - \mu(V_{pX_i} + U_{pX_i})} \\ \dfrac{8\tau\rho W_{pY_i}}{4(U_{pY_i} + V_{pY_i}) - \mu(V_{pY_i} + U_{pY_i})} \end{cases} \tag{4}$$

With:

- τ: Rate defined by interpolator with:

$$\tau = \frac{\varepsilon_i}{\rho} \tag{5}$$

3 Experimental Process

The tool path is defined by three points (**P1**, **P2**, **P3**) and configured as showed in Fig. 3. The coordinates of these three points are given in Table 1.

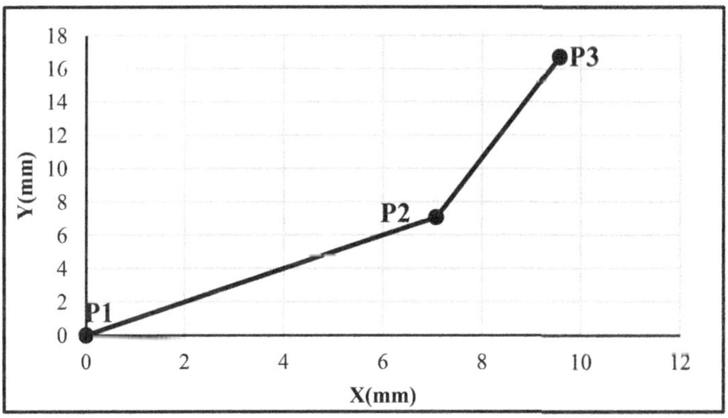

Fig. 3. Programmed linear tool path

Table 1. Coordinates of points

Point	X (mm)	Y (mm)
P1	0	0
P2	7.071	7.071
P3	9.569	16.73

Table 2 summaries the geometric parameters of the smoothing block corresponding to the tool path, as depicted in Fig. 3, with a programmed contour tolerance ρ set to 0.1 mm.

Table 2. Parameters of the smoothing block

μ	τ	Tl_p
2	**0.8**	**0.4**

Two experimental tests were conducted. During the first test, the tool path was programmed using the **G641** function, which considers the computed value of Tl_p. In the second test, the same tool path was programmed using the **G643** function, which incorporates the calculated value of ρ. Motion setpoints and axis feed rates are recorded from the machine.

All tests were carried out on a SPINNER MVC850 3-axis vertical machining (see Fig. 4) center equipped with a SINUMERIK 840D numerical control director. Table 3 summarizes the specifications of NC machine.

Fig. 4. SPINNER MVC850 3-axis vertical machining

4 Results and Discussion

Figure 5 illustrates the programmed trajectory overlaid on the experimental trajectories for the **G641** and **G643** smoothing modes.

The results presented in Fig. 5 show that the **G643** function generates a smoothing block with a higher radius of curvature compared to the **G641** smoothing function. Moreover, the curvature variation of the smoothing block defined by an imposed transition length (**G641**) is greater than that achieved with the programmed contour tolerance (**G643**).

The kinematic behavior of the NC machine at the junction block is influenced by both the geometry of the smoothing block (radius of curvature and its derivative) and the axis maximum kinematic capacities (maximum acceleration and maximum jerk).

Table 3. CNC machine specifications

Parameters	Value
Control	SIEMENS Sinumerik 840D
X-axis	850 mm
Y-axis	510 mm
Z-axis	510 mm
Max. Spindle speed	10000 rpm
Spindle motor power	10 kW
Table clamping surface	1000 × 500 mm
Tool holder	ISO 40
Number of tool places	24 pos
Screen	15″

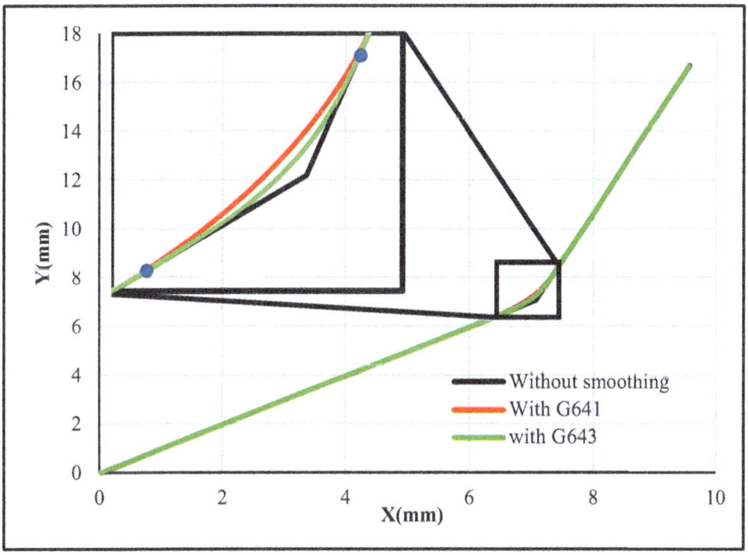

Fig. 5. G641 and G643 effect on tool path

As both experimental tests employ two smoothing blocks with the same transition length Value ($Tl_p = 0.4$ mm) and identical curvature at their junction points (P_1^*, P_3^*), the feed rate is the same at the start and ending of the smoothing block (see Fig. 6).

Due to the inverse relationship between the limit tangential feed rate and the variation in the curvature radius of the element connecting the two programmed linear blocks [15], the smoothing block employing programmed contour tolerance (G643) displays smoother and less varied tangential feed rate compared to that using an imposed transition distance (see Fig. 7).

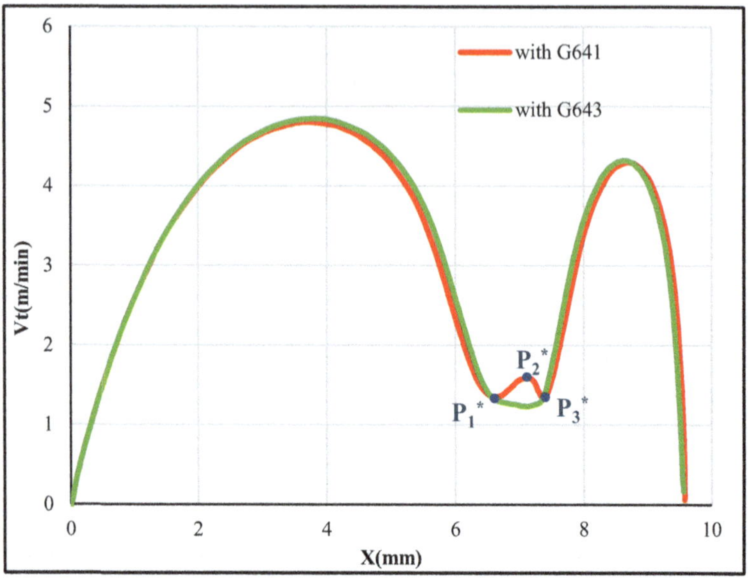

Fig. 6. G641 and G643 effect on the tangential feed rate variation

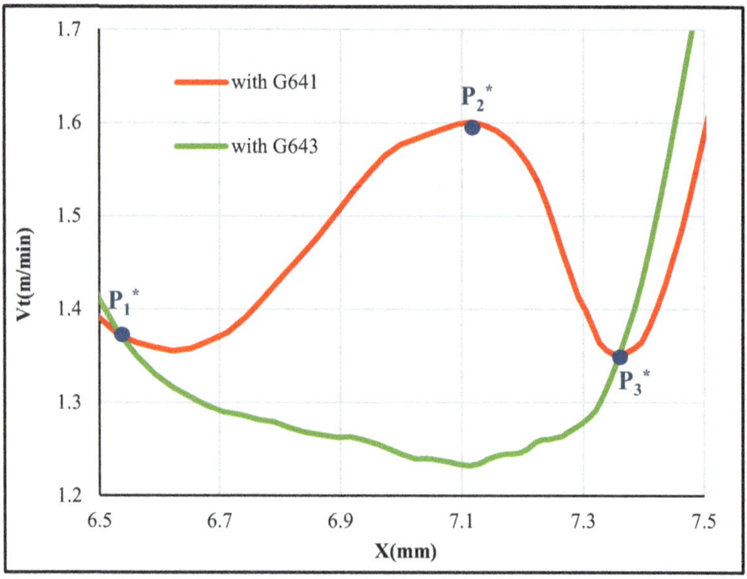

Fig. 7. Feed rate variation along the smoothing bloc

5 Conclusion

A set of experimental tests was performed to explore the effect of the programmed smoothing parameters on the kinematic behavior of the NC machine in HSM. The results show that the **G643** provides a smoother variation in curvature and maintains an almost constant feed rate throughout the smoothing block, making it suitable for finishing operations.

On the other hand, the **G641** presents a greater curvature and a fluctuation in feed speed between the extremums of the smoothing block. It will therefore be preferable to use this function in roughing and semi-finishing operations.

Acknowledgements. This work is carried out thanks to the support and funding allocated to the Unit of Mechanical and Materials Production Engineering (UGPMM/UR17ES43) by the Tunisian Ministry of higher Education and Scientific Research.

References

1. Zhang, Y., Wang, T., Dong, J., Peng, P., Liu, Y., Ke, R.: An analytical G3 continuous corner smoothing method with adaptive constraints adjustments for five-axis machine tool. Int. J. Adv. Manuf. Technol. **109**, 1007–1026 (2020). https://doi.org/10.1007/s00170-020-05402-x
2. Siemens: Mold making with 3 to 5-axis simultaneous milling (2016)
3. Sencer, B., Kakinuma, Y., Yamada, Y.: Linear Interpolation of machining tool-paths with robust vibration avoidance and contouring error control. Precis. Eng. **66**, 269–281 (2020). https://doi.org/10.1016/j.precisioneng.2020.04.007
4. Yang, X., You, Y.: Linear tool path-smoothing method in high-speed machining based on error feasible area and curvature optimization. Appl. Sci. **12**, 9443 (2022). https://doi.org/10.3390/app12199443
5. Liu, Y., Wan, M., Qin, X.-B., Xiao, Q.-B., Zhang, W.-H.: FIR filter-based continuous interpolation of G01 commands with bounded axial and tangential kinematics in industrial five-axis machine tools. Int. J. Mech. Sci. **169**, 105325 (2020). https://doi.org/10.1016/j.ijmecsci.2019.105325
6. FANUC: FANUC Series 30i/31i/32i/35i-model B (2016)
7. Siemens: Milling with SINUMERIK, Mold Making with 3- to 5-Axis Simultaneous Miling (2013)
8. Bouchenitfa, H., Chaves-Jacob, J., Linares, J.-M., Sprauel, J.-M., Azzam, N., Boukebbab, S.: Improvement of toolpath quality combining polynomial interpolation with reduction of toolpath points. Int. J. Adv. Manuf. Technol. **78**, 875–883 (2015). https://doi.org/10.1007/s00170-014-6696-4
9. Bi, Q., Huang, J., Lu, Y., Zhu, L., Ding, H.: A general, fast and robust B-spline fitting scheme for micro-line tool path under chord error constraint. Sci. China Technol. Sci. **62**, 321–332 (2019). https://doi.org/10.1007/s11431-018-9374-6
10. Yeh, S.-S., Su, H.-C.: Implementation of online NURBS curve fitting process on CNC machines. Int. J. Adv. Manuf. Technol. **40**, 531–540 (2009). https://doi.org/10.1007/s00170-007-1361-9
11. Essid, M., Gassara, B., Baili, M., Hbaieb, M., Dessein, G., Bouzid, W.: Analytical modeling of the tool trajectory with local smoothing. In: Benamara, A., Haddar, M., Tarek, B., Salah, M., and Fakher, C. (eds.) Advances in Mechanical Engineering and Mechanics, pp. 24–31. Springer International Publishing, Cham (2019). https://doi.org/10.1007/978-3-030-19781-0_3

12. Lei, C., Yusoff, N., Yap, H.J., Abdelrazek, A.H., Luo, M.: Local tool path smoothing based on symmetrical nurbs transition curve with look ahead optimal method: Experimental and Analytical Study (2022). https://www.researchsquare.com/article/rs-2219822/v1. https://doi.org/10.21203/rs.3.rs-2219822/v1
13. Zhao, G., et al.: An analytical local tool path smoothing method with curvature control. Int. J. Adv. Manuf. Technol. **127**, 1355–1375 (2023). https://doi.org/10.1007/s00170-023-11557-0
14. Siemens: SINUMERIK 840D sl/840DE sl/828D. Notion de base. Manuel de programmation, (2018)
15. Essid, M., Gassara, B., Baili, M., Hbaieb, M., Dessein, G., Bouzid Saï, W.: Analytical modeling of the CNC machine axis motion in high-speed milling with local smoothing. Int. J. Adv. Manuf. Technol. **105**, 457–470 (2019). https://doi.org/10.1007/s00170-019-04157-4
16. Sai, L., Sghaier, R.B., Abdelkader, K., Saï, K., Bouzid Saï, W., Laribi, M.A. (eds.): In: Proceedings of the 2nd International Conference on Innovative Materials, Manufacturing, and Advanced Technologies: IMMAT'2022. Springer, Cham (2024)

Experimental Investigation of Cutting Parameters Influence on Surface Integrity Response During Trimming Operation of Hybrid Composite CGFRP

Sirine Stambouli[1], Lamis Allegue[2], Benoit Schrab[1], Anne Collaine[1], Michel Tourlonias[1], and Ated Ben Khalifa[2](✉)

[1] LPMT, ENSISA, University of Haute Alsace, Mulhouse, France
{sirine.stambouli,benoit.schrab,anne.collaine, michel.tourlonias}@uha.fr

[2] LGM, ENIM, University of Monastir, Monastir, Tunisia
ated.benkhalifa@enim.u-monastir.tn

Abstract. This study aims to investigate the trimming operation of carbon/glass/epoxy hybrid composites. We investigate the influence of the cutting parameters, more specifically the radial engagement and the stacking configuration chosen for the laminated composites, on the surface quality obtained. a surface roughness meter was used to obtain the linear roughness parameters.

Keywords: Hybrid composite · CGFRP · trimming · Surface Integrity Response · surface roughness

1 Introduction

The composite materials market has been in a continuous state of evolution for many years now [1]. Composites are characterized mainly by several factors, such as the type of reinforcement and matrix used, and the structure and stacking of the plies. Currently, the market is divided between carbon-fiber-reinforced composites (CFRP) and glass-fiber-reinforced composites (GFRP) [2]. As the production cost of carbon fibers is relatively high compared to other materials, hybridization techniques have been developed to optimize configurations and reduce manufacturing costs without compromising strength and surface integrity [3]. To obtain final composite parts, operation of trimming is one of the most common finishing processes used [4]. The major obstacles we may encounter when machining hybrid composites are the surface quality obtained and its integrity [5]. These can be influenced by the choice of layer stacking in the glass/carbon composite and the cutting parameters.

In the present work, the roughness of the machined surface is investigated by varying the cut parameters, more specifically the radial engagement.

M. Chouchane et al. (Eds.): CMSM 2023, LNME, pp. 391–399, 2024.
https://doi.org/10.1007/978-3-031-65007-9_41

2 Material and Methods

2.1 Material

The composite material was made with E Glass fibers Satin of 4/ Epoxy resin prepreg and High Strength Carbon fibers 4H Satin/ Epoxy resin prepreg reference HexPly® M26T/50%/G939. The material was supplied by SAFRAN Tunisia. The test pieces were made by a method of autoclave molding. This process of composite manufacturing consists of firmly laying down the prepregs in a mold in the desired sequences and then applying heat and pressure. The purpose of this study is to identify the effect of the hybridization of stratified composites on the integrity and quality of surfaces during trimming operations. This is why, two different layering configurations have been chosen to produce 4 mm thick composite plates. Figure 1 shows the two stacks chosen for the two hybrid composites CGFRP:

H1: [C/G/C/G/C/G/G] s and H2: [G/C/G/C/G/C/C] s.

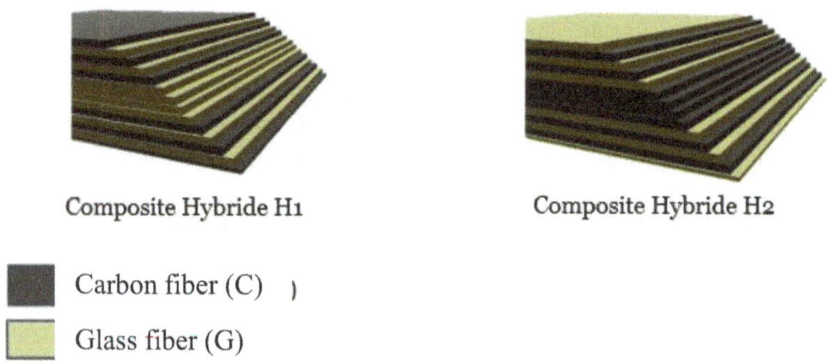

Fig. 1. Hybrid Composites stacking sequences.

To analyze the influence of reinforcement orientation in the two hybrid composites during machining, it was decided to cut two specimens for each hybrid, each with a different orientation. This approach will enable us to compare the results obtained from the two different composites in two different reinforcement orientations for the first round of testing.

The dimensions of plates are 30 mm wide and 4 mm thick. The first specimen in the "perpendicular direction" is 160 mm long, and the second in the "horizontal direction" is 130 mm long.

2.2 Machine and Tools

The trimming tests were carried out using a Charley 2U machine equipped with a spindle rotation up to 24000 rpm, allowing feed rates of up to 100 mm/s, with maximum spindle power of up to 1.5 kW. As for the cutting tool (see Fig. 2), we opted for a SECO carbide milling cutter with a diameter of d = 6 mm whose main characteristics are shown in Table 1.

Table 1. Characteristics of the tool used.

Manufacturer's reference	Teeth number	Coating	Specifications
840060R050Z4.0- DURA	4	Dura	Diamant cut teeth

Fig. 2. Cutting tool.

2.3 Machining Strategy

To carry out trimming tests on hybrid composite plates H1 and H2, the 4 mm thick plates will be positioned and maintained between two flanges tightened by two screws. During machining, the tool will advance at a feed rate Vf (mm/min) with a rotation speed N (rpm). To ensure a balanced distribution of clamping forces on the upper flange, we implemented a clamping strategy by alternating between the two screws and using a torque wrench (see Fig. 3). First, an initial torque of 1 N.m is applied to both screws, then a torque of 3 N.m is applied.

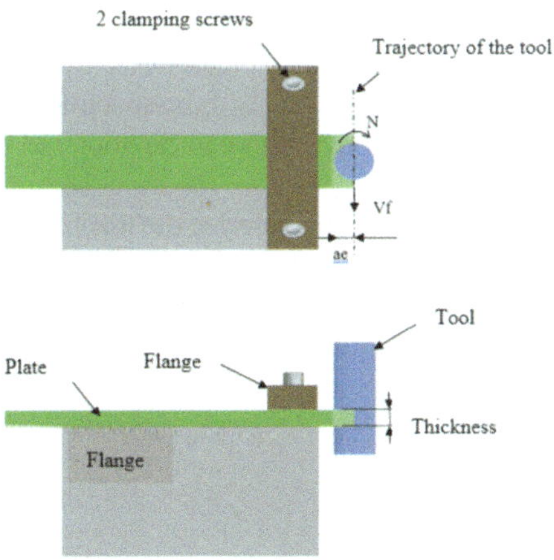

Fig. 3. Trimming strategy.

Since the targeted objective of this study is to analyze the influence of varying cutting parameters during routing on surface quality for both H1 and H2 plate stack types, it is crucial to identify which cutting parameters will be constant in the experiment and which will be varied. The Table 2 shows the cutting parameters used, with their predefined values, and adapted to the composite materials which were previously tested by a PhD researcher from the beforementioned laboratory.

Table 2. Cutting parameters.

Cutting parameters	N (rpm)	Vf (mm/min)	Vc (m/min)	ae (mm)	fz (mm/tooth)	Lpaf (mm)
Value	6101	732	115	0.5	0.012	15.5
				0.75		
				1		

We have chosen to focus our attention solely on the variation of the radial engagement ae (mm) as the key parameter to be developed. This approach is used to quantitatively determine the relationship between radial engagement and surface characteristics after the routing operation on our hybrid composite plates. Three separate and successive tests for each radial engagement value will be carried out, enabling us to acquire data for both types of hybrid composites.

3 Tools and Methods for the Characterization

To form an acquisition system for detecting signals from operations, we have the following equipment at our disposal: Kistler table type 9129AA fixed underneath the lower flange which acts as a force sensor, A type 4508-B three-axis accelerometer mounted on the top flange to analyze the vibrations on the plate caused by routing and an infrared (IR) pyrometer positioned 50 mm from the surface of the specimen to be machined allows to measure the temperature at the interface between the tool and the surface without any direct contact.

The second part of this study is to characterize the machined surface with roughness parameters after each trimming test. A TESA Rugosurf 90G roughness tester is available for post-machining linear roughness measurements. This device makes it possible to assess surface quality in terms of texture and finish by analyzing roughness parameters (Ra, Rz, Rq, etc.).

4 Results and Discussions

The main objective of this study is to analyze the impact of different stacking of the two hybrid composite plates H1 and H2 on the surface quality obtained after trimming tests. Likewise, we analyze the influence of the variation in the cutting parameter: radial engagement ae on the linear roughness parameters, based on data obtained from signals relating to forces during the routing tests. Two testing campaigns will be carried out.

4.1 First Campaign

This campaign aims to see the difference between the orientations of hybrid composite plates; "Perpendicular Direction" and "Horizontal Direction" by comparing forces and roughness parameters. We'll set ae at 0.5 mm.

Histograms in Fig. 4 show the forces for the two plate orientations and for the two types of hybrid composites H1 and H2. Error bars are inserted to give a general idea of the dispersion of values for the 3 tests.

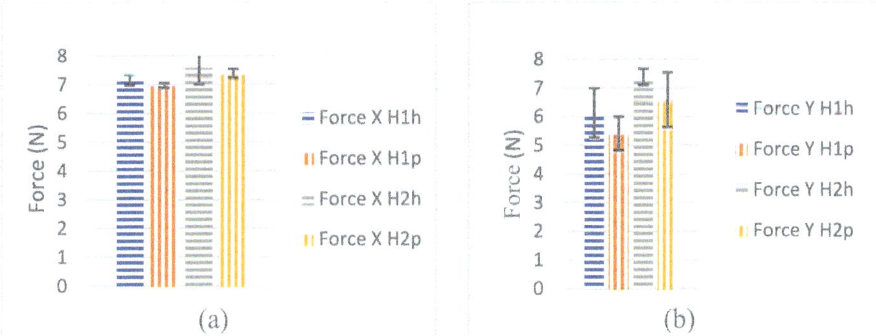

Fig. 4. Force along X and Y direction for composite H1 and H2: (a) Force along X; (b) Force along Y

We can explore and explain these histogramms on two different approaches:

- For all three axes, the average stress values for the "Horizontal" orientation are slightly higher than those for the "Perpendicular" orientation. This observation applies to both H1 and H2 hybrid composites. This difference of around 1N is inconsiderable.
- We find that the three-axis force values for H2 hybrid composite plates are greater than those for H1 hybrid composite plates. This result is justified by the higher carbon content of the H2 hybrid composite, based on ply stacking. Indeed, HS carbon fibers are proven to have higher mechanical properties than E-glass fibers [5], leading to higher force values for the H2 hybrid composite.

The following Fig. 5 shows the mean values of the linear roughness parameter Ra for both H1 and H2 hybrid composites, in "horizontal" and "perpendicular" orientations.

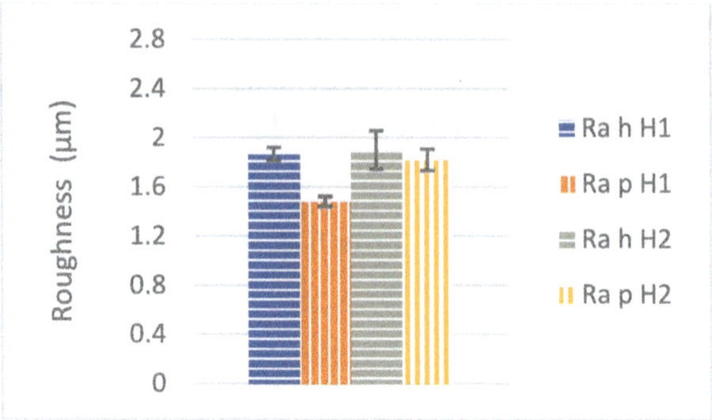

Fig. 5. Roughness for Composite H1 and H2

These results are consistent. Indeed, since the stresses of H2 hybrid composite are greater than those of H1, the roughness values follow the same trend; hence the roughness parameter of H2 hybrid composite is greater than that of H1 hybrid composite.

4.2 Second Campaign

From the first campaign, it can be deduced that the difference between the values for the two orientations "perpendicular" and "horizontal" is not significant, regardless of the stack chosen for machining. This is why, for the second measurement campaign, we consider only the impact of the variation in radial engagement ae for the two types of hybrid composite stack H1 and H2.

Figure 6 shows the evolution of forces with variation in the value of radial engagement for the two hybrid composites H1 and H2.

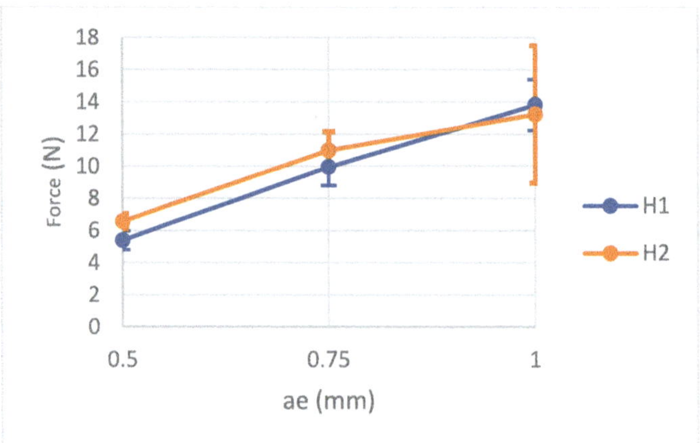

Fig. 6. Evolution of force as a function of ae

We can observe that:

- Forces increase proportionally with radial engagement.
- For ae = 0.5 mm and ae = 0.75 mm, the H2 hybrid composite admits slightly higher force values than the H1 hybrid composite, which is in line with our previous explanations.
- For ae = 1 mm, we notice that the values are reversed; the force values belonging to the H1 hybrid composite surpass those of the H2 hybrid composite.

Error bars based on standard deviation measurements are inserted on the curves. They are used to represent the variability of the measurements, which helps us to better interpret the inversion of the curves. Indeed, the error bars are long for ae = 1 mm, indicating that the measurements are unstable and subject to high variability.

In other words, one of the trimming tests has given values that are too low or too high compared with the others; this may explain the instability of the last part of the curve.

After processing the temperature signals from the infrared pyrometer, Fig. 7 provides an overview of the evolution of the maximum temperature during trimming as a function of radial engagement.

This figure shows:

- The maximum temperature detected increases proportionally with radial engagement ae for hybrid composite stack configurations.
- The maximum temperature during machining of the H2 hybrid composite is higher than that of the H1 hybrid composite. This result is linked to the composition of H2 hybrid composite, which contains more carbon than H1 hybrid composite. Indeed, carbon fibers are stiffer and have a higher thermal conductivity than glass fibers [6], which results in a higher friction rate.

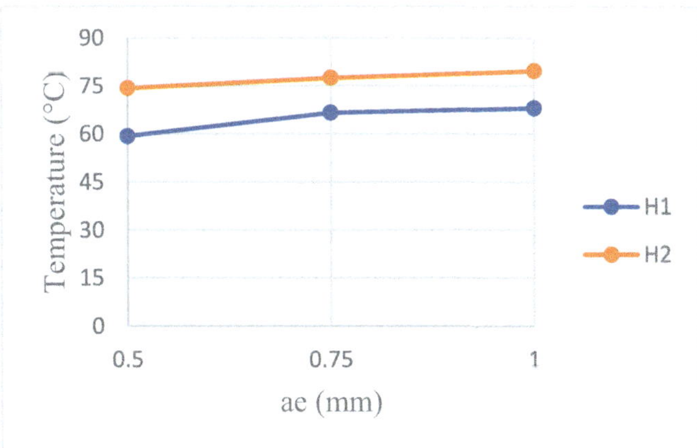

Fig. 7. Temperature as a function of ae

The Fig. 8 below shows the evolution of linear roughness parameter Ra as a function of radial engagement ae.

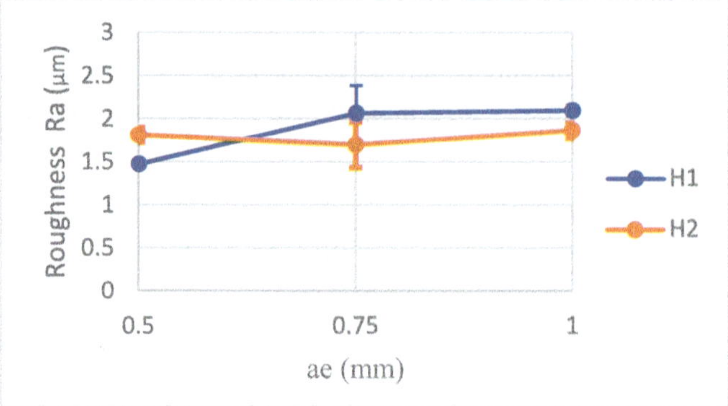

Fig. 8. Evolution of Ra as a function of ae.

It can be seen that:

- the roughness parameter for the H1 hybrid composite increase progressively with radial engagement, which is consistent with the fact that increasing radial engagement leads to an increase in surface roughness [7].
- For radial engagement values equal to ae = 0.75 mm and ae = 1 mm, the values of the roughness parameter of the H1 hybrid composite are higher than those of the H2 hybrid composite. This is because, it is known that carbon fibers generally give lower roughness values than glass fibers [8], and as the H1 hybrid composite contains more glass fiber than the H2 composite, it is obvious that its machined surface is rougher.

5 Conclusion

In this study, we investigated the impact of the cutting parameters and the hybridization of reinforcements on the surface quality of CGFRP composite trimming. Our focus was on varying the radial engagement ae.

Many approaches can be considered to progress in this work, such as improving the experimental protocol, testing other tools, testing other parameter variations to see if the influence is the same.

While the first results are encouraging and the prospects numerous, they highlight the need to quantify vibrations and to have at our disposal a means of characterizing the vibratory behavior of the workpiece during trimming. Vibration measurement will help to better define the operating ranges of a tool in relation to cutting conditions.

References

1. Brooks, G.: Composite Materials: Advances in Materials Science and Engineering. STATES Academic Press (2022)
2. Witten, E., Kraus, T., Kuhnel, M.: Composites Market Report 2016 - Market developments, trends, outlook and challenges (2016)
3. Bahadur Singh, S., Chala, H.: Hybrid effect of functionally graded hybrid composites of glass – carbon fibers (2018)

4. Furet, B., Le Borgne, D., Jolivel, B.: Milling and drilling of composite materials for the aeronautics: feature aeronautics. JEC Compos. Mag. **18**, 41–44 (2005)
5. Azouzi, M., et al.: Characterization of the Surface Roughness of a GFRP by a 3D Profilometer After Trimming. Lecture Notes in Mechanical Engineering, 8th International Congress on Design and Modeling of Mechanical Systems, CMSM 2019 (2020)
6. Ashby, M.F.: Properties of Advanced Composite Materials (2005)
7. Kalpakjian, S., Schmid, S.R.: Machining of Composite Materials (2022)
8. Moyne, J.S., Ammons, M.A. : Composites Machining : Principles, Processes and Applications (2018)

Numerical Investigation of the Hole Saw Tool Geometry Effects on Drilling of CFRP

Amira Hassouna(✉), Slah Mzali, and Salah Mezlini

Mechanical Engineering Laboratory, National Engineering School of Monastir, University of Monastir, Monastir, Tunisia

Hassounaamira0@gmail.com

Abstract. Drilling in Carbon Fiber Reinforced Polymer (CFRP) composites is of paramount importance, especially in industries like aerospace that heavily rely on these advanced materials. The precision and integrity of the drilled holes play a vital role in determining the overall structural performance and reliability of composite components. This study presents a comprehensive numerical investigation into the effect of hole saw tool geometry and cutting parameters on the drilling of Carbon Fiber Reinforced Polymer (CFRP) composites. The analysis aims to optimize the drilling operation by varying the rake angle and the feed rate. To fulfil this, a finite element model is developed using Abaqus explicit software. The outcomes reveal that the rake angle has a negligible effect on the thrust force and the workpiece damage. In contrast, the feed rate significantly affects the thrust force, with a noticeable increase as the feed rate is raised. Finally, it can be observed that the workpiece damage is not influenced by the feed rate.

Keywords: CFRP · Rake angle · feed rate · Thrust force · workpiece damage

1 Introduction

Over the past decade, there has been a notable increase in the utilization of composite materials across various sectors, including aerospace, marine, civil engineering, and electronics. This surge in popularity can be attributed to their advantages, which include reduced weight, enhanced stiffness, and resistance to chemical factors. Despite extensive research on composite materials, challenges persist in several aspects of composite part manufacturing, particularly in terms of machinability. It is important to highlight that drilling constitutes a significant portion of material removal processes. However, characteristics of composites, such as anisotropy, heterogeneity, and the presence of highly abrasive fibers, make drilling operations a complex endeavor, resulting in various issues such as fiber pull-out, burrs, and delamination. To mitigate the occurrence of drilling defects, it is essential to carefully select parameters affecting the drilling process. In this regard, numerous research studies have delved into the impact of cutting parameters and drill geometry on the drilling process. Particularly, Ficini et al. (Ficici et al. 2017) experimentally investigated the effect of the cutting speed and the feed rate on the drilling

M. Chouchane et al. (Eds.): CMSM 2023, LNME, pp. 400–409, 2024.
https://doi.org/10.1007/978-3-031-65007-9_42

quality of 30% glass-fiber reinforced PPA matrix composites using different drills (carbide, ARS, ARS + TiN). They showed that the delamination and the roughness decreased with increasing the cutting speed and decreasing the feed rate. The same results were confirmed by Bosco et al. (Bosco et al. 2013), and Yardimeden et al. (Yardimeden et al. 2014). Similarly, Khashaba (Khashaba 2004) demonstrated the decrease of the delamination with increasing the cutting speed when drilling woven glass/epoxy composites. Franke (Franke 2011) studied the influence of twist drill edge radius on the quality of drilled holes when machining unidirectional glass fiber reinforced thermoplastics. They showed that increasing the edge radius increases the thrust force. Furthermore, Using orthogonal cutting of CFRP, Koplev et al. (Koplev et al. 1983) investigated the effect of rake and clearance angles on the cutting force and the thrust force. They showed that the cutting force decreased with increasing the rake angle, while there was no clear trend towards the thrust force. Concerning the clearance angle, it exerts a substantial impact on the thrust force, while its influence on the cutting force is minimal. They attributed the decrease of the thrust force with the clearance angle to the reduction of the contact zone between the tool and the workpiece.

Other studies have been devoted to the design of new tool geometries. Hocheng & Tsao (Hocheng and Tsao 2006) studied the effect of the drill geometry on the delamination factor. Different types of tools were tested (Candle stick drill, Saw drill, Core drill, Step drill, twist drill). They found that the trepan drill offered the highest critical feed rate and thrust force (high-speed machining without delamination). Mathew et al. (Mathew et al. 1999) compared the geometry of the trepan tool with that of the twist drill when drilling Glass fiber reinforced polymers (GFRP). They concluded that the total volume of material removed by the trepan tool is less than that produced by the twist drill, resulting in a reduction in the cutting force and the thrust force. A 50% reduction in the thrust force and a 10% reduction in the torque were observed, resulting in improved drilling quality.

Within the existing body of literature, numerous tool geometries have been explored with the aim of minimizing drilling defects and elevating the overall quality of drilling processes. In the pursuit of this objective, the present study introduces a novel tool geometry known as the "hole saw tool." The primary focus of this research is to meticulously optimize both geometric parameters and cutting parameters associated with the hole saw tool, with the explicit goal of mitigating drilling defects. To accomplish this task, a finite element model is developed using Abaqus Explict code. The initial part involves the validation of the developed numerical model. In the subsequent section, an investigation into the influence of the rake angle is conducted, specifically focusing on its effects on the thrust force, the torque, and the workpiece damage. Lastly, the study explores the repercussions of varying feed rates on the thrust force, the torque, and the workpiece damage.

2 Finite Element Model

2.1 Geometric Modeling, Meshing and Boundary Conditions

A finite element model simulating the drilling process is developed using ABAQUS-Explicit software. The hole-saw tool exhibits a diameter of 10.5 mm and a thickness of 0.5 mm. It is equipped with 12 consecutive teeth positioned on two circles with varying diameters. These teeth feature a rake angle of 0°, a clearance angle of 5°, and a nose radius measuring 0.01 mm. The hole saw tool is assumed to be rigid, as the Young's modulus of the tool is higher than that of the composite plate. The workpiece, measuring 17 mm in length, 17 mm in width, and 3 mm in thickness, is regarded as deformable. The nodes on the lateral sides of the part are clamped along the x-direction (XSYMM) and the nodes on the other sides are embedded (Fig. 1). A rotational speed of 6000 rpm and a feed rate of 300 mm/min are applied to the tool reference point. In this model, the part is meshed with SC8R (continuum shell) elements. A minimum element size of 2 μm is applied in the contact zone, while a coarse mesh is applied elsewhere. The total number of elements in the workpiece is equal to 2,353,950. Rigid quadrilateral elements of type (R3D4) are used for the discretization of the tool with a total number equal to 20,684.

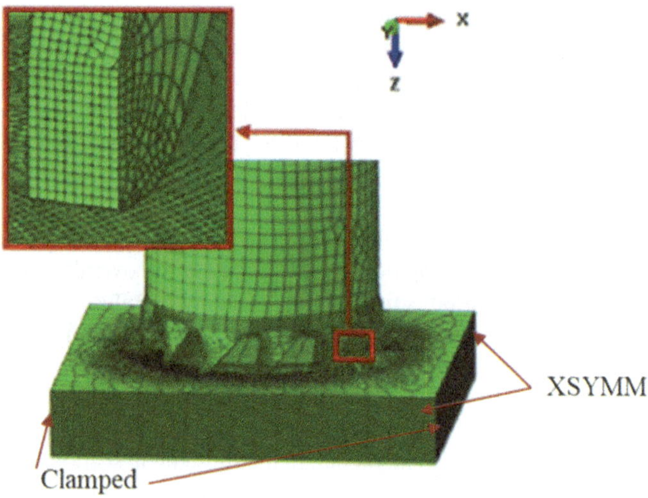

Fig. 1. Finite element model

The contact between the tool and workpiece is simulated using a general contact algorithm available in abaqus, with a fixed friction coefficient of 0.5.

2.2 Material Properties

The composite material used in this study is an epoxy matrix reinforced with unidirectional carbon fibers. This material exhibits orthotropic elastic behavior. Its damage is described using the Hashin criterion. The material behavior and damage law parameters are shown in Table 1.

This criterion recognizes four failure modes, encompassing fiber failure in tension and compression, as well as matrix failure in tension and compression (Hassouna et al. 2020).

Table 1. Material properties (Santiuste et al. 2010)

Parameters	E_1	E2	G_{12}	υ_{12}	
Values (GPa)	126	11	6.6	0.28	
Parameters	X^T	X^C	Y^T	Y^C	S
Values (GPa)	1.95	1.48	0.048	0.2	0.079

3 Results and Discussion

3.1 Validation of the Finite Element Model

The validation of the numerical model involves assessing its ability to predict experimental results. Two criteria have been chosen to validate the developed model.

The initial criterion focuses on the thrust force, and to assess it, the evolution of the thrust force over time, with a fiber orientation angle of 90° is presented in Fig. 2. It can be observed that the thrust force curve can be divided into two phases: (i) the first phase represents the transient part, and (ii) the second phase constitutes the steady-state portion. The value of the thrust force increases over time, reaching an average value of approximately 160 N. The experimental value is around 150 N. An average relative error

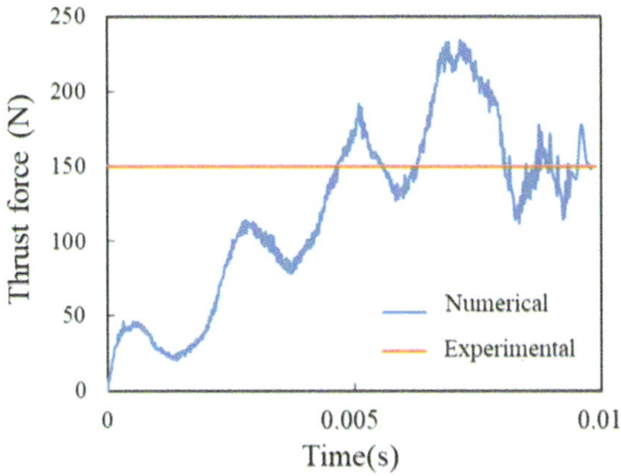

Fig. 2. Comparison between the thrust force of the numerical model and that measured experimentally.

of 6% is revealed. Notably, the numerically measured thrust force closely aligns with the experimentally determined value.

The second validation criterion pertains to damage assessment. Figure 3 displays the damage outcomes from the numerical model and the experimentally determined damage at the hole entrance. The numerically estimated affected area (indicated by the red boundary) spans approximately 12.92 mm, whereas the experimentally observed affected diameter measures approximately 13.56 mm, resulting in a mean relative error of 4%. This indicates a high degree of reliability in the numerical model, with a strong agreement between the experimental and numerical results being achieved.

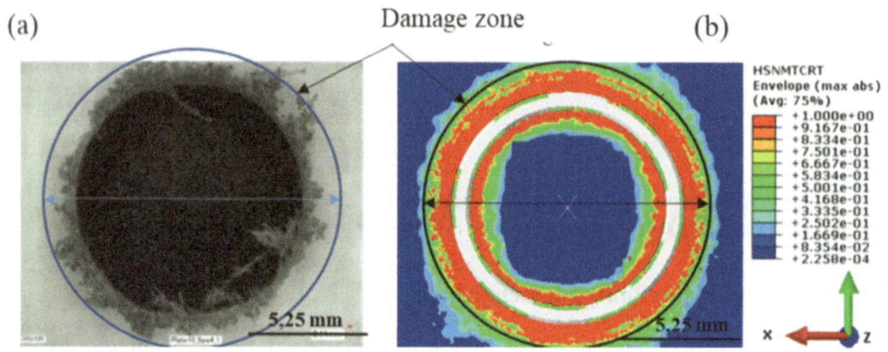

Fig. 3. Damage distribution in the workpiece for (a) experimental observation and (b) numerical observation.

3.2 Damage Analysis of different Failure Modes of the Hashin Criterion

As the Hashin's criterion distinguishes among four modes of failure, namely tensile and compressive fiber failure, as well as tensile and compressive matrix failure, it becomes essential to compare these various failure modes. Figure 4 illustrates the damage surrounding the hole for these distinct failure modes. It is seen that the damage resulting from tensile matrix failure is more significant than that arising from the other modes. In fact, the Hashin criterion's various fracture modes are inversely proportional to the fracture strengths, and since the tensile strength of the matrix is the lowest, the associated damage will be the greatest. Furthermore, it is seen that the damage induced by the fiber compression failure is negligible.

3.3 Effect of the Rake Angle

Cutting Forces Analysis

In order to study the sensitivity of the rake angle to the thrust force and the torque, three trials are carried out with three rake angles 0, 10 and 20°, using a spindle speed of 6000 rpm and a feed rate of 300 mm/min. Figure 5 shows the evolution of the thrust force and the torque for different rake angles. It is seen that the thrust force and the

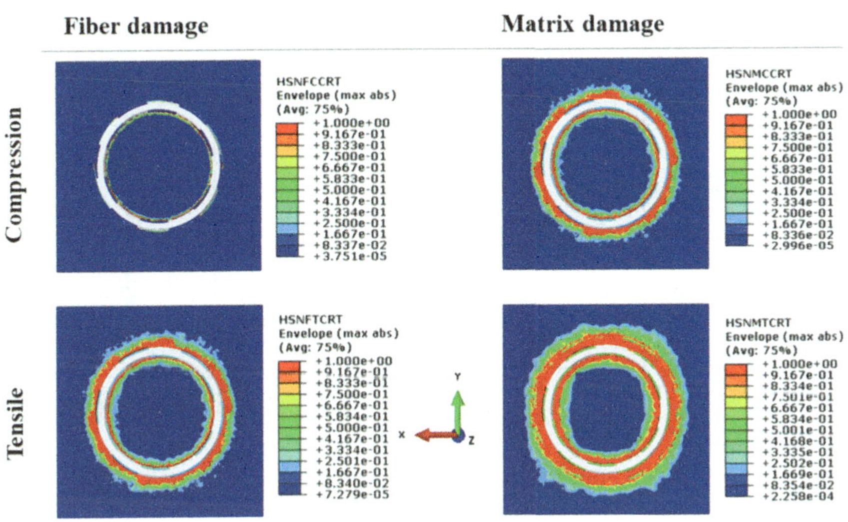

Fig. 4. Damage distribution in the workpiece for different failure modes of the Hashin criterion.

torque are slightly influenced by the rake angle. Indeed, as the rake angle increases from 0 to 10°, the torque decreases by around 10%. Thereafter, as the rake angle increases from 10 to 20°, the torque increases by around 9%. The minimum torque value is seen at a rake angle of 10°. As far as the thrust force is concerned, we note that its evolution is almost stable. A slight increase of around 8% is observed by varying the rake angle from 0 to 20°. The small variation in thrust force with the rake angle can be attributed to the small amount of material removed per tooth during the drilling process. The same result was demonstrated by Yang et al. (Yang et al. 2018) during the orthogonal cutting of the graphite/polymer composite. They showed that the thrust force is insensitive to the positive cutting angle.

Damage Analysis

Figure 6 displays the distribution of damage in the workpiece at various rake angles. It's apparent that the rake angle has a minimal influence on the workpiece damage. For all three angles, the damaged diameter measures approximately 12.92 mm. Similar results were found by Nayak & Bhatnagar (Nayak et al. 2005) during the orthogonal cutting of GFRP composite. They revealed that the rake angle has a non-significant effect on the subsurface damage.

3.4 Effect of the Feed Rate

Cutting Forces Analysis

To investigate the effect of the feed rate on the thrust force and the torque, three feed rates (200, 300, 400 mm/min) are employed. As illustrated in Fig. 7, the feed rate exerts a notable influence on both thrust force and torque. As the feed rate ranges from 200 to 400 mm/min, there is an approximately 80% uptick in the thrust force and a roughly 46% increase in the torque. This surge can be attributed to the fact that raising the feed

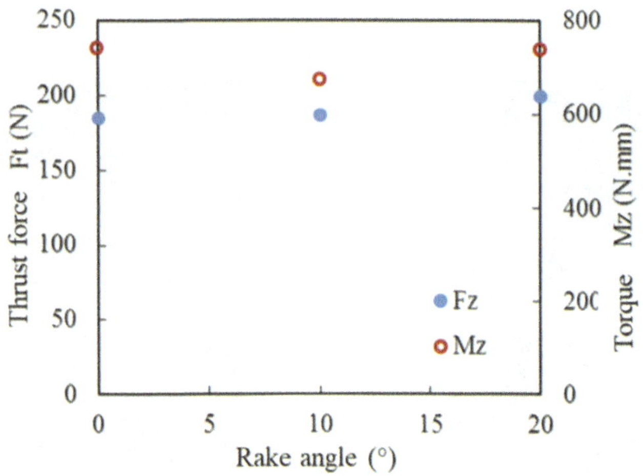

Fig. 5. Effect of rake angle on thrust force and torque.

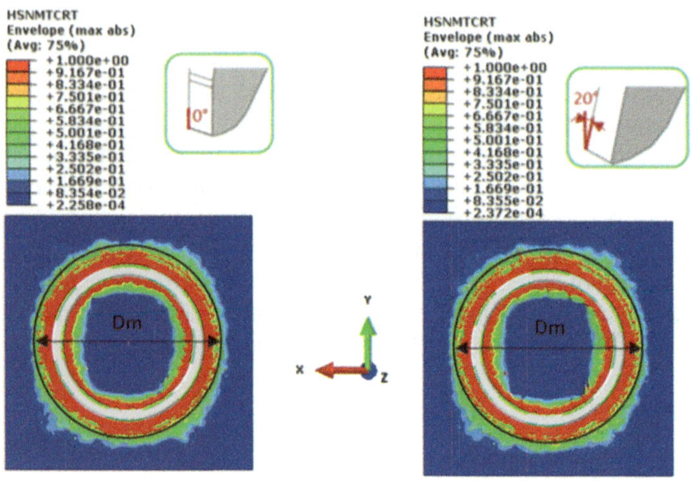

Fig. 6. Damage distribution in the workpiece for different rake angles (a) $\gamma = 0°$ and (b) $\gamma = 20°$.

rate results in a higher feed per tooth and per revolution, which increasing the amount of material that needs to be removed and generates more effort. Eneyew et al. (Eneyew and Ramulu 2014) found the same trend when drilling CFRP using a PCD (polycrystalline diamond) twist drill.

It's important to highlight that the primary cause of defects observed at the hole exit is the thrust force. To mitigate these defects, it's essential to use a lower thrust force (Tsao and Hocheng 2003).

Damage Analysis

The distribution of damage in the workpiece for different feed rates is shown in Fig. 8. It is clear that the effect of the feed rate on the workpiece damage is considered negligible.

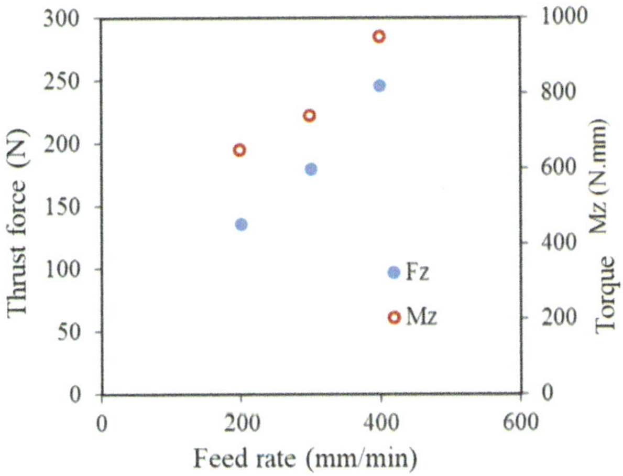

Fig. 7. Effect of feed rate on thrust force and torque for 0° rake angle.

Damaged diameters are of the order of 12.72, 12.92 and 12.82 mm for feed rates of 200, 300 and 400 mm/min respectively. The minimum damage is observed for a feed rate of 200 mm/min.

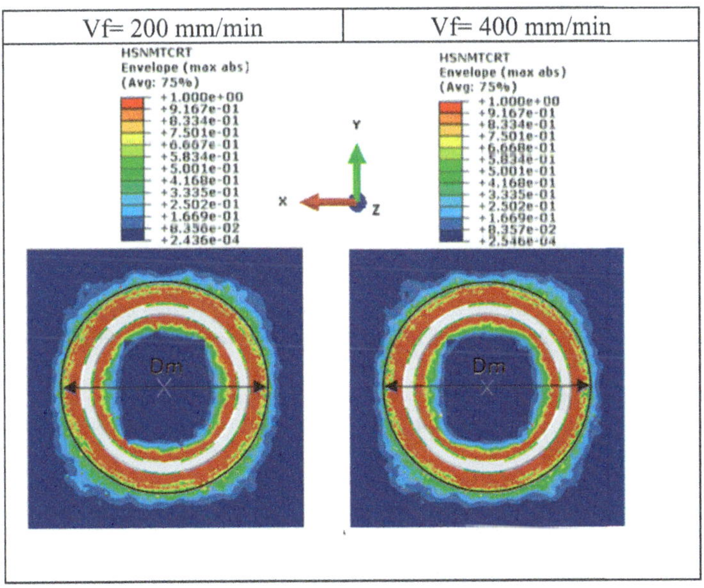

Fig. 8. Damage distribution in the workpiece at different feed rates

4 Conclusion

In this work, a numerical study is carried out to examine the sensitivity of the thrust force, the torque and the damage around the hole to the rake angle and the feed rate. To achieve this, a 3D numerical model simulating the drilling of unidirectional carbon fiber reinforced epoxy matrix composite (CFRP) is developed. The results revealed that the thrust force, the torque and the damage around the hole derived from the numerical model are in good agreement with experimental results. The thrust force, the torque, and the damage around the hole exhibit low sensitivity to changes in the rake angle. However, the feed rate has a remarkable influence on the thrust force and the torque. These two outputs exhibit a notable increase as the feed rate varies from 200 to 400 mm/min. Nevertheless, the effect of the feed rate on the damage around the hole is deemed to be negligible.

In this study, we recommend an average cutting angle of 10° and a low feed speed of 200 mm/min to minimize drilling defects.

References

Bosco, M.A.J., Palanikumar, K., Prasad, B.D., Velayudham, A.: Influence of machining parameters on delamination in drilling of GFRP-armour steel sandwich composites. Procedia Eng. **51**, 758–763 (2013). https://doi.org/10.1016/j.proeng.2013.01.108

Eneyew, E.D., Ramulu, M.: Experimental study of surface quality and damage when drilling unidirectional CFRP composites. J. Market. Res. **3**(4), 354–362 (2014). https://doi.org/10.1016/j.jmrt.2014.10.003

Ficici, F., Ayparcasi, Z., Unal, H.: Influence of cutting tool and conditions on machinability aspects of polyphthalamide (PPA) matrix composite materials with 30 % glass fiber reinforced. Int. J. Adv. Manuf. Technol. **90**(9–12), 3067–3073 (2017). https://doi.org/10.1007/s00170-016-9626-9

Franke, V.: Drilling of long fiber reinforced thermoplastics - influence of the cutting edge on the machining results. CIRP Ann. Manuf. Technol. **60**(1), 65–68 (2011). https://doi.org/10.1016/j.cirp.2011.03.078

Hassouna, A., Mzali, S., Zemzemi, F., Mezlini, S.: Orthogonal cutting of UD-CFRP using multiscale analysis : finite element modeling. J. Compos. Mater. 1–14 (2020). https://doi.org/10.1177/0021998319899129

Hocheng, H., Tsao, C.C.: Effects of special drill bits on drilling-induced delamination of composite materials. Int. J. Mach. Tools Manuf **46**(12–13), 1403–1416 (2006). https://doi.org/10.1016/j.ijmachtools.2005.10.004

Khashaba, U.A.: Delamination in drilling GFR-thermoset composites. Compos. Struct. **63**(3–4), 313–327 (2004). https://doi.org/10.1016/S0263-8223(03)00180-6

Koplev, A., Lystrup, A., Vorm, T.: The cutting process, chips, and cutting forces in machining CFRP. Composites **14**(4), 371–376 (1983). https://doi.org/10.1016/0010-4361(83)90157-X

Mathew, J., Ramakrishnan, N., Naik, N.K.: Investigations into the effect of geometry of a trepanning tool on thrust and torque during drilling of GFRP composites. J. Mater. Process. Technol. **91**, 1–11 (1999)

Nayak, D., Bhatnagar, N., Mahajan, P.: Machining studies of Uni-directional glass fiber reinforced plastic (UD-GFRP) composites part 1: effect of geometrical and process parameters. Mach. Sci. Technol. **9**(4), 481–501 (2005). https://doi.org/10.1080/10910340500398167

Santiuste, C., Soldani, X., Miguélez, M.H.: Machining FEM model of long fiber composites for aeronautical components. Compos. Struct. **92**(3), 691–698 (2010). https://doi.org/10.1016/j.compstruct.2009.09.021

Tsao, C.C., Hocheng, H.: The effect of chisel length and associated pilot hole on delamination when drilling composite materials. Int. J. Mach. Tools Manuf **43**(11), 1087–1092 (2003). https://doi.org/10.1016/S0890-6955(03)00127-5

Yang, D., Wan, Z., Xu, P., Lu, L.: Rake angle effect on a machined surface in orthogonal cutting of graphite/polymer composites. In: Advances in Materials Science and Engineering, (2018). https://doi.org/10.1155/2018/3601918

Yardimeden, A., Kilickap, E., Hisman, Y.: Effects of cutting parameters and point angle on thrust force and delamination in drilling of CFRP. Materialpruefung/Mater. Testing **56**(11–12), 1042–1048 (2014). https://doi.org/10.3139/120.110666

Predictive Modeling of Part Quality in the WAAM Process Using PCA Model Reduction and Machine Learning

Amir Guizani[1](✉), Moncef Hammadi[2], Larbi Yousfi[1], and Mohamed Haddar[1]

[1] LA2MP, National Engineering School of Sfax, University of Sfax, 3038 Sfax, Tunisia
amir.guizani@live.fr, {larbi.yousfi,mohamed.haddar}@enis.tn
[2] QUARTZ Lab. EA7393, ISAE-SUPMECA, 3 Rue Fernand Hainaut, 93400 Saint-Ouen, France
moncef.hammadi@isae-supmeca.fr

Abstract. The Wire Arc Additive Manufacturing (WAAM) technology is a directed energy deposition process that uses an arc welding equipment to melt and continuously deposit material in wire form. This work focuses on the study of the WAAM process using an approach based on machine learning and statistical data analysis to solve the problem of weld quality classification. In this study, experiments were carried out to explore the essential parameters of this process, such as welding speed, wire feed rate and electric current, with the aim of producing high-quality weld layers in terms of geometry. The resulting data from the experiments were rich in information. In order to efficiently manage the data collected during these experiments, Principal Component Analysis (PCA) was applied to reduce the dimensionality of the data while preserving its relevance. Then, several machine learning models, including Random Forest, Logistic Regression, Decision Tree, Naive Bayes Gaussian and SVM, were used to predict welding quality based on parameter data generated by PCA. To evaluate and compare the performance of these models, evaluation metrics such as precision, recall, F1-score and confusion matrix were used. Following this comparative evaluation, we identified that the Random Forest model is the most accurate, capable of reliably predicting part quality based on various combinations of parameters. This approach was implemented in python and was also successfully applied to accurately predict the results of the WAAM process.

Keywords: WAAM process · PCA · Machine Learning · Statistical Data Analysis · Weld Quality Classification

1 Introduction

Arc Additive Manufacturing (WAAM) plays a key role in the field of additive manufacturing. Its fundamental principle is based on the progressive creation of three-dimensional parts, layer by layer. This approach offers several significant advantages, including increased design flexibility, reduced material waste and lower production costs (Taşdemir et al. 2021).

M. Chouchane et al. (Eds.): CMSM 2023, LNME, pp. 410–419, 2024.
https://doi.org/10.1007/978-3-031-65007-9_43

While the advantages of WAAM in additive manufacturing are evident and have been gaining traction in industry applications, there remains a crucial aspect that requires further exploration, which is understanding the influence of various parameters on the welding quality. Addressing this gap not only solidifies the benefits of WAAM but also enhances its potential applications. Consequently, our aim in this study is to carry out an in-depth study of this process involving a large number of experiments. Each of these experiments aims to explore different combinations of parameters and study in detail their influence on the quality of the parts to be obtained. To effectively manage the data collected during these experiments, we propose in this paper a structured methodology aimed at solving the problem of welding quality classification in the WAAM process, exploiting the advantages of machine learning and statistical data analysis. This methodology is based on dimensionality reduction techniques, notably Principal Component Analysis (PCA), thus contributing to simplifying the modeling of this innovative additive manufacturing process.

The remainder of this paper is organized as follows: in the next section, we present an overview of PCA-based work in manufacturing. In Sect. 3, we describe our approach based on machine learning and statistical data analysis to predict part quality in the WAAM process. In Sect. 4, we detail the experimental part carried out to collect the necessary data from the WAAM process. In Sect. 5, we present and discuss the results obtained to illustrate the validity and effectiveness of our approach. Finally, Sect. 6 presents the main conclusions of this study.

2 Synthesis of PCA-Based Work in the Manufacturing Field

Principal Component Analysis (PCA) is a statistical analysis method widely used in manufacturing. It reduces data complexity and reveals trends and hidden relationships within large sets of production data. It is also frequently used to optimize manufacturing processes. It helps identify the key variables influencing production quality and efficiency. By focusing on these variables, companies can make significant improvements to their manufacturing processes (Hwang et al. 2022), (De Freitas Gomes et al. 2012) and (Tong and Wang 2002).

Another essential application of PCA is real-time quality control. Manufacturing plants use PCA to continuously monitor production parameters. It quickly detects deviations from quality standards, enabling immediate intervention to maintain high standards (Straat et al. 2022). This technique is also used in the product design phase. It helps engineers understand the complex interactions between design variables. This understanding is crucial to designing products that meet performance standards while remaining economically viable (Jagadish and Ray 2016). Furthermore, in the field of maintenance, PCA is used to detect early signs of failure in production equipment. This enables maintenance to be planned before major breakdowns occur, thus reducing unplanned downtime and maintaining production continuity (Amruthnath and Gupta 2018).

Another key feature of PCA is its ability to reduce the dimensionality of data. It extracts essential information from large data sets, simplifying the management and analysis of complex data generated by machines and sensors (Kong et al. 2018) and (Bai et al. 2019). In addition, Companies also use PCA to evaluate the performance of their suppliers. It examines data on product quality, delivery times and costs, enabling suppliers to be ranked according to their efficiency and reliability (Matta and Tayal 2017).

Finally, PCA is used to determine optimal combinations of manufacturing parameters. By analyzing experimental results, it identifies the settings that lead to optimum performance, thus contributing to the production of high-quality products while reducing production costs (Moshat et al. 2010) and (Madhavi et al. 2017).

PCA is a versatile technique that brings significant benefits to all aspects of manufacturing, from process optimization and predictive maintenance to quality control and complex data analysis. It remains an essential tool for improving efficiency and quality in the manufacturing sector.

3 Description of the WAAM Process Reduction Approach

Our approach follows a structured process to solve a welding quality classification problem for the WAAM process. It begins by loading data from a file/database containing information about welding operations. This data is then divided into two categories: characteristics and class labels. Characteristics represent specific measurements related to the welding process, such as welding speed, wire feed rate and electric current. Class labels indicate the quality of these operations, classifying each operation as Good “G”, Average “A” or Bad “B”.

Once the data is loaded, it is normalized to ensure comparability. Normalization involves adjusting the features so that they have a mean of zero and a standard deviation of 1. This step is essential to prepare the data for processing by the machine learning models. Then, Principal Component Analysis (PCA) is applied to reduce the complexity of the data. The aim is to simplify the data while retaining important information. By reducing the number of variables, this step improves model efficiency.

The data is then divided into two distinct sets: the training set and the test set. The training set is used to teach the models how to perform classification, while the test set is used to evaluate the models’ performance on unknown data. This division is crucial for assessing the generalizability of the models.

Several classification models such as Random Forest, Logistic Regression, Decision Tree, Naive Bayes Gaussian and SVM are created and trained using the training set. These models are designed to learn how to classify data according to its characteristics. Once the models are trained, they are used to predict the weld quality on the test set. The predictions obtained are then compared with the actual class labels. Various evaluation metrics, such as precision, recall, F1-score and confusion matrix, are used to quantify the quality of each model’s predictions, providing an in-depth assessment of their performance.

Precision measures the proportion of positive observations correctly identified by the model, recall quantifies the model's ability to detect the majority of positive observations, and F1-score combines precision and recall into a single metric to balance false positives and false negatives. The confusion matrix provides a detailed view of prediction results by comparing true positives, false positives, true negatives, and false negatives. These metrics help assess a model's ability to classify data correctly, and guide the choice of the most appropriate model for a specific classification task.

The proposed approach follows a structured workflow from loading data to preparing, splitting, training and evaluating classification models. The main objective is to determine which model provides the best performance for classifying welding quality as a function of process characteristics. This approach has been successfully applied to accurately predict WAAM process results based on parameter combinations generated by PCA. The approach was implemented using Python, due to its ease of use and robustness in the field of machine learning and data analysis.

4 Design of Experiments DOE and Experimental Studies

The electric arc additive manufacturing setup comprises the following components: A six-axis robotic arm: CRP robot (Fig. 1), TIG welding station, Argon gas tank, Filler wire: aluminum alloy ER 5356, Substrate: aluminum alloy plates 5083 and Non-consumable tungsten electrode. A full factorial design 3^3 was created using Minitab 19 to study the effect of heat input on weld quality, and to select appropriate welding parameters. Three factors were considered: electric current, wire feed speed and welding speed, each at three levels (low, medium, high). A total of 27 weld beads were deposited in accordance with the experimental design DOE.

There are several reasons why Design of Experiments (DOE) was chosen for this study of electric arc additive manufacturing. Firstly, DOE allows for the simultaneous evaluation of multiple factors and levels, which facilitates the analysis of interactions between these factors. Additionally, it reduces the number of tests required while maintaining the validity of the results, saving time and resources. Furthermore, DOE offers the possibility of statistically analyzing the collected data, enabling an objective interpretation of the results and facilitating easier decision-making regarding the optimum welding parameters. Therefore, the application of DOE in this study ensures a systematic, efficient, and stable approach to welding analysis and parameter selection.

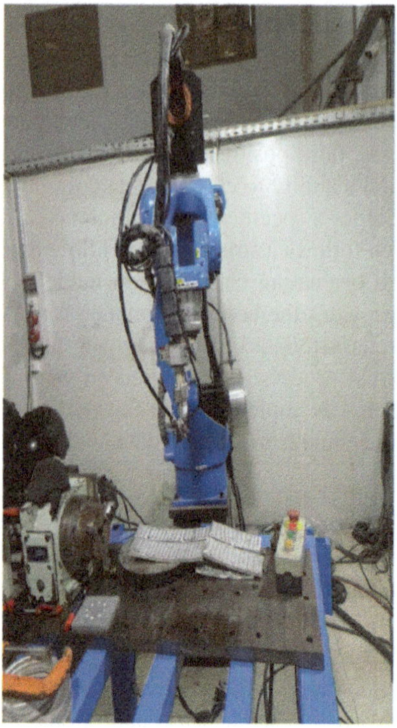

Fig. 1. CRP Robot for Automated Arc Welding in TIG Technology

The values of the selected parameters and the obtained results for each bead are presented in Table 1.

Table 1. Parameter values used for the DOE.

Level	Welding current (A)	Wire Feed Rate (m/min)	Welding Speed (mm/s)
Low	140	1.2	4
Medium	150	1.5	4.5
High	160	1.8	5

The weld beads produced according to the DOE conditions are shown in the Fig. 2.

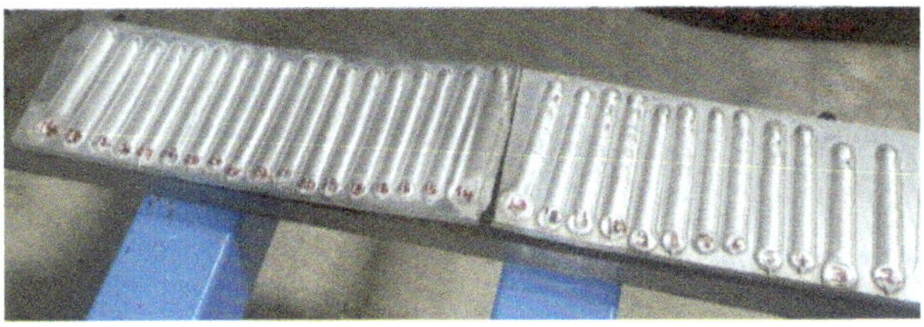

Fig. 2. Experimental Plan Execution

Table 3 presents the characteristics of the tests carried out according to the DOE for the first layer of weld beads deposited on a flat plate.

5 Results and Discussion

The results obtained from Principal Component Analysis (PCA) provide crucial information on data structure and dimension reduction, as presented in Table 2. Firstly, the variance explained by each principal component values indicate that the three principal components (PCs) extracted from the dataset each explain around 33.33% of the total variance. This means that these principal components capture significant aspects of the original data.

Table 2. Results of Principal Component Analysis.

PCs	Explained Variance	Cumulative Variance	Attribute Loads
PC_1	33.33%	33.33%	[0 −1 0]
PC_2	33.33%	66.67%	[−1 0 0]
PC_3	33.33%	100%	[0 0 −1]

Looking at the cumulative variance explained, we see that the first two CPs together explain around 66.67% of the total variance, while the three CPs together cover the entire variance of the data, i.e. 100%. This information is crucial in assessing the usefulness of dimension reduction, as it determines how much of the data variance is retained.

With regard to the attribute load on each principal component, the results show how each attribute (welding speed, wire feed speed and current) contributes to each CP. For example, for the first CP (CP_1), wire feed speed dominates, while the second CP (CP_2) is mainly influenced by welding speed. The third CP (CP_3) is essentially determined by current. These loadings provide insights into the relationships between attributes and CPs, which can be useful when interpreting results.

The PCA results show that dimension reduction retained most of the variance in the data, while identifying the attributes that contribute most to each principal component. This simplifies the data set while retaining relevant information, facilitating more efficient analysis and a deeper understanding of the underlying structure of the data.

Table 3. Characteristics of experimental tests

weld bead	Welding Speed (mm/s)	Wire Feed Rate (m/min)	Welding current (A)	Layer width (mm)	Layer thickness (mm)	Welding quality
1	4.5	1.2	160	9.71	1.2	A
2	4.5	1.8	160	9.66	2.1	A
3	4	1.5	160	10.49	1.72	B
4	4.5	1.5	150	9.80	1.5	A
5	4	1.5	150	9.82	1.82	A
6	4	1.2	140	8.81	1.39	A
7	5	1.5	140	8.70	1.58	G
8	5	1.5	160	10.12	1.40	A
9	4.5	1.5	140	9.05	1.62	G
10	5	1.8	160	10.03	1.79	B
11	4.5	1.2	150	10.43	1.48	A
12	5	1.2	140	8.93	1.30	G
13	5	1.2	160	10.57	1.22	B
14	4	1.2	160	10.94	1.05	B
15	5	1.8	150	8.76	1.72	G
16	4.5	1.8	140	7.74	2.06	A
17	4.5	1.8	150	8.72	1.96	A
18	4	1.2	150	9.41	1.42	G
19	4	1.8	150	9.04	2	G
20	4	1.8	160	9.39	1.97	A
21	5	1.5	150	8.55	1.69	A
22	5	1.2	150	8.81	1.25	A
23	5	1.8	140	7.67	1.70	B
24	4.5	1.2	140	9.05	1.32	A
25	4	1.5	140	8.78	1.71	G
26	4.5	1.5	160	10.36	1.84	A
27	4	1.8	140	8.98	2.03	G

The evaluation results of the different classification models with PCA reveal variable performance, each with its own strengths and weaknesses. We have therefore examined these results in detail to better understand the performance of each model, as shown in Table 4.

Table 4. The results of evaluating classification models with PCA using python.

Models	Evaluation metrics	Welding quality			Accuracy	Macro avg	Weighted avg	Confusion matrices
		G	A	B				
Random Forest	Precision	1	1	1	–	1	1	[[1 0 0] [0 1 0] [0 0 5]]
	Recall	1	1	1	–	1	1	
	F1-score	1	1	1	1	1	1	
Logistic Regression and Decision Tree	Precision	0.33	1	0	–	0.44	0.76	[[1 0 0] [1 0 0] [1 0 4]]
	Recall	1	0.80	0	–	0.60	0.71	
	F1-score	0.5	0.89	0	0.7142	0.46	0.71	
Naive Bayes Gaussian and SVM classification	Precision	0.5	1	0	–	0.50	0.79	[[1 0 0] [1 0 0] [0 0 5]]
	Recall	1	1	0	–	0.67	0.86	
	F1-score	0.67	1	0	0.8571	0.56	0.81	

Where:

- Precision = TP / TP + FP
- Recall = TP / TP + FN
- F1-score = 2 * (Precision * Recall) / (Precision + Recall)
- Macro Average: calculates the metric independently for each class and then takes the average. It treats all classes equally and doesn't consider class imbalance.
- Weighted Average: calculates the metric for each class, but it takes into account class imbalance by weighting the average based on the number of samples in each class.
- Confusion Matrices: are used in machine learning to describe the performance of a classification algorithm. They provide a breakdown of TP, FP, TN, and FN for each class in a classification problem.
- TP: true positives are the cases where the model correctly predicted the positive class.
- FP: false positives are the cases where the model incorrectly predicted the positive class.
- TN: true negatives are the cases where the model correctly predicted the negative class.
- FN: false negatives are the cases where the model incorrectly predicted the negative class.

The Random Forest model clearly emerged as the best performing model. Its precision, recall and F1 scores all reach 100% for each of the "G", "A" and "B". This exceptional performance means that the model correctly classified each test sample into the appropriate categories without any errors. Furthermore, the confusion matrix confirms these results by showing the total absence of classification errors. In short, the Random Forest model stands out for its ability to make perfect predictions for all classes.

In contrast, the Logistic Regression and Decision Tree models perform models present less convincing performances. Although they display an overall accuracy of 71,42%, indicating correct classification of around 71,42% of the test samples, their performance varies considerably by class. In particular, their ability to classify the "G" class is limited, with an accuracy of just 33%. Furthermore, they have significant difficulty identifying the "B" class, as evidenced by zero recall for this class. The Logistic Regression and Decision Tree models also show a precision of 0% for the "B" class, indicating that they failed to correctly classify the samples into this class. The Confusion matrix confirms these limitations by revealing classification errors, particularly in the "G" class.

Regarding, the Naive Bayes Gaussian model and the SVM classification model, they obtain similar performance with an accuracy of 85,71%. This accuracy is mainly attributed to their ability to correctly classify the "A" class. However, these models encounter difficulties in classifying the "B" class, with zero recall for this class. The confusion matrix confirms that they managed to correctly classify the two classes "G" and "A", but were unable to do the same for the "B" class.

In this study, the Random Forest model stands out as the best choice in terms of precision, recall, F1-score and confusion matrix, with impeccable performance for all classes. However, it is essential to note that model performance may depend on the specific characteristics of the data and the particular nature of the classification problem. Consequently, the choice of the ideal model must take into account these performances, as well as the specific requirements of the classification task to be solved.

6 Conclusion

In this study, we presented a structured approach that integrates machine learning and statistical data analysis to predict part quality in the electric arc additive manufacturing (WAAM) process. Our study focused on three key parameters: weld speed, wire feed rate and electric current, all of which are crucial in ensuring the production of high-quality weld layers. Using a meticulously designed experimental framework, we conducted numerous experiments to assess the impact of these parameters on part quality. The experimental data obtained provided valuable information, which was refined and simplified using principal component analysis (PCA), making it easier to manage and interpret.

Based on the experimental data, we applied different machine learning models to perform classifications. After thorough evaluation, the Random Forest model proved to be the most robust, capable of accurately predicting part quality based on combinations of parameters. These results hold great promise for improving production quality and efficiency in additive manufacturing, contributing to more accurate and cost-effective manufacturing processes.

Future work includes the development of data-driven reduced models for other manufacturing processes, such as molding and machining. This phase also includes the development of simplified physical models and surrogate models for the simulation of various aspects of manufacturing. The ultimate aim of this work is to create hybrid models, combining both physical and data aspects, in order to achieve a more accurate and complete representation of manufacturing processes.

Acknowledgements. We would like to express our gratitude to SIAF company, located in Sfax, Tunisia, for allowing us to carry out the experimental part of this work. Additionally, we thank the

Ministry of Higher Education and Scientific Research of Tunisia for funding our research project 22PEJC-D6P7.

References

Amruthnath, N., Gupta, T.: A research study on unsupervised machine learning algorithms for early fault detection in predictive maintenance. In: 2018 5th international conference on industrial engineering and applications (ICIEA), pp. 355–361. IEEE, April 2018

Bai, Y., et al.: A comparison of dimension reduction techniques for support vector machine modeling of multi-parameter manufacturing quality prediction. J. Intell. Manuf. **30**, 2245–2256 (2019)

De Freitas Gomes, J.H., Salgado Junior, A.R., de Paiva, A.P., Ferreira, J.R., da Costa, S.C., Balestrassi, P.P.: Global criterion method based on principal components to the optimization of manufacturing processes with multiple responses. J. Mech. Eng./Strojniški Vestnik **58**(5), 345–353 (2012)

Hwang, J.R., Zheng, J.Y., Kuo, P.C., Huang, C.D., Fung, C.P.: Process optimization of Inconel 718 alloy produced by laser powder bed fusion. Metals **12**(9), 1494 (2022)

Jagadish, Ray, A.: Optimization of process parameters of green electrical discharge machining using principal component analysis (PCA). Int. J. Adv. Manuf. Technol. **87,** 1299–1311 (2016)

Kong, X., Chang, J., Niu, M., Huang, X., Wang, J., Chang, S.I.: Research on real time feature extraction method for complex manufacturing big data. Int. J. Adv. Manuf. Technol. **99**, 1101–1108 (2018)

Madhavi, S.K., Sreeramulu, D., Venkatesh, M.: Evaluation of optimum turning process of process parameters using DOE and PCA Taguchi method. Mater. Today: Proc. **4**(2), 1937–1946 (2017)

Matta, P., Tayal, A.: Supplier performance evaluation for manufacturing industries: re-exploring with big data analysis. In: Advances in Computing and Data Sciences: First International Conference, ICACDS 2016, Ghaziabad, India, 11–12 November 2016, Revised Selected Papers 1, pp. 516–526. Springer, Singapore (2017)

Moshat, S., Datta, S., Bandyopadhyay, A., Pal, P.: Optimization of CNC end milling process parameters using PCA-based Taguchi method. Int. J. Eng. Sci. Technol. **2**(1), 95–102 (2010)

Straat, M., Koster, K., Goet, N., Bunte, K.: An industry 4.0 example: real-time quality control for steel-based mass production using Machine Learning on non-invasive sensor data. In: 2022 International Joint Conference on Neural Networks (IJCNN), pp. 01–08. IEEE, July 2022

Taşdemir, A., Nohut, S.: An overview of wire arc additive manufacturing (WAAM) in shipbuilding industry. Ships Offshore Struct. **16**(7), 797–814 (2021)

Tong, L.I., Wang, C.H.: Multi-response optimization using principal component analysis and grey relational analysis. Int. J. Ind. Eng. Theory Appl. Pract. **9**(4), 343–350 (2002)

WAAM: Optimization of Start and End Zones of the Weld Bead, Case of S235JR Carbon Steel

Kais Fourati[1](✉), Abdessalem Jerbi[2], Elyes Trabelsi[3], and Slim Souissi[1]

[1] Laboratoire des Systèmes Electromécaniques LASEM, École Nationale d'Ingénieurs de Sfax, Université de Sfax, Sfax 3038, Tunisia
kaisfourati1234@gmail.com

[2] Laboratoire de Recherche Optimisation, Logistique et Informatique Décisionnelle (OLID), Institut Supérieur de Gestion Industrielle de Sfax, Université de Sfax, BP-2021 Sfax, Tunisia
abdessalem.jerbi@isgis.usf.tn

[3] Département de Génie Mécanique, Institut Supérieur des Etudes Technologiques de Sfax, Université de Sfax, BP 88 A, 3099 Sfax, Tunisia

Abstract. In recent years, metal additive manufacturing has seen significant growth, offering manufacturers innovative capabilities through sequential layer deposition. Among these techniques, wire-arc additive manufacturing has emerged prominently. However, challenges persist in optimizing welding parameters to control layer dimensions accurately. One overlooked area in existing literature is the initial distortions in walls during wire arc additive manufacturing, especially in weld bead initiation and extinction zones, which can compromise component quality. While some studies propose overlaying start zones onto extinction zones to address these issues, this approach may introduce structural problems like insufficient penetration and potential cracking, affecting the final product's integrity. Our research project aims to optimize parameters at both the start and end of the weld bead to ensure consistent production of defect-free weld lines and ideal walls, particularly those made of S235JR carbon steel. Employing the Design of Experiments method, we optimized weld bead initiation and extinction zones, with a focus on measuring widths and allowances. Although initial findings are promising, further validation through additive wall manufacturing is necessary to confirm the effectiveness and reliability of the optimized parameters for high-quality metal additive manufacturing processes.

Keywords: S235JR carbon steel · robotic welding · wire-arc additive manufacturing · optimization

1 Introduction

Metal additive manufacturing is a new technology that has seen rapid growth in recent years due to the emergence of new processes. It enables the manufacture of three-dimensional parts through successive layer deposition, providing new design and manufacturing freedoms over more traditional methods. Wire-arc additive manufacturing is one of the most common methods of metal additive manufacturing (Shah et al. 2023;

M. Chouchane et al. (Eds.): CMSM 2023, LNME, pp. 420–436, 2024.
https://doi.org/10.1007/978-3-031-65007-9_44

Frazier 2014; Dan et al. 2021; Thompson et al. 2016). This method is based on a system comprising a multi-axis robot and a welding generator with associated equipment (torch, wire feeder, shielding gas supply system, etc.).

Several challenges confront the arc-wire additive manufacturing technique. While the choice of welding parameters to optimize the dimensions of the wall (width and thickness of each layer deposited) is the challenge most discussed in the literature, the deformations experienced by the wall at its start and end are no less important. Indeed, the areas where the weld bead is initiated and extinguished during wall fabrication generate geometric irregularities (bumps and hollows, respectively) (Fig. 1). These irregularities become noticeable after several layers are stacked (Fig. 2a).

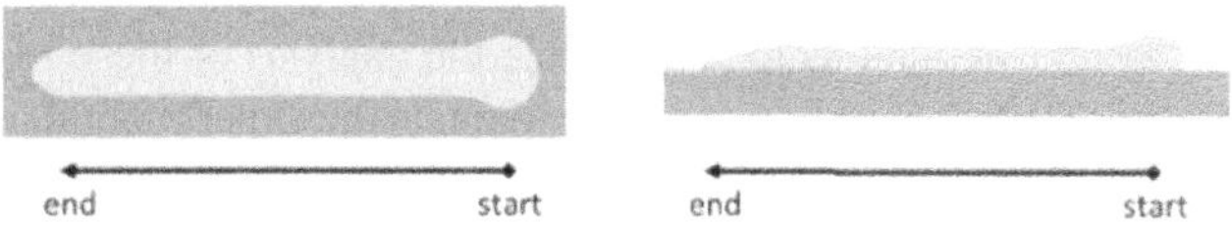

Fig. 1. Weld bead

To address this issue, some researchers propose removing the two zones at the start and end of the wall. Others advocate various deposition strategies (Dunlavey 1983; Park and Choi 2000; Rajan et al. 2001).The most common one is superimposing a start zone on an extinction zone. As a result, the new deposit bump should cover the depression caused by the previous deposit (Fig. 2b). Both methods have several drawbacks. Indeed, the first causes economic losses because of material elimination, whereas the second causes structural problems in the wall because of the superimposition of welding defects, manifested by a lack of penetration and thus a bump in the bead initiation zone, and a crater, which can cause a crack in the bead extinction zone.

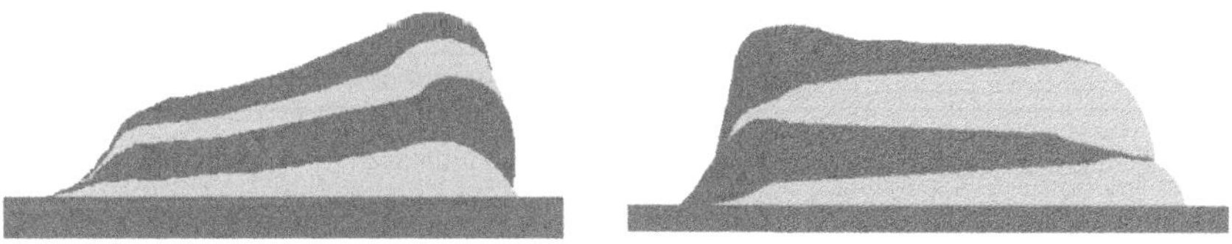

a: identical layers b: alternate layers

Fig. 2. Effect of additive manufacturing strategy

We will attempt to address the issue at its root by attempting to optimize the parameters at the start and end of the weld bead in order to produce uniform, defect-free weld lines, allowing perfect walls to be produced without the use of specific strategies or scenarios.

2 Method

The following approach is used to solve the problem:

1. Choice of bead parameters (material, thickness, width, and allowance, as well as welding parameters).
2. Optimization of arc ignition and extinction zones. The optimization method used is that of experimental design.

The experiment's design method addresses the "black box" phenomena that we seek to clarify in order to better understand their operation and optimize their performance. The method is experimental: data on the observed phenomenon is gathered from planned tests. The goal of experimental design is to structure experimentation and the collection of experimental results, as well as to standardize the analysis of these results (Antony 2023; Montgomery 2019). The following figure depicts the experimental design process (Fig. 3):

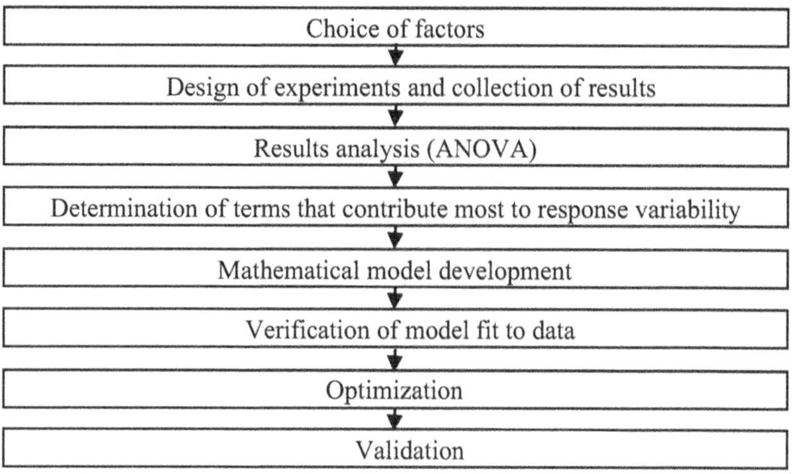

Fig. 3. Experimental design approach

3 Material

All experiments were conducted on a robotized welding system that included a 6-axis "CRP RH14-10-W" robot connected to a "MEGMEET Artsen plus 500" MIG/MAG welding generator, a wire feed system with a filler wire bobbin, and a shielding gas cylinder. Figure 4 depicts the experimental platform.

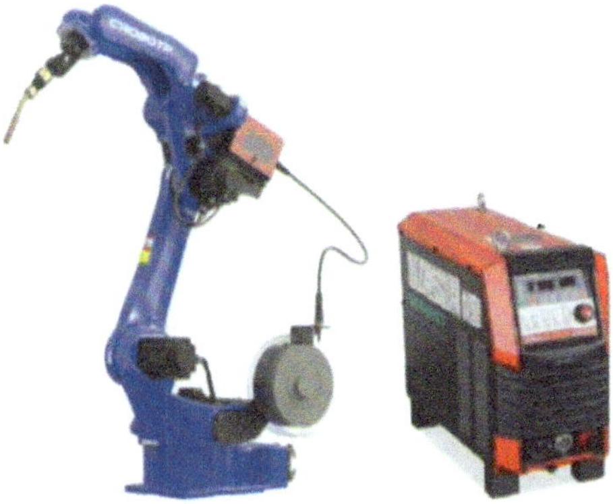

Fig. 4. Robot welding platform

4 Selecting Cord Parameters

S235JR carbon steel was used in this experiment in the form of an 8 mm thick slab with a 1mm welding wire diameter. The welding current and voltage were set to Is = 110A and Us = 22V. The feed and welding wire speeds were set to Va = 8.7 mm/s and V = 9 m/min, respectively. Finally, the stick-out was set to L = 15 mm, and the shielding gas used was ARCAL MAG P. The resulting bead width ranged between 4.9 and 5 mm, with a thickness allowance of 2 to 2.1 mm.

5 Optimization of Arc Ignition and Extinction Zones

5.1 Choice of Factors

The factors considered in this optimization phase are welding energy (E) and welding time (T) (Table 1). We change these two parameters at the start and end of our welding operation. The following equation describes welding energy:

$$E = (60xUxI) / (1000\ xVa) \tag{1}$$

With U and I the welding voltage and current in (Volt and Ampere) and Va in cm/min.

Table 1. Field of experimentation

Factor	Start of the cordon		End of the cordon	
	E(KJ/cm)	T(s)	E(KJ/cm)	T(s)
Level 1	3.582	0.2	2,114	0.2
Level 2	3.779	0.6	2,287	0.6
Level 3	3.978	1	2,463	1
Level 4	4.181	1.4	2,642	1.4
Level 5	4.387	1.8	2,824	1.8

5.2 Design of Experiments and Collection of Results

Since the two factors are taken into account at five levels each, 25 experiments must be taken into account in the experimental design (Table 2) (Pillet 2011). The outcomes are expressed as the widths L (mm) the excess thicknesses S (mm) at the beginning and end of the bead.

Table 2. Design of Experiment and results

			Start of the cordon				End of the cordon			
	Factor		Cordon 1		Cordon 2		Cordon 1		Cordon 2	
Exp	E	T	L	S	L	S	L	S	L	S
1	1	1	5.3	1.8	5.1	2.1	7.3	1.6	6.9	1.3
2	1	2	4.5	2.4	4.4	2.4	6.1	1.4	5.9	1.6
3	1	3	5.7	2.8	6.1	2.9	6	2.8	5.9	3
4	1	4	6.3	2.5	5.9	2.3	6.5	2.9	6.4	3.1
5	1	5	6.2	3.1	6.1	2.8	7.1	3.7	7	3.4
6	2	1	6.8	2.4	6.4	2.5	5.5	1.1	5.4	1.2
7	2	2	6.1	2.9	5.8	3.1	5.8	1.5	5.9	1.5
8	2	3	7.5	3.1	7.4	2.8	7.1	3.1	7.4	3.2
9	2	4	6.6	3.3	6.3	3	6.4	4.1	6.4	4.1
10	2	5	6.5	3.4	6.5	3.5	6.3	4.1	6.5	3.9
11	3	1	5.9	3	6.2	3.1	6.5	4.4	6.7	4.4
12	3	2	7.3	2.6	7	2.6	6.1	4	5.9	4.2
13	3	3	5.7	2.8	5.9	2.9	3.9	2.6	3.8	2.4

(continued)

Table 2. *(continued)*

			Start of the cordon				End of the cordon			
	Factor		Cordon 1		Cordon 2		Cordon 1		Cordon 2	
Exp	E	T	L	S	L	S	L	S	L	S
14	3	4	7.1	3.2	7.3	3	6.7	3.4	6.5	3.4
15	3	5	6.1	2.9	6.2	2.9	7	3.5	6.9	3.8
16	4	1	6.4	3	6.4	2.9	5.6	1.5	5.5	1.6
17	4	2	6.1	3.5	6	3.6	5.5	2.7	5.5	2.6
18	4	3	4.4	3.1	4.3	3.2	6.1	2.4	6.2	2.5
19	4	4	5.8	3.6	5.9	3.7	8.1	4.1	8.3	4.2
20	4	5	6.4	3.3	6.7	2.9	6.5	3.1	6.8	3
21	5	1	7	2.9	7.2	2.9	5.2	2.1	5.1	2.1
22	5	2	6.2	3.1	5.9	3	5.8	2.8	5.9	2.4
23	5	3	6.8	3.5	6.9	3.4	6.7	3.1	6.5	3.1
24	5	4	6.3	3	6.6	2.9	7.4	3	6.9	2.9
25	5	5	7.6	3.6	7.7	3.3	8.2	3.7	8.3	3.5

5.3 Results Analysis (ANOVA)

ANOVA, or Analysis of Variance, plays a crucial role in experiment planning and analysis. It enables the identification of significant effects among various factors, the detection of interactions between these factors, and the assessment of data variability (Tabachnick and Fidell 2007). The following table summarizes the analysis of the variance of the E and T terms and their interaction with the responses. According to this analysis, all the terms under consideration have statistically significant associations with the responses (p 0.05) (Table 3).

Table 3. Analysis of variance

L (Start)					
Term	Degree of Freedom	Som of Square	Mean of Square	F	p
E	4	11.3292	2.83230	104.13	0.000
T	4	2.8312	0.70780	26.02	0.000
ExT	16	16.3628	1.02268	37.60	0.000
Error	25	0.6800	0.02720		

(continued)

Table 3. (*continued*)

L (Start)					
Term	Degree of Freedom	Som of Square	Mean of Square	F	p
Total	49	31.2032			
S (Start)					
Term	Degree of Freedom	Som of Square	Mean of Square	F	p
E	4	3.4960	0.87400	51.41	0.000
T	4	1.5140	0.37850	22.26	0.000
ExT	16	2.4700	0.15438	9.08	0.000
Error	25	0.4250	0.01700		
Total	49	7.9050			
L (End)					
Term	Degree of Freedom	Som of Square	Mean of Square	F	p
E	4	2.2028	0.55070	27.81	0.000
T	4	14.3248	3.58120	180.87	0.000
ExT	16	25.4592	1.59120	80.36	0.000
Error	25	0.4950	0.01980		
Total	49	42.4818			
S (End)					
Term	Degree of Freedom	Som of Square	Mean of Square	F	p
E	4	7.1268	1.78170	112.77	0.000
T	4	16.1748	4.04370	255.93	0.000
ExT	16	20.3532	1.27208	80.51	0.000
Error	25	0.3950	0.01580		
Total	49	44.0498			

5.4 Determination of Terms that Contribute Most to Response Variability

The Pareto charts supported the ANOVA analysis findings in Figs. 5 and 6. The contributions of each term do, in fact, surpass the level of statistical significance. The ranking of the terms' contributions reveals that term E makes a greater contribution to the outcome than the other terms.

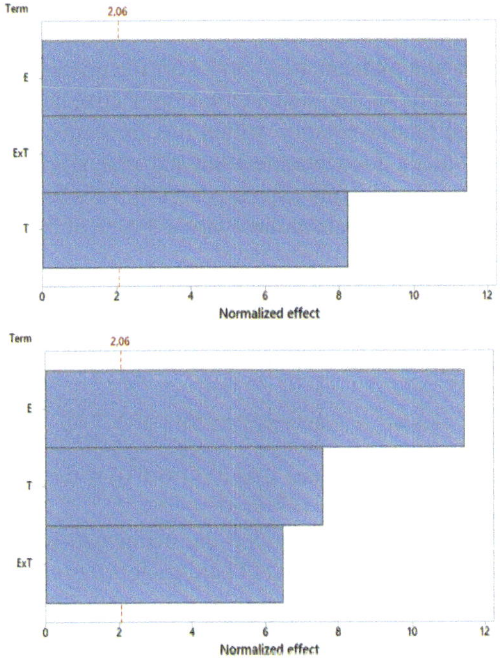

L, start of the cordon

S, start of the cordon

Fig. 5. Pareto chart start of the cordon

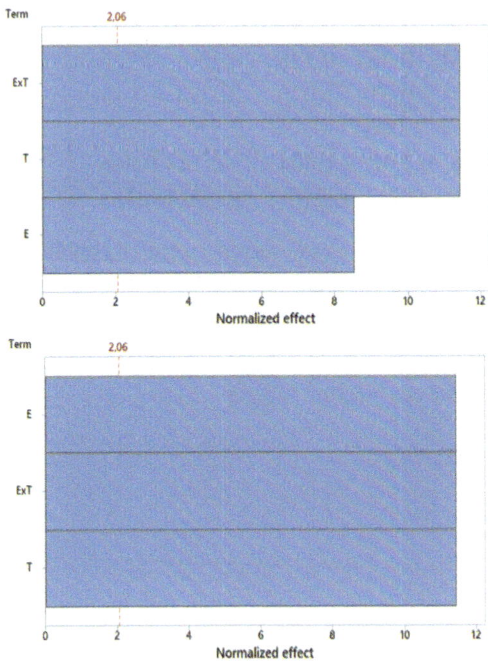

L, end of the cordon

S, end of the cordon

Fig. 6. Pareto chart end of the cordon

5.5 Mathematical Model Development

Embarking on the development of a mathematical model for the process, our aim is to enhance the comprehension of factor impacts and facilitate predictive analysis of experimental results. The adoption of matrix notation within this mathematical framework offers a significantly simplified method for interpretation and analysis (Pillet 2011).

In developing the mathematical models are expressed as follows where factors E and T are expressed as vectors defining the levels of these terms. If one of the factors [A] is at level 1, then the vector would be:

$$[\mathrm{A}] = \begin{bmatrix} 1 \\ 0 \\ 0 \\ 0 \end{bmatrix} \tag{2}$$

t[A] is the is the transpose vector of [A]. Hence t[A] = [1 0 0 0] if the factor [A] is at level 1.

All terms must be taken into account to develop the mathematical model. These models can be expressed as follows where [E] and [T] are vectors defining the levels of these terms.

$$\begin{aligned} \widetilde{\mathrm{L(start)}} &= 6.256 \\ &+ \begin{bmatrix} -0.696 & 0.334 & 0.214 & 0.416 & 0.564 \end{bmatrix}[\mathrm{E}] \\ &+ \begin{bmatrix} 0.014 & -0.326 & -0.186 & 0.154 & 0.344 \end{bmatrix}[\mathrm{T}] \\ &+ \mathrm{t[E]} \begin{bmatrix} -0.374 & -0.784 & 0.526 & 0.386 & 0.246 \\ -0.004 & -0.314 & 1.046 & -0.294 & -0.434 \\ -0.434 & 1.006 & -0.484 & 0.576 & -0.664 \\ 0.546 & 0.536 & -1.304 & -0.144 & 0.366 \\ 0.266 & -0.444 & 0.216 & -0.524 & 0.486 \end{bmatrix} [\mathrm{T}] \end{aligned} \tag{3}$$

$$\begin{aligned} \widetilde{\mathrm{S(Start)}} &= 2.970 + [-0.460 \quad 0.030 \quad -0.070 \quad 0.310 \quad 0.190][\mathrm{E}] \\ &+ [-0.310 \quad -0.050 \quad 0.080 \quad 0.080 \quad 0.200][\mathrm{T}] \\ &+ \mathrm{t[E]} \begin{bmatrix} -0.250 & -0.060 & 0.260 & -0.190 & 0.240 \\ -0.240 & 0.050 & -0.130 & 0.070 & 0.250 \\ 0.460 & -0.250 & -0.130 & 0.120 & -0.200 \\ -0.020 & 0.320 & -0.210 & 0.290 & -0.380 \\ 0.050 & -0.060 & 0.210 & -0.290 & 0.090 \end{bmatrix} [\mathrm{T}] \end{aligned} \tag{4}$$

$$\widetilde{L(End)} = 6.358 + \begin{bmatrix} 0.152 & -0.088 & -0.358 & 0.052 & 0.242 \end{bmatrix}[E]$$
$$+ \begin{bmatrix} -0.388 & -0.518 & -0.398 & 0.602 & 0.702 \end{bmatrix}[T]$$
$$+ t[E] \begin{bmatrix} 0.970 & 0.008 & -0.162 & -0.662 & -0.162 \\ -0.432 & 0.098 & 1.378 & -0.472 & -0.572 \\ 0.988 & 0.518 & -1.752 & -0.002 & 0.248 \\ -0.472 & -0.392 & 0.138 & 1.188 & -0.462 \\ -1.062 & -0.232 & 0.398 & -0.052 & 0.948 \end{bmatrix} [T] \quad (5)$$

$$\widetilde{S(End)} = 2.902 + [-0.422 \quad -0.122 \quad 0.708 - 0.132 \quad 0.032][E]$$
$$+ [-0.772 \quad -0.432 \quad -0.082 \quad 0.618 \quad 0.668][T]$$
$$+ t[E] \begin{bmatrix} -0.258 & -0.548 & 0.502 & -0.098 & 0.402 \\ -0.858 & -0.848 & 0.452 & 0.702 & 0.552 \\ 1.562 & 0.922 & -1.028 & -0.828 & -0.628 \\ -0.448 & 0.312 & -0.238 & 0.762 & -0.388 \\ 0.002 & 0.162 & 0.312 & -0.538 & 0.062 \end{bmatrix} [T] \quad (6)$$

5.6 Verification of Model Fit to Data

S and R^2 are statistical measures employed to evaluate how effectively a model depicts the response variable. S quantifies the discrepancy between observed data values and predicted values, with a smaller S indicating a more accurate model representation. On the other hand, R^2, which ranges from 0 to 100%, reflects how well the model fits the data; higher R^2 values suggest a stronger fit (Rousseeuw and Yohai in 1984; Chen and Chen in 2021). Table 4 summarizes the statistics used to test the models' fit to the response data. S-statistics are low, while R^2-statistics are high. These values support the fit of the models to the response data.

Table 4. Models fit to data test

L-Lc (start)		S-Sc (start)		L-Lc (end)		S-Sc (end)	
S	R^2	S	R^2	S	R^2	S	R^2
0.165	97.82%	0.130	94.62%	0.140	98.83%	0.125	99.10%

5.7 Optimization

The optimization method is based the desirability function (DF) method (Candioti et al. 2014; Cho et al. 2022). This method is based on two steps. The first defines a desirability function by assigning values to responses that reflect their desirability. This involves transforming each value of the performance measure "j" of experiment "i", y_{ij}, into a partial dimensionless desirability function d_i, where $0 \leq d_i \leq 1$. This function includes the choices of the decision maker when constructing the optimization procedure.

A one-directional desirability transformation occurs when the objective is either to maximize or minimize the response, requiring specification of two values, A and B, as the lower and upper limits. Equations (7) and (8) outline the transformation equations tailored for minimizing and maximizing objectives, respectively.

$$d_i = \begin{cases} 1 A \geq y_{ij} \\ \left(\frac{y_{ij}-A}{B-A}\right)^{\omega j} A \leq y_{ij} \leq B \\ 0 y_{ij} \geq B \end{cases} \tag{7}$$

$$d_i = \begin{cases} 1 y_{ij} \geq B \\ \left(\frac{y_{ij}-A}{B-A}\right)^{\omega j} A \leq y_{ij} \leq B \\ 0 A \geq y_{ij} \end{cases} \tag{8}$$

The parameter ωj can be defined as a weighting factor or power exponent that reflects the researcher's subjective judgment regarding the significance of the response in the overall desirability of the product.

A ωj value of 1 suggests the application of a linear desirability function. When ωj is less than 1, the resulting desirability function indicates that performance does not necessarily need to approach the lower or upper limit, depending on the optimization objective, to achieve a higher desirability value. Conversely, if ωj is greater than 1, the desirability function implies that performance must closely approach the lower or upper limit, depending on the optimization goal, to achieve a higher desirability value. To simultaneously optimize multiple performance sets, the individual desirability is combined using a geometric mean in the composite desirability.

$$DF = \left(\prod_{i=1}^{n} d_i\right)^{\frac{1}{n}} \tag{9}$$

A value of DF different from zero implies that all performances are in a desirable range simultaneously. In addition, a value of DF close to 1 means that the combination of the different criteria is globally optimal and the performances values are near the target values.

First, we optimize the responses studied using single-response optimization (Figs. 7 and 8). Figures 9 and 10 depict the multi-response optimization diagrams for these two zones.

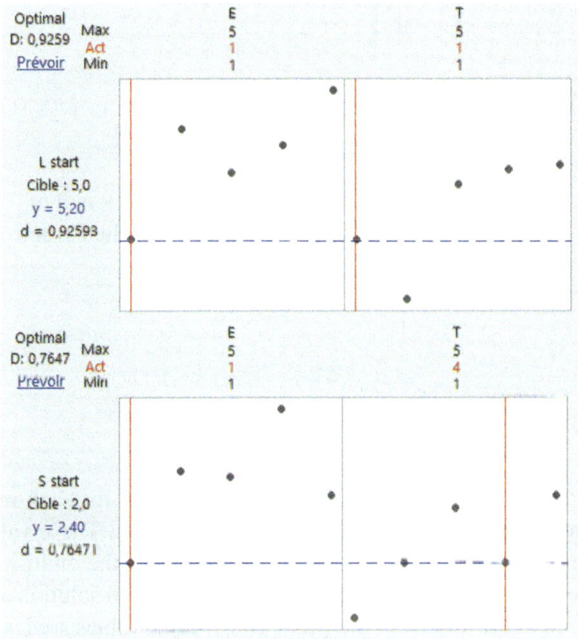

L start of the cordon

S start of the cordon

Fig. 7. Mono-response optimization start of the cordon

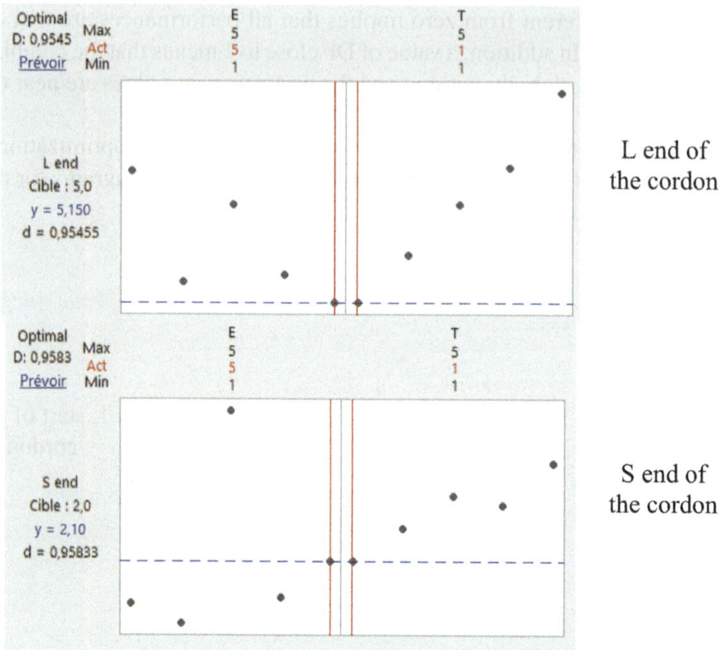

Fig. 8. Mono-response optimization end of the cordon

The optimization is based on calculating individual desirability for each response and weighting each response based on the importance assigned to it. These values are added together to determine the composite or overall desirability of the multi-response system. When composite desirability reaches its maximum, an optimal solution appears. The target values measured in the middle of the bead. These target values are L = 5 mm in width and S = 2 mm in excess thickness, respectively.

The optimum levels of the E and T terms for the mono-response width and thickness of the bead start zone are 1.1 and 1.4, respectively, while the 5.1 levels of the E and T terms generate the optimum levels for the two mono-response width and thickness of the bead end zone. The target value and the optimized value of the mono-responses considered differ slightly. The mono-responses in the early zone have relatively large deviations from the target values (0.2 and 0.4). The deviations in the end zone are smaller (0.15 and 0.10).

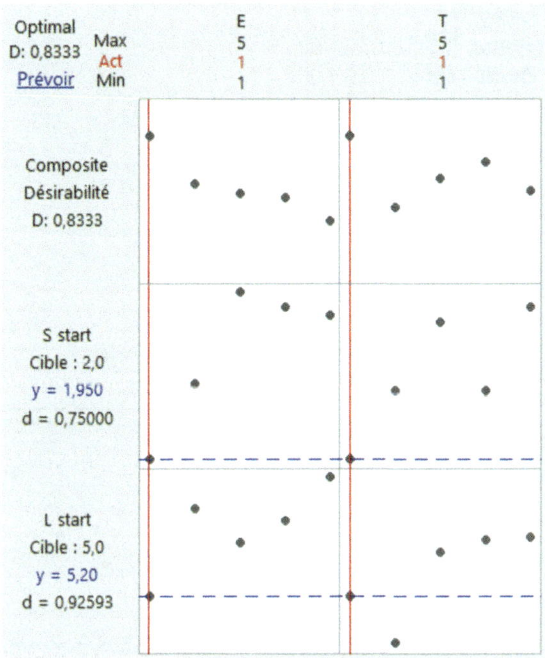

Fig. 9. Multi-response optimization start of the cordon

The multi-response optimization diagrams for these two zones show that the overall desirability of the multi-response system reaches 0.833 and 0.956 for the two zones at the beginning and end of the cord, respectively.

The optimum levels of the E and T terms for multi-responses in the early cord zone are 1 and 1, respectively, whereas the optimum levels of the E and T terms in the late cord zone are 5 and 1. The deviations between the target and optimized values of the considered responses are smaller than in mono-optimization. In fact, the start zone responses deviate by −0.05 and 0.20 from the target values, while the end zone responses deviate by 0.10 and 0.15.

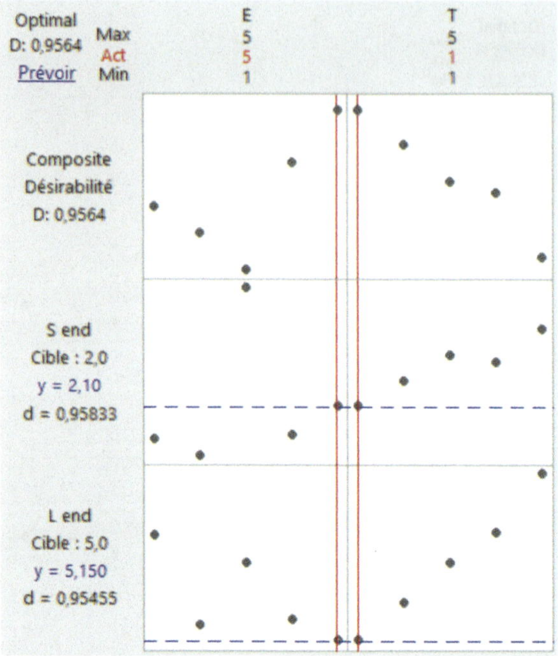

Fig. 10. Multi-response optimization end of the cordon

5.8 Validation

During the validation experiment, we fine-tuned the welding robot parameters using multi-response optimization levels. Specifically, we set E = 3.582 kJ/cm and T = 0.2s for the beginning weld bead zone, and E = 2.824 kJ/cm and T = 0.2s for the end weld bead zone. The results, depicted in the figure below, show that the weld bead has an acceptable appearance. Both the beginning and end zones of the weld bead demonstrate dimensions comparable to those of the middle zone. This was verified through measurements of the two zones, revealing widths and thicknesses of 5.1 mm, 4.9 mm, 2 mm, and 2.05 mm, respectively (Fig. 11).

Fig. 11. Validation welding cordon

6 Conclusion

Metal additive manufacturing has undergone remarkable expansion in recent years, fueled by the development of innovative techniques that enable the production of intricate three-dimensional parts through sequential layer deposition. Among these methods, wire-arc additive manufacturing has emerged as a widely embraced approach, utilizing a system comprising a multi-axis robot, welding generator, and ancillary tools such as a torch, wire feeder, and shielding gas supply. Despite its widespread adoption, wire-arc additive manufacturing encounters several challenges. Chief among these is the predominant focus of research efforts on optimizing welding parameters to control layer dimensions, with insufficient attention given to the initial distortions of walls, particularly in the weld bead initiation and extinction zones. These distortions result in geometric irregularities that become increasingly apparent as layers accumulate.

Some researchers have proposed a deposition approach that overlays start zones onto extinction zones to fill the voids left by previous deposits. However, this method introduces structural concerns, including inadequate penetration leading to protrusions in the initiation zone and potential cracking in the extinction zone. In response to these challenges, our approach aims to optimize parameters at both the start and end of the weld bead to ensure uniform, defect-free weld lines. Unlike previous proposals focused solely on overlaying start and extinction zones, our method addresses structural issues at both ends of the weld bead, with the aim of mitigating problems such as inadequate penetration and potential cracking. Through this comprehensive optimization, we strive to surpass current limitations and produce flawless walls with enhanced structural integrity.

The primary objective of our research project is to enhance the properties of sections made of S235JR carbon steel by optimizing the parameters involved in robotized welding. While our initial results are promising, it is crucial to validate them through additive wall manufacturing processes. This validation step will further confirm the effectiveness and reliability of the optimized welding parameters in enhancing the desired properties of the aluminum alloy sections.

References

Antony, J.: Design of Experiments for Engineers and Scientists. Elsevier (2023)

Candioti, L.V., De Zan, M.M., Cámara, M.S., Goicoechea, H.C.: Experimental design and multiple response optimization. Using Desirab. Funct. Anal. Methods Dev. Talanta **124**, 123–138 (2014). https://doi.org/10.1016/j.talanta.2014.01.034

Cho, J.S., Lee, D.H., Seo, G.J., Kim, D.B., Shin, S.J.: Optimizing the mean and variance of bead geometry in the wire + arc additive manufacturing using a desirability function method. Int. J. Adv. Manuf. Technol. **120**, 7771–7778 (2022). https://doi.org/10.1007/s00170-022-09237-6

Dan, L., Lee, B., Babkin, A., Chang, Y.: Research progress of arc additive manufacture technology. Materials **14**(6), 1415 (2021)

Dunlavey, M.R.: Efficient polygon-filling algorithms for raster displays. ACM Trans. Graph. (Tog) **2**(4), 264–273 (1983)

Frazier, W.E.: Metal Additive manufacturing: a review. J. Mater. Eng. Perform. **23**, 1917–1928 (2014)

Montgomery D.C.: Design and Analysis of Experiments. 9th Edition, Wiley, New York (2019)

Park S.C. Choi, B.K.: Tool-path planning for direction-parallel area milling. Comput.-Aided Des. **32**(1), 17–25 (2000)

Pillet M. (2011). Les plans d'expériences par la méthode TAGUCHI. ISBN - 978–2–9539428–0–4

Rajan, V.T., Srinivasan, V., Tarabanis, K.A.: The optimal zigzag direction for filling a two-dimensional region. Rapid Prototyping J. **7**(5), 231–241 (2001)

Shah, A., Aliyev, R., Zeidler, H., Krinke, S.A.: Review of the recent developments and challenges in wire arc additive manufacturing (WAAM) process. J. Manuf. Mater. Process. **7**, 97 (2023). https://doi.org/10.3390/jmmp7030097

Tabachnick B.G., Fidell L.S. (2007). Experimental designs using ANOVA (Vol. 724). Belmont, CA: Thomson/Brooks/Cole

Thompson, M.K., Moroni, G., Vaneker, T., et al.: Design for additive manufacturing: trends, opportunities, considerations, and constraints. CIRP Ann. **65**(2), 737–760 (2016)

Global Approach for Optimization and Control of CMT-Based WAAM for SS316L

Oussama Trad[1](✉), Ated Ben Khalifa[2], Farhat Zemzemi[1], and Hédi Hamdi[3]

[1] University of Sousse, LMS, ENISo, 4023 Sousse, Tunisia
oussamatrad01@gmail.com

[2] University of Monastir, LGM, ENIM, Avenue Ibn-Eljazzar, 5019 Monastir, Tunisia

[3] Univ Lyon, Ecole Centrale de Lyon, CNRS, ENTPE, LTDS, UMR5513, ENISE, 42023 Saint, France
hedi.hamdi@enise.fr

Abstract. Metal additive manufacturing process represents the most efficient means of producing intricate and compact components with nearly finished shapes. However, a significant drawback of this method is its substantial time and cost requirements. In contrast, wire arc additive manufacturing, commonly known as WAAM, has emerged as a promising alternative for producing large-scale structures due to its cost-effectiveness and reduced fabrication time. Over the past two decades, researchers have dedicated their efforts to investigating WAAM's application with aluminium and its alloys. Nevertheless, there has been limited exploration of the WAAM fabrication process for the stainless steels. This study aims to examine the impact of various process parameters on beads quality and geometry, for that we consider three key process parameters: welding speed (WS), wire feed rate (WFR), gas flow rate (GFR). To streamline our experimentation, we employ the multiple-level factorial design method, which helps reduce the number of trials required to identify optimal input variable ranges. Macro-level structural analysis is conducted on the sixteen weld bead samples to assess their geometric properties. Additionally, we employ ANOVA analysis to investigate the relationship between input process variables and output response variables. To gain a deeper insight into these associations, we employ a multiple regression approach. Using the results obtained, we employ the Teaching Learning-Based Optimization method to determine optimal process parameters and anticipate their corresponding results.

Keywords: Geometry control · DOE · Optimization · TLBO · WAAM-CMT

1 Introduction

As one of the most effective metal additive manufacturing (AM) processing techniques, Wire and Arc Additive Manufacturing (WAAM) achieves high deposition rates, superb material efficiency as well as cos effectiveness (Pattanayak and sahoo 2021). It has been recognised for its capability in AM application for high volume feature production (Jafari et al. 2021).Gas Metal Arc welding (GMAW), Cold Metal Transfer (CMT),Gas Metal

M. Chouchane et al. (Eds.): CMSM 2023, LNME, pp. 437–446, 2024.
https://doi.org/10.1007/978-3-031-65007-9_45

Arc Welding (GTAW) and Plasma Arc Welding (PAW) are some of the energy sources used in AM (Aldalur et al. 2021; Le et al. 2020). This process has gained popularity, notably in the manufacture of 316L components, given its good weld-ability as well as high corrosion resistance, rendering it useful in a variety of industries like nuclear, shipbuilding or even biomedical parts.

Less attention has been paid to optimizing and control this process to achieve the desired geometrical dimensions of weld seams than studying the metallurgical aspects and its mechanical properties mainly for wall printing. To guarantee product with high quality, some elements must be respected like shape stability and uniformity and even material homogeneity. A lot of research studies have sought in this using a variety of methods, the most known one is response surface method, Artificial-Neural-Network and recently Teaching-Learning-Based-Optimization algorithm. Yet it remains an open question how processing factors affect the geometry of weld beads in WAAM.

The goal of this work is to understand the CMT based WAAM, see how far the selected parameters affect the deposit beams and then we aim to optimize and predict the output responses related to the optimum inputs. Bead width (BW), bead height (BH) and bead surface porosity (BP) were measured in an experimental design that included welding speed (WS), wire feed rate (WFR) and gas flow rate (GFR) as process key factors. The Teaching Learning-Based Optimization (TLBO) approach was then used to determine the ideal processing parameters.

2 Experimental Methodology

2.1 Equipment

The A CMT unit operating on a three-axis CNC machine was used to conduct the experiment. For metal deposition, as shown in Fig. 1, the welding torch was plugged into the spindle in the vertical position.

Table 1. Chemical composition of SS316L filler wire (%)

C	Cu	Cr	Mn	Mo	N	Ni	S	Si	Ti	V	P
0.03	0.76	17.83	2.43	1.63	–	10.57	–	0.63	0.267	0.3	–

As the heat source for deposition, a TPS 302i system was used to carry out the experimental trials, based on a cold metal transfer module that provides metal application with reduced heat transfer compared with other arc heat source. The shielding gas used was a mixture of argon (Ar) an Co_2. A commercially available stainless steel filler wire (316L) with a diameter of about 1mm was used for deposition process. The chemical composition of the filler wire is listed in Table 1.

2.2 Research Methodology

We employ a synchronized operating mode allows the current and voltage to be adjusted and controlled according to the wire feed rate (WFR). The WFR stays proportional

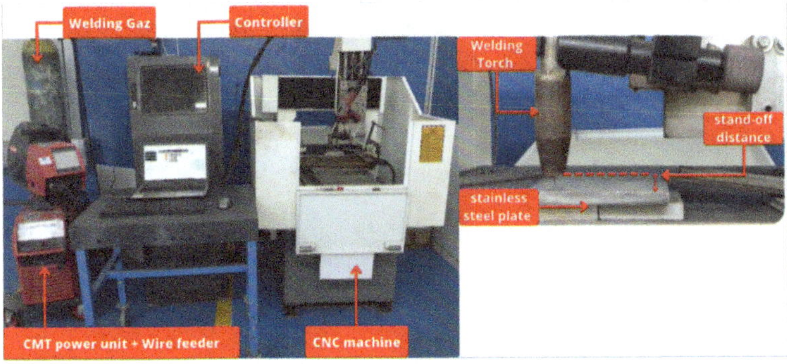

Fig. 1. CMT setup for the WAAM process

to the current and voltage, increasing the WFR causes the current and voltage values to increase, causing an increase in heat input. Four levels of WFR ranging from 3 to 4.5 m/min are used in this study, producing current and voltage levels varying fom 95A-14.5V to 117A-14.8V.

Following a full factorial design, a bead was deposited on the base plate. Wire feed rate (WFR), soldering speed (WS) and gas flow rate (GFR) were chosen as input parameters to the WAAM process for this study. A high wire feed speed results in a high current demand, reflecting a high melting point for the electrode or wire feed. This means that there is sufficient heat available to fuse the wire, which changes the shape of the bead (Dongari and Davidson 2021). Welding speed is desirable for deposition stability as it affects droplet creation, material melting and thermal cycle separation, all of which occur within seconds and influence bead shape. The welding torch is kept away from the deposited surface at a high distance. As a result, the force of the arc decreases and tends to spread, so that the width of the bead and the area of penetration increase with the distance away. The various process parameters and their associated levels are summarised in Table 3 below (Table 2).

Table 2. Main process parameters at different levels applied for the DOE

Parameter	Unit	L1	L2	L3	L4
WS	m/min	0.25	0.5	0.75	1
WFR	m/min	2	3	4	5
GFR	L/min	15	20	–	–

As shown in Fig. 2 the stand-off distance represent the distance between the upper surface of the substrate and the inner nozzle of the CMT welding torch, the stand-off-distance was maintained for 12 mm with an arc length of 9 mm.The purpose of this is to ensure to ensure continued material deposition with good melting zone.

For geometric imperfections examination present in welded beads; an image object detection model was used by Matlab code. Surface defects such as pores, shrinkage

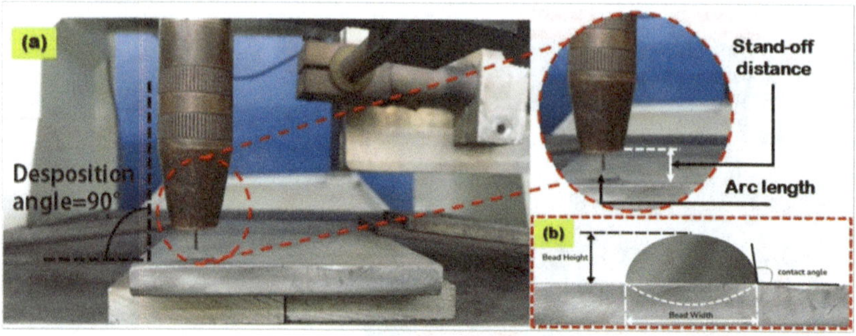

Fig. 2. (a) Setup test, (b) Schematic representation of bead geometry

Table 3. Experimental data relation to weld seam geometry

Trial N°	Corresponding current and voltage		Input parameters			Output response		
	I	V	WFR	WS	GFR	BH	BW	BP
	(A)	(V)	(m/min)	(m/min)	(L/min)	(mm)	(mm)	(%)
S01	102	14.6	3.50	0.75	20	1.41	1.92	2.56
S02	102	14.6	3.50	0.50	15	2.40	2.78	9.96
S03	102	14.6	3.50	0.25	20	2.04	4.06	10.26
S04	110	14.7	4.00	0.75	20	1.31	2.88	35.83
S05	102	14.6	3.50	1.00	15	1.53	2.23	16.11
S06	102	14.6	3.50	0.50	20	1.80	2.54	15.21
S07	110	14.7	4.00	0.50	20	1.44	2.76	2.13
S08	110	14.7	4.00	1.00	15	1.37	2.86	18.67
S09	102	14.6	3.50	1.00	20	1.83	1.55	0.26
S10	95	14.5	3.00	0.50	15	1.97	1.85	15.67
S11	102	14.6	3.50	0.75	15	1.82	3.81	5.72
S12	117	14.8	4.50	0.50	15	1.80	2.23	11.24
S13	95	14.5	3.00	1.00	20	1.03	2.02	38.63
S14	117	14.8	4.50	1.00	15	1.37	3.03	26.71

(continued)

and surface inclusions were manually identified and marked in the resulting images as shown in Fig. 3. The generated values made by the matlab code are promising and offer an application in future work like integrating this code in AI detection system.

Table 3. (*continued*)

Trial N°	Corresponding current and voltage		Input parameters			Output response		
	I	V	WFR	WS	GFR	BH	BW	BP
	(A)	(V)	(m/min)	(m/min)	(L/min)	(mm)	(mm)	(%)
S15	117	14.8	4.50	0.75	20	1.32	3.47	13.79
S16	110	14.7	4.00	0.50	15	1.93	3.12	13.68
S17	117	14.8	4.50	0.25	15	2.43	5.11	6.85
S18	95	14.5	3.00	1.00	15	1.06	2.13	41.47
S19	117	14.8	4.50	1.00	20	1.73	2.91	39.18
S20	110	14.7	4.00	0.25	20	2.47	4.21	9.73
S21	95	14.5	3.00	0.50	20	1.46	2.11	7.43
S22	95	14.5	3.00	0.25	20	2.23	2.86	2.25
S23	110	14.7	4.00	0.75	15	1.86	2.56	8.56
S24	110	14.7	4.00	0.25	15	2.41	4.11	3.98
S25	117	14.8	4.50	0.25	20	2.27	5.29	4.62
S26	95	14.5	3.00	0.75	15	1.37	2.01	33.25
S27	95	14.5	3.00	0.75	20	1.27	2.72	43.13
S28	110	14.7	4.00	1.00	20	1.27	2.45	42.16
S29	95	14.5	3.00	0.25	15	2.23	2.87	2.21
S30	117	14.8	4.50	0.50	20	1.63	4.05	6.69
S31	102	14.6	3.50	0.25	15	2.11	4.11	5.56
S32	117	14.8	4.50	0.75	15	2.17	3.47	11.48

2.3 Parametric Optimization Using the TLBO Method

Ensuring the desired characteristics and minimizing flaws in WAAM relies on the optimization of its parameters, which include variables like welding speed (WS), wire feed rate (WFR), and gas flow rate (GFR). In Fig. 4 we present an application of the Teaching-Learning-Based Optimization (TLBO) algorithm conceived to fine-tune those parameters in the context of WAAM. The attached code depicts the application of the TLBO algorithm to enhance the welding process. Essentially, its objective is to define the most feasible combination of WS, WFR, and GFR for optimizing WAAM. To assess the efficiency of the adopted code, 3 objective functions are derived from statistical analysis using Minitab which provide as three regression functions noted as BH, BW and BP. These functions serve as mathematical models, each addressing specific facets of the process; using WS, WFR, and GFR as input variables and giving a single numeric value, reflecting the quality of the fabricated structure.

The TLBO Algorithm code specifies he next operation values:

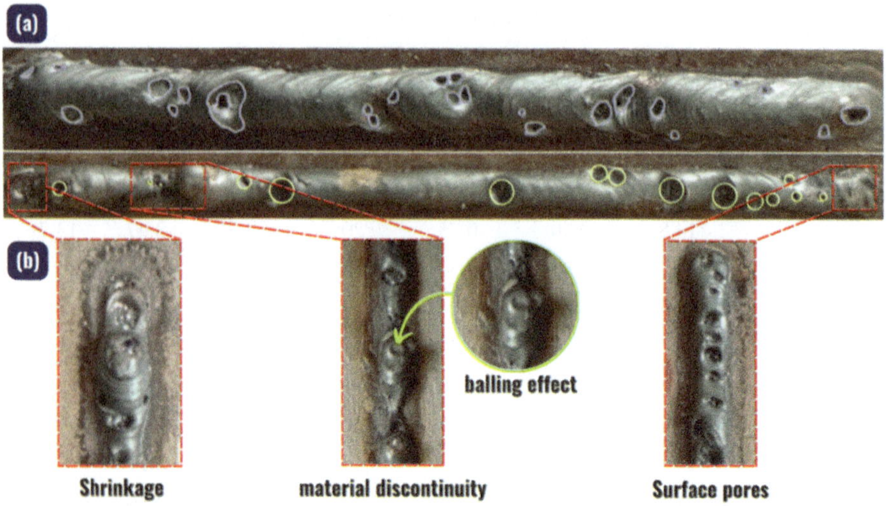

Fig. 3. (a) Porosity detection and measuring using matlab code, (b) Major defects revealed in the WAAM process for single beams

- maximum size of population = 40,
- max-generation = 300, p
- probability of learning = 0.5

Pre-defined limits for GFR, WS, WFR are also given to restrict the searching area during calculation phase. Once the optimization is complete, the code displays the optimum values for GFR, WS and WFR.

3 Results and Discussion

3.1 Evaluating the Impact of Process Parameters on Beads Geometry

The results based on the quality of the weld beads will be discussed in this subsection. In Figs. 5 and 6, specimens are shown in a top view normal to the top surface. For samples S10 and S13, the droplets remain separate, whilst for samples S05, S18, S26 and S27 the droplets appear in pairs. On the other hand, some beads have a good weld shape (S01, S07, S09, S17, S25, S29), particularly sample S09 which has the best geometry, showing the best homogeneity between the height and width of the bead.

While the welding is in progress, the resulting gas inclusions are pushed out as the material melts. Insufficient time for these inclusions to be released during the material solidification stage presents a major problem. At excessively high welding speeds, the weld pool may be insufficiently protected by protective gas. The presence of the gas can create porosity that affects the integrity of the weld. The width of the weld bead is mainly influenced by the wire feed speed and the welding speed. Specifically, a higher wire feed speed results in an excessive supply of material to the weld zone.

The model terms that significantly affect the weld bead width (BW) are welding speed and wire feed rate. In terms of weld bead height (BH), welding speed emerges as

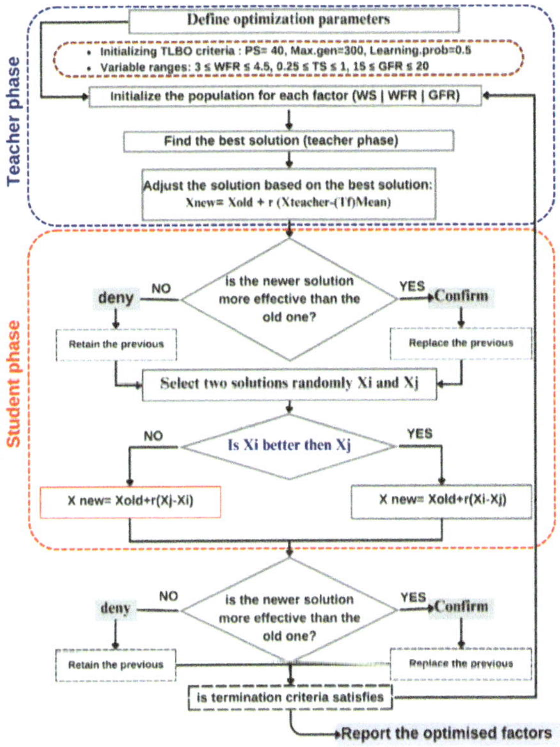

Fig. 4. Optimization methodology via TLBO algorithm

the primary factor, closely followed by gas flow rate (GFR) as secondary factor. Also, WS appears to be the main factor 7affecting the formation of porosity (Fig. 7).

3.2 Validation and Performance of the TLBO Method

Our target is to maximize bead width noted BW and bead height noted BH, and minimizing the porosity (BP). Those functions were extracted from the experimental data in Table 3, using MINITAB 20. These objective functions are represented by the listed equations:

$$^{\text{Maximize}}\text{BW} = 2.43 + 0.26\ \text{WFR} - 0.50\ \text{WS} - 0.122\ \text{GFR} + 0.027\ \text{WFR}^2 + 3.71\ \text{WS}^2 - 0.930\ \text{WFR} * \text{WS} + 0.0562\ \text{WFR} * \text{GFR} - 0.148\ \text{WS} * \text{GFR} \tag{1}$$

$$^{\text{Maximize}}\text{BH} = 0.53 + 1.73\ \text{WFR} - 4.15\ \text{WS} - 0.032\ \text{GFR} - 0.210\ \text{WFR}^2 + 1.160\ \text{WS}^2 + 0.226\ \text{WFR} * \text{WS} - 0.0092\ \text{WFR} * \text{GFR} + 0.0408\ \text{WS} * \text{GFR} \tag{2}$$

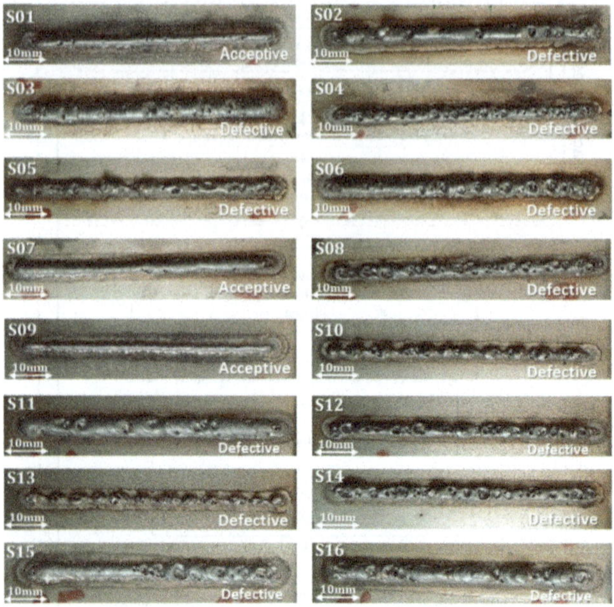

Fig. 5. Geometry results of the deposited beads in the 1st set of experiments

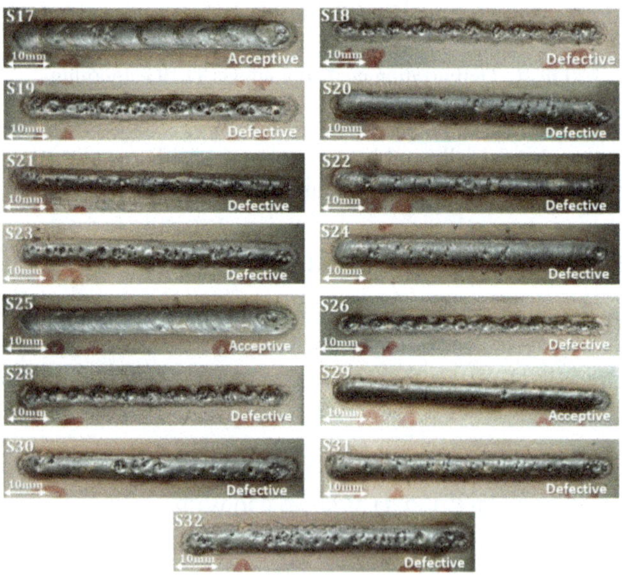

Fig. 6. Geometry results of the deposited beads in the 2nd set of experiments

$$^{\text{Minimize}}\text{BP} = 242 - 111.7\ \text{WFR}$$

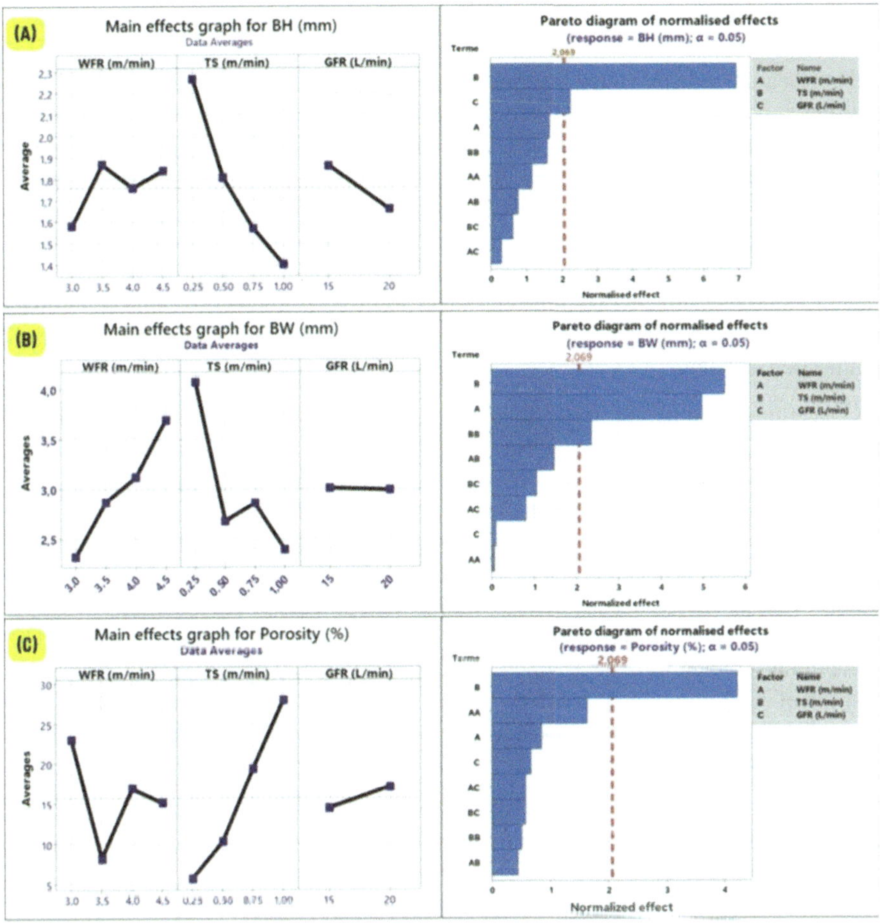

Fig. 7. Main effects graph for (a) Bead Height,(b) Bead Width and (c) Bead Porosity

$$+ 3.1\ WS - 3.56\ GFR + 13.06\ WFR^2$$
$$+ 16.0\ WS^2 - 5.7 WFR * WS + 0.82\ WFR * GFR$$
$$+ 1.63\ WS * GFR \quad (3)$$

The final outcome of our research marks the unveiling of the final optimised parameters, obtained using the TLBO algorithm. These results demonstrate the power of optimisation and the TLBO method to achieve the desired weld bead dimensions within the defined parametric limits.

- Best Solution: WS (m/min) = 0.29 / WFR (m/min) = 3.40 / GFR (L/min) = 16.28
- Best Objectives: BW (mm) = 3.28 / BH (mm) = 2.25 / BP (%) = 5.02

4 Conclusion

In this research, we conducted a series of experiments aimed to examine the influence of different parameters on weld beads dimensions and the presence of surface porosity. These experiments involved four different wire feed rates and welding speeds, as well as two different levels of gas flow. To precisely detect porosity, we implemented a sophisticated MATLAB code, which can measure porosity with a high precision. In our goal to optimise the welding process, we used the Teaching and Learning Based Optimisation method. This technique enabled us to optimize the process parameters, with the ultimate aim of obtaining the ideal weld bead geometry while minimising the presence of porosity in the welded joints. The major findings of this study can be summarised as below:

- Concerning bead height, it is primarily influenced by welding speed (WS), with gas flow rate (GFR) following closely behind. Notably, the gas flow rate has a significant impact, particularly at a value close to 20L/min. This effect can be attributed to the pressure exerted by the gas on the material during the fusion phase.
- Porosity analysis was carried out using image analysis techniques, revealing a noticeable correlation between porosity levels and welding speeds. It should be noted that high welding speeds accelerate the solidification process of the material.
- The utilization of TLBO algorithm enabled the identification of optimum working conditions, which were determined to be a welding speed of 0.29 m/min, a wire feed speed of 3.41 m/min, and a gas flow rate of 16.28 L/min. These optimized parameters yielded the following outcomes: a weld bead width of 3.28 mm, a height of 2.25 mm, and a surface porosity of 5.02%.

Acknowledgement. The authors extend their heartfelt appreciation to the diligent reviewers whose insightful comments and constructive recommendations significantly contributed to the enhancement of this paper.

References

1. Aldalur, E., Suárez, A., Veiga, F.: Metal transfer modes for Wire Arc Additive Manufacturing Al-Mg alloys: Influence of heat input in microstructure and porosity. J. Mater. Process. Technol. **297**, 117271 (2021). https://doi.org/10.1016/j.jmatprotec.2021.117271
2. Dongari, S., Davidson, M.J., 2021. Multi response optimization of Inconel 625 wire arc deposition for development of additive manufactured components using Grey relational analysis (GRA). Metall Mater Eng. https://doi.org/10.30544/586
3. Jafari, D., Vaneker, T.H.J., Gibson, I.: Wire and arc additive manufacturing: Opportunities and challenges to control the quality and accuracy of manufactured parts. Mater. Des. **202**, 109471 (2021). https://doi.org/10.1016/j.matdes.2021.109471
4. Le, V.T., Mai, D.S., Hoang, Q.H.: A study on wire and arc additive manufacturing of low-carbon steel components: process stability, microstructural and mechanical properties. J. Braz. Soc. Mech. Sci. Eng. **42**, 480 (2020). https://doi.org/10.1007/s40430-020-02567-0
5. Pattanayak, S., Sahoo, S.K.: Gas metal arc welding based additive manufacturing—a review. CIRP J. Manuf. Sci. Technol. **33**, 398–442 (2021). https://doi.org/10.1016/j.cirpj.2021.04.010

Investigation of Scanning Strategy Effects on Porosity and Hardness in Selective Laser Melting of H13 Tool Steel

Chams Kisraoui[1], Asma Belhadj[2(✉)], Salma Slama[2], and Noureddine Barka[1]

[1] Mathematics, Software and Engineering Department, University of Quebec at Rimouski, Rimouski, Canada
{chams.kisraoui,noureddine_barka}@uqar.ca

[2] Applied Mechanics and Engineering Laboratory (LMAI-LR-11ES19), University of Tunis El Manar (UTM), National Engineering School of Tunis (ENIT), Tunis, Tunisia
{asma.belhadj,selma.slama}@enit.utm.tn

Abstract. This paper presents partial results of a study aimed at optimizing the selective laser melting (SLM) process applied to H13 tool steel. An experimental investigation was conducted to explore the influence of printing parameters, particularly the laser scanning strategy, on the mechanical properties of SLM-manufactured parts. To achieve this objective, eight experimental conditions with different scanning strategies were studied following an in-depth literature review. In this study, porosity and hardness were determined for parts printed using an SLM machine with two scanning strategies: a simple bidirectional approach and a chess pattern with four rotation angles 0°, 45°, 67°, and 90° for each.

Keywords: Selective laser melting · H13 tool steel · Scanning strategy · porosity · hardness

1 Introduction

Additive manufacturing, also known as 3D printing, has received significant interest from both the industry and the academic world in recent years. It involves adding material to create three-dimensional objects from a digital 3D model, typically layer by layer, in contrast to traditional methods of material-removal manufacturing. 3D printing techniques are gaining increasing attention in the industry, aiming to reduce energy consumption and produce parts with complex geometries that are inaccessible to conventional processes.

The most commonly used method of metal additive manufacturing is Selective Laser Melting (SLM), this technique enables the creation of parts with high precision. The SLM process first appeared in the early 1990s, representing a powder bed fusion technique in which melting is achieved using a laser. The SLM process is currently used for various metallic materials, with a primary focus on enhancing the quality of final parts and reducing defects caused by SLM parameters, such as porosity, considered a major issue (Tang et al 2017, Kruth et al. 2010, Vasquez et al. 2019). This porosity defect is primarily

M. Chouchane et al. (Eds.): CMSM 2023, LNME, pp. 447–455, 2024.
https://doi.org/10.1007/978-3-031-65007-9_46

associated with the Volumetric Energy Density (VED) value, and studies have shown that if this value exceeds a certain threshold, it leads to the formation of keyhole porosities (King et al. 2014). Recently, H13 tool steel has been successfully employed with the SLM process for the fabrication of injection molds (Mazur et al. 2016). However, information available on this material are limited, especially from the perspective of AM, justifying the study presented in this paper. On the other hand, an increasing number of researchers have working to enhance the quality of H13 parts manufactured by SLM through various parameters. Therefore, it is essential to thoroughly the numerous parametric studies conducted by both manufacturers and researchers who have explored this material. The most influential parameters for the production of H13 tool steel parts using the SLM process are scan speed and laser power. A study conducted by Véle et al. (2021) examined the influence of combinations of scan speed and laser power on the porosity and hardness of H13 tool steel. The results of this study revealed that using a low scan speed and high laser power leads to the evaporation of melted material, resulting in the formation of Keyhole pores. An experiment conducted by Aljamal (2019) aimed to study the influence of layer thickness on the microstructure of H13 steel. All samples were prepared using the same laser power, hatch distance, and scan speed, but with different layer thicknesses (30 μm and 60 μm). When the layer thickness was high, the microstructure exhibited large pores and fusion defects. Conversely, when the layer thickness is reduced, material fusion improves. For H13 tool steel, studies on the effect of scanning strategy (SS) are not yet extensive. However, several pieces of research suggest that a well-chosen SS can significantly reduce residual stresses. These studies highlight the influence of scanning vector length and rotation between layers on the quality of parts manufactured by SLM (Jia et al. 2021). Rotating the scanning vector between layers ensures a uniform temperature distribution, thus favoring the production of isotropic parts with improved mechanical properties (Amirjan and Sakiani 2019).

In this study, a series of experimental investigations were undertaken to examine the impact of two distinct scanning strategies and four varied angles of rotation between successive layers on the porosity and hardness of H13 tool steel parts printed with the SLM process. The results of this study provide valuable information for identifying the most advantageous scanning strategy and the optimum angle of rotation for improving the properties of printed parts, particularly in terms of porosity and hardness.

2 Experimental Procedure

2.1 SLM Process and Material

The principle of SLM technology is based on using a laser beam to transfer energy to preheated metal powder. In the SLM process (Fig. 1), a 3D model is created with CAD software and converted to STL format. The process involves preparing a powder bed, melting the top powder layer with a laser in a controlled environment, and building the part layer by layer. After removal from the build platform, some materials may require post-heat treatment for improved properties. In our study, H13 was assessed without post-treatment.

The H13 powder used in this study was supplied by Buffalo Precision Products (BPP). The powder grain size ranges between 15 and 53 μm and we the chemical composition of the powder can be found in Table 1. Additionally, Fig. 2 displays a microscopic image of the "BPP Alloy Powder H13" used in this study.

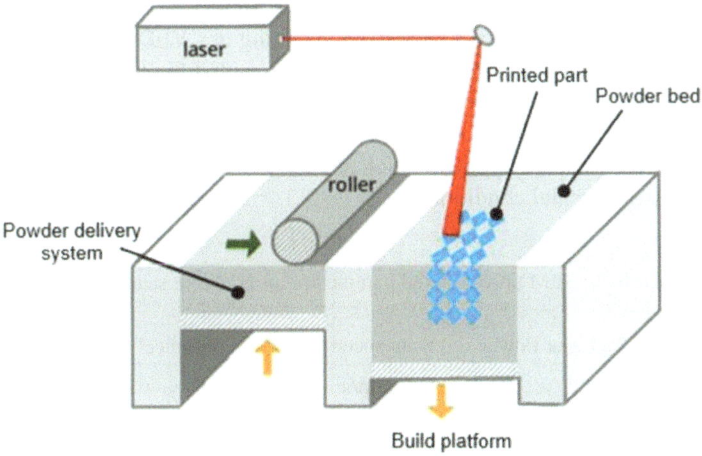

Fig. 1. SLM process

Table 1. Chemical composition of the used H13 powder.

Elements	Fe	Cr	Mn	Mo	Si	V	C	Others
Min (wt %)	Bal	5.0	0	1.0	0.5	0.5	0.3	0
Max (wt %)	Bal	6.0	0.6	2.0	1.0	2.0	0.5	0.05

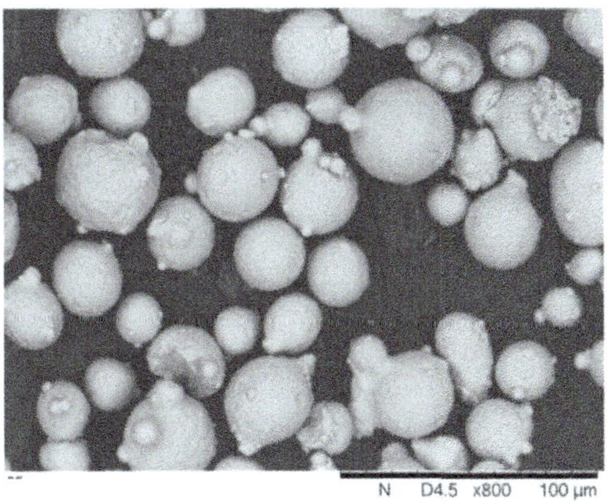

Fig. 2. Microscopic image of "BPP Alloy Powder H13"

2.2 SLM Parameters

The samples for testing were fabricated from H13 powder using an SLM EOS M 290 machine. Based on numerous previous research studies conducted in the UQAR Laboratory, the SLM parameters, such as Laser power, scan speed, hatch distance and layer thickness, used in this study haves been fixed, resulting in a VED value of 68.75 J/mm3 (Table 2). Nevertheless, two scanning strategies, each with different layer rotation angles are explored. The first strategy is a simple bidirectional one (shown in Fig. 3a), while the second involves a chess pattern with 5 mm*5 mm islands (shown in Fig. 3b). Four rotation angles for each are applied, which are 0°, 45°, 67°, and 90°. This resulted in a total of eight experimental conditions.

Table 2. SLM process fixed parameters for H13 tool steel powder.

SLM parameters	Laser Power	Scan speed	Hatch Distance	Layer Thickness
Values	275 W	800 mm/s	0.1 mm	50 μm

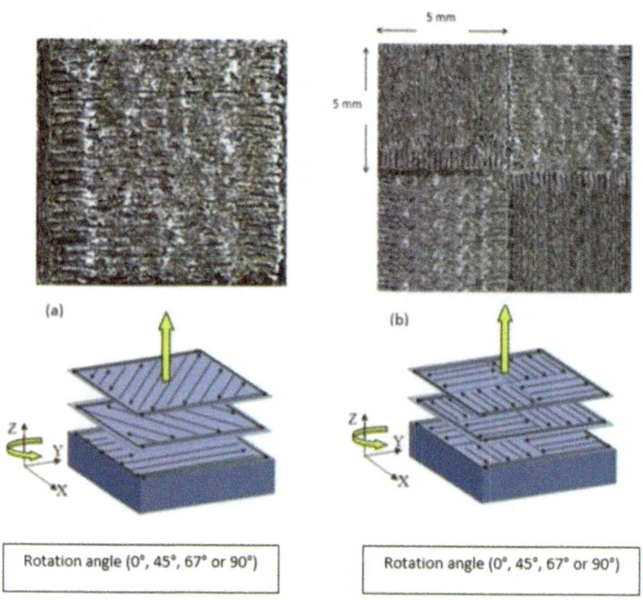

Fig. 3. (a) Simple bidirectional SS (b) Chess SS

2.3 Samples Printing

To conduct the porosity and hardness tests, cubic samples measuring 15 mm × 15 mm × 15 mm (length x width x height) were printed on the SLM machine. Their 3D model was created using SolidWorks software, as shown in Fig. 4, incorporating an 8 mm support structure which will be removed after manufacture. These samples were replicated four times for each experimental condition. Figure 5 illustrates the printed samples corresponding to each SS and angle after they have been separated from the supports and subjected to rectification. A sample coding system has been established.

- Bx: simple bidirectional SS with one of the angles (0°, 45°, 67°, and 90°)
- Cx: Chess SS with one of the angles (0°, 45°, 67°, and 90°).

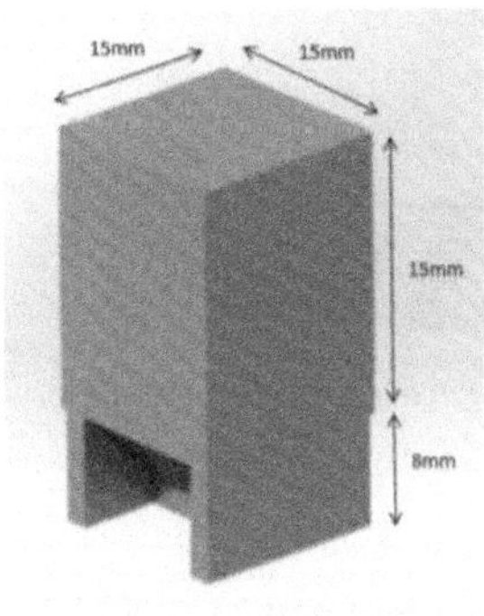

Fig. 4. The cubic sample 3D model.

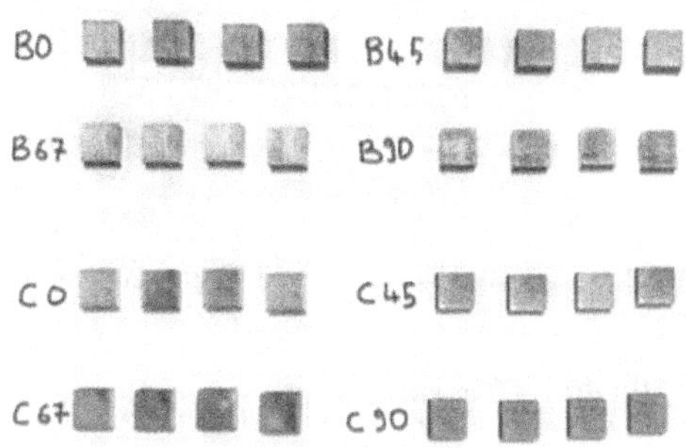

Fig. 5. Printed samples.

2.4 Porosity Measurement

Measuring porosity is directly linked to determining the relative density of samples. For this study, the Archimedes method is employed. This method involves measuring the weight of the sample in two distinct fluids: first in air, and then in water for this study. Dry and wet weight measurements are conducted using a METTLER TOLEDO electronic balance (Fig. 6a) and a density measurement accessory where the cubic sample is immersed (Fig. 6b). The porosity rate in each sample is obtained after calculating its density ρ_{sample} (Eq. 1), its relative density $D_\Gamma(\%)$ (Eq. 2) and its porosity $P(\%)$ (Eq. 3) using the following formulas:

$$\rho_{sample} = \frac{W_{air}}{(W_{air} - W_{Water})} * \rho_{water} \tag{1}$$

$$D_\Gamma(\%) = \frac{\rho_{sample}}{\rho_{theo}} * 100 \tag{2}$$

$$P(\%) = 1 - D_\Gamma(\%) \tag{3}$$

- W_{air}: Weight of sample weighed in air
- W_{water}: Weight of sample submerged in water.
- ρ_{water}: Density of water at 21 °C = 0.99798 g/cm^3
- $\rho_{theo} = 7.8$ g/cm^3: Theoretical density of H13 tool steel

Fig. 6. (a) METTLER TOLEDO electronic balance (b) Density measuring accessory

2.5 Hardness Measurement

Hardness measurement is carried out on cubic samples after cutting them in half parallel to the printing direction, applying a layer of carbon, and ensuring a thorough polish (Fig. 7a). The experimentation took place using was a Clemex hardness tester (Fig. 7b). The choice was made to assess hardness using a 2*2 indentation matrix, as depicted in Fig. 7c, with each point spaced 1 mm apart.

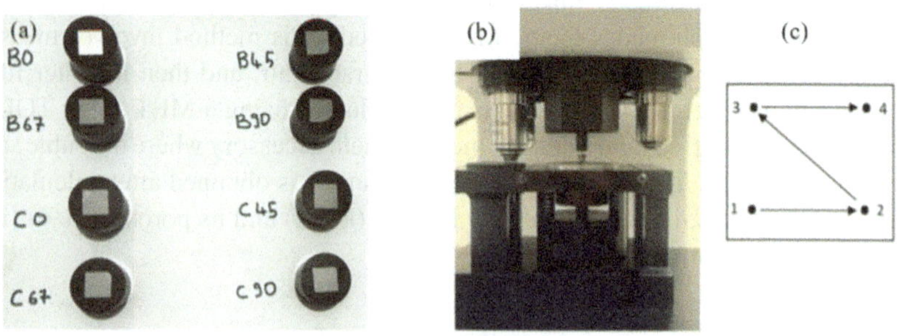

Fig. 7. **(a)** Coated cubes with polished surfaces **(b)** Clemex hardness tester (c) indentation matrix

3 Results and Discussion

3.1 Porosity

Figure 8 shows the results of porosity rate determination for the eight experimental conditions, differing in their SS. This value represents the average of the measurements carried out on the four printed samples for each experimental condition. It can be seen that sample B67, obtained using the simple bidirectional SS with a rotation angle of 67°, exhibits the lowest porosity rate at 0.45%. However, the difference from the other operating conditions is not significant, ranging from 0.45% to 0.69%. This result corroborates the findings of a study by Véle et al. (Véle et al. 2021), which states that the porosity rate depends mainly on the value of the VED, which should be between 80 J/mm3 and 120 J/mm3. In this study, different VED values were employed using a single SS, specifically the fixed bidirectional strategy with a layer rotation angle of 90° (corresponding to sample B90). One of their parameter combinations yields a VED of 69.4 J/mm3, equivalent to that used in this work (68.75 J/mm3). In their case, the recorded porosity rate is approximately 1%, which remains close to the obtained results and confirms our findings.

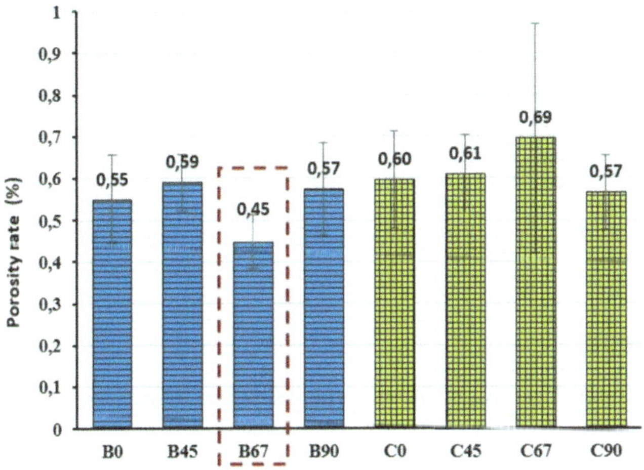

Fig. 8. Average rate of porosity of H13 tool steel for each strategy

3.2 Hardness

Figure 9 presents the results of hardness measurements for the eight experimental conditions, distinguished by their SS. These values represent the average of measurements made on the four printed samples for each experimental condition. It is observed that the simple bidirectional strategy shows a hardness dependent on the rotation angle, reaching its minimum at 67° with a value of 55.7 HRC (B67) and its maximum at an angle of 0°, reaching 59.3 HRC (B0). On the other hand, the chess strategy displays relatively stable hardness values, without major variations related to the angle of rotation.

It should be noted that the hardness of H13 steels formed by SLM, reaching 59,3 HRC, is higher than that of forged steels, whose average hardness is 50 HRC (Dzukey et al. 2020). Hardness variations between the two strategies and the different angles may be due to alternating physical state changes, melting and solidification, as well as the transformation to the solid state that the material undergoes during the SLM process.

This induces microstructural changes influencing its final hardness. It can be seen that the more complex the strategy, the more homogeneous the material. This observation can be validated by additional texture analysis tests.

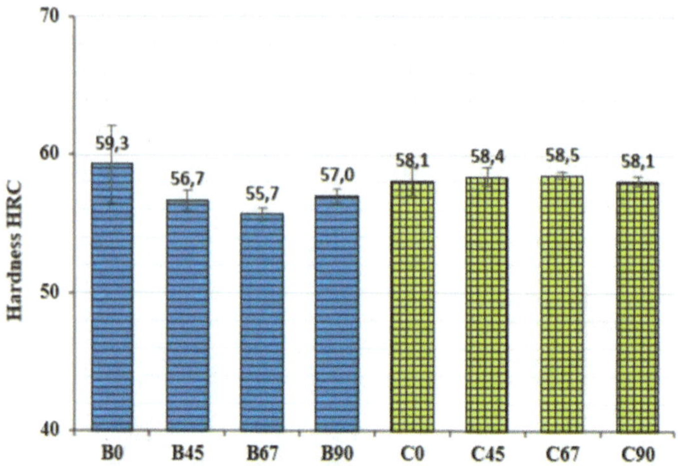

Fig. 9. Average HRC Hardness of H13 tool steel for each strategy

4 Conclusion

This study facilitated the examination of the influence of SS, specifically employing a bidirectional approach and a chess strategy, on the porosity rate and hardness of H13 steel components produced through the Selective Laser Melting (SLM) process. The primary findings from the exploration of eight experimental conditions in this study revealed that:

- The porosity rate of a printed component with a laser power of 275 W, a scanning speed of 800 mm/s, a hatch distance of 0.1 mm and a layer thickness of 50 μm is less than 0.7%, irrespective of the SS employed. Nevertheless, this rate exhibits minimal variation, even with the utilization of a complex strategy (chess with rotation), reaching a minimum of 0.4% with a bidirectional strategy and a rotation angle of 67°.
- The hardness of the components ranges from 55.7 HRC to 59.3 HRC for the bidirectional strategy; however, it remains relatively constant, regardless of the rotation angle, for the chess strategy.

To further explore the effect of the SS, additional tests are imperative to ascertain the potential influence of the strategy choice on tensile properties, as well as on the distribution of residual stresses and the texture of the SLM obtained parts.

References

Aljamal, S.N.H.: Process Optimization of Additive Manufacturing of Tool Steels. Chalmers University of Technology, Sweden (2019)

Amirjan, M., Sakiani, H.: Effect of scanning strategy and speed on the microstructure and mechanical properties of selective laser melted IN718 nickel-based superalloy. Int. J. Adv. Manuf. Technol. **103**, 1769–1780 (2019). https://doi.org/10.1007/s00170-019-03545-0

Dzukey, G.A., Yang, K., Wang, Q., Zhuang, B., Hou, W.: Porosity, hardness, friction and wear performance analysis of H13 SLM-formed samples. J. Mater. Eng. Perform. **29**, 4957–4966 (2020). https://doi.org/10.1007/s11665-020-04999-0

Jia, H., Sun, H., Wang, H., Wu, Y., Wang, H.: Scanning strategy in selective laser melting (SLM): a review. Int. J. Adv. Manuf. Technol. **113**, 2413–2435 (2021). https://doi.org/10.1007/s00170-021-06810-3

King, W.E., Barth, H.D., Castillo, V.M., et al.: Observation of keyhole-mode laser melting in laser powder-bed fusion additive manufacturing. J. Mater. Process. Technol. **214**(12), 2915–2925 (2014). https://doi.org/10.1016/j.jmatprotec.2014.06.005

Kruth, J.P., Badrossamay, M., Yasa, E., et al.: Part and material properties in selective laser melting of metals. In: Proceedings of the 16th International Symposium on Electromachining (ISEM XVI). Shanghai Jiao Tong Univ Press, pp. 3–14 (2010)

Mazur, M., Leary, M., McMillan, M., Elambasseril, J., Brandt, M.: SLM additive manufacture of H13 tool steel with conformal cooling and structural lattices. Rapid Prototyping J. **22**(3), 504–518 (2016). https://doi.org/10.1108/RPJ-06-2014-0075

Tang, M., Pistorius, P.C., Beuth, J.L.: Prediction of lack-of-fusion porosity for powder bed fusion. Addit. Manuf. **14**, 39–48 (2017). https://doi.org/10.1016/j.addma.2016.12.001

Vasquez, E., et al.: Elaboration of oxide dispersion strengthened Fe-14Cr stainless steel by selective laser melting. J. Mater. Process. Technol. **267**, 403–413 (2019). https://doi.org/10.1016/j.jmatprotec.2018.12.034

Véle, F., Ackermann, M., Bittner, V., Šafka, J.: Influence of selective laser melting technology process parameters on porosity and hardness of AISI H13 tool steel: statistical approach. Materials **14**(20), 6052 (2021). https://doi.org/10.3390/ma14206052

Optimization of the Fused Deposition Modeling Parameters Using Definitive screening design

Maroua Jabeur[1](✉), Abdessalem Jerbi[2], Slim Souissi[1], and Ahmed Elloumi[1]

[1] Laboratoire des Systèmes Electromécaniques LASEM, École Nationale d'Ingénieurs de Sfax, Université de Sfax, 3038 Sfax, Tunisia
Marouajabeur30@gmail.com, slim.souissi@isgis.usf.tn

[2] Laboratoire de Recherche Optimisation, Logistique et Informatique Décisionnelle (OLID), Institut Supérieur de Gestion Industrielle de Sfax, Université de Sfax, BP-2021, Sfax, Tunisia
abdessalem.jerbi@isgis.usf.tn

Abstract. Additive Manufacturing offers flexibility in customizing, designing, minimizing waste, faster prototyping and manufacturing complex profiles. The process parameters play a crucial role in the mechanical strength of the 3D printed product. The present research aims to investigate the impact of various 3D printing process settings on the hardness of PLA copper composite. The results of selecting terms with significant effects on the response show that all the factors studied present such effects. Indeed, the analysis of the final screening plans showed that the most influential factors are the filling density and the thickness of the layer for the Flat and Edge positions. The printing speed influences the hardness of the samples in its quadratic form while it itself has no influence. All the factors adopted are influential in their different forms.

Keywords: Additive Manufacturing · fused deposition modeling parameters · optimization · PLA copper composite

1 Introduction

Within the context of integrating novel information technologies, 3D printing is a computer-aided manufacturing process in which components are made by filling material in the design space over time [1–8]. Fused Deposition Modelling (FDM) is thought to be environmentally sustainable [9]. It's one of the most popular types of commercial 3D printing technologies [10]. This popularity is mostly related to the ease of this technology among different additive manufacturing technologies and the feasibility of printing different polymers of which the main ones are PLA (polylactic acid), PETG (polyethylene terephthalate glycol-modified), ABS (acrylonitrile butadiene styrene), nylon (a kind of polyamide) [11]. During the printing process, the mechanical properties can be affected by various parameters of the FDM technology such as raster angle, build plate and nozzle temperature, building orientation, layer thickness, and printing speed [13–15].

To balance manufacturing and the environment, sustainable materials play significant role [16]. 3D printing is crucially more sustainable than conventional processes and

M. Chouchane et al. (Eds.): CMSM 2023, LNME, pp. 456–464, 2024.
https://doi.org/10.1007/978-3-031-65007-9_47

more reasonable for the earth and the atmosphere [17]. Thus, manufacturers must use promising materials to replace non-biodegradable plastics because of their biodegradable and renewable properties. This is in order to maximize the collective environmental impact. PLA is a biodegradable polymer that mainly has good mechanical properties compared to other conventional polymers that aremore easily found in everyday life in the most diverse products [18]. The degradation of PLA is important in terms of the post-consumer period when it dominates the possibility of recycling or use of the material as well in terms of durability during its period of use. [19].

The definitive screening design technique has been proposed recently [20]. It is a new class of three-level experimental design that demonstrates superior performance to the classical experimental design and provides a better estimation of main and interaction effects [21].

The aim of the current study is to investigate the evolution in hardness of printed specimens with different FDM parameters for two building orientations (On flat and On Edge). The main objective is to assess the extent to which variations in several key fabrication parameters of FDM affect hardness of 3D printed parts.

2 Method Description

Definitive screening designs are a new category of experimental designs that have appeared relatively recently used to identify the truly significant factors among a very large number of potential candidate terms.

The traditional trial planning method start with a screening design to identify important main factors, then transit to a resolution IV design to estimate interactions between factors. By retaining statistically significant factors, researchers often obtain a full 2^k design. Sometimes, all necessary trials for a full factorial design are in the initial screening, eliminating the need for additional trials. Adding axial and central points converts the factorial design into a response surface design. Definitive screening plans streamline this process into a single step, making them a definitive solution.

The analysis of definitive screening plans is more intricate compared to other types of plans due to the inclusion of all considered terms, often resulting in completely saturated designs. Consequently, there are usually no degrees of freedom available to estimate the error term. To yield satisfactory results, the principle of parsimony must be applied. This principle involves limiting the number of parameters or variables considered in the statistical representation associated with a given phenomenon. Thus, a Step-by-Step regression is necessary during the analysis stage to adhere to this principle. The next step involves regression where variables are added or removed one at a time based on their significance until all variables not included in the model have a greater p-values than a specified α value. While all variables in the model have p-values less than or equal to the α value must be included the regression model.

The R^2 and adjusted R^2 statistics are essential for assessing the model's fit to the data. R^2 represents the percentage of variation in the response explained by the model, ranging from 0% to 100%. As more predictors are added to the model, the R^2 value always increases. Adjusted R^2 is used to compare models with different numbers of predictors, considering the model's complexity. While these statistics are crucial, residual plots, such as the Henry line and residual value plot, are necessary to verify if the model meets assumptions. As part of this study five factors were chosen. Each of these factors has three levels (Table 1).

Table 1. Studied factors

Factors	Symbol	Levels		
Nozzle temperature (°C)	A	200	210	220
Bed temperature (°C)	B	40	50	60
Manufacturing speed (mm/s)	C	40	50	60
Infill density (%)	D	50	75	100
layer thickness (mm)	E	0.2	0.3	0.4

3 Results and Discussion

The definitive screening plan adopted includes 13 experiments (Table 2). Each of these experiments was replicated five times. The filament of PLA with classic copper (Futura Form) with 1.75 mm diameter was used. Table 2 summarizes the term selection results during the different stepwise regression steps.

The analysis results of the main effects of the factors studied as well as their interactions and their quadratic effects on the hardness of the samples for the two printing positions (Flat and Edge) are illustrated in Figs. 1, 2 and 3.

The Pareto charts in Fig. 1 shows certain factors emerge as primary influencers, including factors A, D, and E across both positions, although with varying degrees of significance between positions. Additionally, interactions such as AxE and BxC for the flat position, and AxE and BxE for the edge position, are observed, despite some factors within these interactions lacking individual significance, as seen with factors B and C for both positions. Lastly, quadratic terms, particularly the CxC term, exhibit notable effects, although not significant as primary factors.

Table 2. Screening plan and results

Exp	Factor					Flat					Edge				
	A	B	C	D	E	Rep1	Rep2	Rep3	Rep4	Rep5	Rep1	Rep2	Rep3	Rep4	Rep5
1	0	+1	+1	+1	+1	62	62.5	62	63.5	63	58	59	58	57	54
2	0	−1	−1	−1	−1	60	58	58	58.5	58.5	59.5	62	62	61	61.5
3	+1	0	+1	−1	−1	58	53	54.5	54	52.5	59	60.5	61	61	60
4	−1	0	−1	+1	+1	62	61.5	60	61.5	58	54.5	52.5	55	56	55

(*continued*)

Table 2. (*continued*)

Exp	Factor					Flat					Edge				
	A	B	C	D	E	Rep1	Rep2	Rep3	Rep4	Rep5	Rep1	Rep2	Rep3	Rep4	Rep5
5	+1	+1	0	+1	−1	64	63.5	61.5	63	63	65	68	68	67	68
6	−1	−1	0	−1	+1	57	60	60	61	60	57	59	62	53	52.5
7	+1	−1	+1	0	+1	62	61	62.5	61	60	58	58	61	61	61.5
8	−1	+1	−1	0	−1	63	57	57	61	60.5	63	63	63.5	65	64.5
9	+1	−1	−1	+1	0	59	62	58.5	59	59	60	60	59	60	60
10	−1	+1	+1	−1	0	63	61	60	60.5	61	58	60	60	57	59
11	+1	+1	−1	−1	+1	58	58.5	58.5	59	59	58	57	56.5	57	58
12	−1	−1	+1	+1	−1	64	60	60	60	63	63	62.5	64	63	61.5
13	0	0	0	0	0	63	62	62	63.5	62	63	61.5	62.5	63	62

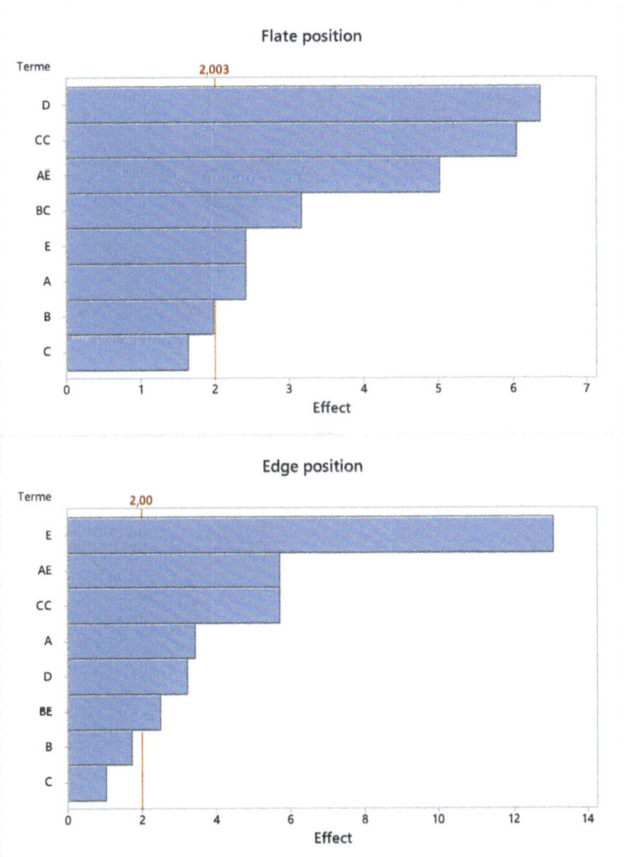

Fig. 1. Pareto diagrams of effects on response ($\alpha = 5\%$)

The graphs based on Henry lines of residual values help verify if these residual values follow a normal distribution (Fig. 2). Indeed, the residual values in both positions approximately follow a straight line.

Residual plots against order are used to check the hypothesis that residual values are independent of each other. The residual values in both positions show no trend or pattern when displayed in chronological order. Indeed, they are randomly distributed around the center line and thus considered independent (Fig. 3).

The step by step analyses of terms confirm the result of the graphic analyses (Tables 3 and 4). Indeed, factors A, D, and E present significant p values. Additionally, signification p-values are observed for the interactions AxE and BxC in the flat position, and AxE and BxE in the edge position even if factors B and C present non significant p-values. Lastly, the quadratic terms CxC exhibit notable p-value effects in both positions.

The R^2 and adjusted R^2 statistics from the selection step show relatively high values for both positions. They reach respectively 64.13% and 59.01% for the flat position

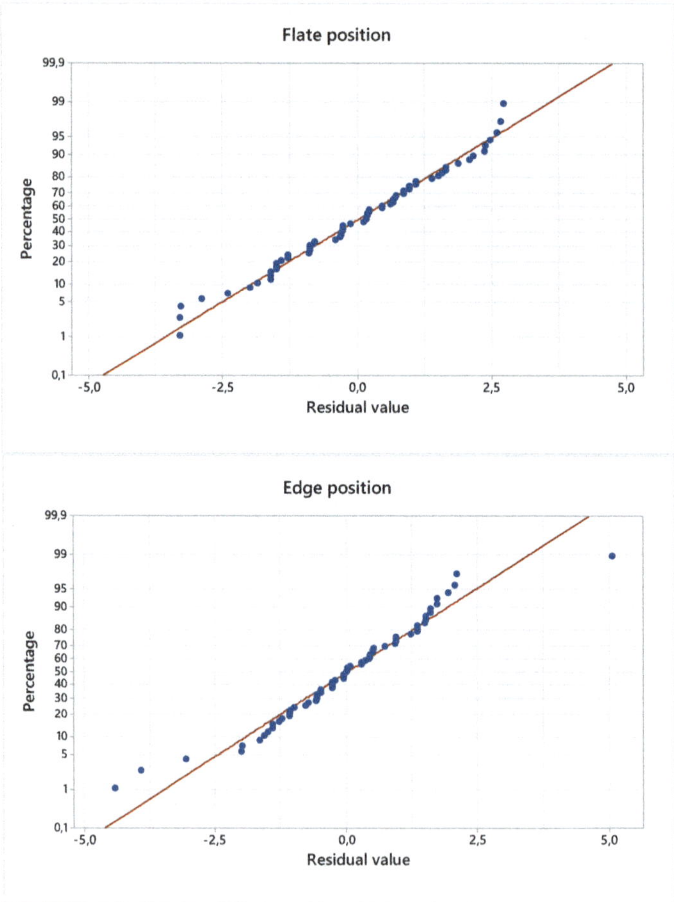

Fig. 2. Henry's lines

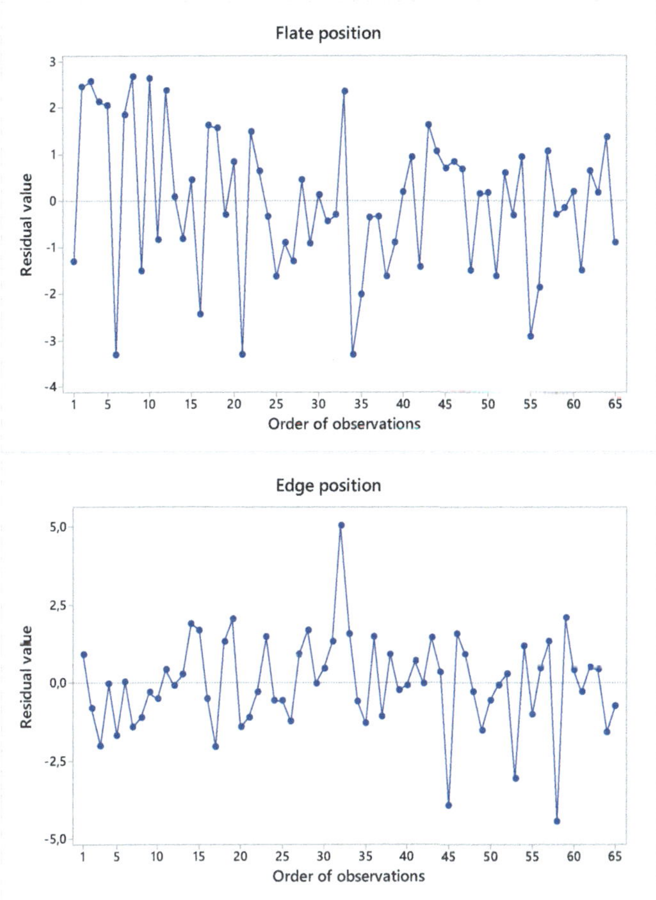

Fig. 3. Residual values in relation to observations

(Table 5) and 82.54% and 80.04% for the edge position (Table 6). These values show the good fit of the model during the regression phase even if they still need to be improved.

In summary, the analysis of the final screening plans showed that the most influential factors are respectively D and E for the Flat and Edge positions. Factor C influences the hardness of the samples in its quadratic form while it itself has no influence. The interaction between factors A and E generates significant effects for both positions.

Table 3. Step-by-step term selection on flat position

	Step 1		Step 2		Step 3		Step 4	
	Coeff	P	Coeff	P	Coeff	P	Coeff	P
Constant	60.223		61.700		62.470		62.894	
D	1.480	0.000	1.480	0.000	1.480	0.000	1.480	0.000
C			0.380	0.199	0.380	0.142	0.380	0.107
CxC			−1.920	0.003	−2.921	0.000	−3.472	0.000
A					−0.560	0.032	−0.560	0.019
E					0.560	0.032	0.560	0.019
A*E					1.155	0.001	1.791	0.000
B							0.460	0.053
BxC							0.995	0.003

Table 4. Step-by-step term selection on edge position

	Step 1		Step 2		Step 3		Step 4		Step 5	
	Coeff	P	Coeff	P	Coeff	P	Coeff	P	Coeff	P
Constant	60.154		62.100		63.170		63.170		62.714	
E	–2.960	0.000	–2.960	0.000	−2.960	0.000	−2.960	0.000	−2.960	0.000
C			0.230	0.469	0.230	0.372	0.230	0.341	0.230	0.313
CxC			–2.530	0.000	−3.921	0.000	−3.921	0.000	−3.329	0.000
A					0.780	0.003	0.780	0.002	0.780	0.001
AxE					1.605	0.000	1.605	0.000	1.636	0.000
D							0.730	0.003	0.730	0.002
B									0.390	0.090
BxE									–0.714	0.015

Table 5. R^2 statistics on flat position

	Step 1	Step 2	Step 3	Step 4
R^2	26.01%	37.83%	55.21%	64.13%
R^2 (ajust)	24.84%	34.77%	50.58%	59.01%

Table 6. R^2 statistics on edge position

	Step 1	Step 2	Step 3	Step 4	Step 5
R^2	53.49%	62.83%	76.41%	79.66%	82.54%
R^2 (ajust)	52.75%	61.01%	74.41%	77.56%	80.04%

4 Conclusion

Additive Manufacturing, particularly Fused Deposition Modelling (FDM), offers unparalleled flexibility and efficiency in customizing designs, reducing waste, and accelerating prototyping while preserving the mechanical integrity of the final product. The present study focused on exploring the impact of various 3D printing process parameters on the hardness of PLA copper composite. The results highlighted significant effects of all studied factors, with filling density and layer thickness emerging as the most influential factors for different printing orientations. Additionally, the quadratic form of printing speed exerted a notable influence on hardness. The adoption of sustainable materials like PLA underscores the environmentally conscious approach of Additive Manufacturing, positioning it as a promising solution for both manufacturing and environmental sustainability. The utilization of definitive screening design techniques facilitated a comprehensive understanding of the interaction between fabrication parameters and hardness, illuminating the optimization of 3D printing processes for enhanced mechanical properties.

Refcrences

1. Bhagia, S., Bornani, K., Agrawal, R., et al.: Critical review of FDM 3D printing of PLA biocomposites filled with biomass resources, characterization, biodegradability, upcycling and opportunities for biorefineries. Appl. Mater. Today **24**, 101078 (2021)
2. Maqsood, N., Rimašauskas, M.: Delamination observation occurred during the flexural bending in additively manufactured PLA-short carbon fiber filament reinforced with continuous carbon fiber composite. Results Eng. **11**, 100246 (2021)
3. Ngo, T.D., Kashani, A., Imbalzano, G., et al.: Additive manufacturing (3D printing): a review of materials, methods, applications and challenges. Compos. B Eng. **143**, 172–196 (2018)
4. Palanisamy, C., Raman, R., Dhanraj P.K.: Additive manufacturing: a review on mechanical properties of polyjet and FDM printed parts. Polym. Bull. 79, 7065–7116 (2022). https://doi.org/10.1007/s00289-021-03899-0
5. Tofail, S.A.M., Koumoulos, E.P., Bandyopadhyay, A., et al.: Additive manufacturing: scientific and technological challenges, market uptake and opportunities. Mater. Today **21**, 22–37 (2018)
6. Tümer, E.H., Erbil, H.Y.: Extrusion-based 3D printing applications of PLA composites: a review. Coatings **11**, 390 (2021)
7. Lay, M., Thajudin, N.L.N., Hamid, Z.A.A., et al.: Comparison of physical and mechanical properties of PLA, ABS and nylon 6 fabricated using fused deposition modeling and injection molding. Compos. B Eng. **176**, 107341 (2019)

8. Bergonzi, L., Vettori, M., Stefanini, L., et al.: Different infill geometry influence on mechanical properties of FDM produced PLA. In: IOP Conference Series: Materials Science and Engineering, vol. 1038, p. 012071 (2021)
9. Stefaniak, A.B., Bowers, L.N., Cottrell, G., et al.: Towards sustainable additive manufacturing: the need for awareness of particle and vapor releases during polymer recycling, making filament, and fused filament fabrication 3-D printing. Resour. Conserv. Recycl. **176**, 105911 (2022)
10. Gao, X., Yu, N., Li, J.: Influence of printing parameters and filament quality on structure and properties of polymer composite components used in the fields of automotive. In: Structure and Properties of Additive Manufactured Polymer Components, pp. 303–330. Elsevier (2020)
11. Wickramasinghe, S., Do, T., Tran, P.: FDM-based 3D printing of polymer and associated composite: a review on mechanical properties defects and Treatments. Polymers **12**, 1529 (2020)
12. Naser, A.Z., Deiab, I., Darras, B.M.: Poly(lactic acid) (PLA) and polyhydroxyalkanoates (PHAs), green alternatives to petroleum-based plastics: a review. RSC Adv. **11**, 17151–17196 (2021)
13. Penumakala, P.K., Santo, J., Thomas, A.: A critical review on the fused deposition modeling of thermoplastic polymer composites. Compos. B Eng. **201**, 108336 (2020)
14. Letcher, T., Waytashek, M.: Material property testing of 3D-printed specimen in PLA on an entry-level 3D printer. In: Volume 2A: Advanced Manufacturing. Montreal, Quebec, Canada: American Society of Mechanical Engineers, p. V02AT02A014 (2014)
15. Rybachuk, M., Alice Mauger, C., Fiedler, T., et al.: Anisotropic mechanical properties of fused deposition modeled parts fabricated by using acrylonitrile butadiene styrene polymer. J. Polym. Eng. **37**, 699–706 (2017)
16. Jamwal, A., Agrawal, R., Sharma, M., Kumar, A.: Sustainable material selection for indian manufacturing industries: a hybrid multi-criteria decision-making approach. In: Pratap Singh, R., Tyagi, D.M., Panchal, D., Davim, J.P. (eds.) Proceedings of the International Conference on Industrial and Manufacturing Systems (CIMS-2020).Lecture Notes on Multidisciplinary Industrial Engineering, pp. 31–43. Springer, Cham (2022). https://doi.org/10.1007/978-3-030-73495-4_3
17. Javaid, M., Haleem, A., Singh, R.P., et al.: Role of additive manufacturing applications towards environmental sustainability. Adv. Ind. Eng. Polym. Res. **4**, 312–322 (2021)
18. Sadeghi, B., Sadeghi, P., Marfavi, Y., et al.: Impacts of cellulose nanofibers on the morphological behavior and dynamic mechanical thermal properties of extruded polylactic acid/cellulose nanofibril nanocomposite foam. J. Appl. Polym. Sci. **139**, 51673 (2022)
19. Lv, S., Liu, X., Gu, J., et al.: Effect of glycerol introduced into PLA based composites on the UV weathering behavior. Constr. Build. Mater. **144**, 525–531 (2017)
20. Jones, B., Nachtsheim, C.J.: A class of three-level designs for definitive screening in the presence of second-order effects. J. Qual. Technol. **43**(1), 1–15 (2011)
21. Jones, B., Nachtsheim, C.J.: Definitive screening designs with added two-level categorical factors. J. Qual. Technol. **45**(2), 121–129 (2013)

Characterization of 3D Printed Cellular Structures Based on Hollow Spherical Cells

Mohamed Khlif(✉), Moez Frikha, Naserddine Ben Ali, and Chedly Bradai

Electro-Mechanical Systems Laboratory (LASEM), National Engineering School of Sfax, University of Sfax (ENIS), PB 1173-3038 Sfax, Tunisia

Mohamed.khlif@enis.tn

Abstract. This study investigates the compressive mechanical properties of 3D printed cellular structures based on hollow spherical cells. Utilizing Fused Deposition Modeling (FDM) with polylactic acid (PLA), we fabricated structures with four distinct architectures featuring square-hexagonal stacking and open-closed porosity variations. Each architecture was printed with four different wall thicknesses (0.8 mm, 0.6 mm, 0.4 mm, and 0.2 mm) to analyze the influence of geometric parameters on mechanical performance. Young's modulus of the material was evaluated through a series of carefully controlled quasi-static compression tests, yield strength, and overall stress-strain behavior. The results demonstrate a strong dependence of mechanical properties on the relative density of the structures, which is influenced by both cell morphology and wall thickness. Hexagonal stacking with open porosity and thicker walls exhibited superior strength and stiffness compared to other configurations. The experimental findings align well with the Gibson-Ashby model for cellular structures, confirming the role of relative density in dictating mechanical behavior. This research provides valuable in-sights for the design and fabrication of 3D printed cellular structures with tailored properties for various engineering applications, including tissue engineering and lightweight structural components.

Keywords: hollow-spheres-structures · polylactic acid (PLA) · mechanical properties

1 Introduction

The ability to produce lightweight parts with intricate geometries without incurring exorbitant costs makes FDM, an additive manufacturing technology, a compelling choice for engineers. This freedom in design stems from the non-linear relationship between geometric complexity and production expenses inherent to the FDM process [1, 2, 3].

Therefore, the family of ordered foam structures is considered as a new group of multifunctional materials with specific characteristics. It is possible to configure their application features depending not only on the cells morphology but also on material used during a structural fabrication process, [4, 5].

M. Chouchane et al. (Eds.): CMSM 2023, LNME, pp. 465–473, 2024.
https://doi.org/10.1007/978-3-031-65007-9_48

In recent years, a broad range of biomaterials including biopolymers (and biocomposites), metallic biomaterials, ceramics, etc. [6, 7] along with latest additive fabrication techniques are more and more implemented to manufacture controlled structures used for various applications, such as repairing artificial bone scaffolds[8, 9]. AM technologies offer good opportunity to create controlled structures with differences in morphology of cell shapes, sizes and porosities from a CAD model allowing the possibility to modify the geometry as required [10].

Biopolymers are widely used for the creation of additively manufactured structures including PLA (Poly-Lactic-Acid) and PLGA (Poly-Lactic-co-Glycolic-Acid) [11, 12]. It is commonly believed and was experimentally investigated in [13, 14, 15] that polymeric cellular structures display basic mechanical properties and depend on the relative density, which is obtained from the architectural morphology and the constituent material. A combination of bending and stretching occurs when the structure is loaded [16, 17].

Generally, a stretching-solicited structure is stiffer and stronger than a bending-dominated structure for the same density [18]. The mechanical properties of porous structures are determined by their architecture and morphology, when the material and density are fixed. Usually, while the zone of densification occurs at lower strains [19], the mechanical characteristics of cellular structures (specially the collapse strength and plateau strength) in-crease with increasing relative density.

Compression characterization tests allow assessment of the mechanical behaviors of porous materials or structures, including *(E)*Young's-modulus, ultimate strength, energy dissipation, etc. The relation of these properties with the structure morphology (i.e. Geometric parameters) allows the future design of structures with different topology for a large range of applications. Besides, the experimental study of these properties helps to understand and evaluate the impact of geometric factors, not only on mechanical performance but also on manufacturing defects.

Cellular structures offer many study opportunities for various engineering applications and can be created by using different 3D printing techniques. Consequently, cell structures have received a lot of attention to find the most optimal shape that can enlarge the stiffness and strength-to weight ratio [20].

Newly, the focus has shifted towards developing porous structures with light weight and controlled morphologies, such as hollow spherical cells structures. The porous structure, featuring hollow spheres, displays a dual porosity characteristic, combining both open and closed porosity features. This kind of structures offers uniform cell distribution, which provides more predictable mechanical and physical properties when compared to structures with random cell size distribution. Due to their novel morphological features, hollow spherical cells have been used in various fields of engineering, such as, tissue engineering and structural engineering.

The most used analytical model to examine the mechanical properties of parts made additively is described by Gibson [21] and Ashby [22]. They showed the deformation as a function of the effective properties of cell structures. Mechanical properties and damage mechanism are demonstrated on geometric configurations designed with different morphologies [23], such as 3D honeycomb like structure, cube truncated [24, 25], closed-cell structure [26], hollow spherical cells [27], etc. Several other studies have also focused on process parameter optimization while using commercially available materials. Zhen [28] investigated the angle and wall thickness impact of honeycomb geometry on mechanical properties. These two geometric parameters show a significant effect on mechanical behavior and failure mechanisms. The structures generated by the wall angle typically often have a lower stress threshold than that induced by the cell wall thickness. Maskery [29] studied the com-pressive mechanical behaviour of 3D matrix structures manufactured via SLM technology. Failure mechanisms are linked to the size cell, and the authors are recommended to use small cells to eliminate low strain structural failure caused by stress concentration, crack propagation and localized fracture. Nevertheless, the small cell size presents a limitation for AM techniques. Yang E [30] evaluated the influence of geometric factors on the mechanical behavior of Ti6Al4V structures made in 3D. The wall thickness and cells number have been shown to strongly affect the mechanical behavior of these structures (young's modulus and compressive strength).

2 Materials and Method

2.1 Specimen Materials and Manufacturing Processes

As mentioned in the literature, A fundamental properties of 3D cellular structures is the ability to structurally suit to the specific and intrinsic mechanical properties [13]. In this paper, cellular structures are proposed according to four varied architectures. The structures morphologies are present-ed in Fig. 1. Cellular structures were manufactured by Raise3D machine with a (PLA) Poly-lactic acid. This biopolymer is a material for all FDM printers. The mass per unit volume of the PLA is almost $\rho s = 1240\ kg/m^3$.The cellular structures has a dimension of 40*40*40 mm with 10 mm cell-diameter. It can be mentioned that the center-to-center distance of the hollow spheres with open porosity is 8 mm which is significantly smaller than the diameter of the spheres.

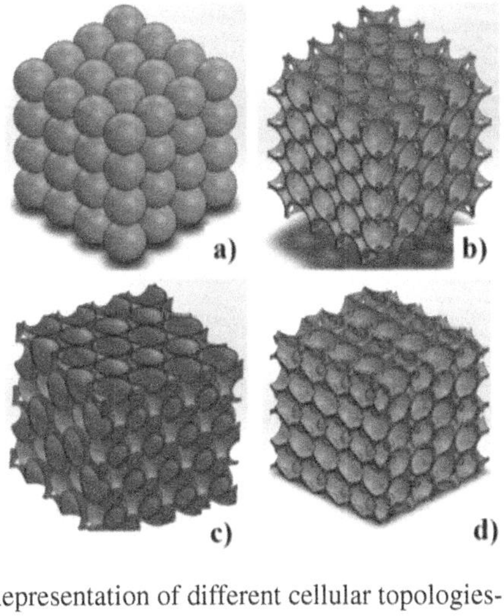

Fig. 1. Representation of different cellular topologies- CAD file.

Furthermore, this was produced using a layer height of 0.15 mm with different wall thicknesses while changing the printer nozzle to create cellular structures (namely 0.2 mm-0.4 mm-0.6 mm and 0.8 mm), as shown in Fig. 2. The nozzle and bed temperature are set to 190° and 60 °C, as well as a print speed to approximately 60 mm/s.

The technology of FDM-3D Printing is an additive manufacturing process that builds objects by layering and fusing materials together in a specific pattern. Just by exceeding its glass transition temperature, the process starts by melting PLA filament. The molten material is then carefully deposited in layers, gradually forming the complete 3D object [27]. The manufactured specimens were arranged according to four spherical unit-cell architectures having square/hexagonal stack and open/closed porosity. Each configuration is made with four wall thickness, as shown in Fig. 2.

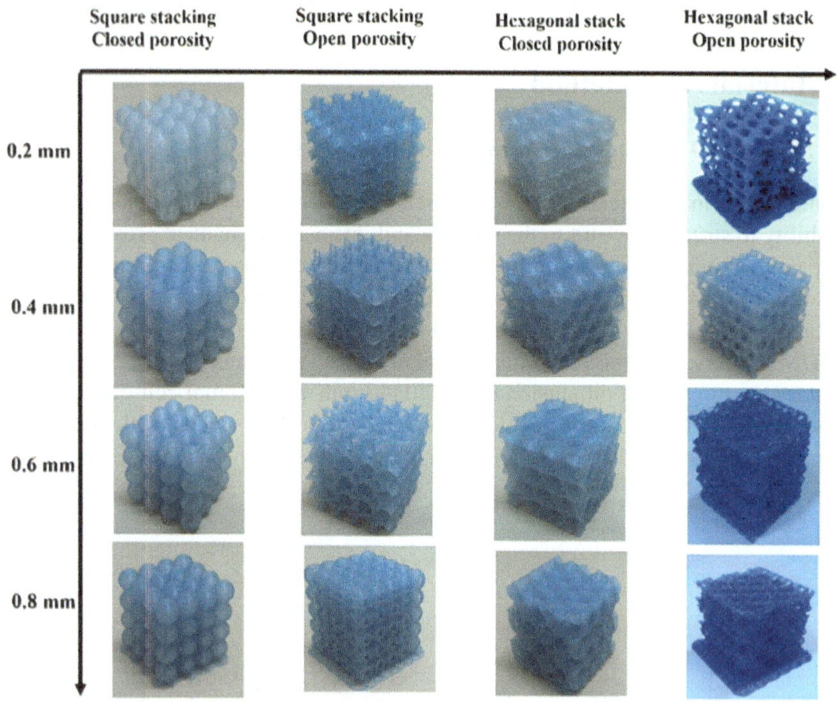

Fig. 2. PLA ordered cell structures with different wall thicknesses.

2.2 Quasi-static Compression Test

Compression testing is a method to characterize the stiffness and compressive strength of specimens. Quasi-static compression tests were applied on the different 3D structures using the same MTS universal machine with a 20 kN load sensor. For the compression tests the displacement rate of 1 mm/min and the load sensor measuring the applied force [27]. During mechanical testing, the loading direction has the same manufacturing direction. The opposite faces of the samples on which the platens of the machine transmit the force without modifying their geometry (Fig. 3).

For each specimen type, 3 tests were at least done, to take into account eventual presence of manufacturing defects. The quality of the manufactured specimen was examined using a digital microscope.

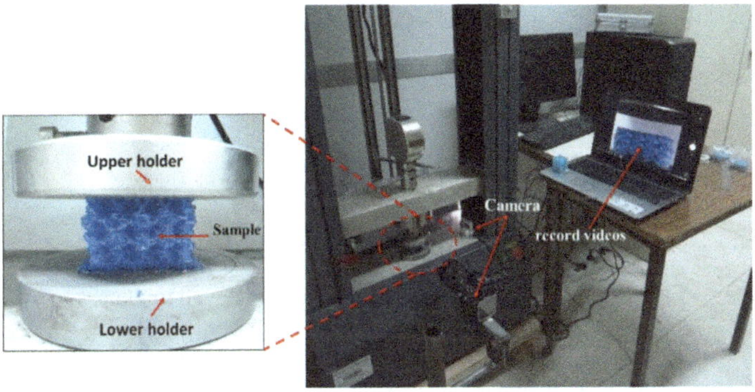

Fig. 3. View of the configuration for compression test along the manufacturing direction.

3 Results and Discussion

3.1 Microstructural Evaluation

To assess the density of each configuration, we weighed the specimens after manufacturing and calculated their relative densities, as presented in Fig. 4. All specimens increased in relative density with increasing thickness. Square based structures and closed porosity exhibited the lowest porosity and the hexagonal stacking morphology with open porosity showed the densest structures exhibit a strong correlation with the surface area of their respective unit cell topologie.

As reported in Table 1, the porosity of the structures varies from 70.7% to 94%. The morphology based on a square stack and a closed cell shows the greatest porosity with a thickness of 0.2 mm. On the other hand, the lowest porosity with a structure of a square stack has a open cell and a thickness of 0.8 mm.

3.2 Compressive Properties

Figure 5 shows the stress-strain curves for the structures tested at varied configurations. Four distinct groups of curves are presented, categorized by the spherical cell thickness: 0.2 mm, 0.4 mm, 0.6 mm, and 0.8 mm. Every obtained compressive stress-strain curve presented the characteristics of a typical cell structure. The stress-strain curves for all configurations display a characteristic pattern, including an initial Young's modulus, a plateau region, and a densification phase, where stress increases with strain as the curve progresses.

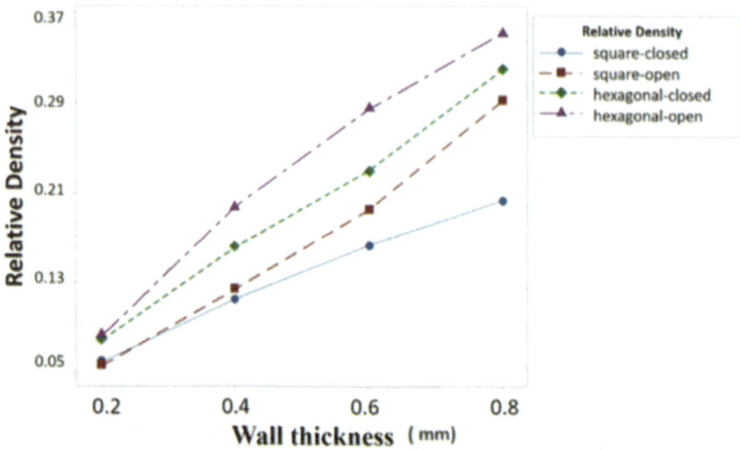

Fig. 4. Density variation with respect to cell wall thickness.

Table 1. Percentage porosity of different cellular structures

wall thicknesses	0.2 mm	0.4 mm	0.6 mm	0.8 mm
morphology	Porosity (%)	Porosity (%)	Porosity (%)	Porosity (%)
square–stack/closed–porosity	94±0.18	88.5±0.2	83.6±0.17	79.6±0.11
square–stack/open–porosity	93.4±23	87.6±15	80.4±0.16	70.7±0.14
hexagonal–stack/closed–porosity	92.06±0.21	83.8±0.18	77±0.21	71.2±0.18
hexagonal–stack/open–porosity	93.1±0.19	85±0.2	83.6±0.19	79.9±0.13

The stress-strain curve initially exhibited a linear relationship, indicating an elastic response. However, as the strain increased, the stress fluctuated around the mean value, likely due to the failure of the spherical cells. The third region is densification where the deformation continues to occur until the rupture of the cells of the structure. During this phase, structures begin densifying and the stress-strain response starts to increase remarkably denoting the densification stage of the spherical cells.

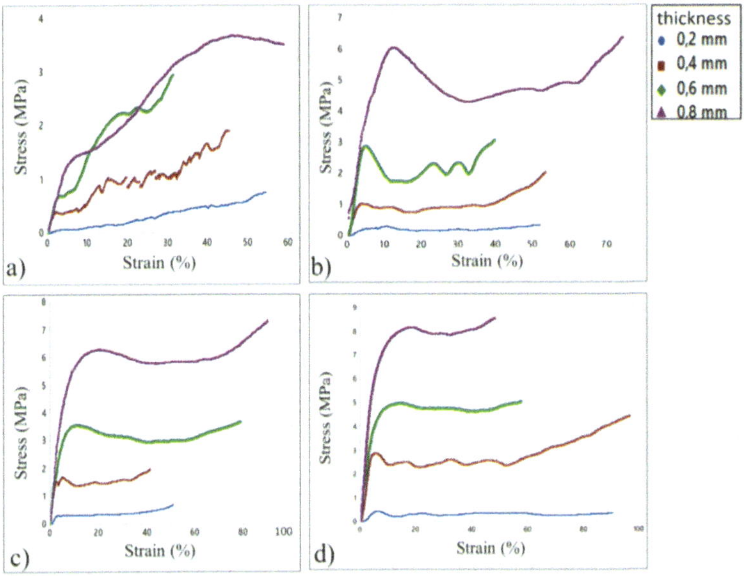

Fig. 5. Mechanical behavior of cellular structures with varying thicknesses: square-stack and closed-porosity *(a)*, square-stack and open-porosity*(b)*, hex-agonal stack and closed-porosity*(c)*, hexagonal stack and open-porosity*(d)*.

4 Conclusion

This work is the first of its kind to explore the impact of density on the mechanical properties of ordered cellular structures with hollow-spheres cells. In this article, the experimental results of mechanical properties for diverse morphology of spherical cells with a square_hexagonal stack withopen_closed porosity are presented. The wall thicknesses of the spherical cells are made with a single contour of value 0.2, 0.4, 0.6 and 0. 8 mm.The experimental study was based on the quasi-static compression tests. The specimens were manufactured from PLA polymer with the use of FDM technology. It was seen that by applying density classification to the varied cell morphologies, the behavior of structures and mechanical properties are significantly modified. The results show that increasing the wall thickness of the spherical cells improves the mechanical properties. Hexagonal stacking structure and open porosity has higher mechanical properties than other morphology. A good agreement between the evaluated mechanical properties of cellular structures with Gibson-Ashby model was achieved. A comparative study was performed, and it was found that the structure with a hexagonal stack and open porosity possessed the highest properties.

References

1. Zhang, Y., Bernard, A., Gupta, R.K., Harik, R.: Evaluating the design for additive manufacturing: a process planning perspective. Procedia CIRP **21**, 144–150 (2014)

2. Thompson, M.K., et al.: Design for additive manufacturing: trends, opportunities, considerations, and constraints. CIRP Ann. **65**(2), 737–760 (2016)
3. Lebaal, N., Zhang, Y., Demoly, F., Roth, R., Gomes, S., Bernard, A.: Optimised lattice structure configuration for additive manufacturing. CIRP Ann. **68**(1), 117–120 (2019)
4. Mitschke, H., Schury, F., Mecke, F.K., Wein, F., Stingl, M., Schröder-Turk, G.E.: Geometry: the leading parameter for the Poisson's ratio of bending-dominated cellular solids. Int. J. Solids Struct. **100**, 1–10 (2016)
5. Gao, Y., Shang, K., Chang, N., Li, L.: Structural analysis of corrugated board furniture design based on polymer material. Acta Microscopica **28**(5) (2019)
6. Wang, W., Yeung, K.W.: Bone grafts and biomaterials substitutes for bone defect repair: a review. Bioact. Mater. **2**(4), 224–247 (2017)
7. Pierantozzi, D., et al.: 3D printed Sr-containing composite scaffolds: Effect of structural design and material formulation towards new strategies for bone tissue engineering. Compos. Sci. Technol. **191**, 108069 (2020)
8. Chocholata, P., Kulda, V., Babuska, V.: Fabrication of scaffolds for bone-tissue regeneration. Materials **12**(4), 568 (2019)
9. Touri, M., Kabirian, F., Saadati, M., Ramakrishna, S., Mozafari, M.: Additive manufacturing of biomaterials– the evolution of rapid prototyping. Adv. Eng. Mater. **21**(2), 1800511 (2019)
10. Rider, P., Kačarević, ŽP., Alkildani, S., Retnasingh, S., Schnettler, R., Barbeck, M.: Additive manufacturing for guided bone regeneration: a perspective for alveolar ridge augmentation. Int. J. Mol. Sci. **19**(11), 3308 (2018)
11. Sharma, P., Pandey, P.M.: Morphological and mechanical characterization of topologically ordered open cell porous iron foam fabricated using 3D printing and pressure less microwave sintering. Mater. Des. **160**, 442–454 (2018)
12. Gregor, A., et al.: Designing of PLA scaffolds for bone tissue replacement fabricated by ordinary commercial 3D printer. J. Biol. Eng. **11**(1), 31 (2017)
13. Jiang, H., Ziegler, H., Zhang, Z., Meng, H., Chronopoulos, D., Chen, Y.: Mechanical properties of 3D printed architected polymer foams under large deformation. Mater. Des. 108946 (2020)
14. Abou-Ali, A.M., Al-Ketan, O., Rowshan, R., Al-Rub, R.A.: Mechanical response of 3D printed bending-dominated ligament-based triply periodic cellular polymeric solids. J. Mater. Eng. Perform. **28**(4), 2316–2326 (2019)
15. Hindy, A., et al.: Synthesis and characterization of 3D-printed functionally graded porous titanium alloy. J. Mater. Sci. 1–13 (2020)
16. Khaderi, S., Deshpande, V., Fleck, N.: The stiffness and strength of the gyroidlattice. Int. J. Solids Struct. **51**, 3866–3877 (2014)
17. Maconachie, T., et al.: The compressive behaviour of ABS gyroid lattice structures manufactured by fused deposition modelling. Int. J. Adv. Manuf. Technol. **107**(11–12), 4449–4467 (2020)
18. Deshpande, V., Ashby, M., Fleck, N.: Foam topology: bending versus stretching dominated architectures. Acta Mater. **49**, 1035–1040 (2001)
19. Del Rosso, S., Iannucci, L.: On the compressive response of polymeric cellular materials. Materials **13**(2), 457 (2020)
20. Brandt, M. (Ed.): Laser Additive Manufacturing: Materials, Design, Technologies, and Applications. Woodhead Publishing, UK (2016)
21. Gibson, L.J., Ashby, M.F.: Cellular Solids: Structure and Properties. Press Syndicate of the University of Cambridge, Cambridge, UK, p. 183e231 (1997)
22. Ashby, M.F., Evans, A., Fleck, N.A., Gibson Hutchinson, J.W., Wadley, H.N.G., Delale, F.: Metal foams: a design guide. Appl. Mech. Rev. **54**, B105 (2001)

23. Benali, N., Hammami, D., Khlif, M., Bradai, C.: Optimization of FDM manufacturing parameters of a biodegradable thermoplastic (PLA). In: Aifaoui, N., et al. (eds.) Design and Modeling of Mechanical Systems - IV. CMSM 2019, pp. 355–362. LNME. Springer, Cham (2020). https://doi.org/10.1007/978-3-030-27146-6_38
24. Compton, B.G., Lewis, J.A.: 3D-printing of lightweight cellular composites. Adv. Mater. **26**, 5930–5935 (2014)
25. Yavari, S.A., Ahmadi, S.M., Wauthle, R., Pouran, B., Schrooten, J., Weinans, H.: Relationship between unit cell type and porosity and the fatigue behavior of selective laser melted meta-biomaterials. J. Mech. Behav. Biomed. Mater. **43**, 91–100 (2015)
26. Wang, S., Zheng, Z., Zhu, C., Ding, Y., Yu, J.: Crushing and densification of rapid prototyping polylactide foam: meso-structural effect and a statistical constitutive model. Mech. Mater. **127**, 65–76 (2018)
27. BenAli, N., Khlif, M., Hammami, D., Bradai, C.: Mechanical and morphological characterization of spherical cell porous structures manufactured using FDM process. Eng. Fract. Mech. 106527 (2019)
28. Li, Z., Jiang, Y., Wang, T., Wang, L., Zhuang, W., Liu, D.: In-plane crushing behaviors of piecewise linear graded honeycombs. Compos. Struct. **207**, 425–437 (2019)
29. Maskery, I., Aboulkhair, N.T., Aremu, A.O., Tuck, C.J., Ashcroft, I.A.: Compressive failure modes and energy absorption in additively manufactured double gyroid lattices. Addit. Manuf. **16**, 24–29 (2017)
30. Yang, E., et al.: Effect of geometry on the mechanical properties of Ti-6Al-4V gyroid structures fabricated via SLM: a numerical study. Mater. Des. 108165 (2019)

Numerical Analysis of the Mechanical Behavior of 3D-Printed Parts

Ghalia Nhassa(✉), Romdhane Ben khalifa, and Ali Snoussi

Laboratory of Applied Thermodynamics, National Engineering School of Gabes, ENIG, Street of Omar Ibn El Khatab, Zig Eddakhlania, 6029 Gabes, Tunisia
ghalianh52@gmail.com

Abstract. Additive manufacturing is one of the growing trends in manufacturing processes. This technology enables the creation of complex forms and geometries that were previously impossible or too challenging to achieve with existing technologies. The consideration for the characteristics of the manufacturing process, the objective of this work is to numerically simulate the mechanical properties of FDM components using a CAD model. We simulated the FDM process using a numerical environment and CAE software, and then employed the finite elements technique. In truth, additive manufacturing (FDM) process simulations have shown very positive outcomes on a number of fronts. For instance, their mechanical characteristics make them suitable for use in the design of industrial products. Furthermore, FDM process simulations can provide valuable insights into the effectiveness of the production process, such as stress levels and Elastic modulus. Indeed, FDM process simulations can also optimize the design and production process, reducing cost and time.

Keywords: additive manufacturing · finite element modeling · mechanical behavior · 3D printing

1 Introduction

Determining the proper shape of structural components is a significant and critical issue for engineers. In all areas of structural mechanics, the correct design of parts is essential to verify their strength and use. Indeed, this challenge appears daily in mechanical manufacturing, such as aviation, automobiles, naval, precision machinery, and civil engineering structures. Currently, allowed manufacturing processes, additive manufacturing has many environmental advantages: fewer raw materials consumed and a significant reduction in energy requirements. Indeed, this process technology is a viable alternative to these challenges. Additive manufacturing is a prototyping or rapid manufacturing process that involves stacking layers one by one to build a complete part from the digital data of a 3D model. It allows the production of objects without shape constraints: hollow parts, thin-walled parts, undercuts, etc. These technological performances open up horizons for developing this technique in terms of materials, design flexibility, precision, and execution speed [1]. Among additive manufacturing technologies, Fused Deposition

M. Chouchane et al. (Eds.): CMSM 2023, LNME, pp. 474–486, 2024.
https://doi.org/10.1007/978-3-031-65007-9_49

Modeling (FDM) technology was developed by Scott Crump in 1988 [2]. It consists of controlling a heating nozzle according to a manufacturing algorithm generated by the machine's CAM software to extrude a filament. The extruded thermoplastic filament is deposited at a temperature just above the melting point. Two servomotors control the nozzle to move in two horizontal directions, while the support table is driven by a servomotor to lower and allow the stacking of the layers [3]. The development of additive manufacturing and the field of application, which is becoming increasingly vast thanks to the several parameters of the FDM process, influence the mechanical properties of printed parts. Among these parameters, we can cite: the orientation of the construction defines the direction in which the part is oriented along three axes [4, 5] the thickness of the layer defines the height of each addition of material, and it presents one of the essential characteristics of each printer [6], the temperature of the plate [7], the width of the filament [8]… etc.

Additive manufacturing, like fused deposition modeling (FDM), has become more popular because it can make intricate shapes and designs that are hard to produce with traditional methods. Many studies have looked at the strength, precision, and surface finish of 3D printed parts and simulating the FDM process using computer-aided engineering software.

Caminero et al. [9] investigated the mechanical properties, dimensional accuracy, and texture of PLA-based composites manufactured using fused filament fabrication. The authors highlighted the challenging nature of achieving the ideal combination of process parameters and the influence of graphene nan platelet reinforcement on these properties. This study sheds light on the influence of process parameters on FDM component mechanical properties, emphasizing the need for further research to optimize the manufacturing process.

Furthermore, the study by Samykano et al. [10] focused on the influence of printing parameters on the mechanical properties of FDM-printed ABS. The authors reported that the ideal combination of parameters is challenging, suggesting the need for advanced techniques to optimize the printing process and enhance the mechanical properties of FDM parts.

In addition, Caminero et al. [9] investigated the effect of three important process parameters, namely infill density, printing speed, and layer thickness, on the tensile properties of PLA specimens. This study highlights the significance of process parameters in determining the mechanical properties of FDM parts and indicates the need for further research to optimize these parameters for enhanced mechanical performance.

In a study by Rosenthal et al. [11] the mechanical properties and microstructure of AlSi10Mg components made using laser beam additive manufacturing (AM) technology were examined. The results of this research offer valuable information on the mechanical characteristics of parts created with various additive manufacturing methods, highlighting the ability to replicate mechanical properties and improve the manufacturing process for better performance.

Kitayama et al. [12] investigated the multi-objective optimization of injection molding process parameters. They highlighted the influence of bed temperature, primary layer thickness, and infill patterns on manufactured parts' tensile and flexural strength.

This study's findings emphasize the significance of process parameters in determining the mechanical properties of manufactured parts, indicating the need for advanced optimization techniques to enhance the mechanical performance of FDM parts.

Various printing parameters influence the mechanical characteristics of 3D-printed PLA structures. Serra et al. [13] highlighted the importance of incorporating PEG in PLA-based blends for tissue engineering scaffolds, showing increased surface topography, wettability, and elastic modulus with the addition of PEG. Senatov et al. [14] focused on the mechanical characteristics and shape memory effect of porous scaffolds made of 3D-printed PLA. Anisotropic mechanical properties related to build orientation were discussed by Torrado et al. [15], emphasizing the effect of specimen geometry on mechanical anisotropy. Zou et al. [16] investigated the elasticity and yielding performance of 3D-printed ABS material, emphasizing the impact of printing orientation on mechanical properties. Hinchcliffe et al. [17] demonstrated the enhancement of PLA mechanical properties through additive manufacturing and post-tensioning of natural fiber reinforcement. Liao et al. [18] proposed a heat treatment post-3D printing to optimize crystallinity and overcome anisotropy in 3D-printed PLA objects. Finally, Hsueh et al. [19] studied the impact of printing settings on the mechanical and thermal characteristics of PETG and PLA 3D printed materials, showing increased mechanical properties at higher printing temperatures. These studies collectively emphasize the significance of printing parameters in influencing the mechanical characteristics of 3D-printed PLA structures.

To sum up, the research shows that simulations for additive manufacturing, especially FDM, have shown positive outcomes regarding printed parts' mechanical properties. However, finding the perfect mix of process parameters is still a hurdle, underscoring the importance of more studies to fine-tune the manufacturing process and improve the mechanical properties of FDM parts. Future research could explore advanced optimization methods, real-time process control, and monitoring to achieve the desired mechanical performance in additive manufacturing.

The literature review suggests that there are still gaps in our understanding that need to be explored further. Studies show that finding the right balance of process parameters and using advanced optimization techniques is a significant challenge. Future research should concentrate on improving optimization methods, real-time control, and monitoring to improve manufacturing processes and enhance the strength of FDM parts.

Investigating how different materials and additive manufacturing techniques impact the mechanical properties of parts is essential. Filling these knowledge gaps can help improve additive manufacturing processes and create high-quality parts with better mechanical properties in the future.

The focus of this study is on examining the behavior of components created through printing. By utilizing the element method, researchers can replicate the stress distribution within a printed component. This analysis aids in pinpointing points and refining the overall design. Additionally, these simulations offer insights into the characteristics of 3D-printed parts, including their durability and rigidity.

The research delves into attributes of 3D-printed components, such as durability and rigidity. It also explores the factors that impact these attributes significantly, such as material composition and printing settings. Furthermore, it explores strategies for enhancing the features of printed parts. The influence of printing settings on the properties of 3D-printed components is substantial. Variables like layer thickness, infill density, and print speed can substantially alter the strength and rigidity of the end product. Hence, comprehending and optimizing these factors are essential for achieving the desired characteristics of printed components.

2 Material and Methods

3D printing begins with the design of an object on CAD software. Then, the 3D model is saved as a stand-for file. Stereo lithography (. STL). This STL format has established itself as the standard for any 3D printing process. The last step before printing is to cut the STL file. Software called a slicer completely cuts the digital model into a series of individual layers and sends the machine all the instructions necessary to produce the print (Fig. 1).

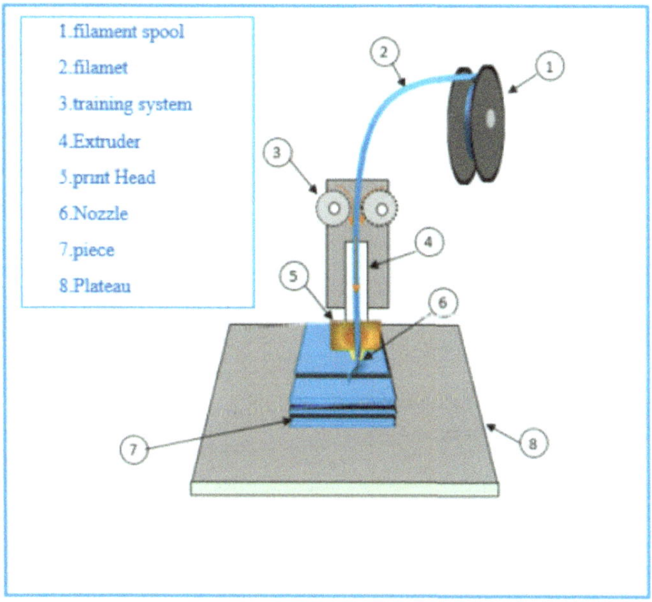

Fig. 1. A schematic representation of the FDM process

The FDM machine admits several parameters that have dominant roles in obtaining a part of good quality and good mechanical properties (filling pattern, printing speed, number of contours, raster angle, etc.). These parameters are explained in the figure below (Fig. 2).

When manufacturing specimens for testing the mechanical properties of materials, particularly in the context of 3D printing, there are a few things to keep in mind.

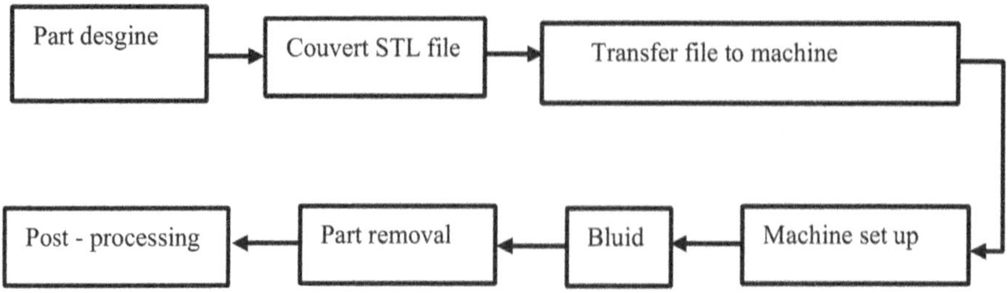

Fig. 2. Deposition parameters of the FDM process

2.1 Material and Specimen Manufacture

The material used for model making in this study was thermoplastic filament polylactide (PLA). The tensile modulus and tensile strength of the filament are 60MPa and 3.5MPa respectively.

ASTM D638–10 is a standard test method for determining the tensile properties of plastics. Type IV specimens are specifically used for testing thin plastic films. These specimens in Fig. 3 have a rectangular shape with a uniform width and a gauge length marked in the center.

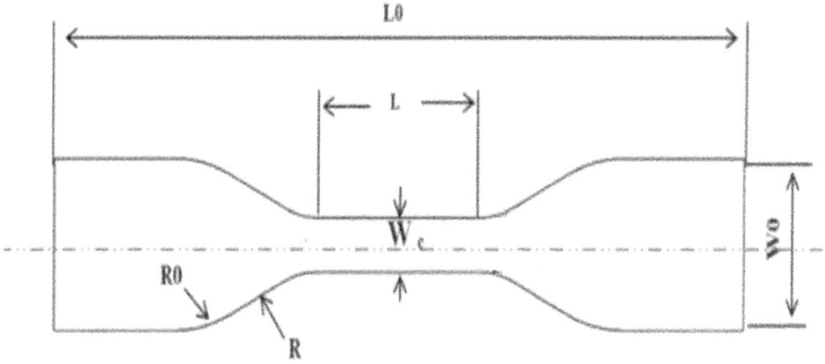

Fig. 3. Tensile test specimen (ASTM D638–10 Type IV)

During the tensile test, the specimen is gripped at the more expansive grip areas and subjected to a controlled stretching force until it fractures. The test measures various mechanical properties. The properties of interest include the elastic modulus, elongation at break, tensile strength, and critical mechanical properties that provide insights into the material's behavior under tension.

It is important to note that while ASTM D638–10 provides guidelines for the Type IV specimen, specific dimensions and requirements may vary depending on the version of the standard in use. For precise specifications, it's always advisable to refer to the latest version of the standard (Table 1).

Table 1. Specimen size

Type	L_0	L	Wc	W_0	R_0	e
D638	115	33	6	19	14	3.2

2.2 Taguchi Experimental Design

Taguchi experimental design is a statistical method developed by Genichi Taguchi for designing experiments and optimizing processes. It is widely used in various fields, including engineering and manufacturing, to study the effects of different factors on the performance of a system or product. Regarding the numerical analysis of the mechanical behavior of 3D-printed parts, Taguchi's experimental design can be applied to optimize the printing parameters and identify the most significant factors affecting the mechanical properties of the parts.

By applying the Taguchi experimental design to the numerical analysis of the mechanical behavior of 3D-printed parts, the key factors can be systematically identified, and the printing process optimized to achieve the desired mechanical properties. This strategy may enhance the quality and performance of 3D-printed parts while reducing the time and cost of trial-and-error simulation.

After studying the literature reviews, it is found that the most influencing parameters on the quasi-static mechanical response of FFF parts are infill density, printing velocity, raster orientation (Table 3) and outline overlap. These four parameters were selected as factors for the experiment (Table 2).

Table 2. Factors and control levels used for the design of experiments (DOE)

Parameters	Code	Level			Unit
		1	2	3	
Infil density	F1	10	80	100	%
Printing speed	F2	5	30	80	mm/s
Raster orientaion	F3	0,45	0,−45	−45,45	°
Ouline overlap	F4	1	3	6	–

After studying the literature reviews, it was found that the parameters most influencing the quasi-static mechanical response of FFF parts are infill pattern, infill density, printing velocity, raster orientation, and outline overlap. These five parameters were selected as factors for the experiment.

The infill pattern refers to the inside arrangement of the construction printed part that can impact its strength and stiffness. Infill density dictates the quantity of material used to occupy a component. It is affecting its weight and durability. Printing velocity affects the part's speed, potentially impacting the final product's quality. Raster orientation dictates the direction of the printed layers, which influences the part's mechanical

Table 3. Type of raster orientation of printing specimens

Raster orientaion	Specimens
0,45°	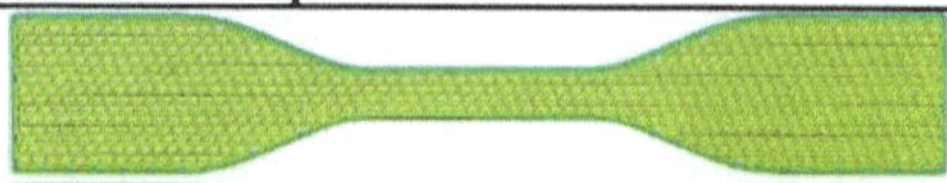
0,-45°	
-45°,45°	

properties. In 3D printing, outline overlap plays a vital role in the strength and integrity of a printed part. Factors such as infill density, printing speed, and outline overlap must be considered to achieve the desired outcome. By finding the right balance between these variables, manufacturers can create durable and high-quality 3D-printed parts that meet their specific application requirements.

Table 3 presents the quantity of 3D-printed test specimens as specified in the Taguchi L8 table. Each test specimen was printed using different printing parameters, such as temperature, speed, and layer height. The results of each test specimen will be analyzed to determine the optimal combination of printing parameters that will produce the highest-quality 3D prints (Table 4).

Table 4. The number of test specimens according to the Taguchi L9 table.

N°	F1	F2	F3	F4
1	10	5	0,45	1
2	10	30	0,−45	3
3	10	80	−45,45	6
4	80	30	0,−45	6
5	80	80	−45,45	1
6	80	5	0,45	3
7	100	80	−45,45	3
8	100	5	0,45	6
9	100	30	0,−45	1

The Taguchi L9 table allows for efficient testing involving multiple parameters and a limited quantity of test specimens. The numerical simulation was carried out on the CAE Abaqus software, allowing for detailed structural behavior analysis under different loading conditions. The simulation results indicated that the design met all safety requirements and could withstand the anticipated stresses and strains. This information was crucial in finalizing the production plans and ensuring the final product's structural integrity. Furthermore, the simulation also provided insights into potential areas of improvement in the design, allowing engineers to make any required modifications before proceeding with production. With the design's reliability confirmed, manufacturers could move forward confidently, ensuring that the final product meets the required standards and specifications (Fig. 4).

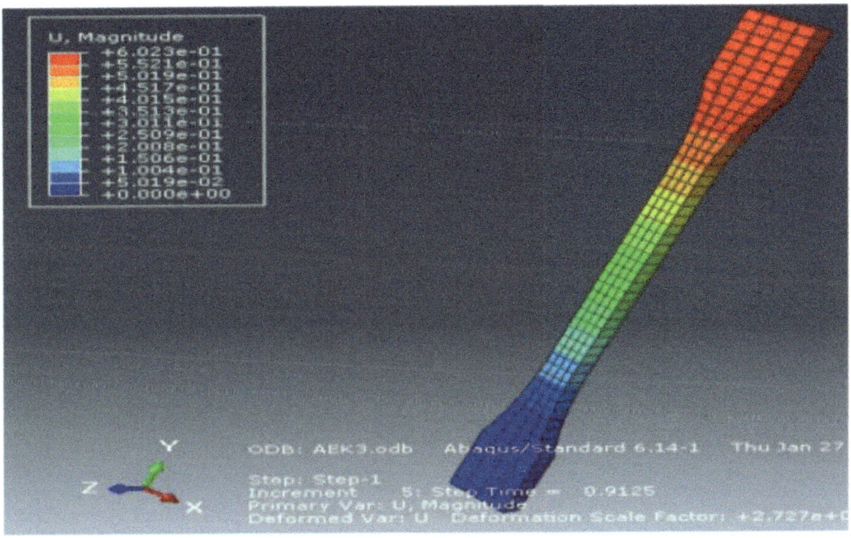

Fig. 4. Maximal displacement

The CAE Abaqus software played a crucial role in the development process, providing an indispensable tool for ensuring the project's success.

The software is thorough analysis enabled engineers to pinpoint any possible weaknesses in the design and make adjustments to enhance the overall structure. This proactive approach saved time and resources and prevented potential issues that could have surfaced during testing or after the product's release.

3 Results and Discussion

Mechanical properties of printed PLA samples determined by tensile testing according to ASTM D638–10-TypeIV standard are shown graphically. The figure compares the behavior of different printed and carefully tested tensile samples according to different selected parameters. The test specimen (SP5) performs best, coloring 111.5MPa and an elastic modulus of 3666.6 MPa. (Fig. 5) shows the stress evolution depending on deformation during a tensile test for the PLA specimen. PLA is characterized by a nonlinear elastic behavior during monotonous loading; thanks to this, the curve gives the elongation at the break of each specimen. Results may vary considerably depending on the choice of manufacturing parameters that modify the filling and the internal structure of the specimen. Young's modulus, the elastic modulus, measures a material's stiffness. It is determined through mechanical tests that analyze how a material responds to stress. The primary way to calculate Young's modulus is by performing a tensile test.

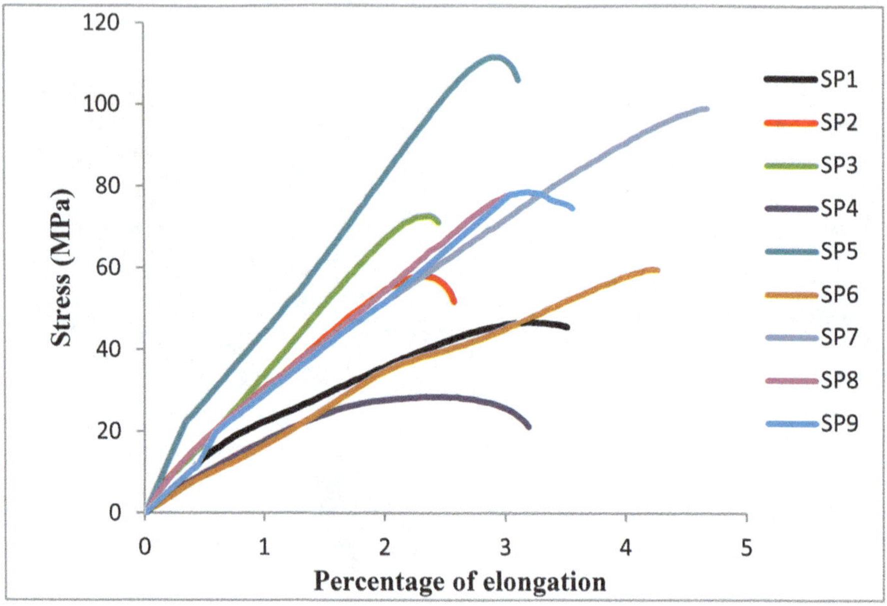

Fig. 5. The stress evolution as a function of strain during a tensile test for the PLA specimen.

To assess the impact of variables such as infill density, printing velocity, raster orientation, and outline overlap on the observed Young's modulus, the stress for each factor was analyzed in the figures below (Fig. 7 and Fig. 8). The infill density was found to significantly impact both the apparent Young's modulus and stress, with higher infill densities resulting in higher values for both parameters (Fig. 6).

Printing velocity also played a role, with faster printing speeds leading to lower apparent Young's modulus and stress values. Raster orientation and outline overlap were found to have minimal effects on these mechanical properties. Overall, it is clear that infill density and printing velocity are vital factors to consider when optimizing 3D printing parameters for desired mechanical properties.

The study of how variations in various factors affect stress given that the second level of this factor fixes the percentage of filling (F1) at 100%; on the other hand, printing speed (F2) has no effect on stress value, for the third factor (F3), the first level of this factor determines the maximum value of stress; and about the effect of the number of contours, we observe that the second level (6 contours) carries out the constraint. The number of contours plays a significant role in determining stress levels. The percentage of filling and the maximum stress value are also critical factors in this study. Understanding how these variables interact and affect stress levels is crucial for further research and

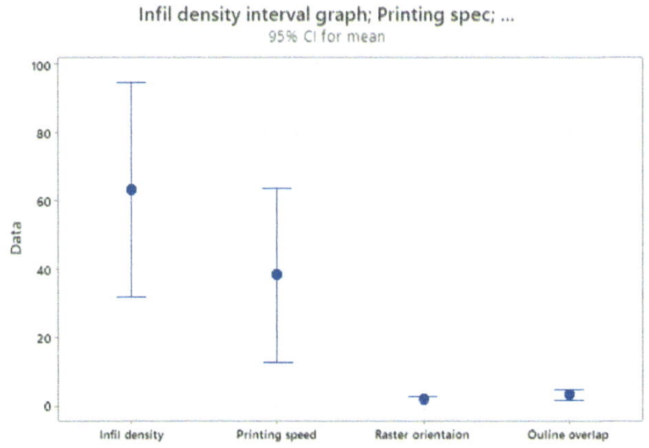

Fig. 6. 3D printing parameter interval chart

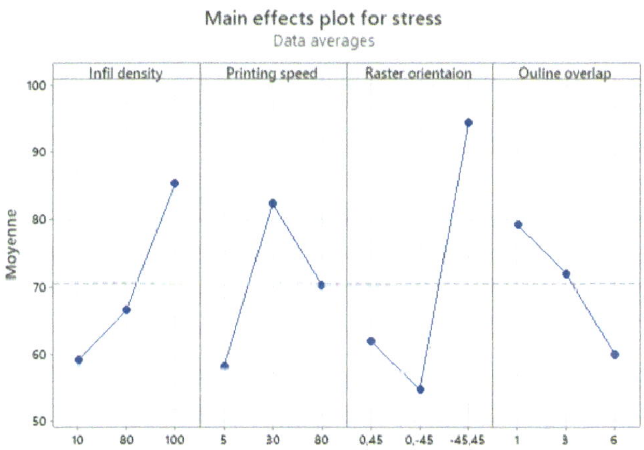

Fig. 7. Main effects for stress

development. By analyzing the impact of each factor, researchers can make informed decisions to optimize stress levels in various applications.

In the same way, the study of the effects of the variation of different factors on the modulus of elasticity shows that the percentage of the filling (F1) affects the first level (10%%); on the other hand, the printing speed (F2) also does not influence the value of the modulus of elasticity, for the third factor (F3) the maximum value of the constraint is given by the first level (0°,45°) of this factor and concerning the effect of several contours we notice that the constraint is carried out by the second level (6 contours). These results indicate that the percentage of filling and number of contours significantly impact the modulus of elasticity, while printing speed does not. Additionally, the first level of the third factor, which is the angle of the constraint, has the most significant effect on the

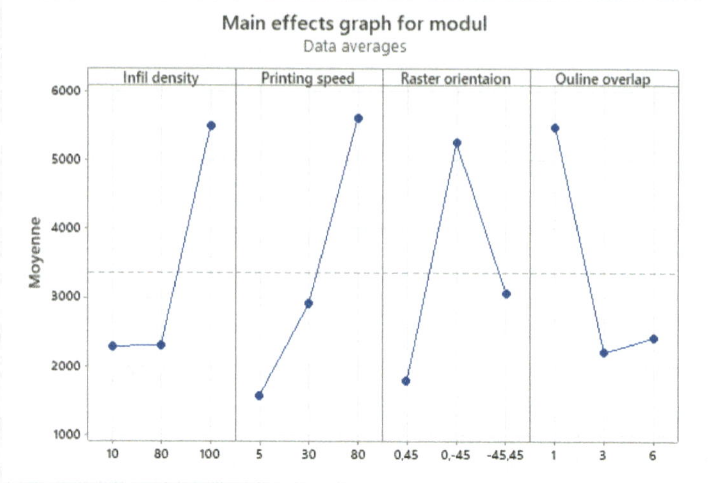

Fig. 8. Main effects for modulus of elasticity

modulus of elasticity. Overall, it is clear that certain factors play a more crucial role in determining the modulus of elasticity in the printing process.

4 Conclusion

3D printing plays a vital role in all mechanical manufacturing industries. It is instrumental in rapid prototyping and has proven capable of significantly reducing design and production times for new models of mechanical parts. The numerical study that started in this paper makes it possible to show the importance of the mechanical characteristics of PLA by determining the effects of the variables of the 3D printing of the PLA Material. Through our research, we used finite element methods to calculate CAD parts obtained by simulating an additive manufacturing (FDM) process. This results are very encouraging on several levels. For example, we can modify some printing parameters to obtain a structure with better mechanical properties and modify it again. The choice of mesh size and type of solver is essential to reducing calculation time while maintaining an accurate and acceptable calculation. When we look at how factors like infill density, printing speed, raster orientation and outline overlap affect the properties of printed PLA, it is clear that these variables have an impact on the final characteristics of the material. Manufacturers and 3D printing enthusiasts need to consider these factors to achieve their desired mechanical properties in their printed PLA. Additionally, delving into parameters could offer perspectives on enhancing the mechanical properties of printed PLA for different uses.

References

1. Sakly, A., Laksimi, A., Kebir, H., Benmedakhen, S.: Experimental and modelling study of low velocity impacts on composite sandwich structures for railway applications. Eng. Fail. Anal. **68**, 22–31 (2016)

2. Othmani, M., Chouaf, A., Zarbane, K.: Modeling and numerical analysis of the mechanical behavior of parts obtained by the FDM type additive manufacturing process. In: Proceedings of the Mediterranean Symposium on Smart City Application, pp. 1–4 (2017)
3. Abderrafai, Y., Hadi Mahdavi, M., Sosa-Rey, F., Hérard, C., Otero Navas, I., Piccirelli, N., et al.: Additive manufacturing of short carbon fiber-reinforced polyamide composites by fused filament fabrication: formulation, manufacturing and characterization. Mater. Des. **214**, 110358 (2022)
4. Wang, K., Li, S., Rao, Y., Wu, Y., Peng, Y., Yao, S., et al.: Flexure behaviors of ABS-based composites containing carbon and Kevlar fibers by material extrusion 3D printing. Polymers **11** (2019)
5. Hanon, M.M., Alshammas, Y., Zsidai, L.: Effect of print orientation and bronze existence on tribological and mechanical properties of 3D-printed bronze/PLA composite. Int. J. Adv. Manuf. Technol. **108**, 553–570 (2020)
6. Aveen, K.P., Bhajathari, F.V., Jambagi, S.C.: 3D Printing mechanical characteristion of polylactic acid and bronze filled polylactic acid components. In: IOP Conference Series: Materials Science and Engineering, vol. 376, p. 012042 (2018)
7. Sneha, P., Balamurugan, K., Kalusuraman, G.: Effects of fused deposition model parameters on PLA-Bz composite filament. In: IOP Conference Series: Materials Science and Engineering, vol. 988, p. 012028 (2020)
8. Fountas, N.A., Kostazos, P., Pavlidis, H., Antoniou, V., Manolakos, D.E., Vaxevanidis, N.M.: Experimental investigation and statistical modelling for assessing the tensile properties of FDM fabricated parts. Procedia Struct. Integr. **26**, 139–146 (2020)
9. Caminero, M., Chacón, J.M., García-Plaza, E., Núñez, P., Reverte, J.M., Bécar, J.: Additive manufacturing of PLA-based composites using fused filament fabrication: effect of graphene nanoplatelet reinforcement on mechanical properties, dimensional accuracy and texture. Polymers **11** (2019)
10. Samykano, M., Selvamani, S.K., Kadirgama, K., Ngui, W.K., Kanagaraj, G., Sudhakar, K.: Mechanical property of FDM printed ABS: influence of printing parameters. Int. J. Adv. Manuf. Technol. **102**, 2779–2796 (2019)
11. Rosenthal, I., Stern, A., Frage, N.: Microstructure and mechanical properties of AlSi10Mg parts produced by the laser beam additive manufacturing (AM) technology. Metallogr. Microstruct. Anal. **3**, 448–453 (2014)
12. Kitayama, S., Miyakawa, H., Takano, M., Aiba, S.: Multi-objective optimization of injection molding process parameters for short cycle time and warpage reduction using conformal cooling channel. Int. J. Adv. Manuf. Technol. **88**, 1735–1744 (2016)
13. Serra, T., Ortiz-Hernandez, M., Engel, E., Planell, J.A., Navarro, M.: Relevance of PEG in PLA-based blends for tissue engineering 3D-printed scaffolds. Mater. Sci. Eng. C **38**, 55–62 (2014)
14. Senatov, F.S., Niaza, K.V., Zadorozhnyy, M.Y., Maksimkin, A.V., Kaloshkin, S.D., Estrin, Y.Z.: Mechanical properties and shape memory effect of 3D-printed PLA-based porous scaffolds. J. Mech. Behav. Biomed. Mater. **57**, 139–48 (2016)
15. Torrado, A.R., Roberson, D.A.: Failure analysis and anisotropy evaluation of 3D-printed tensile test specimens of different geometries and print raster patterns. J. Fail. Anal. Prev. **16**, 154–164 (2016)
16. Zou, R., Xia, Y., Liu, S., Hu, P., Hou, W., Hu, Q., et al.: Isotropic and anisotropic elasticity and yielding of 3D printed material. Compos. B Eng. **99**, 506–513 (2016)
17. Hinchcliffe, S.A., Hess, K.M., Srubar, W.V.: Experimental and theoretical investigation of prestressed natural fiber-reinforced polylactic acid (PLA) composite materials. Compos. B Eng. **95**, 346–354 (2016)
18. Liao, Y., Liu, C., Coppola, B., Barra, G., Di Maio, L., Incarnato, L., et al.: Effect of porosity and crystallinity on 3D printed PLA properties. Polymers **11** (2019)

19. Hsueh, M.-H., Lai, C.-J., Wang, S.-H., Zeng, Y.-S., Hsieh, C.-H., Pan, C.-Y., et al.: Effect of printing parameters on the thermal and mechanical properties of 3D-printed PLA and PETG, using fused deposition modeling. Polymers **13** (2021)

Study of Fatigue Life of Structures Obtained by 3D Printing Subjected to Rotating Fatigue Test

Amina Remadi[1](✉), Sami Chatti[2], and Cyrine Lifti[2]

[1] National Engineering School of Sousse ENISo, Mechanical Laboratory of Sousse LMS, University of Sousse, Sousse, Tunisia
amina.remadi1@gmail.com

[2] Higher Institute of Applied Science and Technology of Sousse, Tahar Ben Achour Sousse, 4003 Sousse, Tunisia

Abstract. Many studies have consistently highlighted the influence of printing parameters on the mechanical properties of FDM-printed polymer components as they play a significant role in the mechanical performance of these parts. This paper is focused on experimental investigations using a rotating fatigue testing methodology. In order to determine the effect of 3D printing parameters, including infill pattern, infill density, and pattern orientation, impact the fatigue characteristics of test specimens subjected to rotating bending fatigue tests. This paper outlines the utilization of a design of experiments approach to not only evaluate the significance of these printing parameters and their interactions, but also to establish regression equations giving fatigue life as a function of the 3D printing parameters.

Keywords: additive manufacturing · fatigue test · regression equation · errors

1 Introduction

Additive Manufacturing (AM), also known as 3D printing, is a revolutionary manufacturing process that has used in a new area of production and design. Unlike traditional subtractive manufacturing methods, where material is removed to create a final product, AM builds objects layer by layer from the ground up [1, 2]. This innovative approach enables the transformation of digital designs, often in the form of 3D Computer-Aided Design (CAD) models, into physical objects with unparalleled precision and complexity [3].

Additive Manufacturing (AM) is a diverse field encompassing several distinct methods that share a common principle. Among these 3D printing techniques, Fused Deposition Modeling (FDM) stands out as the most widely adopted. FDM operates by melting a thermoplastic polymer material, such as Polycarbonates (PC), polylactic acid (PLA), acrylonitrile butadiene styrene (ABS), nylon, Thermoplastic Polyurethane (TPU), and others [5]. The melted material is then extruded in a precise pattern, layer by layer, to form a tangible object.

M. Chouchane et al. (Eds.): CMSM 2023, LNME, pp. 487–500, 2024.
https://doi.org/10.1007/978-3-031-65007-9_50

Numerous studies have been conducted to study the effect of these process parameters on the final properties of an FDM printed part and to optimize them. Investigation on fatigue life of a 3D printed part is important because they determine the real-life of the part.

While extensively discussed in the existing literature [4, 5], it's noteworthy that there is a like of comprehensive investigations into the influence of 3D printing parameters on fatigue performance. Whereas numerous studies have presented overviews of fatigue data for additively manufactured components, a detailed examination of how specific 3D printing settings impact fatigue behavior remains relatively scarce. For instance, Safai et al. [6] offered a valuable synthesis of fatigue data in additively manufactured parts, including a brief discussion of Fused Filament Fabrication (FFF) technology. However, their work did not delve deeply into the effect of various 3D printing parameters. Similarly, Shanmugam et al. [7] conducted a commendable review focusing on the fatigue behavior of printed parts, emphasizing material characteristics. However, a thorough analysis of the multifaceted relationship between 3D printing parameters and fatigue performance was not a primary focus.

The purpose of this study is to investigate the influence of 3D printing parameters on the results of the rotating bending fatigue test using a design of experiments (DOE) approach. This study aims, among other, to obtain a regression equation giving the fatigue life in function of the process parameters. The Minitab software is used in order to study a full experimental plan requiring experimental results of 54 rotating fatigue tests. The factors affecting service life include the fill rate A, the layer orientation B, the fill pattern C, and the applied force D. This work is interested on examining how changes in these parameters, as well as their interactions, affect the number of cycles to failure. Regression's equations are then established giving the fatigue life as a function of the aforementioned factors. The differences between these equations are studied. Finally, a validation test is used in order to validate the selected.

2 Materials and Methods

2.1 PLA

Polylactic acid (PLA) is a biodegradable thermoplastic polymer made of renewable resources, most derived from corn starch or sugarcane [8]. It has a relatively low melting temperature, typically in the range of 150 °C to 220 °C. The PLA has a density of 1.31g/cm^2, a tensile strength of 20.9 MPa, a tensile modulus of 1882 Mpa and a Charpy impact strength of 5.7 kJ/m^2. It is a convenient choice for rapid prototyping and manufacturing. It can be blended with a variety of materials such as wood, carbon fiber, metal, and stone to create PLA composites. The filament, given in Fig. 1, has a diameter of 1.75 mm.

2.2 Specimen Design

The fatigue specimens are designed as illustrated in Fig. 2. To create a part using a 3D printer, it requires the geometric specifications, which are obtained from slicing software in the form of an STL file. Subsequently, the file is opened using the IDEA MAKER software and then transferred to the printer.

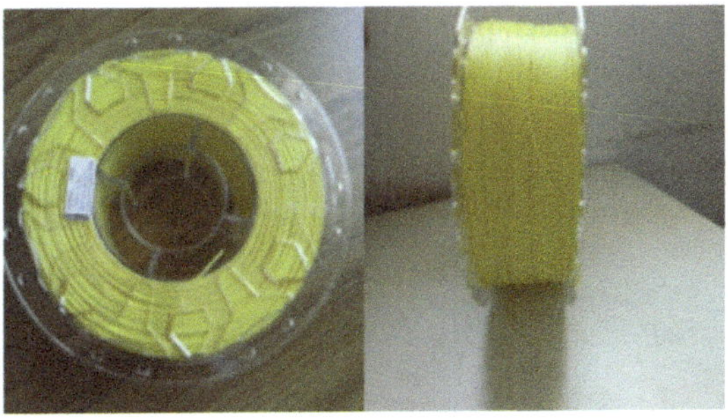

Fig. 1. Filament spool of PLA

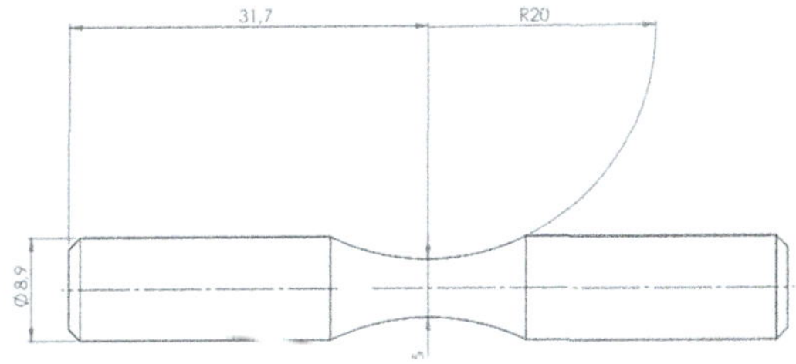

Fig. 2. Specimen dimensions

2.3 FDM Process

The PLA is heated and extruded through a nozzle to 3D printer and the specimen is made layer by layer using the Gcode instructions. With each layer, the tray is lowered automatically, and the process is repeated until the specimen is printed completely. The 3D printer used in the experiment is Raise3DN2.

2.4 Printing Parameters

The effect of 3D printing parameters - Fill Density (A), Build Orientation (B), Fill Pattern (C), and the applied flexural force - is investigated on the fatigue life of printed specimens. With

- Build orientation refers to the direction in which a part is positioned relative to the platform, along the X, Y, and Z axes. In this work, three orientations are considered: 0°, 45°, and 90°.
- Infill Density: This refers to the internal structure of a 3D printed part, determined by the proportion of infill volume with filament. It ranges between 0% and 100%. We consider two infill densities: 60% and 80%.
- Infill Pattern: Different infill patterns are utilized in 3D printed parts. In our study, we investigate three infill patterns: Grid (G), Triangular (T), and Line (L), as depicted in Fig. 5. For the applied force, we applied 15 N, 20N, and 25N.

Three levels are used for each parameter, except for fill density, for which only two levels (60% and 80%) are considered, as displayed in Table 1.

Table 1. 3D printing factors

Coded factor	−1	0	1
Factor			
Applied Force (D)	15 N	20 N	25 N
Fill Pattern (C)	Grid	Triangular	Line
Fill Pattern Orientation (B)	0	45°	90°
Fill Density (A)		60%	80%

2.5 Rotating Bending Test

The fatigue testing setup, designated as HSM19D-E-14-032 and given in Fig. 3, is located at the ISSAT Sousse. This device is equipped with a digital display interface, which offers real-time data concerning the applied force magnitude and the cycle count. Figure 4, gives a schematic illustration of the fatigue testing setup showing the applied force and the boundary conditions.

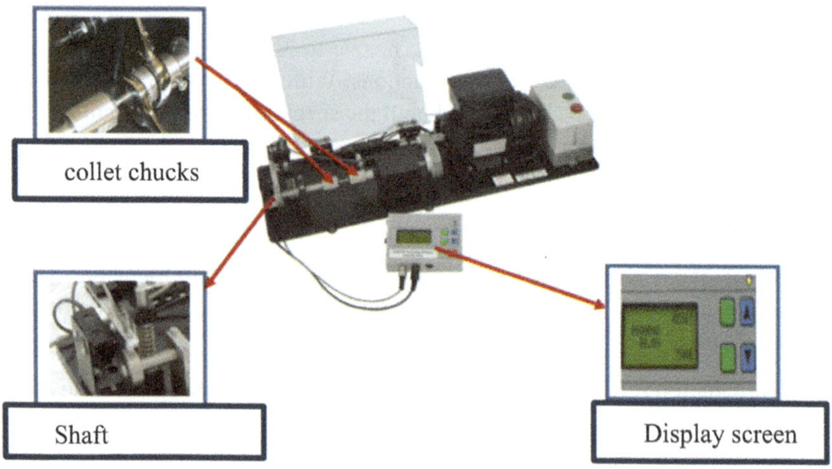

Fig. 3. Set up of the rotating bending test

The fatigue printed specimens are shown in Fig. 5. It is important to notice that there are no contours in order to get reliable results.

In Fig. 6, the specimens after-fatigue test are displayed. For the majority of specimens, fractures occur predominantly at the midpoint where the cross-section of the specimen is smallest.

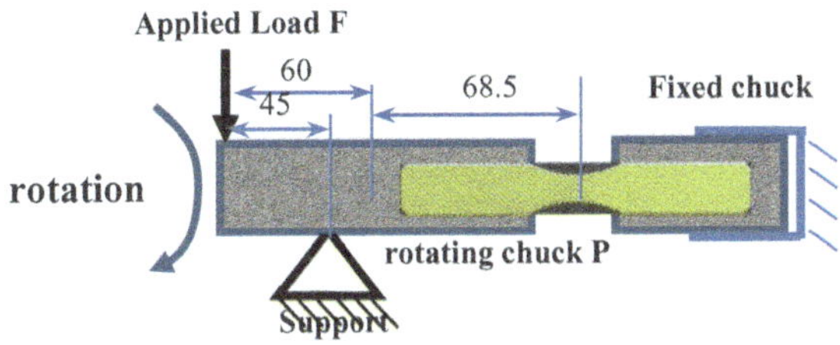

Fig. 4. Schematic illustration of the rotating bending test

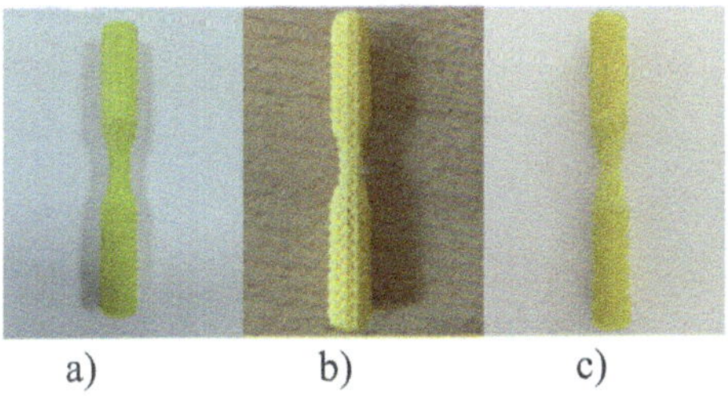

Fig. 5. Different infill patterns: a) grid, b) triangle, and c) line

Fig. 6. Specimen fractured by rotating bending fatigue test

2.6 Mixed Full Factorial Design of Experiments

The design of experiments (DOE) approach is a systematic method for determining the relationship between factors of a process and the process's outputs. DOE is a versatile data collection and analysis tool that can be used in a variety of studies. It enables the manipulation of many input variables to determine their influence on the desired results (responses). By changing many inputs at the same time, DOE can uncover critical interactions that might otherwise go undetected when testing with one component at a time. In this study, a mixed full factorial design of experiments is used needing 54 tests displayed in Table 2. Notice that the considered factors are detailed in Table 1.

Table 2. Domain of experiments

Factor / Test	A	B	C	D	N_{cycle}
1	0	−1	1	−1	96
2	0	1	0	−1	2256
3	0	1	−1	0	248
4	1	0	0	1	1274
5	1	1	1	1	2526
6	0	1	0	1	1437
7	0	0	−1	−1	874
8	1	−1	0	0	1483
9	1	0	1	1	263
10	0	−1	−1	0	248
11	1	0	−1	0	6088
12	0	1	0	0	1619
13	1	1	−1	−1	5128
14	0	0	0	0	374
15	0	0	1	1	13
16	1	0	−1	−1	14178
17	1	1	0	−1	4801
18	1	−1	−1	0	2736
19	0	0	0	1	109
20	1	1	1	−1	9193
21	1	1	−1	0	2736
22	0	0	−1	1	466
23	1	1	1	0	3633
24	1	1	−1	1	1740
25	1	−1	1	−1	7783
26	1	−1	1	0	638
27	1	−1	−1	1	1740
28	1	1	0	0	2347
29	1	0	0	−1	9091
30	0	−1	0	1	169

(continued)

Table 2. (*continued*)

Factor / Test	A	B	C	D	N_{cycle}
31	0	0	0	−1	531
32	1	1	0	1	2069
33	0	0	1	−1	65
34	1	0	1	0	401
35	0	1	−1	1	198
36	0	−1	−1	−1	618
37	0	1	1	0	271
38	1	−1	−1	−1	5128
39	0	−1	1	1	23
40	0	0	1	0	22
41	1	−1	0	−1	4654
42	1	0	−1	1	3995
43	0	−1	0	0	287
44	0	−1	0	−1	580
45	0	−1	−1	1	198
46	1	−1	0	1	769
47	0	0	−1	0	722
48	1	0	1	−1	502
49	1	−1	1	1	483
50	1	0	0	0	1407
51	0	1	1	−1	424
52	0	−1	1	0	44
53	0	1	−1	−1	618
54	0	1	1	1	135

3 Results and Discussion

3.1 Pareto Charts

A Pareto chart is a statistical tool that allows determining the impact degree of the considered factors on the response. Figure 7 shows the Pareto chart for the different parameters. it is noticed that the infill density followed by the applied load are the most important factors then by the infill cell orientation and lastly the infill pattern. Accordingly, the minimum fatigue life is obtained for the test number 15 (with density of 60% and force of 25 N) and the maximum fatigue life for the test number 16 (density of 80% and applied force of 15 N).

In addition, Fig. 7 shows that the interaction between the studied parameters has a significant effect on the fatigue life.

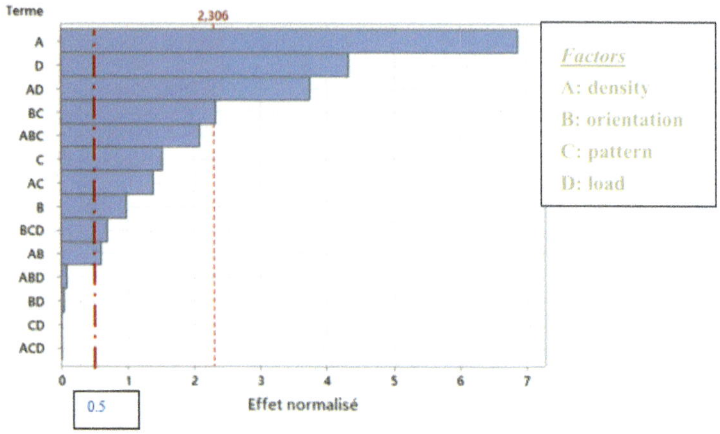

Fig. 7. Pareto chart for number of cycles

3.2 Regression Equations

ANOVA analysis provides many regressions equations that predict the response. The differences between these equations are dictated by considering a reference line position given in Fig. 7. This line specifies factors and their interactions considered in the regression equation. In the case of the default value of the reference line of 2.306, adopted by Minitab, the number of cycles until failure is expressed by:

$$\mathrm{Log}\,(\mathrm{Ncycle}) = 2.39 - 0.25\,\mathrm{D} + 0.97\,\mathrm{A} - 0.058\,\mathrm{AD} + 0.16\,\mathrm{BC} \tag{1}$$

Figure 8.a provides a comparison between the experimental values and those provided by Eq. 1. Figure 8.b displays the errors for each test with a maximum error of 92% and an average value of 14%. In order to reduce these errors, more interactions are considered. For this purpose, a parameter of 0.5 is used as the reference line value (see Fig. 7). The regression equation becomes:

$$\begin{aligned}\log(\mathrm{Ncycle}) &= 2.39 + 0.979\ \mathrm{A} - 0.25\,\mathrm{D} - 0.39\,\mathrm{C} + 0.25\,\mathrm{B} \\ &- 0.06\,\mathrm{D} + 0.15\,\mathrm{AC} - 0.13\,\mathrm{AB} + 0.18\,\mathrm{CB} - 0.06\,\mathrm{ABC} + 0.05\,\mathrm{CD}\end{aligned} \tag{2}$$

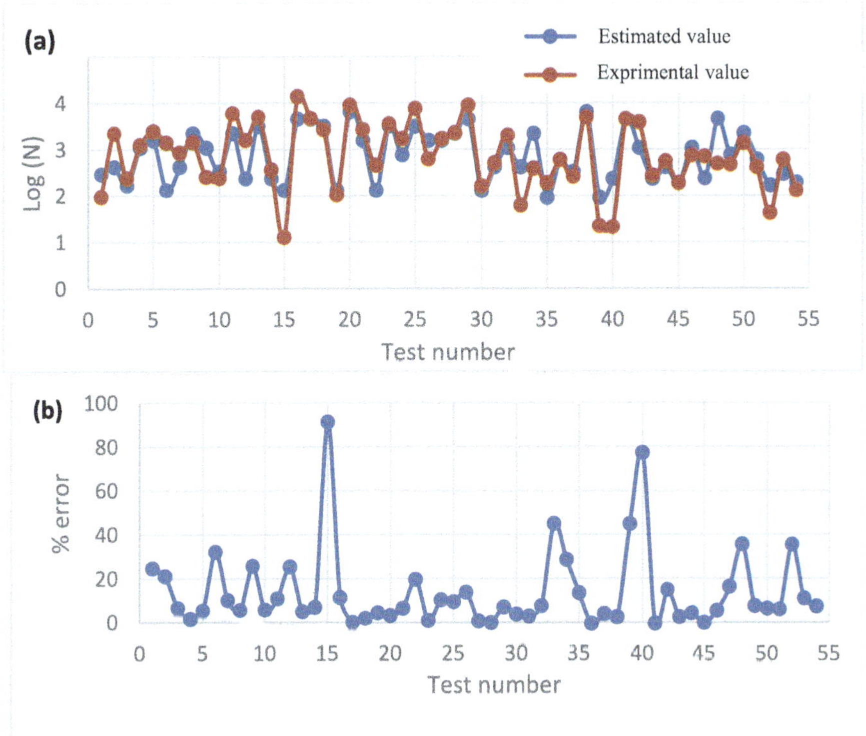

Fig. 8. a) Comparison between experimental values and estimated values for regression Eq. (1); b) Errors for each test

Figure 9 presents the error diagram. In this case, both the maximum error and the average errors decreased to 57% and 9.8% respectively.

Minitab provides the following complete equation containing all interactions with non-zero coefficients:

$$\begin{aligned}\text{Log (Ncycle)} &= 2.5512 - 0.256\,\text{D} - 0.7564\,\text{C} + 0.3292\,\text{B} + 0.634\,\text{A} + 0.0259D^2 \\ &- 0.5072C^2 + 0.2314B^2 - 0.0462\,\text{DC} + 0.0386\,\text{DB} - 0.0579\,\text{DA} + 0.1836\,\text{CB} \\ &+ 0.1501\,\text{CA} - 0.127\,\text{BA} + 0.0204D^2\text{C} - 0.0343D^2\text{B} + 0.1239D^2\text{A} + 0.0093\,D^2\text{C} \\ &+ 0.0154\,\text{DCB} - 0.0004\,\text{DCA} - 0.0002\,\text{D}B^2 + 0.0499\,\text{DBA} - 0.0866C^2B \\ &+ 0.4924C^2\text{A} + 0.5321B^2\text{C} - 0.0547\,\text{CBA} - 0.1077\,B^2\text{A} + 0.0653\,\text{DCBA} \qquad (3)\end{aligned}$$

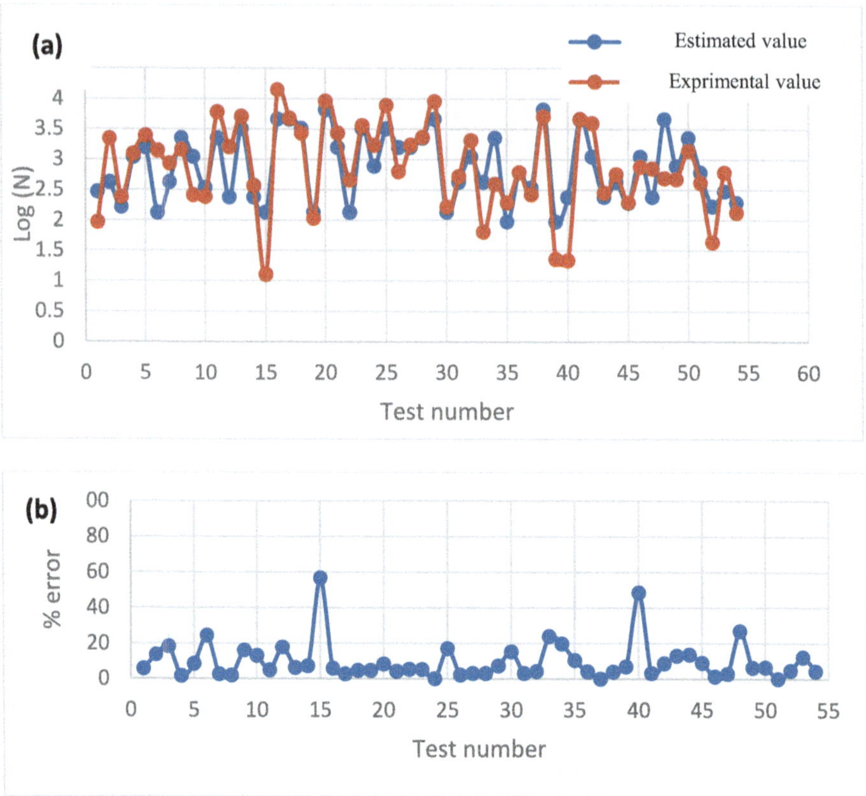

Fig. 9. a). Experimental values and estimated values by the regression Eq. (2) for the reference line value of 0.5; b) Errors for each test

This equation consists of 28 coefficients. Figure 10 shows that the maximum error is 14% and the average error is 3%.

Figure 7 shows that the interactions (C D, B D and ABD) are relatively weak. By removing these interactions, the Minitab software provides reduced regression Eq. (4) with only 14 coefficients. As shown in Fig. 11, the maximum error is then 34% and the average error is 6.2%.

$$\begin{aligned} \log \text{Ncycle} = {} & 2.57 + 0.62\ A - 0.28\ D - 0.74\,C + 0.25\ B \\ & - 0.51\ C^2 + 0.23\ B^2 + 0.15\,AC - 0.13\,AB + 0.16\ C\ B \\ & + 0.15\,(AD)^2 + 0.49\,(AC)^2 - 0.11\,AB^2 + 0.53\ CB^2 \end{aligned} \tag{4}$$

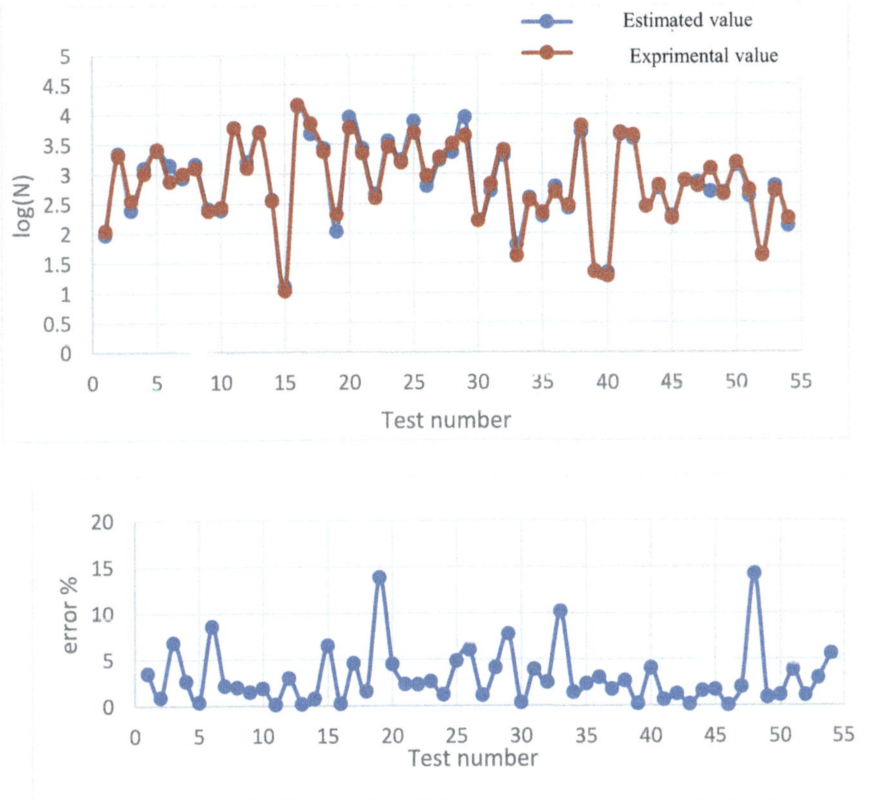

Fig. 10. a). Comparison between experimental values and estimated values by the full regression Eq. (3); b). Errors for each trial

Table 3 gives the errors provided by the four previous regression equations. It is true that Eq. (4) is the most accurate, however it is quite complex. An alternative is to use (3) which contains only 14 factors and allows an average error of 6.2% with a prediction capability of 86.07%.

3.3 Validation

To verify the validity of the model derived from the experimental plan, a specimen made by AM with 70% fill density and the grid pattern orientation of 45° was tested, and subsequently, the response values for the fatigue test were re-evaluated using the regression (3). Table 4 shows that the experimental fatigue lives are close to that given by the analytical model with a mean error of 5.2% which is within the mean error obtained from the (4).

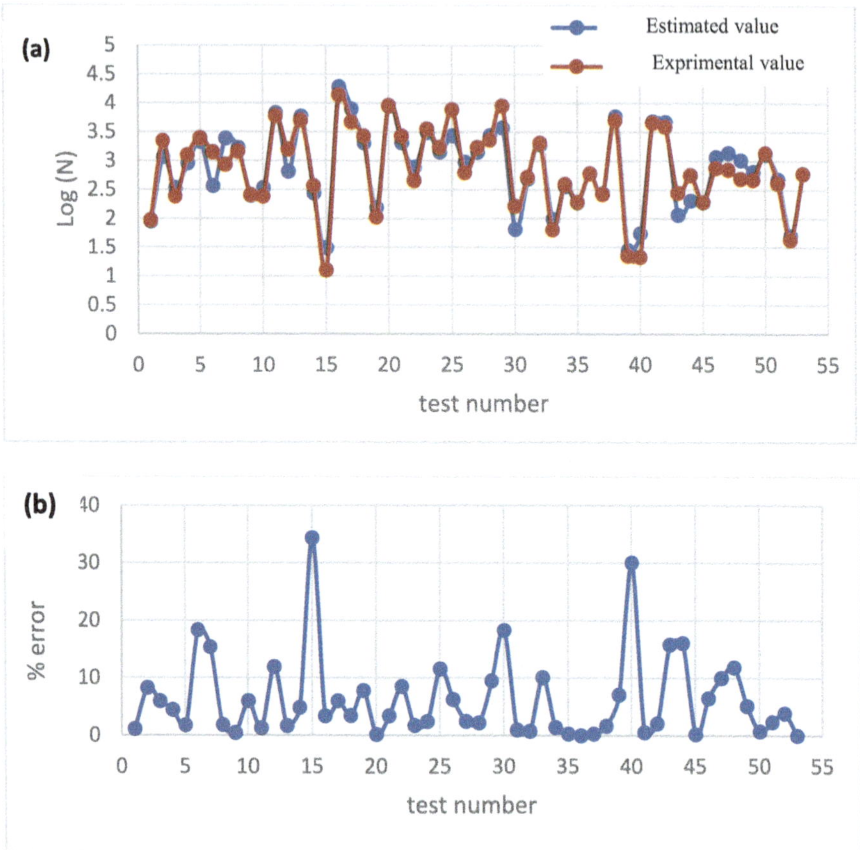

Fig. 11. a). Comparison between experimental values and estimated values by the simplified regression Eq. (4); b) Errors for each test

Table 3. Comparison of results given by the different regression equations

	Number of factors	Max Error (%)	Mean Error (%)	Prediction capability (%)
(1)	5	92	14	53
(2)	11	57	10	69
(3)	28	14	3	93
(4)	14	35	6	86

Table 4. Experimental and estimated results of validation specimen

Applied load	Experimental Log(N)	Estimated Log(N)	Error (%)
15 N	3.7	3.8	2.7
20 N	3.2	3.4	6.3
25 N	3.0	3.2	6.7

4 Conclusion

In this work we carried out a rotating bending fatigue test campaign on samples adapted to the fatigue machine. The four factors: applied bending force, fill density, fill orientation and fill pattern were experienced. Fully design of experiments (54 tests) was used to study the effect of these factors on the fatigue life. As expected, it was found that the applied force and the fill density have the main effect on the fatigue life. Many regressions equations were obtained by considering different relevant factors and their interactions. An optimized regression equation (with 14 coefficients) was adopted leading to a good fit with experiments with mean error of 6.2% and prediction capability of 86%. A validation test was experienced giving an error of 5.2% when compared with the adopted regression equation.

As perspective, other factors can be considered to study their effects on the fatigue life of 3D printed samples subjected to fatigue test in order to obtain a more general regression equation. In addition, the Wohler curve can be obtained and optimization approach can be conducted to establish the best factors combination in order to improve the fatigue life of 3D printed samples.

References

1. Stavropoulos, P., Foteinopoulos, P., Papacharalampopoulos, A., Bikas. H.: Addressing the challenges for the industrial application of additive manufacturing: towards a hybrid solution. Int. J. Lightweight Mater. Manuf. **1**(3), 157–168 (2018)
2. Wong, K., Hernandez, A.: A Review of Additive Manufacturing. ISRN Mechanical Engineering, pp. 1–10 (2012)
3. Durakovic, B.: Design for additive manufacturing: benefits. Trends and challenges. Period. Eng. Nat. Sci. (PEN) **6**(2), 179 (2018)
4. Wiang, C.-P., Cheng, Y.-C., Lin, H.-W., Chang, Y.-L., Pasang, T., Lee, S.Y.: Optimization of FDM 3D printing parameters for high strength PEEK using the Taguchi method and experimental validation. Rapid Prototyp. J. **28**, 1260–1271 (2022)
5. Wickramasinghe, S., Do, T., Tran, P.: FDM-based 3D printing of polymer and associated composite: a review on mechanical properties. Defects and treatments. Polymers **12**, 1529 (2020)
6. Safai, L., Cuellar, J.S., Smit, G., Zadpoor, A.A.: A review of the fatigue behavior of 3D printed polymers. Addit. Manuf. **28**, 87–97 (2019)
7. Shanmugam, V., et al.: Fatigue behaviour of FDM-3D printed polymers. polymeric composites and architected cellular materials. Int. J. Fatigue **143**, 106 (2021)

8. Bagsik, A, Schöppner, V.: Mechanical properties of fused deposition modeling parts manufactured with ULTEM. In: ANTEC, pp. 1294–1298 (2011)

3D-Printed Bio-Sourced Meta-material: Tensile Properties

A. Hamrouni[1,2(✉)], J. L. Rebiere[1], A. ElMahi[1], Moez Beyaoui[2], and Mohamed Haddar[2]

[1] Acoustics Laboratory of Le Mans University (LAUM), UMR CNRS 6613, Le Mans University, Av. O. Messiaen, Le Mans, France
anis_hamrouni@yahoo.com, {jean-luc.rebiere, abderrahim.elmahi}@Univ-lemans.fr

[2] Laboratory of Mechanics Modeling and Production (LA2MP), National School of Engineers of Sfax, University of Sfax, Sfax, Tunisia
Mohamed.haddar@enis.rnu.tn

Abstract. Lightweight materials are in high demand across several industries that require materials capable of withstanding various loads. One of the most common ways to achieve lightweight materials is through the use of architectural sandwiches that involve printing two skins onto a honeycomb structure. These sandwiches provide a versatile solution for providing specific mechanical and dynamic properties that are suitable for different industrial sectors. The current study focuses on the tensile properties of auxetic and non-auxetic materials made from three different honeycomb structures, namely rectangular, re-entrant, and hexagonal lattice. The study uses a bio-based material consisting of flax fibers and polylactic acid, which is a sustainable and environmentally friendly option. The specimens fabricated using a Raise3DPro2 Plus printer, and various tensile tests performed on honeycomb structures with different relative densities of the unit cell to determine the effect of honeycomb lattices on the tensile properties of the material. The results of the study show that the lattice structure of the honeycomb material significantly affects the structural Young's modulus and Poisson's ratio. Additionally, the study includes the development of a numerical model that demonstrates good agreement between the experimental results and numerical predictions. This model can used to predict the tensile properties of honeycomb structures for different relative densities of the unit cell, providing a valuable tool for engineers and designers. Overall, this study findings could apply to various industrial sectors and highlights the potential of using bio-based materials and honeycomb structures for creating lightweight materials with specific mechanical and dynamic properties.

Keywords: Poisson's ratio · Tensile · metamaterial · Additive manufacturing · PLA/Flax material

M. Chouchane et al. (Eds.): CMSM 2023, LNME, pp. 501–509, 2024.
https://doi.org/10.1007/978-3-031-65007-9_51

1 Introduction

Additive manufacturing is a highly innovative technology that allows for quick prototyping of complex structures and has gained popularity due to its advantages over traditional manufacturing methods. These advantages include smaller setups, faster production times, rapid design changes, reduced production costs (Mathur et al. 2022). The proliferation of 3D printing in the market has seen a swift surge in recent years, surpassing projected estimates (Kabir et al. 2020). The mechanical properties of materials produced through 3D printing are largely influenced by the choice of base materials (Wang and Zou 2022). Recent advancements in 3D printing have included using natural fibers to reinforce the matrix, resulting in composite materials with low-density, high-energy absorption capacity, and good static properties. Polylactic acid (PLA) is another sustainable material that has been used in 3D printing. The mechanical properties of composite materials are influenced by the honeycomb geometry of the structure. The mechanical properties of 3D printed materials have been shown to vary based on the properties of the base materials by multiple authors (Liu et al. 2021; García-Collado et al. 2022). Recent developments in 3D printing have resulted in the use of natural fibers to reinforce the matrix, which enhances the mechanical properties of the composite materials. Flax fibers are a popular choice for this purpose due to their low density, high energy absorption capacity, good static properties, and durability (Scarpa and Tomlinson 2000; Hamrouni et al. 2023a; Hamrouni et al. 2023b). They are also considered a durable alternative due to their availability (Khalfallah et al. 2014). Additionally, polylactic acid (PLA) is a biodegradable and recyclable material that is derived from sustainable resources, making it an attractive option for use as a matrice (Okada 2002; Satyanarayana et al. 2009). Studies have been conducted on short flax fibers-reinforced polylactic acid (PFF), which has been found to have low density, high energy absorption capacity, and high specific strength (Hamrouni et al. 2023c; 2023d; 2023e). The structure design of a composite material, particularly the honeycomb geometry, has a significant impact on its static properties as demonstrated by (Mousanezhad et al. 2016; Zamani et al. 2022). Researchers have also evaluated the strength of conventional and auxetic structures in comparison, finding that the tensile performance of auxetic structures was particularly noteworthy (Imbalzano et al. 2018). In this study, three honeycomb designs made from PFF material are developed using additive manufacturing, and their static properties are tested using experimental tensile tests. The Poisson's ratio and Young's modulus of the structures are also compared to numerical predictions.

2 Materials and Methods

2.1 Honeycomb Structures

Figure 1 displays the geometries of the structures designed for this work. The samples are designed and fabricated by using three cell shapes (Table 1). Their relative densities are determined by expression (1) where ρ is the cell density and ρ_s the Flax/PLA material density. The cells are inserted into a square with an H dimension of 25/4 mm.

Derived from Fig. 1, $l = H/(2\cos\theta)$ and $h = (H/2)\,(1\text{-tg}\theta)$. The cells plane directions are align with the (X, Y) directions, with the width extending along the Y-axis and the

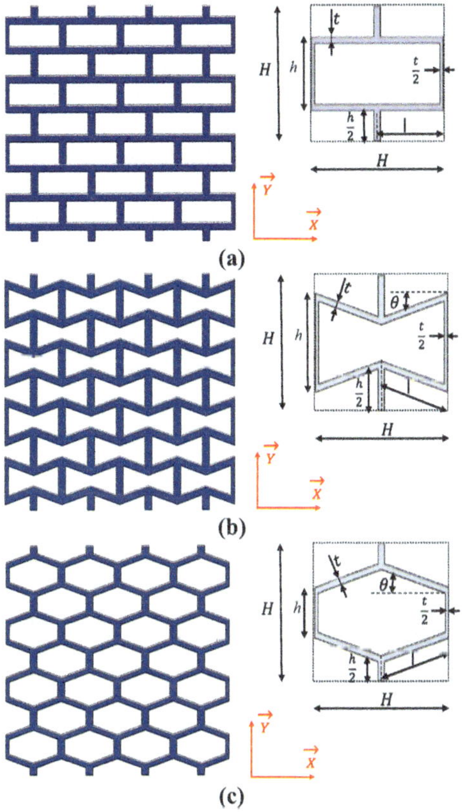

Fig. 1. (a) Honeycomb structures (a) Rectangular (b) Re-entrant (c) Hexagonal.

Table 1. Structure parameters.

Structure	θ(°)	$\rho/\rho s$ (%)	h (mm)	l (mm)
Rectangular	0°	28.8	3.12	3.12
Re-entrant	−20°	33.52	4.26	3.32
Hexagonal	+ 20°	26.53	1.98	3.32

thickness by Z-direction. The specimen's dimensions are set with a width and thickness of 25 mm and 5 mm, respectively.

$$\frac{\rho}{\rho_s} = \frac{t}{H\cos\theta}(2 + \cos\theta + \sin\theta) \quad (1)$$

2.2 Additive Manufacturing Process

In the present study, the fabrication of meta-materials is carried out using a 3D printer known as the "RAISE3D Pro2 Plus" (Fig. 2). This device is part of the latest generation of professional 3D printers and is equipped with numerous features that enable the production of high-quality, highly precise components. The process involves a power supply block drawing a filament from the feed coil and feeding it into a heating block. The material, once melted, is dispensed onto the printing platform via an extrusion nozzle, ensuring a steady flow of viscous composite in accordance with the parameters specified in Table 2. To generate the desired shapes, a CAD model of the meta-materials is required. This model is subsequently translated into a set of instructions that the RASE3DPro2 Plus can interpret, facilitated by the specialized Idea Maker software.

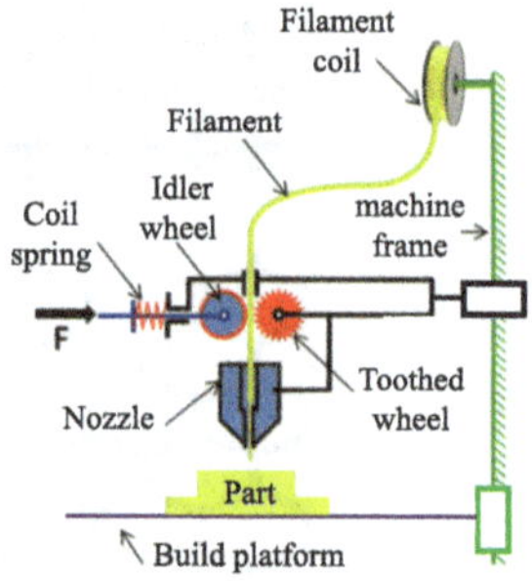

(a) (b)

Fig. 2. (a) RAISE3D Pro2 Plus (b) 3D printing process.

Table 2. 3D printing parameters

Nozzle diameter (mm)	Nozzle temperature (°C)	Platform temperature (°C)	Print speed (mm/s)	Layer thickness (mm)	infill density (%)
0.4	220	60	60	0.2	100

2.3 Experimental Tensile Test

Tensile tests are carried out to determine the static properties of the architectural cores. The antitrichiral cores are 3D printed and tested in accordance with ASTM D638 (International 2014). A standard INSTRON hydraulic machine equipped with a 1 kN load cell and a speed of 1 mm/min, is used to perform tensile tests. The structures are printed with an effective length of 100 mm, as shown in Fig. 3. Two 25 mm long blocks are printed with the architectural cores to hold the sample during testing. An extensometer is used to measure transverse displacement. The position of the movable grips is used to determine longitudinal displacement. To ensure repeatability, five specimens are printed and tested for each configuration.

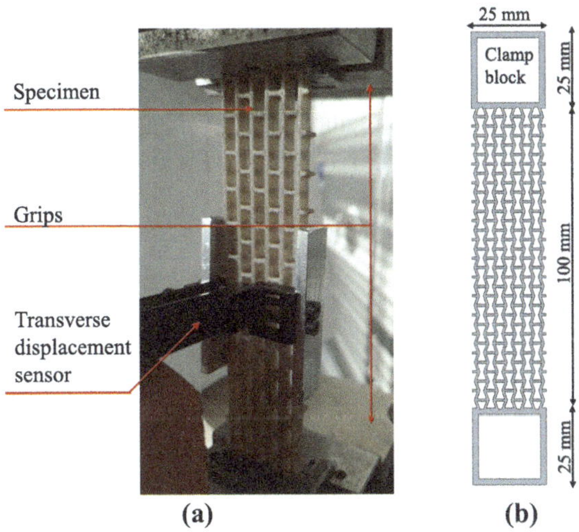

Fig. 3. (a) Experimental tensile test equipment (b) Tensile specimen

2.4 Numerical Model

Numerical simulations were conducted to assess the tensile behavior based on experimental results. This model corroborates the experimental data, allowing us to ascertain the structures static properties. The core was crafted using CAD software. The design from Solidworks was subsequently translated into commands for use with ABAQUS/Standard computation software. It is presumed that the raw material demonstrates elastic behavior. The geometry, along with the boundary conditions for the tensile test are presented in Fig. 4. The cores tested had dimensions of 100 mm in length, 25 mm in width, and 5 mm in thickness.

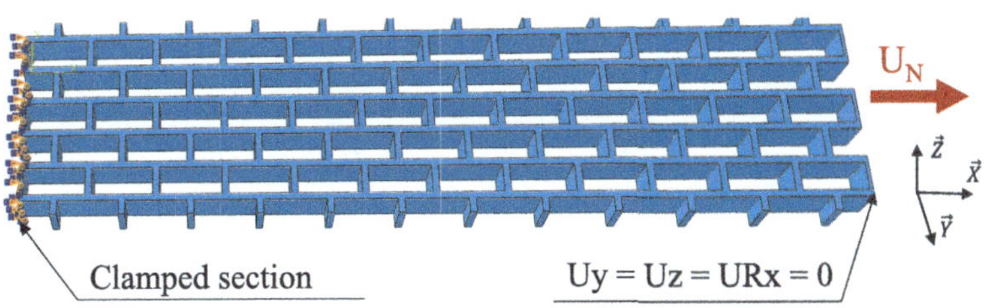

Fig. 4. FE model of tensile test with boundary and loading conditions.

3 Results

Tensile tests are conducted on the three structures. The structural Young's modulus and Poisson's ratio of each structure may be calculated using the experimental stress/longitudinal strain (ε_L) and transverse strain (ε_L) curves as shown in Fig. 5.

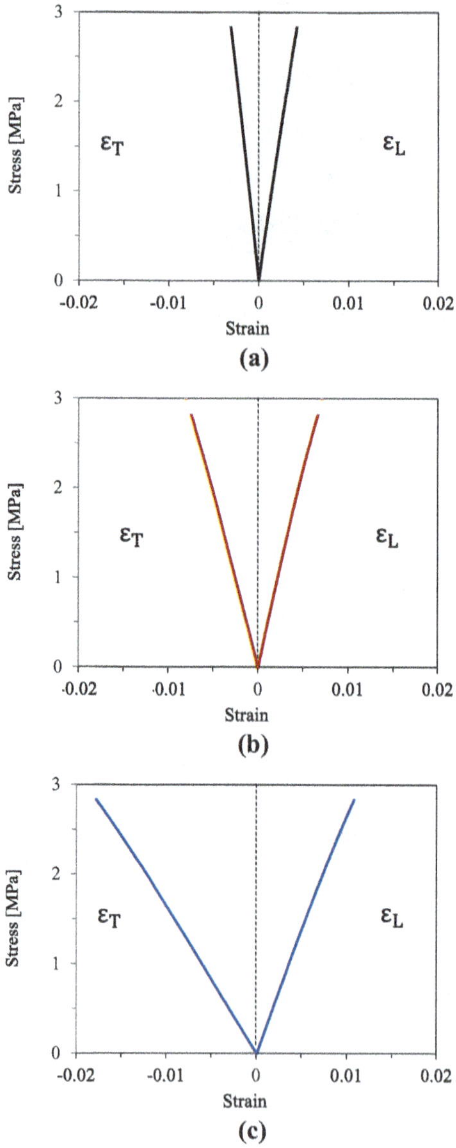

Fig. 5. Stress/strains evolution of structures (a) Rectangular (b) Re-entrant (c) Hexagonal.

It is found that the transverse deformation is positive only for the reentrant honeycomb due to the swelling of the structure during the tensile test, whereas it is negative for the hexagonal and rectangular honeycombs due to the structure shortening during the test. This is because these materials may thin or thicken due to the influence of the tensile load on each structure's cell walls.

Figure 6 shows the Young's modulus of different structures. The rectangular lattice structure has the highest modulus, followed by that of the reentrant honeycombs. Indeed, the hexagonal structure has the lowest Young's modulus value. The reason for this is that the structure doesn't have cell walls that lean in the applied load direction, which diminishes the structure's flexibility during tensile test, thereby demonstrating good stiffness. Conversely, hexagonal and re-entrant cell walls contribute to the flexibility of the structure when subjected to tensile stress. Finite element modelling was used to calculate the Young's modulus of the structure. The results show a close correlation between the experimental and numerical methods.

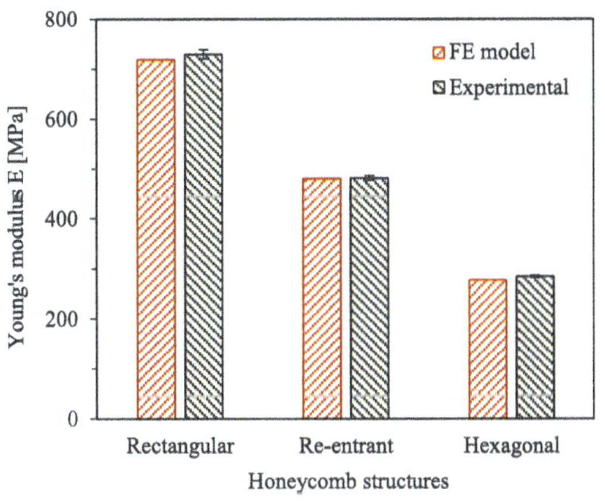

Fig. 6. Experimental and numerical Young's modulus evolution of structures.

Table 3. Poisson's ratios of honeycomb structures.

Structure	Experimental	FE Model
Rectangular	0.73 ± 0.02	0.71
Re-entrant	-1.18 ± 0.05	−1.3
Hexagonal	1.68 ± 0.05	1,6

Moreover, the Poisson's ratio is determined by using transverse/longitudinal strain curves in the field of linear elastic deformation. Table 3 shows the experimental and numerical Poisson's ratio for each structure lattice. The results show that the ratio is positive for the rectangular and hexagonal honeycombs which shows their non-auxetic behavior. However, an auxetic behavior is noticed for the re-entrant structure.

4 Conclusion

The use of 3D printing to create architectural materials is currently under investigation. Specifically, the materials being studied are composed of polylactic acid (PLA) reinforced with flax fibers, and researchers have selected three honeycomb lattices to analyze their impact on the composite material's tensile properties. Through experimentation, they discovered that the Poisson's ratio varied between -1.18 and 1.68 depending on the lattice structure, and they also determined that the rectangular structures demonstrated good stiffness. The results of the finite element (FE) model aligned with the experimental findings regarding static properties.

References

García-Collado, A., Blanco, J.M., Gupta, M.K., Dorado-Vicente, R.: Advances in polymers based multi-material additive-manufacturing techniques: state-of-art review on properties and applications. Addit. Manuf. **50**, 102577 (2022)

Hamrouni, A., Rebiere, J.L., El Mahi, A., Beyaoui, M., Haddar, M.: Damping behavior of bio-based anti-trichiral materials made with additive manufacturing. In: Akrout, A., Abdennadher, M., Feki, N., Abbes, M.S., Chaari, F., Haddar, M. (eds.) Advances in Acoustics and Vibration IV. ICAV 2022. Applied Condition Monitoring, vol. 22, pp. 214–223. Springer, Cham (2023a). https://doi.org/10.1007/978-3-031-34190-8_24

Hamrouni, A., Rebiere, J.-L., El Mahi, A., Beyaoui, M., Haddar, M.: Static study of bio-based architectural materials made with 3D printing technology. In: Akrout, A., Abdennadher, M., Feki, N., Abbes, M.S., Chaari, F., Haddar, M. (eds.) Advances in Acoustics and Vibration IV, ICAV 2022. Applied Condition Monitoring, vol. 22, pp 224–233. Springer, Cham (2023b). https://doi.org/10.1007/978-3-031-34190-8_25

Hamrouni, A., Rebiere, J.-L., El Mahi, A., Beyaoui, M., Haddar, M.: Experimental and finite element analyses of a 3D printed sandwich with an auxetic or non-auxetic core. J. Sandwich Struct. Mater., 10996362231151454 (2023c). https://doi.org/10.1177/10996362231151454

Hamrouni, A., Rebiere, J.-L., El Mahi, A., Beyaoui, M., Haddar, M.: Experimental and numerical investigation of the static behavior of a 3D printed bio-based anti-trichiral sandwich. J. Compos. Mater. **57**(13), 2161–2178 (2023). https://doi.org/10.1177/00219983231169332

Hamrouni, A., Rebiere, J.-L., El Mahi, A., Beyaoui, M., Haddar, M.: Poisson's ratio influence on structural dynamic properties of 3D printed bio-based sandwich with an anti-trichiral core. J. Vibr. Control, 10775463231185914 (2023e). https://doi.org/10.1177/10775463231185914

Imbalzano, G., Linforth, S., Ngo, T.D., Lee, P.V.S., Tran, P.: Blast resistance of auxetic and honeycomb sandwich panels: comparisons and parametric designs. Compos. Struct. **183**, 242–261 (2018). https://doi.org/10.1016/j.compstruct.2017.03.018

International A. Standard test method for tensile properties of plastics. ASTM international (2014)

Kabir, S.F., Mathur, K., Seyam, A.-F.M.: A critical review on 3D printed continuous fiber-reinforced composites: history, mechanism, materials and properties. Compos. Struct. **232**, 111476 (2020)

Khalfallah, M., et al.: Innovative flax tapes reinforced Acrodur biocomposites: a new alternative for automotive applications. Mater. Des. **64**, 116–126 (2014)

Liu, G., Xiong, Y., Zhou, L.: Additive manufacturing of continuous fiber reinforced polymer composites: design opportunities and novel applications. Compos. Commun. **27**, 100907 (2021)

Mathur, K., Kabir, S.F., Seyam, A.-F.M.: Tensile properties of 3D printed continuous fiberglass reinforced cellular composites. J. Textile Inst. **113**(1), 60–69 (2022)

Mousanezhad, D., Haghpanah, B., Ghosh, R., Hamouda, A.M., Nayeb-Hashemi, H., Vaziri, A.: Elastic properties of chiral, anti-chiral, and hierarchical honeycombs: a simple energy-based approach. Theor. Appl. Mech. Lett. **6**(2), 81–96 (2016)

Okada, M.: Chemical syntheses of biodegradable polymers. Prog. Polym. Sci. **27**(1), 87–133 (2002)

Satyanarayana, K.G., Arizaga, G.G., Wypych, F.: Biodegradable composites based on lignocellulosic fibers—an overview. Prog. Polym. Sci. **34**(9), 982–1021 (2009)

Scarpa, F., Tomlinson, G.: Theoretical characteristics of the vibration of sandwich plates with in-plane negative Poisson's ratio values. J. Sound Vib. **230**(1), 45–67 (2000)

Wang, P., Zou, B.: Improvement of heat treatment process on mechanical properties of FDM 3D-printed short-and continuous-fiber-reinforced PEEK composites. Coatings **12**(6), 827 (2022)

Zamani, M.H., Heidari-Rarani, M., Torabi, K.: A novel graded auxetic honeycomb core model for sandwich structures with increasing natural frequencies. J. Sandwich Struct. Mater. **24**(2), 1313–1339 (2022)

Effect of Waste Forest Content on the Mechanical and Thermal Properties of 3D Printing PLA Parts

Sarra Helaoui[1,2](✉), Ahmed Koubaa[1], Hedi Nouri[2], Martin Beauregard[1], and Sofiane Guessasma[3]

[1] UQAT, 445, boul. de l'Université, Rouyn-Noranda, QC J9X 5E4, Canada
helso2@uqat.ca

[2] Laboratory of Electromechanical Systems (LASEM), National Engineering School of Sfax, University of Sfax, 3038 Sfax, Tunisia

[3] UR, Equipe Paroi Végétale et Polymères Pariétaux, 1268 Biopolymères Interactions Assemblages, Site de la Géraudière BP71627, Cedex 03, 44316 Nantes, France

Abstract. Poly (lactic acid) is a promising biopolymer in 3D printing applications, offering a potential eco-friendly alternative to petrochemical plastics. The modification of PLA by incorporating fillers such as forest industrial residues is used to enhance its properties in line with the sustainable goals. This study aims to investigate the potential of biocomposites made of PLA with forest industrial residues such as wood ashes and wood sawdust up to 20% weight fraction and to understand the effect of fiber type and weight fraction on their thermal properties, mechanical performance and microstructure characteristics. For the thermal properties, it has been found that adding fiber decreased the thermal stability, the glass transition temperature, and the cold crystallization temperature with adding fibers but did not significantly change the melting point temperature of PLA. The crystallinity rate of PLA increased with adding filler and increasing the weight fraction up to 20%. The tensile properties highlighted a decrease in strain and tensile strength, while Young's modulus showed no clear trend. The microstructural analysis revealed that a higher filler content resulted in increased porosity, roughness, and visible pull-out fiber on both the filament and printed parts' surface.

Keywords: PLA · biocomposites · Biodegradable · forest industrial residues · FDM

1 Introduction

Additive manufacturing has transformed the manufacturing industry's ability to produce complex designs. Its advantages in material efficiency and post-processing elimination reduced manufacturing time, waste, and cost. Fused Deposition Modeling (FDM) is a promising method for producing biocomposites due to its cost-effectiveness and user-friendly operation.

M. Chouchane et al. (Eds.): CMSM 2023, LNME, pp. 510–521, 2024.
https://doi.org/10.1007/978-3-031-65007-9_52

There is a growing interest in developing biodegradable alternatives in response to the excessive use of petroleum-based plastics and their negative environmental impact. Poly-lactic acid (PLA) is a fully biodegradable polymer derived from microbial synthesis with attractive properties such as renewability, biocompatibility, and cost-effectiveness[1]. However, its high brittleness, poor thermal stability, and low crystallization kinetic have limited its applications[2]. On the other hand, using forest industrial residues (FIR) has become increasingly important in the forest industry, as it offers a sustainable solution to the disposal of these materials and reduces the negative environmental impact [3–5]. With its abundance, affordability, and renewability, FIR has the potential to be a valuable source of raw materials for developing wood-plastic composites (WPC). Recent studies focused on incorporating these lignocellulosic fibers into the PLA matrix to improve its mechanical and thermal properties. The resulting biocomposites offer an opportunity to utilize FIR for sustainable and eco-friendly 3D printing, providing an environmentally conscious solution for material production.

Limited research explored the use of FIR, such as wood ashes [6, 7] and sawdust [8, 9] or cellulose [10–13] as reinforcement in biocomposites production. The combination of PLA and natural fiber poses a significant challenge, given the hydrophilic nature of wood and the hydrophobicity of PLA, which leads to inadequate interfacial bonding [14, 15] and results in strength properties reduction in the 3D-printed structures. Kariz et al. incorporated wood powder into PLA pellets at 10% to 50% fractions. The tensile strength increased from 55 MPa to 57 MPa upon adding 10% wood fiber. However, with a larger fraction (50%), a decline in tensile strength was observed at higher concentrations of wood fiber [16]. This decrease in strength was attributed to the challenge faced by polymer binders in fully encapsulating the wood particles, resulting in poor bonding and weak stress transfer.

This study investigated the potential of PLA WPC with forest industrial residues such as wood ashes and wood sawdust up to 20% weight fraction and to characterize the effect of fiber type and weight fraction on the thermal properties, mechanical performance, and microstructure of these biocomposites for FDM technology.

2 Materials and Methods

2.1 Experimental Procedure

PLA (Ingeo 4032D grade) was procured from Nature Works LLC and dried to eliminate moisture. Rettenmaier Canada supplied cellulose fibers as a reference fiber, Wood ashes were provided by the company West Fraser, La Sarre, Q, Canada and the biomaterial laboratory generated wood sawdust. Fibers were ground and sieved to achieve a fiber size $< = 150\ \mu m$, then oven-dried at 70 °C to reach a moisture content lower than 3%. PLA and fibers were mixed using different portions of fibers (Table 1).

The samples were prepared following a three-step extrusion process (Table 2 and Fig. 1). The first extrusion homogenized the polymer and fiber mixture. The second extrusion manufactured filament spools while the third extrusion is the printing step.

2.2 Characterization of Wood Polymer Composites

2.2.1 Differential Scanning Calorimetry (DSC)

Filament sample characterization was analyzed using a Q20 device equipped with a cooling accessory (TA Instrument, Newcastle, TX, USA) and an aluminum crucible. The cycle was repeated twice, consisting of a heating run from 10 °C to 250 °C at a rate of 10 °C/min, followed by an isothermal period of 1 min to stabilize the thermal history, and then a cooling phase from 250 °C to 10 °C at a rate of 10 °C/min. The resulting scans provided data for determining the melting temperature (T_m), glass transition (T_g), and crystallization temperature (T_c). The crystallinity $X_{PLA-Fiber}$ of the samples was calculated using Eq. (1) [17]:

$$X_{PLA-Fiber}\% = \frac{\Delta H_m - \Delta H_c}{\Delta H_m^{\circ} \times (w_{PLA})} \times 100 \quad (1)$$

where:

ΔH_m is the total enthalpy of melting,
ΔH_c is the enthalpy of cold crystallization,
ΔH_m° is the theoretical enthalpy of fusion for fully crystalline PLA (93 Jg^{-1}) [18],
w_{PLA} is the mass fraction of PLA in the sample.

Table 1. Composition of the samples.

Formulation	PLA Weight Fraction (%)	Fiber Weight Fraction (%)		
		Wood Ash (WA)	Wood sawdust (GP)	Cellulose Fiber (CF)
PLA	100%	-	-	-
PLA-WA1	95%	5%	-	-
PLA-WA2	90%	10%	-	-
PLA-WA3	85%	15%	-	-
PLA-WA4	80%	20%	-	-
PLA-GP1	95%	-	5%	-
PLA-GP2	90%	-	10%	-
PLA-GP3	85%	-	15%	-
PLA-GP4	80%	-	20%	-
PLA-CF1	95%	-	-	5%
PLA-CF2	90%	-	-	10%
PLA-CF3	85%	-	-	15%
PLA-CF4	80%	-	-	20%

Table 2. Biocomposites elaboration.

Samples preparation steps	Equipment	Parameters
1- Extrusion	Twin-screw extruder (Haake PolyLab OS Rheodrive 7, HAKKE, Germany)	Temperatures were set at 180 °C and 190 °C, respectively The rotation speed is 50 rpm
2- Filament process	3devo Composer 450 desktop filament maker (Utrecht, Netherlands)	The diameter of the filaments is 1,75 ± 1mm The temperatures were set at 170 °C, 175 °C, 180 °C, and 185 °C for zones 1,2,3, and 4 respectively The rotation speed is 4 rpm
3- 3D printing	Printer: CreatBot DX Plus	Nozzle diameter: 0,8 mm Extrusion temperature: 200 °C Bed temperature: 60 °C Infill speed: 30 mm/s, Infill percentage: 100% Layer thickness: 0,4 mm Raster angle: ± 45°

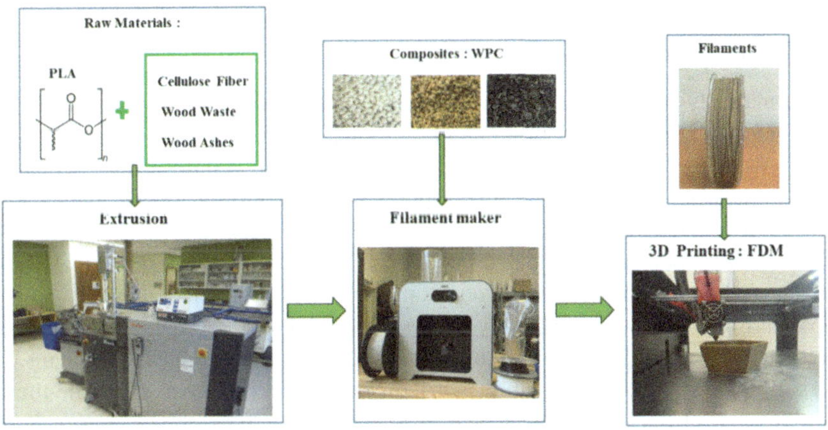

Fig. 1. Samples preparation using Fused Deposition Modeling.

2.2.2 Thermogravimetric Analysis (TGA)

The thermal stability of samples was conducted using thermogravimetric analysis with a Q50 device from a TA instrument controlled by TA Analysis software. Neat PLA, fibers, and filaments weighing approximately 10–20 mg were subjected to a heating cycle from 10 °C to 600 °C at 40 °C/min in a nitrogen atmosphere.

2.2.3 Tensile Test

The tensile properties of 3D-printed structures were assessed using a Zwick Universal Testing Machine equipped with a 20 kN load cell (Z020, Zwick Roell Group, Germany). The crosshead speed was fixed at 2 mm/min at ambient temperature. The reported results represent the average of at least four measurements under ASTM D 638 using Type 1 samples.

2.2.4 Scanning Electron Microscopy (SEM)

This test was used to investigate the morphology of filaments and cross-sections of tensile samples using a scanning electron microscope (ThermoFisher, Quattro S, USA), and an acceleration voltage of 15kV was applied.

3 Results and Discussion

3.1 Thermogravimetric Analysis (TGA)

The thermal anlysis was conducted using TGA. Table 3 provides the onset temperature (T_{onst}), the maximum degradation rate (T_{max}) temperature, and the residual weight at 600 °C. Wood sawdust started decomposition earlier than cellulose fiber, which was expected because lignocellulosic fiber's thermal decomposition started with lignin's degradation at the lowest temperature, followed by hemicellulose and then cellulose [19]. However, wood ashes remained stable up to 600 °C with a higher residual mass of 90% due to the pre-combustion of wood components. PLA filament indicated an onset temperature of 295 °C with a relatively low residual weight of 0,4%. The addition of fibers decreased $theT_{onst}$ and T_{max} of the biocomposites with increasing fiber content [20]. Fibers influenced the thermal stability of the composites, mainly because the thermal decomposition of fibers was lower than PLA. The PLA-CF exhibited greater thermal stability. However, PLA-WA degraded earlier than the other samples. Increasing fiber content increased the composite's residual mass, mainly because the fibers' residual mass was higher than the matrix.

3.2 Differential Scanning Calorimetry (DSC)

In the DSC analysis summary of PLA and filaments, two behaviors were presented in Fig. 2. There was a significant reduction inT_g with the addition of fiber weight from 5% to 20%, mainly attributed to weak interfacial adhesion and interaction between fillers and PLA matrix [21]. This reduction also suggested that the composites crystallize

Table 3. Thermogravimetric analysis measurements.

Formulation	T_{onst} (°C)	T_{max} (°C)	Residual weight (%)
PLA	295,1	368,9	0,4
Cellulose Fiber	222,4	363,7	18,7
PLA-CF1	293,2	370,7	1,4
PLA-CF2	275,6	365,1	2,3
PLA-CF3	267,2	357,2	3,5
PLA-CF4	261,6	355,2	3,9
Wood sawdust	215,1	358,9	21,1
PLA-GP1	275,6	370,0	2,8
PLA-GP2	266,5	350,9	3,2
PLA-GP3	267,4	346,1	4,6
PLA-GP4	256,1	339,6	6,6
Wood Ashes	0,0	448,9	88,9
PLA-WA1	236,4	341,6	4,7
PLA-WA2	231,7	307,5	8,6
PLA-WA3	220,4	284,3	13,8
PLA-WA4	217,9	262,7	21,4

earlier than neat PLA. This finding was confirmed by the fact that cold crystallization temperature T_{cc} of neat PLA at 131 °C shifted to a lower temperature as the fiber content increased. Regarding the melting temperature T_m, adding residues wasn't significantly influenced, maintaining a value of 176 ⊥ 1 °C (Fig. 3).

3.3 Mechanical Properties

The mechanical results were highlighted with the tensile test. Figure 4 presented the tensile strength, Young's modulus, and strain at break for 3D-printed samples with varying fiber content. In Fig. 4 (a), the results for elastic modulus showed no clear trend with increasing filler weight. However, with some biocomposites, Young's modulus exceeded the PLA sample of 3,44 GPa. For PLA-GP4, PLA-CF3, and PLA-WA3, the tensile modulus was higher by 20,9%, 4%, and 11,3%, respectively. Figure 4 (b) showed that the tensile strength of 3D-printed composites was lower than pure PLA (51,64 MPa) and showed different trends.

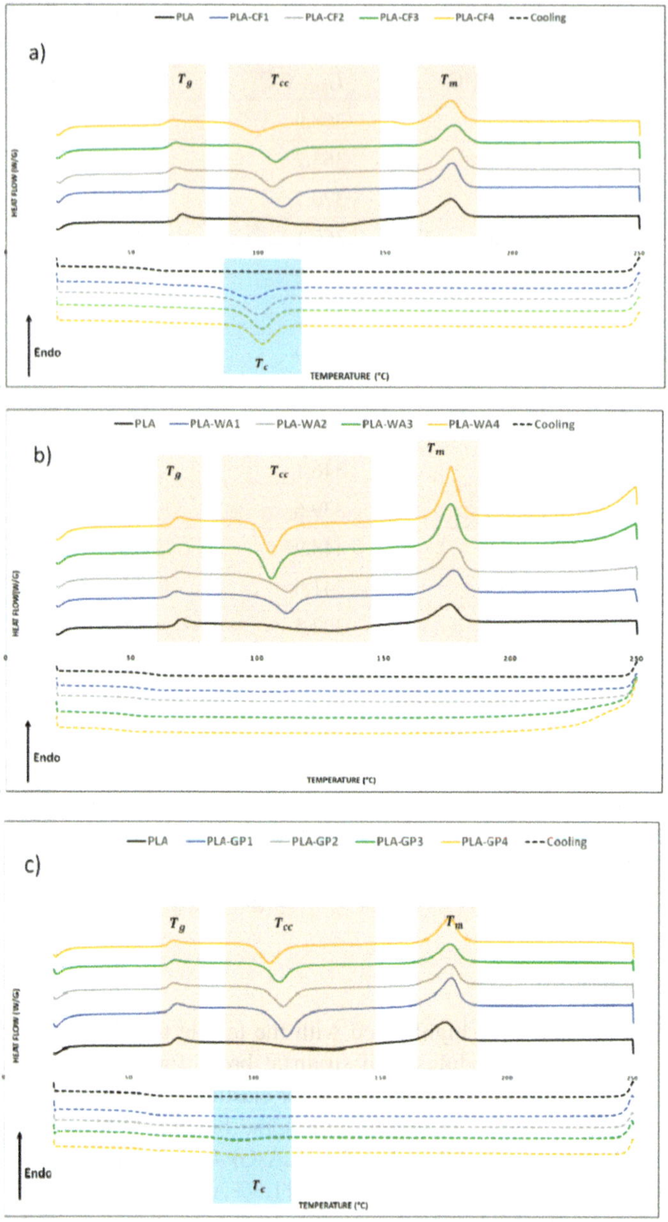

Fig. 2. Differential scanning calorimetry curves of biocomposites: a) PLA-CF, b) PLA-WA, c) PLA-GP.

From Fig. 4(c), it is evident that the strain at break significantly decreased with adding fillers, which was less than pure PLA by 41%, 64%, and 50% with 20% of fiber weight of GP, CF, and WA, respectively. Adding hydrophilic fibers to hydrophobic PLA

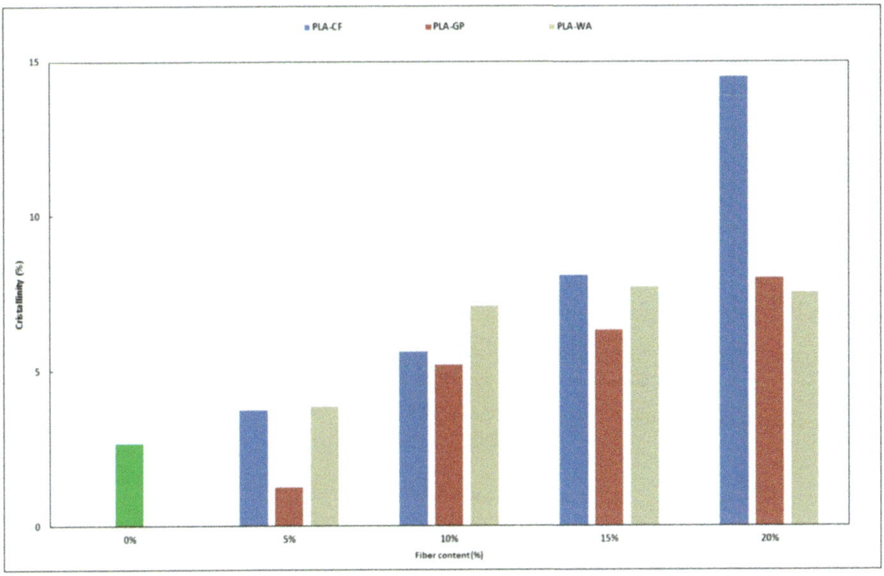

Fig. 3. The crystallinity of biocomposites as a function of fiber content.

reduced tensile elongation and localized stress concentration, making them susceptible to premature failure and a decrease in the polymeric matrix's ductility, resulting in more fragile biocomposites [22].

Adding fibers decreased the mechanical properties, namely the stress and strain at the break, making the material more brittle [22]. In addition, it's important to distinguish between two types of porosity in the printed samples that affected mechanical behavior. The inner layer porosity inside the deposited layers acted as local defects and reduced tensile strength caused by the poor compatibility and bonding between PLA matrix and fillers, pull-out fibers, and fiber agglomeration [23]. The interlayer porosity between layers is caused by the printing process, such as low pressure and heterogeneous temperature distribution [24]. Overall, incorporating forest residue fibers in PLA did not yield the expected reinforcement for structural applications.

3.4 Scanning Electron Microscopy

Cross-sectional SEM analysis was conducted to examine the surface of the composite filaments and 3D printed parts, as illustrated in Fig. 5, Fig. 6, and Fig. 7.

In Fig. 5 and Fig. 6, the pure PLA highlighted a smooth fracture surface; with the incorporation of forest residues, the roughness of the filament became more prominent with visible assembled fibers on the surface. While some areas encapsulated fibers, gaps between the PLA and fiber interfaces were apparent. The addition of fibers led to the debonding of the filler and pull-out from the matrix, along with observed fiber agglomeration, indicating reduced fiber dispersion [25], mainly due to a decrease in interfacial adhesion resulting from the different polarities between the hydrophobic matrix PLA and hydrophilic fibers [26].

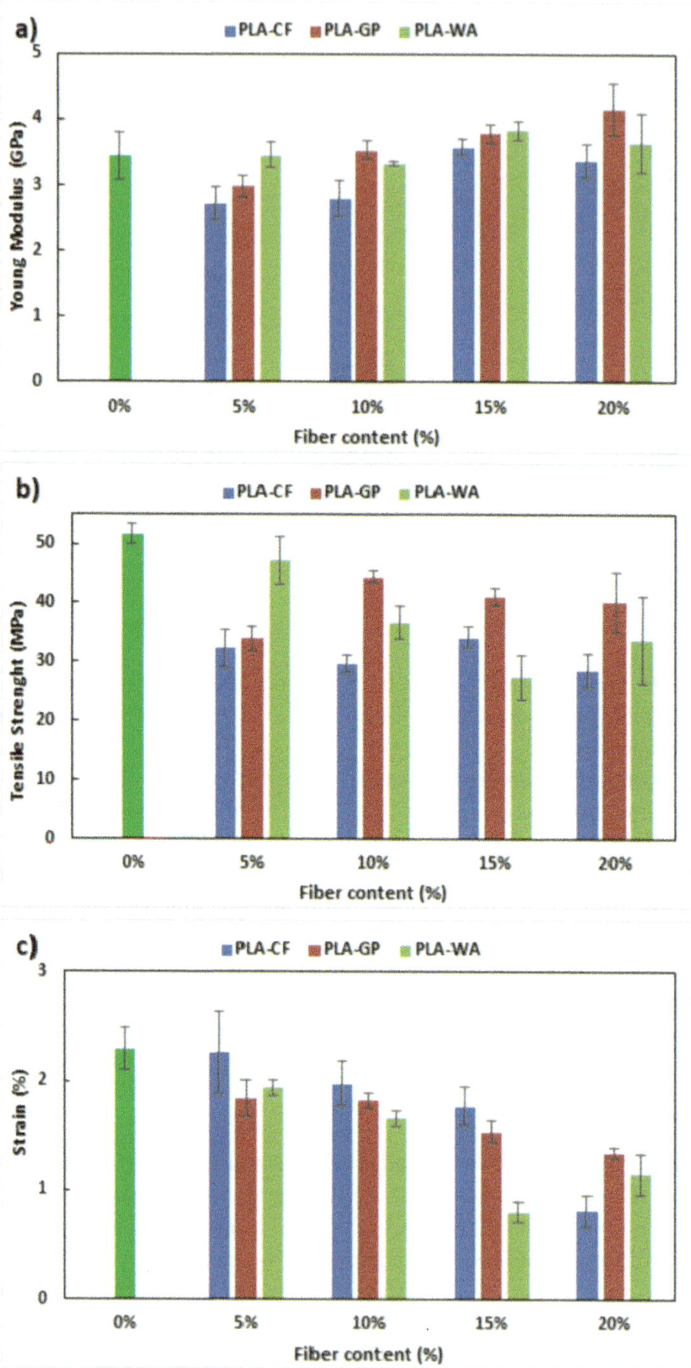

Fig. 4. Tensile properties of 3D printed specimens: a) Young Modulus, b) Tensile Strength, and c) Strain at the break as a function of fiber content.

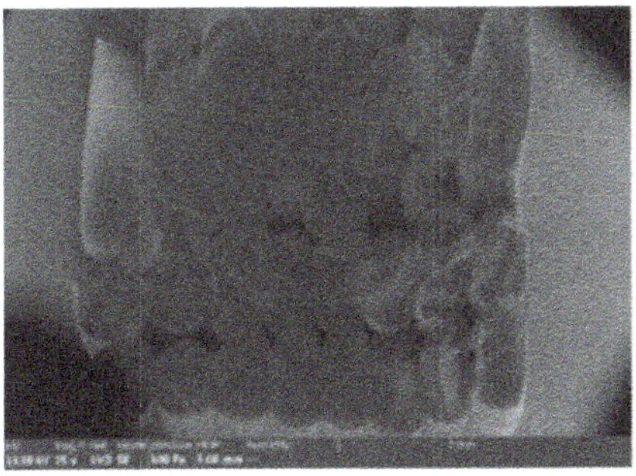

Fig. 5. SEM analysis of the cross-section of pure PLA 3D printed samples.

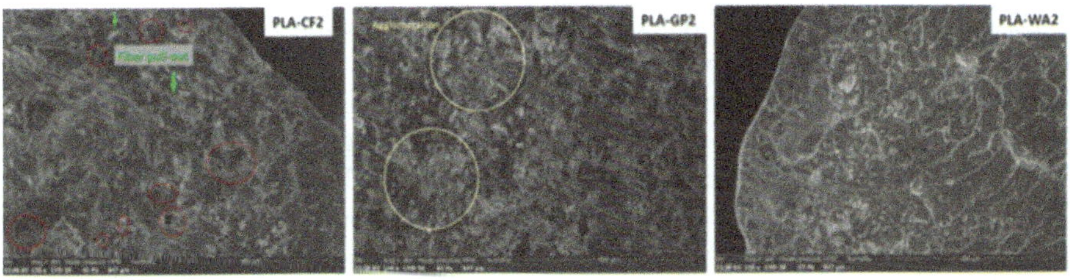

Fig. 6. SEM analysis of the cross-section of filament samples at 10% of weight fraction at 400μm.

Figure 7 revealed a significant increase in porosity compared to filament samples, emphasizing the porosity generated during the FDM process. These irregular voids, larger than in-layer porosities, resulted from the layering process and could be attributed to insufficient melting and blending pressure in the FDM device. Furthermore, lower print quality, inadequate flow, and misalignments between adjacent filaments further contributed to the observed voids, consistent with prior research [27]. The volume and number of voids on the fracture surface increased with higher fiber content, resulting in inadequate fiber-polymer stress transfer and a subsequent reduction in the mechanical

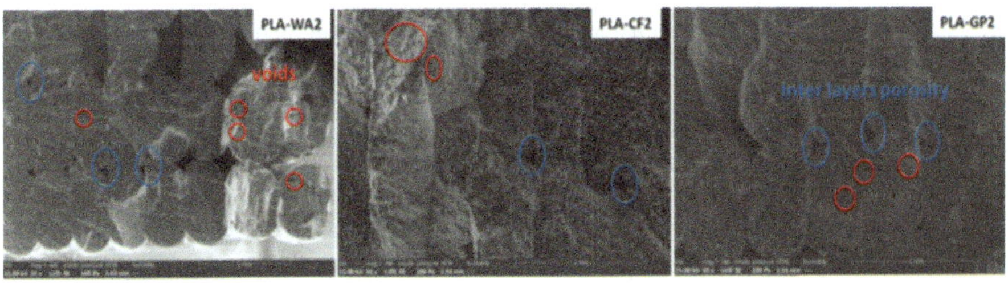

Fig. 7. SEM analysis of the cross-section of 3D printed samples.

reinforcement efficiency of the fiber in the 3D-printed part [28]. Additionally, fiber dispersion deteriorated as fiber content increased from 5% to 20%, and noticeable agglomeration generated various defects in the samples. These results indicated that the incorporation of waste fibers in PLA polymer primarily served a filler role, emphasizing the need to improve the adhesion and compatibility of the composites.

4 Conclusion

Incorporating forest residues into PLA composites has shown potential for contributing to the circular economy. Combining wood sawdust, wood ashes, and cellulose fiber as filler materials, 3D printable samples were successfully created. Fiber enhanced the crystallinity of biocomposites and affected their thermal stability and mechanical properties. A higher filler content increased porosity, roughness, and visible pull-out fiber on the filament and printed parts' surface. Despite some limitations in its mechanical properties, the biodegradability and biobased nature of these biocomposites make them suitable for various applications in line with the circular economy goals. Using adhesives could improve the biocomposites' compatibility and mechanical properties, making it a key area for future investigation. Finally, using renewable resources to produce composites highlights our modern society's growing emphasis on sustainability.

References

1. Wróblewska-Krepsztul, J., Rydzkowski, T., Borowski, G., Szczypiński, M., Klepka, T., Thakur, V.K.: Recent progress in biodegradable polymers and nanocomposite-based packaging materials for sustainable environment. Int. J. Polym. Anal. Charact. **23**(4), 383–395 (2018)
2. Ren, Q., Wang, J., Zhai, W., Lee, R.E.: Fundamental influences of induced crystallization and phase separation on the foaming behavior of poly (lactic acid)/polyethylene glycol blends blown with compressed CO2. Ind. Eng. Chem. Res. **55**(49), 12557–12568 (2016)
3. Sang, L., Han, S., Li, Z., Yang, X., Hou, W.: Development of short basalt fiber reinforced polylactide composites and their feasible evaluation for 3D printing applications. Compos. B Eng. **164**, 629–639 (2019)
4. Awad, S., et al.: Evaluation of characterization efficiency of natural fiber-reinforced polylactic acid biocomposites for 3D printing applications. Sustain. Mater. Technol. **36**, e00620 (2023)
5. Tanase-Opedal, M., Espinosa, E., Rodríguez, A., Chinga-Carrasco, G.: Lignin: A biopolymer from forestry biomass for biocomposites and 3D printing. Materials **12**(18), 3006 (2019)
6. Daramola, O., et al.: Influence of bamboo stem ash on some properties of polyester matrix. SVOA Mater. Sci. Technol. **2**, 1–8 (2020)
7. Sanusi, O.M., Oyinlola, A.K., Akindapo, J.O.: Influence of wood ash on the mechanical properties of polymer matrix composite developed from fiberglass and epoxy resin. Int. J. Eng. Res. Technol **2**, 344–352 (2013)
8. Estakhrianhaghighi, E., Mirabolghasemi, A., Zhang, Y., Lessard, L., Akbarzadeh, A.: 3D-printed wood-fiber reinforced architected cellular composites. Adv. Eng. Mater. **22**(11), 2000565 (2020)
9. Liu, L., Lin, M., Xu, Z., Lin, M.: Polylactic acid-based wood-plastic 3D printing composite and its properties. BioResources **14**(4), 8484–8498 (2019)

10. Wang, Z., et al.: Preparation of 3D printable micro/nano cellulose-polylactic acid (MNC/PLA) composite wire rods with high MNC constitution. Ind. Crops Prod. **109**, 889–896 (2017)
11. Ren, Q., et al.: Cellulose nanofiber reinforced poly (lactic acid) with enhanced rheology, crystallization and foaming ability. Carbohyd. Polym. **286**, 119320 (2022)
12. Ambone, T., Torris, A., Shanmuganathan, K.: Enhancing the mechanical properties of 3D printed polylactic acid using nanocellulose. Polym. Eng. Sci. **60**(8), 1842–1855 (2020)
13. Murphy, C.A., Collins, M.N.: Microcrystalline cellulose reinforced polylactic acid biocomposite filaments for 3D printing. Polym. Compos. **39**(4), 1311–1320 (2018)
14. Le Guen, M.-J., et al.: Influence of rice husk and wood biomass properties on the manufacture of filaments for fused deposition modeling. Front. Chem. **7**, 735 (2019)
15. Tao, Y., Wang, H., Li, Z., Li, P., Shi, S.Q.: Development and application of wood flour-filled polylactic acid composite filament for 3D printing. Materials **10**(4), 339 (2017)
16. Kariz, M., Sernek, M., Obućina, M., Kuzman, M.K.: Effect of wood content in FDM filament on properties of 3D printed parts. Mater. Today Commun. **14**, 135–140 (2018)
17. Lim, L.-T., Auras, R., Rubino, M.: Processing technologies for poly (lactic acid). Prog. Polym. Sci. **33**(8), 820–852 (2008)
18. Pilla, S., Gong, S., O'Neill, E., Rowell, R.M., Krzysik, A.M.: Polylactide-pine wood flour composites. Polym. Eng. Sci. **48**(3), 578–587 (2008)
19. Bouafif, H., Koubaa, A., Perré, P., Cloutier, A., Riedl, B.: Wood particle/high-density polyethylene composites: thermal sensitivity and nucleating ability of wood particles. J. Appl. Polym. Sci. **113**(1), 593–600 (2009)
20. Daver, F., Lee, K.P.M., Brandt, M., Shanks, R.: Cork–PLA composite filaments for fused deposition modeling. Compos. Sci. Technol. **168**, 230–237 (2018)
21. Zhang, Q., et al.: Improvement on the properties of microcrystalline cellulose/polylactic acid composites by using activated biochar. J. Clean. Prod. **252**, 119898 (2020)
22. Petchwattana, N., Channuan, W., Naknaen, P., Narupai, B.: 3D printing filaments prepared from modified poly (lactic acid)/teak wood flour composites: an investigation on the particle size effects and silane coupling agent compatibilization. J. Phys. Sci. **30**(2), 169–188 (2019)
23. Dong, Y., Ghataura, A., Takagi, H., Haroosh, H.J., Nakagaito, A.N., Lau, K.-T.: Polylactic acid (PLA) biocomposites reinforced with coir fibers: evaluation of mechanical performance and multifunctional properties. Compos. Part A Appl. Sci. Manuf. **63**, 76–84 (2014)
24. Le Duigou, A., Correa, D., Ueda, M., Matsuzaki, R., Castro, M.: A review of 3D and 4D printing of natural fiber biocomposites. Mater. Des. **194**, 108911, (2020). https://doi.org/10.1016/j.matdes.2020.108911
25. Tekinalp, H.L., et al.: High modulus biocomposites via additive manufacturing: cellulose nanofibril networks as "micro sponges." Compos. B Eng. **173**, 106817 (2019)
26. Zhang, Q., et al.: Biocomposites from organic solid wastes derived biochars: a review. Materials **13**(18), 3923 (2020)
27. Milosevic, M., Stoof, D., Pickering, K.L.: Characterizing the mechanical properties of fused deposition modeling natural fiber recycled polypropylene composites. J. Compos. Sci. **1**(1), 7 (2017)
28. Gauss, C., Pickering, K., Barbier, M., Miller, T.: Additive manufacturing of hygromorphic structures using regenerated cellulose/PLA biocomposites. Mater. Today Proc. (2023)

Author Index

M. Chouchane et al. (Eds.): CMSM 2023, LNME, pp. 523–525, 2024.
https://doi.org/10.1007/978-3-031-65007-9

SPRINGER NATURE

GPSR Compliance

The European Union's (EU) General Product Safety Regulation (GPSR) is a set of rules that requires consumer products to be safe and our obligations to ensure this.

If you have any concerns about our products, you can contact us on ProductSafety@springernature.com

In case Publisher is established outside the EU, the EU authorized representative is:

Springer Nature Customer Service Center GmbH
Europaplatz 3
69115 Heidelberg, Germany

The manufacturer's authorised representative in the EU is Springer Nature Customer Service Centre GmbH, Europaplatz 3, 69115 Heidelberg, Germany. If you have any concerns regarding our products, please contact ProductSafety@springernature.com

Printed and bound by CPI Group (UK) Ltd, Croydon, CR0 4YY
28/09/2026
02224057-0007